Crushing and Grinding Process Handbook

Crushing and Grinding Process Handbook

C. L. Prasher

Consultant to Chemical
and Mechanical Engineering Industry,
Linora Technical Associates

John Wiley & Sons Limited
Chichester · New York · Brisbane · Toronto · Singapore

Library of Congress Cataloging-in-Publication Data:

Prasher, Charles Lindsay
Crushing and grinding process handbook.

Includes bibliographies and index.
1. Size reduction of materials—Handbooks, manuals, etc.
I. Title.
TP156.S5P73 1987 621.9′14 85-26434
ISBN 0 471 10535 X

British Library Cataloguing in Publication Data:

Prasher, Charles Lindsay
Crushing and grinding process handbook.
1. Size reduction of materials
I. Title
660.2′8422 TP156.S5
ISBN 0 471 10535 X

Printed and bound in Great Britain by
Biddles Ltd, Guildford and King's Lynn

Contents

Each chapter is preceded, if appropriate, by its own NOTATION and concluded by its own REFERENCE LIST

Preface

The comminution literature is very extensive. Bibliographies are available running into hundreds of thousands of references covering the last hundred years. This is an indication of the intense interest which has been shown in the process of particle size reduction and the considerable effort to understand its hidden secrets. Early workers took a simplistic approach, not fully appreciating the complexity of the interrelationships of variables involved. 'Laws' were formulated which only applied over limited ranges of the variables. Thus it was easy to find conflicting conclusions. During the last thirty years particularly, equations have been developed to match the complex nature of the comminution process, and the advent of large computers has provided the means whereby solutions may be obtained, many of which invoke statistical mathematics.

There was therefore a need to show the limitations of the early simplistic approaches, to attempt to harmonize conflicting conclusions and to set down in a logical sequence the most important research advances and the mathematical models which embraced them. The aim of this book is to fill this need by placing into the hands of technical personnel in industry and universities a summary of the underlying principles of comminution as far as modern studies have progressed.

It is admitted that this is no easy task. It was commenced by Alfred Ratcliffe, to whom tribute must be paid as initiator of the concept of this handbook. However, owing to his illness prior to his untimely death in 1981, he was unable to keep abreast with the rapid changes which were taking place in fundamental thinking on the subject and hence in research progress. Thus the initial manuscript had to be rewritten entirely to reach this published form. Chapter 9 has been contributed by Professor Kapur of the Indian Institute of Technology, Kanpur, to whom grateful thanks are given.

The help of the libraries of the Institute of Chemical Engineers, of the Society of Mining Engineers and of Blue Circle Cement at Greenhithe is acknowledged. Above all, the friendly assistance of the West Sussex Library Service has been invaluable.

The typing and editing services of Mrs Clarice Millar are gratefully acknowledged.

Throughout, the understanding assistance of the staff of John Wiley & Sons at Chichester has seen the book through to completion, and to them thanks are due.

C.L.P.

CHAPTER 1

Introduction

1.1 SURVEY OF INDUSTRIAL COMMINUTION

Comminution is a major unit operation in chemical engineering practice. In terms of global quantities of material reduced in size it has been estimated that the annual tonnage is of the order of several thousand million, and in terms of the energy expended the yearly megawatt hours total several hundred million (Hiorns, 1971), of which the USA uses more than thirty million (Kaplan, 1981). These are the measures of the importance of comminution in industry today.

Top product size ranges from a few centimetres in crushing plants to a few microns in very fine grinding equipment, implying a 10,000 to 1 variation in product size.

The products themselves range from homogeneous pure chemicals to the most heterogeneous of minerals, covering the whole gamut of the Moh hardness scale, with the possible exception of diamond. Again, machine capacities may be as high as several hundred tonnes per hour, absorbing several thousand horsepower, or as low as a few kilogrammes per hour for very fine products.

These are measures of the diversity within the industries in which comminution plays a part. High capacity industries include the preparation of road-making materials, cement manufacture, mineral beneficiation and the fine pulverization of coal for combustion in boilers. Smaller capacity plants include the preparation of pigments and paints, pharmaceuticals and materials for the ceramics and pottery industries.

Traditionally crushing applies to size distributions with a top size greater than 5 mm (approximately 80% passing 104 microns) and grinding to finer distributions as shown in Table 1.1. This table also indicates the different types of machinery that are involved as the product becomes smaller. At the same time the specific energy consumption rises steeply as the finer stages of size reduction are reached.

Table 1.1. Classification of comminuted products by size

Stage of reduction	Typical machine	Product size passing 80%* (micron)	Specific energy† (kWh/tonne)
Primary crushing	Compression or impact crusher	10^6–10^5	0.1–2
Secondary crushing	Compression or impact crusher	10^5–10^4	
Coarse grinding	Rod mill	10^4–10^3	2–4
Fine grinding	Ball or vertical spindle mill	10^3–10^2	5–20
Very fine grinding	Tube mill	10^2–10	20–100
Super-fine grinding	Attrition or fluid energy mill	10 –1	100–1000

Reproduced by permission of the American Institute of Mining, Metallurgical and Petroleum Engineers

*Hukki (1961) †Hukki (1975)

1.2 RANGE OF PARTICLE REQUIREMENTS IN INDUSTRY

The word *particle* is used throughout the book for a fragment of comminution regardless of its size.

Within each sector of industry different features of the product assume importance. For example, both size and shape of crushed products are important in aggregates for road-making and crucial to the regular downward flow patterns of solids in shaft kilns, as well as to the upward flow of gases. In the fine grinding of pulverized coal for power-station boilers, the size distribution can be of the order of 60% minus 75 microns for reactive lignites. For bituminous coals, however, 70% minus 75 microns is commonly called for. Anthracites with their low volatile content demand about 80% minus 75 microns. For all these coals, particle shape or specific surface is not specified and the most important size feature is really the limitation of the percentage of coarse material so that all particles will completely burn out within the furnace residence time. Cement clinker needs to be ground finer than anthracite, but the strength of the concrete is found to be strongly related to the specific surface of the clinker, for which the Blaine index has become meaningful within the industry. However, specific surface alone is not a sufficient criterion, and particles in a particular limited small size range should be present in as high a proportion as possible. This illustrates the need for both surface and size measurements in the product of one industry. Taking an example from very finely ground products, the opacity of paint is inversely proportional to the pigment particle size when this is greater than the wavelength of incident light. When the particle size is smaller than the wavelength of light, the reverse is the case. Jefferies (1969). Hence the most careful control of very small particle sizes is important to the paint industry.

Cases could be multiplied: the quality of cosmetics, the behaviour of drugs, the rate of chemical reaction, the effectiveness of a catalyst and the optimum conditions for the separation of an ore from a mineral matrix, all demand restraints to be placed on the size and/or shape and/or specific surface of the ground constituents. Thus considerable space is devoted to these considerations and to the principles of classification, which is the process of removal of particles of the required size distributions from a coarser circulating distribution. Classification is an area, in fact, where industry is calling for higher sharpness of cut, which can only be effected after detailed study of the process.

1.3 ENERGY SAVING IN INDUSTRY

Despite the large volume of literature over the last century, representing considerable effort chiefly by universities and industry, as recently as twenty years ago men of experience were saying that 'the relationship between size distribution and energy is *starting* to be understood' (Callcott in discussion on Lynch and Moore, 1965). Within this twenty-year span there has been an alarming rise in fuel costs, which has catalysed the urgency of reduction in the energy consumption for comminution. For example the US Bureau of Mines, the National Science Foundation and the US Department of Energy have combined to consider a new research organization for this purpose (Kaplan, 1981). A report by the National Material Advisory Board (NMAB, 1981) declares that comparatively little public money has been spent on comminution because the magnitude of the potential saving has not been realized. This is estimated at 10% of the present levels, that is three million megawatts in the USA alone, and globally several tens of millions annually. In the present climate of high incentive to reduce energy costs it is quite possible that more public money will be made available to accelerate the better understanding of the fundamentals of the size reduction process. This is where all such research must start, and a necessary prerequisite is an appraisal of the basic studies already undertaken. It is hoped that this handbook will prove valuable for that purpose.

1.4 PATTERN OF STUDY OF THE BASICS OF COMMINUTION

Basic to the handbook is the recognition of the following areas for progressive study:

(1) Fracture physics in the context of size reduction.
(2) Controlled fracture tests involving the measurement of size distributions (or surface) and the energy for breakage with (a) single particles or (b) particle beds.
(3) Size reduction and classification in laboratory and pilot scale mills.
(4) Scale-up from small to large mills.
(5) Mathematical simulation of (a) the comminution process *per se*, and (b) associated systems such as particle transport in moving beds and classification.
(6) New methods of comminution.

Each of these areas is briefly considered below.

(1) In this section solids are studied in relation to their material properties such as strength, hardness, toughness, brittleness and yield. The stress–strain patterns within particles as they approach and undergo fracture are measured and assessment made of the energy associated with microcracks within the material. Particles tested are both symmetrical in shape, such as the sphere and irregular as found in practical comminution.

The materials studied for the most part have been associated with the major industries that use crushing and grinding:

(a) cement industry—quartz, limestone, cement clinker
(b) power industry—coal
(c) ceramics industry—glass

and this applies also to other levels of study. Fracture physics is the subject of Chapter 3.

(2) Many specially designed breakage tests have been undertaken, but not all may claim full control of all the variables. Greater credence has been given to those tests in which control has been maintained, whether single particles or beds of particles have been involved. It has been convenient to separate size distribution studies (Chapter 4) from energy measurements (Chapter 6) but they are clearly interrelated and their links are not overlooked.

(3) Since most commercial methods of grinding, as distinct from crushing, include a classification loop, the basics of classification are spelt out in Chapter 5 in preparation for inclusion in simulation of closed circuits.

(4) Two aspects of scale-up have been distinguished. Firstly, the 'end result' scale-up has been treated in Chapter 7 in which detailed variations of mill internal parameters are ignored. Secondly, the effect of scale-up on such items as ball and powder charge movement within a tumbling mill, for example, has to be known if optimum conditions for mill operation are to be assessed. This aspect of scale-up is treated in Chapter 10.

Experiments with full-scale plant are important relative to:

(a) comparison with laboratory or pilot-scale mills
(b) improvement of existing plant
(c) design of new plant.

The results of such experiments are introduced as appropriate throughout the book.

(5) Mathematical modelling of mill systems is proving to be so advantageous that recent years have seen a strong swing towards simulation of each zone of operation, including grinding, mass transport of powder and classification. Hence Chapters 8, 9 and 10 are devoted to this important aspect of comminution.

(6) A watch must always be kept on new methods of comminution which

have potential for the future, and several are selected for review in Chapter 11. Throughout the work, the bearing of basic studies on practical size reduction is constantly kept to the forefront of discussion.

1.5 REFERENCES

Hiorns, F. J. (1971). 'Wear measurements in size reduction machinery', *Chem. Proc. Eng.*, January, 47–49.

Hukki, R. T. (1975). 'The principles of comminution: an analytical summary', *Eng. Min. J.*, **176**, 106–110.

Hukki, R. T. (1961). 'A proposal for a Solomonic settlement between the theories of Rittinger, Kick and Bond', *Trans. AIME*, **220**, 403–408.

Jefferies, H. D. (1969). In *Dispersion of Powders in Liquids* (Parfitt, C. D., Ed), Elsevier, Ch. 8.

Kaplan, L. J. (1981). 'R&D on size reduction is urged to save energy', *Chem. Eng.*, 28 December, 33–34.

Lynch, A. J. and Moore, D. E. (1965). 'The mechanical breakage of mineral particles', *Proc. 1st Tewksbury Symp.*, Melbourne, 26–30 August 1963. Published by Univ. of Melbourne, 1965, 344–362.

NMAB (1981). 'Comminution and energy consumption', Report of the National Material Advisory Board of the US National Research Council.

CHAPTER 2

Review of process development

2.1 INTRODUCTION

The purpose of this chapter is threefold:

(1) to summarize recent progress in comminution by highlighting selected facets: this to encourage mill operators and designers to make the maximum use of these advances for the better optimization and control of existing plant and the more accurate design of new plant;

(2) to expose areas where research is lacking: this to encourage further basic studies by universities and manufacturing industry;
(3) to review up-to-date requirements as comminution-oriented industry develops so that the best type of mill may be selected to fulfil these requirements.

Throughout, reference is made to the appropriate sections of the book.

2.2 FRACTURE PHYSICS

2.2.1 Preview

Fracture physics was mainly developed in areas of fracture avoidance, such as airframe manufacture and bridge construction. It is relatively recently that it has been applied to comminution where fracture is welcomed! However, its introduction has seen considerable improvement in the understanding of the mechanics of breakage.

2.2.2 Methods of force application

The method of application of force to a particle is found to affect the breakage pattern and four basic methods are recognized:

(1) *impact*—particle concussion by a single rigid force
(2) *compression*—particle disintegration between two rigid forces
(3) *shear*—the wrenching apart of a particle by the action of a rigid force or the swirling of a fluid
(4) *attrition*—the scraping of one surface on another, be it particle against a rigid face or particle against particle.

Although in many practical mills it may not be easy to ascertain the proportion of each dominant mode, in basic studies it is important to distinguish these methods of force application, because each may affect the fracture pattern differently.

2.2.3 Single- versus multiple-particle breakage

The main thrust in fracture physics has been with single particles, and that down to 1 micron diameter. This has demanded highly sophisticated equipment, such as microcompressors capable of applying loads of 50 tonnes on to such small particles; the measurement of very high transient temperatures associated with crack formation in solids; and the exploration of stress patterns. Much of the initial study has been with spheres, but it has also been with irregularly shaped particles. An important feature has been the application of statistical analysis to the results of tests in sufficient number to make the analysis valid.

By contrast, multiple-particle breakage has received less intensive study than that of single particles. In the light of the fact that the majority of practical mills operate with beds of particles, this seems to be a gap in research. It is strongly suggested that considerable reward would result from further controlled tests of particle assemblages (see §§6.2.3 and 6.2.4.)

2.2.4 Efficiency versus energy utilization

Formerly great efforts were made to overcome the difficult problem of the measurement of the efficiency of the grinding process by calorimeter methods. From more recent studies in fracture physics it is known that efficiency can only be low—of the order of 3% at the maximum—and that energy utilization is a much more meaningful concept, which is used to good advantage in comparison of energy test data. (see Chapter 6.)

2.2.5 Additives

For many years it was thought that additive adsorption on the surface of a crack within a particle could hasten crack propagation (the Rehbinder effect) or that absorbed molecules could pin dislocations near the crack surface (the chemomechanical effect). However, one of the rewards from single-particle studies has been experimental evidence that neither of these effects is valid and that the true reason for grinding improvement by additives lies in the increased particle mobility produced thereby (see §3.4.2) which allows more particles than otherwise to be in a position to be nipped by the grinding media in a tumbling mill, for example.

2.2.6 Grind Limit

Single-particle measurements of the critical particle size at which the transition from brittle to completely plastic behaviour occurs (see §3.2.6.4) suggests that a grind limit is likely, because further comminution action would then only change shape and not size. Evidence has been produced for tumbling mills that size distributions on a log–log grid, which are straight lines at a given angle over most of the range, tend to become asymptotic to a particle size as comminution proceeds (see §4.3.7). On the other hand, other workers have produced size distributions for the same material down to a size lower than the asymptotic size. Evidence has also been found for chemical change to occur as particle sizes become very small, which may partly account for different results by some experimenters. Others again have proposed an apparent grind limit associated with a particular type of mill. This is clearly an area for further examination, demanding very specialized skills for the resolution of the problem.

2.3 PARTICLE SHAPE

Particle shape can be industrially important as explained in §4.2.1.

Because of the large numbers of particles involved no two of which are exactly alike, the assessment of particle shape has always been a difficult task. The advent of image analysers with back-up computer equipment has simplified the problem to some extent and has given a new impetus to research in what is admittedly a complex field. (see §4.2.5.)

2.4 PARTICLE SIZE DISTRIBUTION

2.4.1 Changes as grinding proceeds

The changes which occur as grinding proceeds have the following features:

(1) Primary breakage results in a bi- (or tri-) modal distribution.
(2) Continued further breakage sees the move toward a mono-modal distribution.
(3) Whatever the feed size, after sufficient grinding the distributions tend to coincide.
(4) Repeated fracture produces more fines than single fracture.

It is heartening to record that, resulting from single-particle studies, all these features could be forecast. Item 1, for example, results from a cone of high stress intensity immediately below the contact area of the applied force, giving many small fragments, whereas the rest of the particle breaks up into a fewer number of larger pieces: i.e. a bimodal distribution. With respect to item 2, as grinding proceeds, the larger pieces from primary breakage are preferentially selected, not only because they are weaker than smaller ones but the smaller will tend to be shielded in the interstices between the larger. Hence the peak in the distribution associated with larger particles gradually disappears, but that of the fines mounts higher, giving a monomodal distribution.

Similar explanations are available for features 3 and 4, as given in §§4.3.5.4 and 4.3.5.5 respectively.

2.4.2 Mathematical representation in terms of parameters linked with physical properties

For many years the search has continued for mathematical expressions for size distributions containing parameters which are meaningful in terms of particle physical properties. Only very recently, there have been conducted in Australia drop shatter tests on brown coal char (easy to shatter) and bituminous coal (more difficult to shatter). The analysis of the results of these tests has been made in terms of volume flaws, assumed to be Poisson distributed. It is shown that for these materials there is a linear relation between the number of volume activated flaws and the breakage probability. This can only be regarded as a hypothesis until it has been repeated for a wider range of both materials and particle sizes, but it is the first glimmer of hope in an otherwise abortive scene. Hence every encouragement should be given to the furtherance of this interesting excercise. (see §4.3.8.4.)

2.4.3 Mathematical representation as arrays

The representation of a size distribution as a numerical array introduced fifty years ago in matrix form, has proved to be a powerful concept. Today the general term notation typified in equation (4.34) is more usual than the matrix format, but either allows comuterized handling of size distributions in full. Attempts to represent a distribution by a single digit, such as x_{80} or by a single surface area value, have obvious disadvantages compared with the complete size information contained in an array, which is therefore being used increasingly.

2.5 CLASSIFICATION

2.5.1 Representation of classifier performance

Advances in the representation of classifier performance take cognisance of several types of deviation from the classification (Tromp) curve, such as (i) an apparent bypass effect due to fines carried in the interstices between large particles in the coarse stream, (ii) the effect of comminution which may occur within a classifier, (iii) a shortfall of data in the smaller size range, and (iv) a fish-hook effect due, for example, to the recoating of separated particles by still finer particles. All these are embodied in a selectivity curve (function) as distinct from the classification curve (function), and lead to a better understanding of the inner workings of a classifier and hence of the total system of which the classifier is but one unit.

Similarly, greater provision than formerly is now made for measurement errors which affect the mass balance over a classifier in each size fraction. (see §5.2.7.)

2.5.2 Sharpness of cut

The importance of sharpness of cut in a classifier is being increasingly recognized and new designs which incorporate improved sharpness are appearing on the market. An example for dry systems is the O'Sepa classifier (see §5.4.2.2(iii)) which has done much to improve the overall performance of cement plants.

There appears to be no hydraulic classifier with a sharpness of cut equivalent to that of the O'Sepa in a dry system, and this represents a considerable challenge to some enterprising organization interested in wet systems.

2.6 LIMITATIONS OF ENERGY LAWS

Since the turn of the century, battles have raged between the proponents of Kick and Rittinger's laws relative to the energy required for comminution. Fifty years ago Bond entered the arena with his concept of energy being inversely proportional to the square-root of particle size. More recent work has demonstrated that such simple expressions only apply over limited ranges of specific energy as judged by Figure 6.27. The same is true of the laws of Holmes and Charles, and hence a better system must be looked for.

2.7 GRINDABILITY UNCERTAINTIES

The disadvantages of single grindability indices, whether Hardgrove or Bond, are now appreciated (see §6.8) and their use in mill design has been seen to give no grounds for complacency. This is demonstrated for the Bond equations in Figures 6.35 and 6.36. Similar problems have arisen with mill design based on the Hardgrove index—e.g. vertical spindle mills. Hence a need has been recognized for a method which takes into account the inner workings of a mill system and does not rely on a single point value of a grindability.

2.8 SHORTCOMINGS OF END-RESULT SCALING LAWS

The scaling of capacity and energy requirements of a mill for several decades has been accomplished by end-result scaling laws, derived from dimensional analysis which produces relationships between dimensionless groups of variables, as discussed in Chapter 7 for various mill types.

The drawback of this overall process is that the characteristics of specific operational zones within the mill are masked. Hence there is a need for a system which allows for the scaling of these zones according to their own criteria. The price paid, of course is greater complexity, but the reward is greater accuracy (see §7.4.)

2.9 THE SIZE–MASS BALANCE METHOD OF MILL DESIGN

To compensate for the uncertainties of grindability, the limitations of simplistic energy laws and the shortcomings of end-result scaling laws, there has emerged during the last twenty years a more comprehensive method of mill design, variously called the size–mass balance (SMB) or the population balance model.

Fundamental to this model are two basic functions:

(1) the specific rate of breakage S of a particle in a given size range (see §8.2.2);

(2) the breakage distribution function B for that size range (see §8.2.3).

Incorporating these functions it is possible to write an equation for the mass balance in a batch process for particles in each size range (see §8.3) and to find a solution to the equation (see §8.4). Methods of determination of the breakage functions *S* and *B* are listed in §8.5. There has developed a considerable literature on the SMB method, particularly for tumbling mills, which are the most widely used mills in the world, and Chapter 9 is devoted to a study of the same for batch processes.

Continuous mill operation demands the introduction of the concept of residence time distribution (RTD), and the measurement of the RTD function and its mathematical representation have been documented in Chapter 8 relative to open circuit mills.

However, as most plants operate on closed circuit, which involves a classifier and recirculated streams, these added features have to be incorporated into the SMB equations. Models at various levels of sophistication have been developed and these are presented for each type of mill in Chapter 10. The models may be applied to the control and optimization of existing plant or to the design of new plant, and examples of each are incorporated into that chapter. From the summary at the close of Chapter 10 it is easy to see the areas where adequate simulation is still missing.

The future is expected to produce models of even greater sophistication than those currently available, which will involve greater complexity of computer programs but will result in more accurate prediction of mill performance both in respect of the product size distribution and the power requirements at a given capacity. However, until secure relations between material properties and breakage functions are established, even the SMB method will remain an empirical system: it will not attain the status of a science. Hence the importance of linking breakage probability with some physical feature of the product, in which a small start has been made, as discussed in §2.4.2. The furtherance of this avenue of research presents a tremendous challenge to those interested in the advancement of comminution.

2.10 NEW COMMINUTION METHODS

Side by side with advance of fundamental comminution knowledge must proceed the development of new comminution methods and the modification of existing mills for specialized duties.

As industrial needs change, so the demands for comminuted products change. A recent example is the need for coal slurries which will replace high-cost oil, and which may be pumped and atomized just like oil. This calls for very fine grinding of coal at a cost acceptable to the industry.

In the very fine grinding regime (see Table 1.1) a special form of planetary mill (Szego) has been developed, and ultrasonic equipment is of interest (see §11.2) In the super-fine regime, the USBM attrition mill has proved its worth (see §11.3).

It is always essential to keep an open mind on unconventional size reduction means such as:

(1) the sudden expansion of steam to rupture particles, as used in the Snyder process (§11.4);
(2) electrohydraulic crushing (§11.5);
(3) thermal decrepitation (§11.6);
(4) chemical comminution (§11.7).

At present none of these unconventional systems appears to be an economic proposition.

2.11 EPILOGUE

At some time in the future it may be possible to write a book entitled 'The science of size reduction'. Then the basic equations will be available whereby a secure calculation may be made, for example, to assess the energy needed to produce a given size distribution from a given material with a coarser distribution, without a prior empirical test. Until then, empiricism is the order of the day.

However, the considerable advances in the understanding of the fundamentals of comminution which have taken place in the last twenty years will undoubtedly proceed apace, having as a long-term goal the elimination of empiricism.

In the meantime it can only be beneficial to mill designers and operators to take full advantage of existing knowledge by using the SMB methods for both optimization of existing plant and the design of new plant.

CHAPTER 3

The physics of breakage

NOTATION FOR CHAPTER 3

$2a$ Crack length
a^* Threshold value of crack length in change from a deformation-controlled to a fracture-controlled situation
A New surface produced by an elemental fracture unit
B Crack width

d	Particle size or diameter of sphere
d_c	Critical particle size for brittle/plastic transition
$e(v)$	Kinetic energy related to an elemental fracture unit
E^*	Young's modulus
F	Force applied to a particle or indenter test specimen
F^*	Threshold value of force in change from a deformation-controlled to a fracture-controlled situation
H	Hardness of a material
K_c	Toughness of a material
L	Length
p_0	Maximum compressive component of stress in the force contact area
R	Radius of force contact area on a sphere
dU	Energy received by an elemental fracture unit from the surrounding elastic stress field
U_f	Fracture energy
v	Crack velocity
V	Volume
Y	Yield stress
α_0	Indenter constant
β	Energy consumed by an elemental fracture unit
ε_θ	Longitudinal strain
$\varepsilon\phi$	Latitudinal strain
$\xi(a)$	Function depending on test-piece geometry in an indenter test
σ_0	Uniform uni-axial stress applied to a middle notched specimen
σ_{ta}	Maximum tensile stress in force contact area
τ_{max}	Maximum shear stress in force contact area

3.1 INTRODUCTION

3.1.1 Preview

The physics of breakage is a vast subject, as demonstrated for example by Bilby's (1979) review which has nearly three hundred references. Only a very limited survey is made in this chapter, without much mathematical detail, but with sufficient coverage to highlight the main features relative to comminution.

3.1.2 Atomic attraction and repulsion in solids

It has been noted in Chapter 1 that the basic equations for size reduction calculations are not yet available. One of the reasons lies in the fact that comminution deals with solids in which complex cohesive bonds exist between atom and atom and between molecule and molecule, which are arranged in ordered arrays, unlike in gases, for which the atoms and molecules are disordered. Not only are there cohesive bonds, but there must also be repulsive forces or a solid would tend to shrink and so increase in density at constant temperature, which is not the case.

Following Marshall (1974), an example is taken of sodium chloride, for which the attractive forces between a sodium ion and a chloride ion are due to electrostatic influences. If, due to these, the ions approach each other, the attractive force will increase; but equally, because of the electron cloud which surrounds the nucleus, a repulsive force exists which will neutralize the attractive force, resulting in a state of equilibrium. This equilibrium is not static but dynamic, and the interatomic distance fluctuates because of the thermal motions of the atoms; but it has a mean value which increases with temperature. Rupture is possible because the net force of attraction between two atoms becomes very small at a distance as little as 10% greater than the mean atomic distance at room temperature.

Rupture takes place clearly by tensile stress. Compressive stress just pushes the atomic nuclei closer together. When compressive forces applied to a particle produce rupture this is because they generate tensile stresses within the particle. When shear accompanies compression there is a tendency for layers of atoms to slide over each other, and with sufficient shear, particle rupture can occur. Fracture can be produced by thermal means alone because increase in temperature implies an increase in mean atomic distance as well as increased amplitude of atomic vibrations. These manifest themselves in thermal expansion, and differential thermal expansion leads to rupture. Similarly rupture can be aided by chemical means because stressed bonds between atoms and molecules are more subject to chemical attack than are unstressed bonds.

3.1.3 Elastic and plastic properties of solids

3.1.3.1 Uniform metal bar

Before rock materials are discussed, the well-documented case of a uniform metal bar is taken.

If the bar is supported at one end and loaded in the direction of its length with a force which puts the bar in tension, then the stress within the material is defined as the force per unit area of cross-section. The elongation produced per unit length is the resulting strain. Considerations of the equilibrium between attraction and repulsive forces of adjacent atoms lead to the relation that stress is proportional to strain (Hooke's law) and the proportionality constant is known as Young's modulus. In fact this is only true of small strains (in practice about 1% of the mean interatomic distance) and hence the straight-line relation between stress and strain, where the material is behaving elastically, applies only over AB in Figure 3.1

Along with the extension of the bar, the cross-sectional area is reduced and the ratio of lateral strain to longitudinal strain is known as Poisson's ratio.

Application of increasing force beyond the value at B in Figure 3.1 causes the metal to 'flow' as shown by BE, where the solid displays ductile qualities.

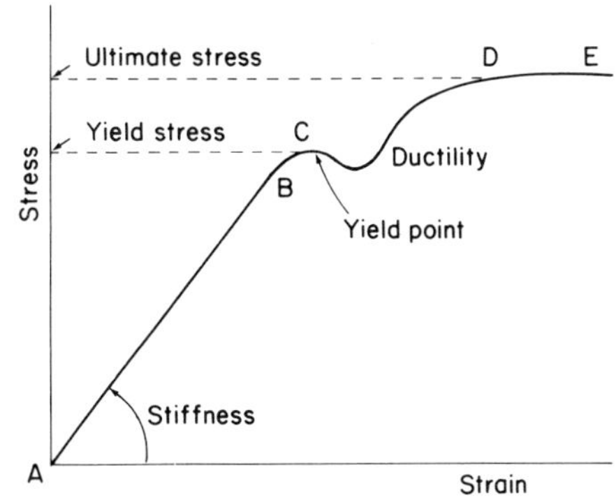

Figure 3.1. Features of a stress–strain curve for a metal bar

The solid is behaving plastically. The stress at point C, where yielding begins, is termed the 'yield stress' and is related to layers of atoms sliding over each other. Eventually the rod breaks and the stress at which this occurs is called the 'ultimate stress'.

3.1.3.2 Rock materials

When a stress–strain curve is produced for rock material such as granite, its characteristics are rather different from those of metals. Figure 3.2 is typical. Four stages to brittle fracture have been noted by Brace and Byerlee (1967):

I and II The behaviour is primarily elastic, but in stage II gains not only distort elastically, but also slide relative to each other.

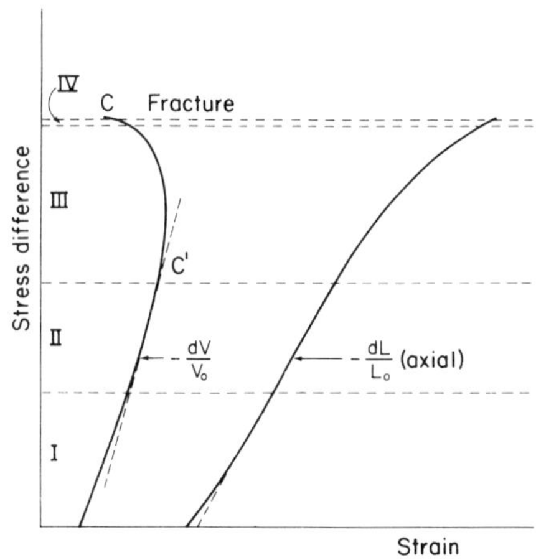

Figure 3.2. Stages in the stress–strain behaviour preceding fracture of rock in compression (Brace and Byerlee, 1967). *Reproduced by permission of the American Institute of Mechanical Engineers*

III The process of II continues, together with the formation of open axial cracks which cause dilation of the rock, indicated by the strong change of shape in the volume stress–strain curve of Figure 3.2.

IV Growth of cracks from grain boundaries occurs, with fault formation leading to rapid fracture.

The straight-line portion associated with Young's modulus in stages I and II is well-defined.

3.1.3.3 Stiffness versus strength

In Figure 3.1 the greater the angle made by AB with the strain axis, the greater the stiffness of the material. Hence Young's modulus is a measure of material stiffness. This is quite different from the strength of the material, and Table 3.1 shows that stiffness and strength do not correlate well. Usually the most easily ground materials (e.g. limestone) have relatively high Young's modulus together with low strength.

Table 3.1. Comparison of material stiffness and strength (Marshall, 1974)

Substance	Young's modulus ($Nm^{-2} \times 10^{-9}$)	Ultimate tensile stress ($Nm^{2} \times 10^{-6}$)
Rubber	0.007	28
PVC	1.4	30
Limestone	35	7
Glass	70	80
Mild steel	210	420
Carbon whiskers	800	7000

Reproduced by permission of the Institution of Chemical Engineers.

3.1.4 Strain energy

The product of stress and strain is energy per unit of cross-sectional area, known as strain energy. It is given by the area between the curve AD in Figure 3.1 and the strain axis. The area under AB only is called the elastic strain energy.

It must not be thought, however, that the strain energy is uniformly distributed throughout a particle. Inglis (1913) was aware that strain energy is concentrated around holes, particularly where the radius of curvature is small, and he postulated the well-known concept of a stress concentration factor, which is the number of times the local stress is greater than the mean stress value in the material.

3.1.5 Hardness, toughness and brittleness

Hardness is described in relative terms by the Moh scale, in which a solid will scratch another placed below it in the list given in Table 3.2.

Quantitively, hardness is decided by indentation with a hard profile as follows:

Profile	*Test*	*British Standard*
Ball	Brinell	BS 240 (1962)
Pointed cone	Vickers	BS 427 (1961)
Pyramid	Rockwell	BS 891 (1961)

Thus hardness H is resistance to deformation. On the other hand, toughness K_c is resistance to fracture. Recently Lawn and Marshall (1979) have defined an index of brittleness as the ratio H/K_c,
where $H = F/\alpha_0 R^2$ and $K_c = F\xi(a)$. Hence:

$$\text{Index of brittleness} = \frac{1}{\alpha_0 R^2 \xi(a)} \tag{3.1}$$

which is independent of the applied load, and can be obtained by means of an indentation test (Vickers). Typical values are given in Table 3.3.

The above study thus offers a valuable approach to material classification for ease of comminution. Although the concepts have been developed in the context of a Vickers hardness test, the brittleness index has more general application. Materials with high brittleness index will be easy to comminute. Those with low brittleness index will be difficult to comminute because of their toughness, and plastics are a good example.

The whole area of indentation fracture is becoming of increasing importance and students are referred to review papers such as Bilby (1979) and Lawn and Wilshaw (1975).

Table 3.2. Moh hardness scale (Marshall, 1974)

Hardness number	Substance	Common practical test
1	Talc or graphite	Marks paper
2	Rocksalt or gypsum	2.5 is finger nail hardness
3	Calcite	
4	Fluorspar	4 is copper coin hardness
5	Apatite	
6	Felspar	6 is window glass hardness
7	Quartz	6.5 is pocket knife blade hardness
8	Topaz	
9	Sapphire	
10	Diamond	

Reproduced by permission of the Institution of Chemical Engineers.

Table 3.3. Typical hardness, toughness and brittleness values (Lawn and Marshall, 1979)

Material	Deformation/fracture parameters: Hardness H *(GPa)*	Deformation/fracture parameters: Toughness K_c *(MPa $m^{1/2}$)*	Deformation/fracture parameters: Brittleness H/K_c ($\mu m^{-1/2}$)	Threshold parameters: Critical load F^* (N)	Threshold parameters: *Critical crack* length a^* (μm)
Medium-strength steel	5	50	0.1	800,000	12,000
NaCl (monocrystal)	0.24	0.4	0.6	30	330
Al_2O_3 (MgO doped)	12	4	3	2	12
SiC (hot-pressed)	19	4	5	0.6	5
SiO_2 (glass)	6.2	0.7	9	0.02	1.5
Si (monocrystal)	10	0.6	17	0.002	0.4

Reproduced by permission of the American Ceramic Society.

3.1.6 Crack propagation in solids

Most solids have some parts weaker than others owing to the presence of flaws, dislocations, microcracks, crystal boundaries and impurities. Griffith (1920) used a simplified system of a small elliptical flaw in a large thin plate, because the mathematics of the three-dimensional case is difficult, but it has been treated by Sack and Wills (1941) and Elliott (1947). Griffith followed up the Inglis (1913) concept of stress concentration factors in the regions of the tips of the flaws and showed that cracks will be initiated therefrom. By performing an energy balance on the moving crack tip, Griffith further showed that the overall stress round the flaw must reach a critical value. Once the crack starts, if the strain is not rapidly relieved, the system becomes unstable and crack propagation accelerates to high velocities, reaching 40% of the speed of sound in the solid.

During this rapid progress, the strain energy relaxed will exceed the energy associated with the new surface produced, and the surplus energy can concentrate on other cracks in the material and propagate these too. Hence once a crack begins to spread, multiple fracture can ensue.

Paul and Gangal (1967) may be referred to for treatment of initial and subsequent fracture curves for bi-axial compression of brittle materials.

Brace and Byerlee (1967) have also shown that the simple process described by the Griffith theory does not apply for rocks under compression, since, as seen in the volume change curve of Figure 3.2, crack growth must begin much earlier than the fracture at C. In fact it must begin at C′, when there commences the dilation referred to in §3.1.3. This is in keeping with two features of rock breakage under compression which are otherwise difficult to explain:

(1) The absolute fracture strength of a rock is very reproducible from sample to sample compared with the variation usually observed with brittle fracture of materials in tension.
(2) Formation of the final fault requires almost as much stress (C to C′ in Figure 3.2) as that required to initiate crack growth, which is the stress C′.

Materials which behave in the manner described above have a high brittleness index; that is, the strain energy within them is relaxed by crack propagation rather than the alternative of deformation, as discussed in §5.1.

To return to Lawn and Marshall (1979), further analysis of the data they assembled was made using the following normalized parameters (to make them dimensionless):

For the deformation and fracture process	$F/(K_c/H)^2$
For the indenter load	$F/K_c(K_c/H)^3$

A universal diagram, produced as Figure 3.3, enables the response of either crack propagation or deformation to be predicted for any material for which the hardness and toughness are found by standard tests. The curve is essentially an 'equilibrium' curve in the sense that cracks corresponding to a point

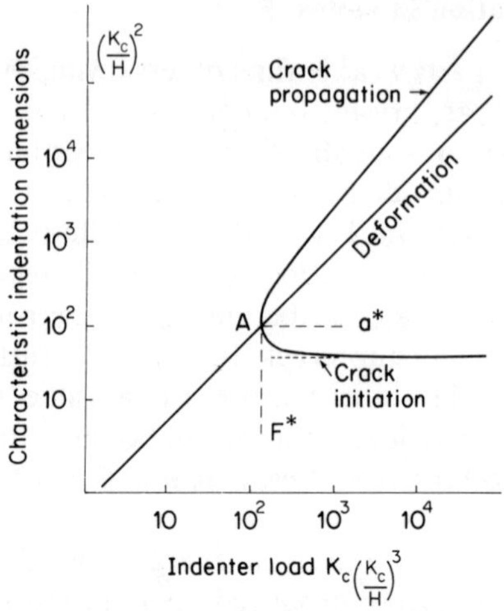

Figure 3.3. Universal deformation/fracture diagram (Lawn and Marshall, 1979). *Reproduced by permission of the American Ceramic Society*

on the curve represent a configuration that is on the point of extension. In the initiation region the curve indicates that for smaller crack nuclei a higher indentation load is required to maintain a state of equilibrium. The equilibrium is unstable in the initiation region, so that once a nucleus extends, the crack grows spontaneously at constant load to the propagation (stable) band of the equilibrium.

The intersection of the deformation and fracture curves at A indicates also the threshold condition for which F^* is the maximum load that a solid might sustain in a contact event without the onset of fracture and a^* is the critical crack length corresponding to F^*. When $F^* > F$ conditions are such that the process is deformation-controlled, and conforming materials—for example metallic compounds and polymers—are difficult to comminute. Similarly when $F^* \ll F$ the damage is essentially fracture and conforming materials are relatively easy to comminute.

3.1.7 Review

It is hoped that the basic concepts covered in this introduction may be considered a sufficient background to the physics of breakage. The subject is further detailed in subsequent sections by consideration of the factors on which the stress pattern round a crack depends, and of the energy required to produce a crack. The effects of chemical additives are also discussed.

3.2 PARTICLE STRESS PATTERN

3.2.1 Features for discussion

Following Schönert (1979) the three-dimensional stress pattern produced in a particle under the influence of applied forces is dependent on

(1) the number and direction of the forces;
(2) the deformation behaviour (elastic or plastic);
(3) the rate of deformation;
(4) the particle size and shape.

These are now considered in turn.

3.2.2 Effect of number of contact forces

As indicated by Jaeger (1967), from any one contact point cracks develop in the direction of another force contact point as shown in Figure 3.4.

Breakage probability within an assembly of spheres increases with the number of contact points per particle up to a limited number of the order of ten. If more than this, the breakage probability decreases and the greater is the tendency for cracks to run near the periphery of the solid. These probability features were demonstrated by Hoffmann and Schönert (1971) who loaded a binary mixture of 2 mm and 1 mm glass spheres with a compression of 28 N/mm^2. Figure 3.5 shows that with only 2 mm spheres, about 80% of them were disintegrated. As the proportion of 1 mm spheres increased, the breakage probability fell rapidly. With the addition to the 2 mm spheres of 0.5 mm spheres instead of the 1 mm spheres, the fall was even more rapid owing to the further increase of the number of contact points.

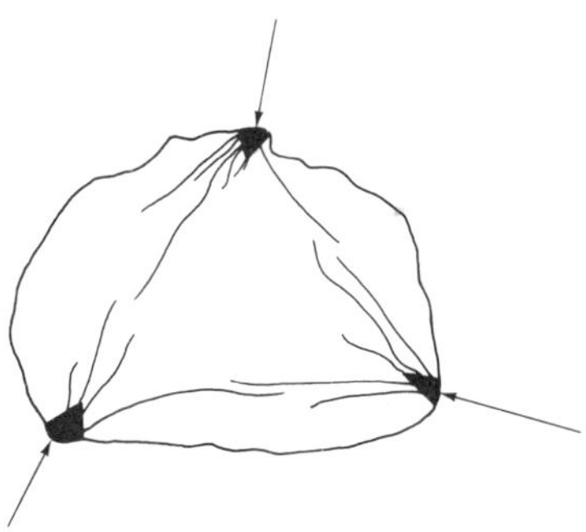

Figure 3.4. Pattern of crack development within a particle under the influence of multiple forces

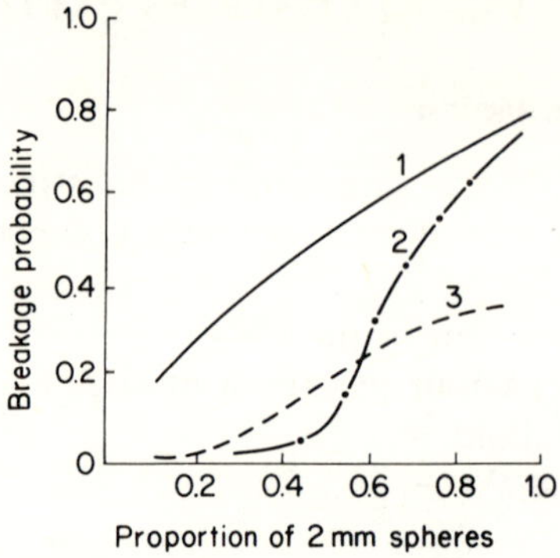

Figure 3.5. Breakage probability of a binary mixture of spheres (Schönert, 1979). *Reproduced by permission of the author*
(1) 2 mm spheres in mixture of 1 mm and 2 mm
(2) 2 mm spheres in mixture of 0.5 mm and 2 mm
(3) 0.5 mm spheres in mixture of 0.5 mm and 2 mm

This has a direct application to a bed of particles as exists in a ring-roll or ring-ball mill, and shows the importance of removal from the bed of fine particles of output size, otherwise the probability of larger particles being broken is reduced by the increased number of contact points per particle.

3.2.3 Deformation behaviour—radial forces

3.2.3.1 Theoretical studies

This section deals with spheres acted on by radial forces.

It is essential to differentiate between materials which behave predominantly elastically or plastically. For the elastic case with point force applications, theoretical treatments are available from Hertz (1881) and Huber (1904) for the volume within the sphere in the force contact area and its immediate surroundings, and by Lurge (1963) for parts distant from the contact zone. There is no known theory for the case of plastic behaviour.

3.2.3.2 Contact zone considerations

Glass is a material which normally behaves elastically. Consider a glass sphere acted on by a weighted horizontal hard flat plate, producing a total load F. The sphere will deform, giving a circle of contact of radius R. Thus the applied pressure is $F/\pi R^2$. Inside the sphere (Hamilton and Goodman, 1966) the maximum compressive stress p_0 arising from the deformation occurs at the centre of contact and it has a value of $3F/2\pi R^2$, which is equal to 1.5 times the

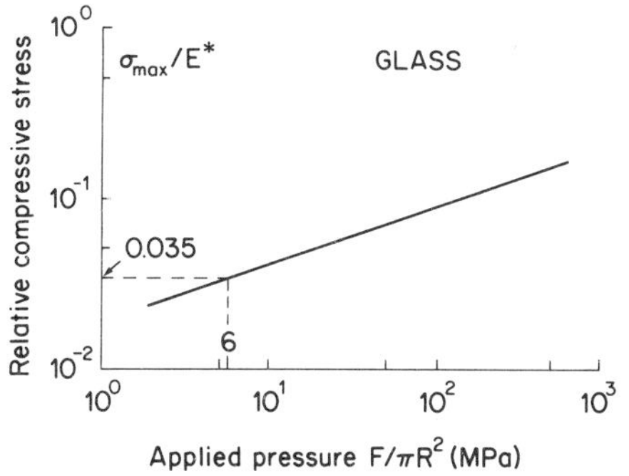

Figure 3.6. Relation between the applied pressure and the resultant compressive component of stress in the force contact area of a glass sphere (Schönert, 1979). *Reproduced by permission of the author*

applied pressure. The relation between p_0 and the applied pressure is given in Figure 3.6. For an applied pressure of only 6 MPa the graph indicates that $p_0/E^* = 0.035$.

Taking Young's modulus for glass as 62,040 MPa (9×10^6 psi), then $p_0 = 0.035 \times 62{,}040 = 2110$ MPa. But at 2000 MPa glass begins to deform plastically (Bridgeman, 1952). Under these conditions glass is behaving plastically at normal temperature.

Hamilton and Goodman (1966) also show that, as the Hertz–Huber theory predicts, the highest tensile stress σ_{ta} occurs round the edge of the contact area, giving rise to a ring crack observed by many (e.g. Rumpf and Schönert, 1972). Within this ring in a cone-shaped volume, lesser tensile stresses also occur.

Further, Hamilton and Goodman (1966) indicate the presence of shear stresses, the maximum τ_{max} of which is found at a distance $R/2$ below the centre-point of the contact area and within the cone volume.

Following Schönert (1979) the relative values of these three maximum stresses for glass with a Poisson's ratio of 0.3 are:

$$p_0 \colon \tau_{ta} \colon \tau_{max} = 1{:}0.11{:}0.5$$

Since p_0 is itself 1.5 times the applied pressure, the shear stresses can be up to 0.75 of the applied pressure, which illustrates the intensity of the stress just below the contact area—much greater, in fact, than in the rest of the particle. Hence, for the conditions stated, a more intense size reduction would be predicted for the material in the cone below the contact area than in the rest of the sphere. This was confirmed in practice for glass spheres by Arbiter *et al* (1969). The fines from the contact zones they described as the 'complement' to the 'residue' from the rest of the sphere. In fact, a bimodal size distribution (Snow and Paulding, 1973) resulted from the overall fracture, which is discussed in greater detail in §4.3.5.2. Under certain conditions trimodal distributions are obtained (Hukki, 1943; Rumpf, 1965).

When a sphere of plastic material such as PMMA (a low molecular weight acrylic resin) is treated in a similar manner to the glass spheres, it is observed (Rumpf and Schönert, 1972; Gildemeister and Schönert, 1976) that a cone-shaped volume is driven into the sphere by the applied force, pushing the surrounding material sideways as described in §3.2.3.3(ii). The contact cone remains undestroyed, in contrast to the case of glass, for which the contact cone breaks into very fine fragments.

3.2.3.3 Consideration of parts distant from the contact zone

To study the effect of applied pressure in parts other than the contact zone, 50 mm spheres of steel and of PMMA were each equipped with two sets of stain gauges. One set of four was secured on a meridional arc in the *yz* plane and lying in the direction of the meridian, to give the longitudinal strain ε_θ. The other set of four was placed on a meridian in the *xz* plane and lying in the longitudinal direction to give ε_ϕ, the latitudinal strain, as shown in Figure 3.7. Only one-quarter of the circumference was treated, as the strain pattern was symmetrical. A compressive load of 2500 kg was applied.

The resulting strain patterns are shown in Figure 3.8, in which compressive strains are marked inside the circumference and tensile strains outside it. In each case the lengths of the radial lines show relative magnitudes of the strains, but note that the scale is different in the two cases.

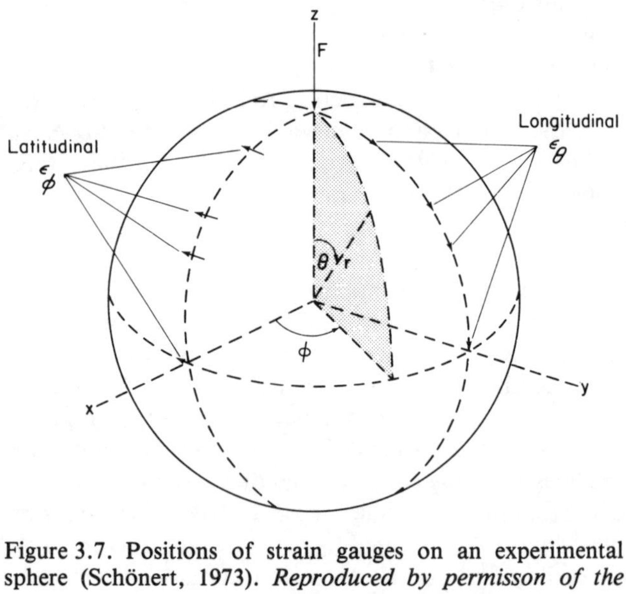

Figure 3.7. Positions of strain gauges on an experimental sphere (Schönert, 1973). *Reproduced by permisson of the author*

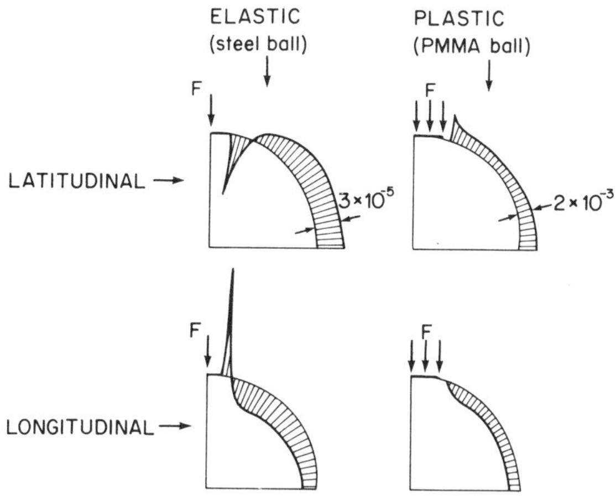

Figure 3.8. Strain pattern resulting from a radially applied force on elastic and plastic spheres. Shaded areas indicate the degree of strain, tensile shown outside and compressive shown inside the quadrants (Schönert, 1973). *Reproduced by permission of the author*

(i) Taking the *elastic* case first (steel ball) and looking at the latitudinal strain ε_ϕ, it may be noted that the strain just outside the contact area is compressive and so large that, if it were drawn to scale, it would go off the diagram. At an angle at the centre of about 16° with the vertical, the strain changes to tensile and evens out to a level of 3×10^{-5} around the circumference.

Similarly the longitudinal strain in the elastic case is exceedingly large but tensile on the edge of the contact circle. At an angle at the centre of about 11° the strain becomes compressive and evens out to a similar magnitude to the latitudinal strain around the circumference.

From Hertzian theory, the stresses corresponding to these strains can be calculated and typical values are:

Compressive stress at centre of contact area = 600 kg/mm^2
Longitudinal tensile stress on edge of contact area = 88 kg/mm^2
Maximum latitudinal tensile stress = 0.6 kg/mm^2

Since cracks develop perpendicular to the maximal tensile stresses, it would be predicted that:

(1) Ring and cone cracks would develop round the contact circle because of the high longitudinal tensile stress displayed there.
(2) Cracks would run round curved surfaces like onion peel following surfaces of constant strain, which are in this pattern.
(3) No meridional cracks would appear because of the low latitudinal tensile stress.

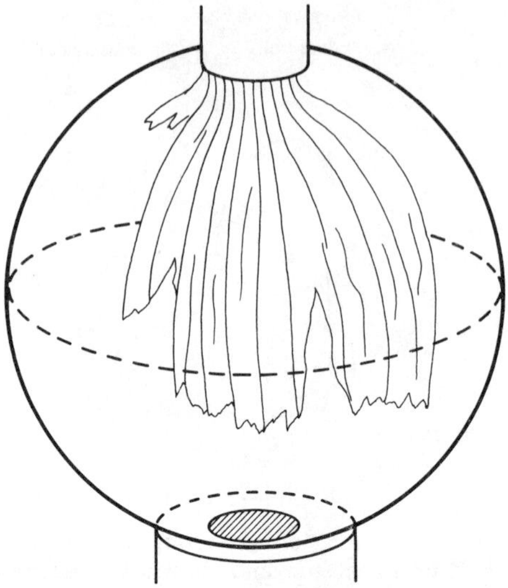

Figure 3.9. Crack pattern within an elastic sphere under an applied radial compressive force (Schönert, 1973). *Reproduced by permission of the author*

All these expectations are realized in practice. Evidence of the ring crack has already been mentioned in §3.2.3.2, and Figure 3.9 shows the onion peel effect for a 6 mm diameter glass sphere.

(ii) Coming now to the *plastic* sphere, the longitudinal strain is compressive and the latitudinal is tensile, as seen in Figure 3.8, but the order of magnitude is 100 times greater than for the steel sphere with the same applied force. Because PMMA is a visco-elastic material, it is not possible to apply Hertzian theory to calculate the stresses near the contact zone. However, it is observed that high latitudinal tensile strains occur around the contact area because beneath it a cone-shaped volume is driven into the sphere by the applied force, as noted in §3.2.3.2, pushing the surrounding material sideways and upwards along logarithmic spirals, as shown in Figure 3.10. The outer shell is thus strained as if under internal pressure.

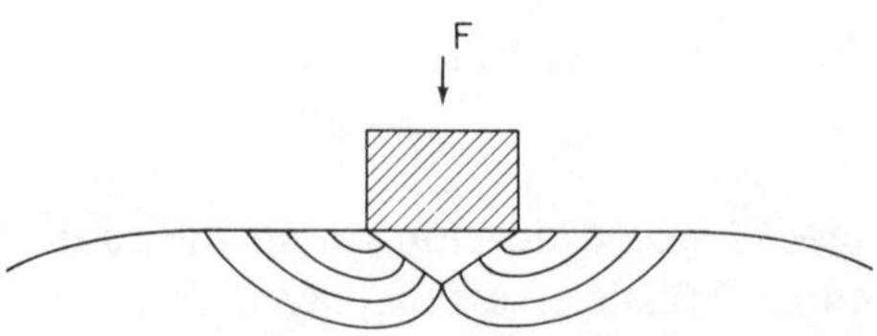

Figure 3.10. Effect of a radially applied force on a plastic sphere near to the area of force contact (Schönert, 1979). *Reproduced by permission of the author*

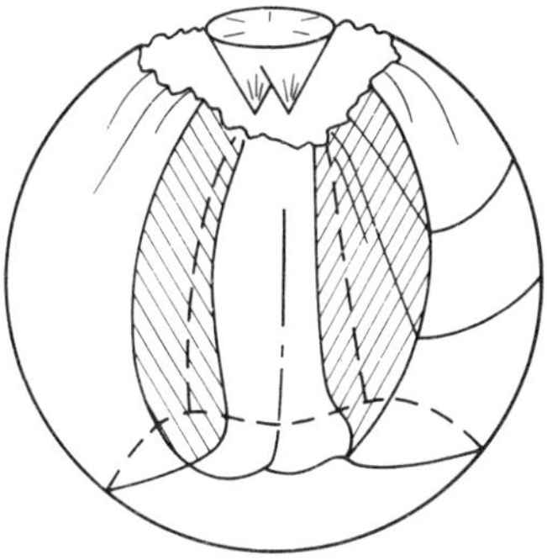

Figure 3.11. Crack pattern within a PMMA sphere under a radially applied force at $-196°C$ (Gildemeister and Schönert, 1976). *Reproduced by permission of DECHEMA, Frankfurt*

Associated with this lateral tensile strain will be lateral tensile stress, and hence cracks parallel to meridional planes would be forecast, dividing the sphere into orange lith like fragments. This expectation is realized with visco-elastic materials. The contact cones remain undestroyed, but the rest fragments as described by Gildemeister and Schönert (1976) for impact and by Rumpf and Schönert (1972) for compression. Figure 3.11 applies to the breakage of a PMMA sphere at $-196°C$ and shows both the solid cone and the meridional cracks.

3.2.4 Deformation behaviour—non-radial forces

This section deals with a sphere acted on by a non-radial force, which can be considered as having two components, a radial and a tangential.

For a sphere behaving elastically, the stress pattern with and without shear is shown in Figure 3.12, calculations having been performed on the basis of the Hertz–Huber theory, with a coefficient of friction of 0.25, by Hamilton and Goodman (1966). By the application of shear, the stress pattern becomes asymmetrical and the tensile stress is considerably increased on one side of the force contact area. Thus it would be expected that the sphere would fail under increasing proportion of shear with a decreasing compressive load, and this has been demonstrated experimentally for elastic behaviour, as shown in the glass sphere case in Figure 3.13.

For a sphere behaving plastically in the contact zone, the application of shear does not greatly affect the solid cone which is formed beneath the contact area and which is responsible for the disintegration of the rest of the sphere. Hence superimposed shear does not strongly alter the breakage pattern in the plastic case, as seen for cement spheres in Figure 3.13. Although hardened cement is normally thought of as a brittle material, here it is acting plastically

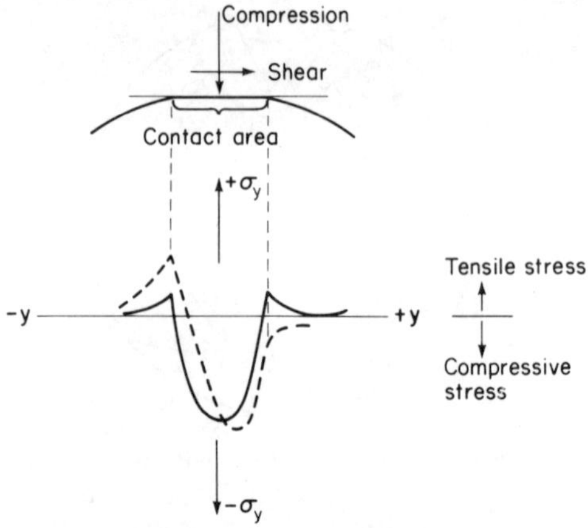

Figure 3.12. Comparison of stresses within a sphere near the force contact area, with and without shear (Schönert, 1979). *Reproduced by permission of the author*
—— Without shear
- - - With shear

in the force contact area. Because most materials behave in this way in the contact volume, attrition, which implies shear forces, must be regarded as wasted energy.

There is, however, a secondary effect of shear. After the particle has been

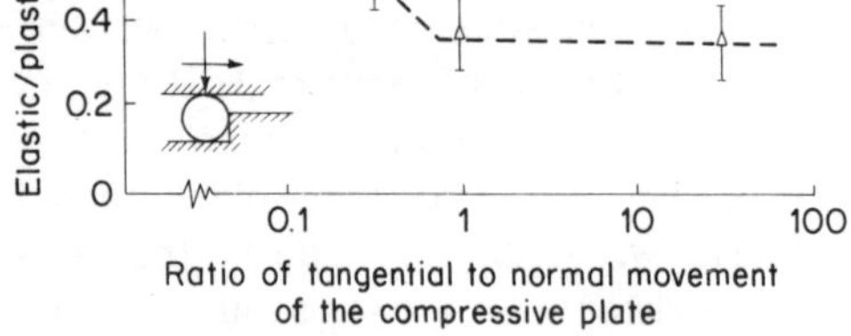

Figure 3.13. Ratio of breakage force for the elastic to the plactic case for spheres, versus ratio of the tangential to the normal components of force (Schönert, 1979). *Reproduced by permission of the author*
—— 5 mm diameter cement spheres without shear at 18 N/mm^2 (inelastic)
- - - 1 mm diameter glass spheres without shear at 500 N/mm^2 (elastic)

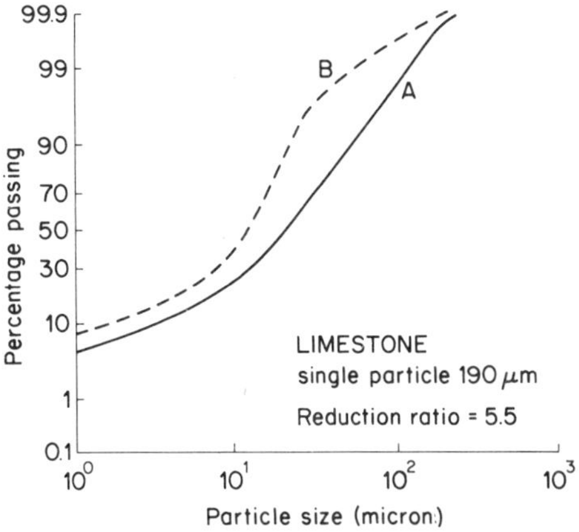

Figure 3.14. Size distributions of limestone produced by crushing rolls with relative roll speeds of 1 : 1 and 10 : 1 (Schönert, 1979). *Reproduced by permission of the author*
—— Speed ratio 1 : 1
--- Speed ratio 10 : 1

broken, shear can help to distribute the fragments and so aid the comminution process. Thus (Ohe, 1967) a single 190 micron particle of limestone crushed between rolls moving at equal speeds produced the size distribution A in Figure 3.14. When the ratio of the rolls was put at 10:1, size distribution B resulted for the same reduction ratio because of the favourable dispersion of the particles. Again, although limestone is usually a brittle material, in the zone of force contact it is acting plastically.

3.2.5 Effect of deformation rate

It is well known that toffee is easily broken by a sharp tap with a hammer, but attempts to sever by slow bending are usually abortive. This is a common example of the effect of deformation rate on fracture.

Again impact and compression are, conveniently, thought of as distinct modes of force application, but they really only differ in the rate at which the force is applied.

Practical deformation velocities vary considerably with the mode of comminution as follows:

Machine	*Velocity* (m/s)
Crusher, roll mill, ball mill	0.1–10
Impact mill	20–200
Fluid energy mill	up to 500

There appear to be two main effects of change of velocity and its consequent influence on the particle deformation rate.

The first is the change in the stress field in an elastic body. Each distortion caused by a contact force is distributed in the body as elastic waves, which on internal reflection meet oncoming waves to change the stress values associated with the wavefronts. Under favourable reflection conditions it has been found by Gildemeister and Schönert (1972) that, for spheres, cracks can be caused by such waves. However, for irregularly shaped particles it is highly improbable that the reflection conditions would ever favour crack formation.

The second is a change in deformation behaviour in a visco-elastic body caused by the change in the deformation rate. The greater the velocity the less the deformation but the higher is the stress. This is a characteristic of such materials as polystyrene, which are therefore better crushed by impact than by slow compression.

3.2.6 Effect of particle size

3.2.6.1 Survey

Small particles have internal stress patterns different from larger ones and thus they behave differently. Three reasons have been advanced relating to (i) flaw size, (ii) stored elastic energy and (iii) plastic deformation effects.

3.2.6.2 Flaw size

Clearly flaw size must decrease with decreasing particle size. Hence the stress levels for breakage have to be increased accordingly.

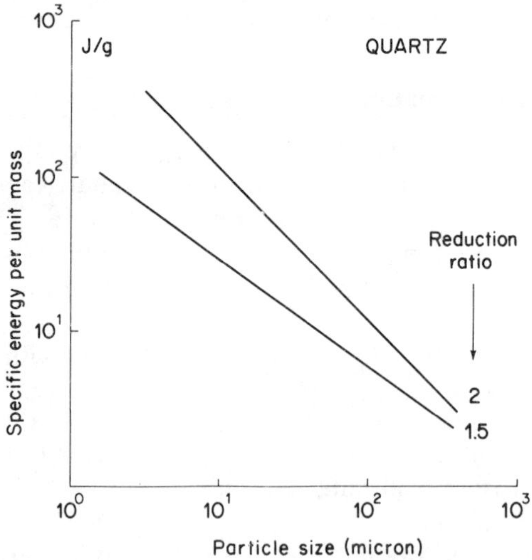

Figure 3.15. Specific energy per unit mass versus particle size for quartz (Steier and Schönert, 1972). *Reproduced by permission of DECHEMA, Frankfurt*

On a force–displacement diagram, Steier and Schönert (1972) have shown for both quartz and limestone that with 100 micron particles, the generally rising curve is punctured by many peaks and troughs, each indicating internal fractures, before final breakage takes place. By contrast, the force–displacement curves for particles in the 1–3 micron size range for the same material show smooth curves, indicating no internal fractures up to final breakage.

The same researchers have plotted mass related energy (in joules per gram) versus particle size (in microns), as shown in Figure 3.15, for quartz at reduction ratios of 1.5 and 2. There is a marked increase in the work of crushing related to mass as the size decreases, so indicating increase in particle strength. There is inherent in this experiment an increase in the relative volumes of the contact zones with smaller particles, which Hess and Schönert (1981) later recognized as a relevant factor in the increased plasticity as particle size is reduced.

3.2.6.3 Stored elastic energy

The second reason is associated with the elastic energy stored within a particle at the moment of crack release, which energy decreases proportionally with particle volume. However, the energy needed for crack propagation diminishes proportionally with the cross-section. As particle size is reduced, it can happen that less energy is stored than is required and the crack will stop. Then more energy must be fed in before final breakage ensues. This phenomenon is specially found in plastic materials, which may have to be impacted many times before they break.

3.2.6.4 Plastic deformation effects

Thirdly, plastic deformation effects become increasingly important as particle size is reduced. For example, Steier and Schönert (1972) compressed single quartz and limestone particles, varying in size from 100 to 1 micron, and observed the fracture pattern in an electron microscope. Down to about 5 microns the number of radial cracks were seen to increase with reduced particle size, owing to increased plastic deformation in the middle of the particle. Below 5 microns the plastic deformation spread out further, so there was no further increase in radial cracks. Finally at the 1 micron size the whole particle deformed without cracks. Its behaviour was plastic throughout. For polystyrene, however, the transition to unfractured deformation occurred at the 10 micron size.

This phenomenon was apparently first observed by Klemm and Smekal (1941) in their work on the scratching of glass and quartz, and they called it 'athermal melting'.

For coal, Parish (1967) reported on the plastic deformation critical size observed earlier by Boddy (1943). Based on experiments with cubes of coal, Parish was able to relate the critical particle size for complete plastic deformation with dry mineral matter free carbon content of the coal over a size range

Table 3.4. Critical particle sizes for brittle/plastic transition

Material	d_c (micron)	Reference
Cement clinker	10–20	Hess and Schönert (1981)
Limestone	4.4	Steier and Schönert (1972)
Crystalline boron	3	Hess and Schönert (1981)
Quartz	1.7	Steier and Schönert (1972)
Glass	0.5	Kendall (1978)

from 500 to 10 microns. For irregularly shaped coal particles, since the compressive force is initially applied to a few specific points, rather than the face of a cube, the critical sizes for plastic deformation are smaller, ranging from 100 to 1 micron. Hess and Schönert (1981) report a critical size of 35 microns for a coal specimen on which they worked, which is in keeping with Parish's results.

The critical particle size d_c for the transition to completely plastic behaviour of brittle materials has been given by Kendall (1978) as:

$$d_c = 32\ U_f E^* / 3Y^2 \tag{3.2}$$

Such an equation, of course, is developed with simplifying assumptions and can only give an order of magnitude for the particle size. It takes no account of particle shape nor of crack initiation considerations nor of temperature and rate effects, but it does serve a useful function in comminution. For example, for calcium carbonate it predicts a limiting particle size of about 1 micron, which compares favourably with the known average particle size of 0.8 microns from prolonged milling of the material (Gregg, 1968). Thus it has a direct bearing on the grind limit discussed in §4.3.7.

Other approximate critical particle sizes for the transition to completely plastic behaviour of brittle materials are shown in Table 3.4.

Under the headings of plastic deformation must also be mentioned a residual stress effect in sufficiently small particles which can effect breakage, not when the load is applied, but after it has been removed. Schönert (1972) refers to a 21 micron diameter quartz glass sphere, photographed after compression loading and showing shadow regions underneath the contact areas. The shadows indicate that the material in these regions is compressed to a higher density than the rest of the sphere. The stress field created under these conditions remains even after unloading, and can result in particle rupture after unloading. When this happens, the fragments are much coarser than those produced when conditions are such that rupture occurs during loading, for the same sized sphere.

3.2.7 Spherical versus other shaped specimens

Early workers such as Carey and Bosanquet (1933), Smekel (1938) and Gaudin and Yavaska (1946) employed cubes of material in singe-particle crushing tests, thinking this was the simplest shape. However, the load may be concentrated at any irregularity on the platen surface supplying the force or on the cubic particle surface, causing longitudinal cleavage. Another difficulty is that particle and platen expand laterally by different amounts under applied force, causing a friction force between platen and particle. Obert *et al.* (1946) showed that this effect causes the apparent compressive strength to vary with height–width ratio for cylindrical samples, which are therefore not favoured. For cylinders (or rectangular parallelopipeds with square ends) it has been found necessary to make the height–width ratio at least 2:1 for most materials, to avoid a fracture plane within the specimen intersecting the ends receiving the load. This has been explored for compacted powders by Bhattacharya (1966), but the same principles hold good for any solid material.

Later workers have bypassed the above problems by using spheres, which Hiramatsu and Oka (1966) have shown to have similar internal stress distribution patterns under compression as irregularly shaped specimens, as long as the local curvature of the specimen is not very different from that of a sphere. Snow and Paulding (1973) have gone further, by producing density–size distributions from irregularly shaped particles which are not dissimilar from those obtained from the breakage of spheres under similar conditions. (see §4.3.8.3(c).)

Hence no single particle test data based on the crushing of cubes or cylinders are considered valid.

3.3 CRACK PROPAGATION ENERGY

The Griffith criteria for crack propagation, introduced in §3.1.6, are that (a) it is energetically possible, i.e. that the energy of the new surface produced is less than the energy released by relaxation of strain in the surrounding material and (b) that there is a mechanism whereby it can occur, i.e. that plastic flow is inadequate to relieve the stress, which is the case with brittle materials.

Originally Griffith (1920) considered only the loss of energy from the elastic strain field. Later Irwin (1947) and Orowan (1949) incorporated a correction for plastic deformation, and more recently Rumpf (1961) has considered other sources of energy such as residual elastic stresses due to mechanical or thermal pretreatment, thermic energies of atoms, ions or molecules and chemical reaction or adsorption at the crack tip. Rumpf also considered ways of energy dissipation other than crack propagation, such as the kinetic energy of material just after the crack passes through a section; the plastic deformation near the tip; and electric charges on the new surface produced.

Many of these items are very difficult to evaluate, and as a simplification, the elastic stress field U is considered the only energy source. Following Schönert (1972) the energy consumption mechanisms are denoted by β, a function of temperature and crack velocity v. β is called the fracture surface energy related to an elementary fracture unit. Let $e(v)$ be the kinetic energy related to the unit and dA the differential increase of fracture surface. Then for a propagating crack:

$$dU/dA = \beta + e(v) \tag{3.3}$$

or the energy from the strain field must be equal to or greater than the energy consumed by the crack. At the starting point, the kinetic energy is negligible, so:

$$-\mathrm{d}U/\mathrm{d}A = \beta \tag{3.4}$$

Fracture energy U_f is equal to the energy offered from the surrounding strain field to propagate the crack of width B and length $2a$, and is associated with half the new crack surface, or:

$$U_f = -\mathrm{d}U/2B\mathrm{d}a \tag{3.5}$$

Since $\mathrm{d}A = B\,\mathrm{d}a$ this leads to:

$$-\mathrm{d}U/\mathrm{d}A = U_f/2 \tag{3.6}$$

For a middle-notched specimen in a fracture test, with uniform uniaxial stress σ_0 applied:

$$U_f = \pi\sigma_0 a/E^* \tag{3.7}$$

Then from equations (3.4), (3.6) and (3.7):

$$\beta = \pi\sigma_0 a/2E^* \tag{3.8}$$

Thus, if σ_0, a and E^* are measured in a fracture test, it is possible to calculate β, the fracture surface energy, and this has been done for the following materials (Schönert, 1972):

Glass	10^3–10^4 ergs/cm^2
Plastics	10^4–10^6
Metals	10^6–10^8

By contrast, the free surface energy of solids is of the order of 10^2–10^3 ergs/cm^2, which is quite insignificant in fracturing.

Clearly comminution cannot occur without expenditure of energy commensurate with β. In fact, the plastically deformed zone at the crack tip is a moving heat source and the resulting temperature rise from the very considerable energy expenditure has been measured by Wells (1953) using thermocouples. The temperature attained depends on crack resistance, width of crack zone and on crack velocity. Theoretical treatment results in a solution with width of crack zone as the only unmeasurable quantity, as shown by Weichert and Schönert (1978), but plausible values can be estimated as by Dugdale (1960) giving calculated temperatures in degrees Kelvin as follows:

Plastics	200–800
Calcite	up to 1300
Glass	1000 to 4000
Quartz	up to 4400

For PMMA, Fuller *et al.* (1975) have actually measured temperatures of approximately 500° K, which falls within the calculated range for plastics.

These high temperatures are for very short time durations, of course, of the order of 10^{-12} to 10^{-7} seconds, but they could well be the reason why quartz changes from the crystalline to the amorphous state during vibration milling, as indicated by Rehbinder and Chodakow (1962). Steier *et al.* (1974) have reported the same change during compression stressing of quartz, and Roth (1977) for sugar during impact milling.

3.4 INFLUENCE OF ADDITIVES

3.4.1 Early theories

There is an extensive literature on additives which Rose and Sullivan (1958) summarized up to that date. More recently Somasundaran and Lin (1972) and Fuerstenau (1979) have reviewed the subject, revealing proposed explanations of the observed environmental effects which have not been substantiated by later, more carefully controlled experiments.

The immediate environment of a crack has been elegantly explored by Schönert *et al.* (1969) by their measurement of crack velocities in the special case of slow crack propagation in glass samples, and of the effect of moisture and temperature on those velocities. From Figure 3.16, a plot of crack velocity versus fracture energy, it is clear that the presence of moisture promotes such

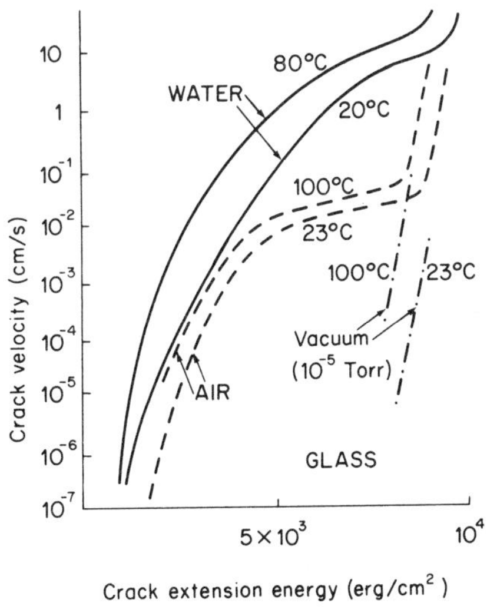

Figure 3.16. Crack velocity versus fracture energy for glass under different crack environmental conditions (Schönert *et al.*, 1969). *Reproduced by permission of Chapman & Hall*

slow crack propagation. Similarly, increase in temperature gives a shift to lower fracture energy levels. Above a velocity of only 10^{-1}cm/s, however, all the curves coincide, and the moisture and temperature effects are slight.

As Klimpel and Austin (1982) point out, such slow crack propagation is associated with surface flaws in the specimens tested and is therefore framed in a context rather remote from conditions of industrial breakage, where volume flaws predominate. For these practical conditions, crack propagation can approach the speed of sound in the material being broken, and it is known that molecules of additive cannot diffuse down a crack at such rates. Hence it would not be expected that they could influence the energy of a propagating tip. Hence such mechanisms as the Rehbinder effect (Rehbinder and Kalinkovskaya, 1932), invoking adsorption of grinding aids on to the surface of a flaw are most unlikely to be valid. Similarly are doubted the chemomechanical effects proposed by Westbrook (1966) in which absorbed molecules are thought to pin dislocations near the crack surface, thus preventing easy movement of dislocations under stress gradients. In fact, Locher and von Seebach (1972) have given strong experimental evidence against chemomechanical effects in the dry grinding of cement clinker using additives.

3.4.2 Fluidity/flowability theory

The truth lies in another direction, that of increased fluidity of slurries and increased flowability of dry powders with the addition of grinding aids. By chemi-sorption of the additive on to particle surfaces, they are given greater mobility, increasing the probability of a particle being nipped between tumbling balls in a ball mill, for example. For the same reason, Opoczky (1981) has demonstrated a more uniform grain size distribution with than without grinding aids, and the delay of the onset of both aggregation and agglomeration by keeping particles well dispersed. In fact, Locher and von Seebach (1972) have asserted that chemicals such as ethylene glycol and butylamine do not affect the dry breakage of coarse material but only become a factor in improved comminution as fine material begins to build up. In the context of wet grinding, Meloy and Crabtree (1967) examined the effect of a variety of liquids, such as water, glycerine, sugar solution, chloroform, acetone, *n*-hexane etc., and concluded that the additive action was to coat colliding surfaces with particles, thus reducing metal-to-metal ball contact which converts energy to heat without doing any breakage work. Thus, for a given power expenditure, increased capacity of the order of 10–20% at constant fineness, or alternatively increased fineness at constant capacity, is possible with the use of a correctly selected grinding aid (Opoczky, 1981).

3.4.3 Viscosity control in wet grinding

Klimpel and colleagues in the 1970s (e.g. Klimpel and Manfroy 1978) set out to identify specific industrial grinding aids based on the fluidity theory. As indicated by Klimpel and Austin (1982) they rightly started on basic experiments without aids and found that:

(1) increasing the solids loading of a slurry ground for a given time produced a gradually increasing percentage of fines less than a stated size;
(2) particular size distributions corresponding to a closer packing efficiency of particles had a similar effect;
(3) these trends persisted only as long as the slurry viscosity was not too high;
(4) reduction of viscosity by addition of water was self-defeating in respect of (1), (2) and (3) because of inherent volume dilution.

Hence chemical fluidity modifiers were looked for which would allow more dense slurries to interact with tumbling media in a mill, thus extending the effects (1) and (2) by lowering the viscosity limitation. The materials found were generally low-molecular-weight anionic water-soluble polymers (e.g. XFS 4272, a Dow Chemical Co. patented compound) which have been proved effective not only in laboratory tests, but in full-scale plant (Klimpel, 1980). It is important that such chemicals do not adversely affect processes downstream of the comminution plant, such as those of flotation, thickening, pelletization, and tests for this must be made in any particular application. In the end, the decision to use an additive depends on whether or not the gain in production outweighs the additive cost per ton of output. In many industries, additives have been proved advantageous on this basis. The Dow Chemical Co. quote on-site grinding additive cost at 12–30 cents (US 1980) per ton of dry ore feed. This is economically justifiable for a number of operations (Klimpel and Austin, 1982).

One particular field of exploration in which additives are essential is that of the manufacture of coal–water slurry for combustion to replace oil fuel. In this context, a valuable advantage of the grinding aid is the reduction of water content, of a stable slurry. Without this reduction the loss of boiler efficiency owing to the presence of the water would offset the reduced fuel cost, but with the reduction viable economies seem possible (Funk, 1981). (see §11.2.2.2(b).)

3.5 REVIEW OF THE PHYSICS OF BREAKAGE

3.5.1 General findings

Some of the more general findings of this chapter which impinge directly on practical comminution may be summarized as follows:

(1) More energy is needed to break small particles than large.
(2) When a brittle particle is stressed by an applied force, underneath the area of force contact there develops a cone-shaped zone of stress concentration in which particle degradation is intense compared with the space outside the zone. Hence fines are derived chiefly from these cone-shaped zones and a much coarser grade from the rest of the particle. (see §4.3.5.)

(3) The rate of application of breakage forces is more important for visco-elastic materials than for brittle materials. Visco-elastic materials are better fractured by high-speed than by slow-speed applications. Thus plastics are better comminuted by impact than by slow compression.
(4) The size distribution from primary breakage is complex and is usually bimodal. (see §4.3.5.2.)
(5) From considerations of the critical particle size for the transition to fully plastic behaviour for brittle materials, a grind size limit is predicted. (see §4.3.7.2.)
(6) There emerges the importance of fracture energy (several orders of magnitude greater than free surface energy) which it is necessary to attain before any fracture can occur.
(7) Chemomechanical explanations of the advantageous effects of additives on comminution have not been substantiated: the real cause of their success lies in greater fluidity of slurries and flowability of powders induced by the grinding aids. This in turn increases the probability of particles being nipped between balls, in a ball mill for example; and hence, without any power increase, either an increased capacity at a given fineness or an increased fineness at a given capacity have been successfully achieved.

Clearly practical comminution involves stressing as many particles as possible to the fracture point at the highest possible rate. Since many of the forces applied to an individual particle may not reach a value sufficient to attain the necessary fracture energy within the particle, much of the energy will be lost as heat. Hence fundamental studies predict a low efficiency for the grinding process. Further, from item (6) above, it is not sensible to take surface energy as a basis for the estimation of efficiency.

3.5.2 Special applications

Veldkamp (1980), with the assistance of Rumpf, has used the basic equations of fracture physics, with quite drastic simplifying assumptions, and applied them to the design of the type of fluid energy mill in which jets of the material to be ground are fired towards each other (see Figure 3.17). It is beyond the scope of this chapter to detail the equations used: suffice it to say that using hardness data and critical stress intensity factors from indentation fracture tests, for the material concerned, the following were correctly forecast:

(1) It would be impossible to jetmill strontium hexafluoride because of the very high jet velocity required.
(2) It would be possible to jetmill manganese zirconium ferrite.
(3) For all materials it is impossible to reduce size endlessly: a grind limit exists for jetmilling as for other comminution methods.

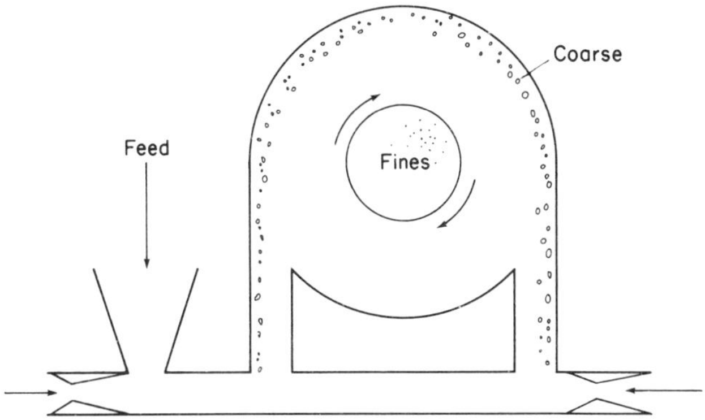

Figure 3.17. Longitudinal section of an opposed jet fluid energy mill

In the above example, the basic fracture process is high-velocity impact, but compression cases have also been studied to advantage. Dowding and Lytwynyshyn (1982) have applied point load deformation and fracture test results on concrete, mortar, two types of limestone, amphibolite and taconite to the improvement of the design of jaw crushers, leading to a reduction in weight of the swing plate by as much as 20%.

These examples serve to illustrate the increasing use of fracture mechanics to enhance the design and elucidate the performance of size reduction machinery.

3.5.3 Single-particle and particle-bed studies

In each of the examples in §3.5.2 where fracture mechanics has been applied successfully to practical comminution, no bed of particles was involved. However, the greater proportion of machines in current use incorporate a bed of particles, and hence particle-bed studies must be important.

It is evident from this chapter that single-particle studies in fracture physics have received more attention than assemblages of fragments, and this is understandable, since a particle buried in a bed is difficult to retain in experimental control.

A further feature of note is that a single particle undergoing primary breakage will not necessarily produce the same fragment distribution or use exactly the same amount of energy as another apparently similar particle of the same material under the same breakage conditions, because the internal flaw pattern may not be the same. Hence, to obtain realistic results, either a sufficient number of successive tests must be made and a statistical average value taken of distribution and of energy, or a sufficiently large number of the single particles must be broken together and their products pooled to obtain the size distribution. It is clearly less time-consuming to follow the latter course, as long as particle interference is avoided.

By contrast, in a bed of particles under compression, for example, there is already a sufficient number of particles present for random variations to be evened out, and a greater likelihood exists of reproducibility of product size distribution and of the energy used.

3.6 REFERENCES

Arbiter, N., Harris, C. C., and Stambolitzis, G. A. (1969). 'Single fracture of brittle spheres', *Trans. SME/AIME*, **244**, 118–133.

Bhattacharya, D. (1966). 'Direct measurement of unconfined yield strengths of powders', Ph.D. thesis, University of Bradford.

Bilby, B. A. (1979). 'Putting fracture to work', *Proc. 4th Tewksbury Symp. Fracture*, Melbourne, Australia, 1:1–1:44.

Boddy, R. G. H. B. (1943). 'Notes on a technique for demonstrating the plasticity of coal', *Proc. Conf. Ultrafine Struct. Coal Coke*, BCURA, London, 326–341.

Brace, W. F., and Byerlee, J. D. (1967), 'Recent experimental studies of brittle fracture of rocks'. In: *Failute and Breakage of Rocks*, Proc. 8th Symp. Rock Mech., University of Minnesota, 15–17 Sept. 1966, AIME, 58–81.

Bridgman, P. W. (1952). 'The resistance of 72 elements, alloys and compounds to 100,000 kg/cm^2', *Proc. Am. Acad. Arts Sci.*, **81**, (4), 166–251.

British Standard 240: *Method for Brinell Hardness Test*, Part 1 (1962) *Testing of Metals*.

British Standard 427: *Method for Vickers Hardness Test*, Part 1 (1961) (1981) *Testing of Metals*.

British Standard 891: *Method for Rockwell Hardness Test*, Part 1 (1961) (1981) *Testing of Metals*.

Carey, W. F., and Bosanquet, C. H. (1933). 'A study of crushing brittle solids', *J. Soc. Glass Technol.*, **17**, 384–410.

Dowding, C. H., and Lytwynyshyn, G. (1982). 'Point load deformation relationships and design of jaw crusher plates', *Powder Technol.*, **31**, 277–286.

Dugdale, D. S. (1960). 'Yielding of steel sheets containing slits', *J. Mech. Phys. Solids*, **8**, 100–104.

Elliott, H. A. (1947). 'An analysis of the conditions for rupture due to Griffith cracks', *Proc. Phys. Soc. (London)*, **59**, 208–223.

Fuerstenau, D. W. (1979). 'Grinding aids in comminution systems; a report prepared for US National Academy of Sciences'.

Fuller, K. N. G., Fox, P. G., and Field, J. E. (1975). 'The temperature rise at the tip of fast moving cracks in glassy polymers', *Proc. R. Soc. London A*, 537–557.

Funk, J. E. (1981). 'Coal-water slurry and methods for its preparation', US Patent 4,282,006.

Gaudin, A. M., and Yavasca, S. S. (1946). 'Principles of comminution—size and surface distribution', *Trans. AIME*, **169**, 88–94.

Gildemeister, H. H., and Schönert, K. (1972). 'Berechnung zur wellenausbreitung in kugeln und bruchphänomene in kreisscheiben bei prallbeanspruchung', Dechema Monograph 69, 233–267.

Gildemeister, H. H., and Schönert, K. (1976). 'Bruchphanomene und spannungsfeld in prallbeanspruchten kugeln', Dechema Monograph 79, 131–149.

Gregg, S. J. (1968). 'Surface chemical study of rupture and flow in solids', *Chem. Ind.*, 11 May, 611–617.

Griffith, A. A. (1920). 'The phenomenon of rupture and flow in solids', *Phil. Trans. Royal Soc. A*, **221**, 163–198.
Hamilton, M., and Goodman, E. (1966). 'The stress field created by a circular sliding contact', *J. App. Mech.*, **33**, 371–376.
Hertz, H. (1881). 'On the contact of elastic solids'. English translation from *J. fur die Reine und Angewandte Math.*, **92**, 156–171. Miscellaneous Papers, Macmillan, NY (1896), 146–162.
Hess, W., and Schönert, K. (1976). 'Die zerkleinerung von kalksteinkugeln bei kombinierter druck- und schubbeanspruchung', Dechema Monograph 79, 151–165.
Hess, W., and Schönert, K. (1981). 'Brittle-plastic transition in small particles', *Inst. Chem. Eng. Symp. Ser.* No. 63, Birmingham, England D2/2/1–9.
Hiramatsu, Y., and Oka, Y. (1966). 'Determination of the tensile strength of rock by a compression test of an irregular test-piece', *Int. J. Rock Mech. Miner. Sci.*, **3**, 89–99.
Hoffmann, N., and Schönert, K. (1971). 'Bruchanteil von glaskugeln in packungen von franktionen and binären mischungen', *Aubereit. Tech.*, **12**, 513–518.
Huber, M. T. (1904). 'Zur theorie der beruhrung fester elastischer korper', *Annalen. Phys.*, **14**, 153–163.
Hukki, R. T. (1943). 'An experimental study of the principles of comminution', D.Sc. thesis, Massachusetts Institute of Technology.
Inglis, C. E. (1913). 'Stresses in a plate due to the presence of cracks and sharp corners', *Trans. R. Inst. Naval Arch.*, **55**, 219–241.
Irwin, G. R. (1947). 'Fracturing of metals', *ASM Symp.*, Chicago, 147.
Jaeger, J. C. (1967). 'Failure of rocks under tensile conditions', *Int. J. Rock Mech. Miner. Sci.*, **4**, 219–237.
Kendall, K. (1978). 'The impossibility of comminuting small particles', *Nature*, **272**, 710–711.
Klemm, W., and Smekal, A. (1941). 'Über den grundvorgang des polierens von gläsern', *Naturwissen*, **29**, 688–690.
Klimpel, R. R. (1980). 'The engineering analysis of dispersion effects in selected mineral processing operations', Proc. Symp. Fine Particles Processg, Las Vegas, 1979. Somasundaran, P. (Editor), AIME, New York, 1129–1154.
Klimpel, R. R., and Austin, L. G. (1982). 'Chemical additives for wet grinding of minerals', *Powder Technol.*, **31**, 239–253.
Klimpel, R. R., and Manfroy, W. (1978). 'Chemical grinding aids for increasing throughput in the wet grinding of ores', *Ind. Eng. Chem. Process Des. Devel.*, **17**, 518–523.
Lawn, B. R., and Marshall, D. B. (1979). 'Hardness, toughness and brittleness: an indentation analysis', *J. Am. Ceram. Soc.*, **62**, 347–350.
Lawn, B. R., and Wilshaw, R. (1975). 'Review—indentation fracture: principles and applications', *J. Mater. Sci.*, **10**, 1049–1081.
Locher, F. W., and von Seebach, H. M. (1972). 'The influence of adsorption on industrial grinding', *Ind. Eng. Chem. Process Des. Devel.*, **11**, (2), 190–197.
Lurge, A. J. (1963). In *Raumliche Probleme der Elastizitatstheorie*, Akademie Verlag, Berlin, 153.
Marshall, V. C. (Ed.) (1974). 'Comminution: a report by the Inst. Chem. Eng. Working Party', Inst. Chem. Eng., London, 5,6,53.
Meloy, T. P., and Crabtree, D. (1967). 'Surface tension and viscosity in wet grinding', Dechema Monograph 57, 405–426.
Obert, L., Windes, S. L., and Duvall, W. I. (1946). 'Standardized tests for determining the physical properties of mine ores', US Bureau of Mines Report of Investigations, 3891.
Ohe, van H. (1967). 'Einzelkornzerkleinerung von kalkstein und zementklinker—

kornern in einem waltzensplatt mit und ohne friktion', *Chemie. Ing. Tech.*, **39**, 357–364.
Opoczky, L. (1981). 'Ways of influencing the particle size distribution of ground products in dry ball mills', *Interceram.*, **4**, 327–389, 403.
Orowan, E. (1949). 'Fracture and strength in solids', *Rep. Prog. Phys.*, **12**, 185–232.
Parish, B. M. (1967). 'The effects of coal rank and particle size on the plastic behaviour of coal', *Br. J. App. Phys.*, **18**, 233–240.
Paul, B., and Gangal, M. (1967). 'Initial and subsequent fracture curves for bi-axial compression of brittle materials', *Proc. 8th Symp. Rock Mech.*, University of Minnesota, 15–17 Sept. 1966, AIME, New York, 113–141.
Rehbinder, P. A., and Chodakow, G. S. (1962). 'Feinmahlung von quarz', *Silikatech.*, **13**, 200–208.
Rehbinder, P., and Kalinkovskaya, N. (1932). Reduction in stability of the surface layer of a solid by the absorption of surface acting substances (Russian). *J. Tech. Phys. USSR*, **2**, 726–755.
Rose, H., and Sullivan, R. (1958). 'Ball, tube and rod mills', Chemical Publishing Co., New York, 236–251.
Roth, D. (1977). 'Das agglomerationsverhalten von frisch gemahlenem zucher', *Zucher.*, **30**, (9), 464–470.
Rumpf, H. (1961). 'Problemstellungen und nuere ergebnisse der bruchtheorie', *Materialprufung*, **3**, 253–265.
Rumpf, H. (1965). 'Die einzelkornzerkleinerung als grundlage einer technischen zerkleinerungswissenschaft', *Chemie. Ing. Tech.*, **37**, 187–202.
Rumpf, H., and Schönert, K. (1972). 'Die brucherscheinungen in kugeln bei elastischen sowie plastischen verformungen durch druckbeanspruchung', Dechema Monograph 69, 51–55.
Sack, R. A., and Wills, H. H. (1941). 'Extension of Griffith theory of rupture to three dimensions', *Proc. Phys. Soc.*, **58**, 729–736.
Schönert, K. (1972). 'The role of fracture physics in understanding comminution phenomena', *Trans. SME/AIME*, **252**, (1), 21–26.
Schönert, K. (1973). 'Single particle crushing of polymers', H. Heywood Memor. Symp., 17–18 Sept., Loughborough University of Technology.
Schönert, K. (1979). 'Aspects of the physics of breakage relevant to comminution', *Proc. 4th Tewksbury Symp. Fracture*, 17–18 Feb., University of Melbourne, Australia, 3:1–3:30.
Schönert, K., Umhauer, H., and Klemm, W. (1969). 'The influence of temperature and environment on the slow crack propagation in glass', *Proc. 2nd Int. Conf. Fracture*, Brighton, 1969, Chapman & Hall, London, 474–482.
Smekel, A. (1938). 'Compression comminution of cubic single particles', *Z. VDI Beinheft verfahrenstechnik*, **6**, 159–165. (English translation US NTC, 70-14798).
Snow, R. H., and Paulding, B. W. (1973). 'Prediction of fragment size distribution from calculations of stress in spheres and discs', H. Heywood Memor. Symp., 17–18 Sept., Loughborough University of Technology.
Somasundaran, P., and Lin, I. J. (1972). 'The effect of the nature of environment on comminution processes', *Ind. Eng. Chem. Process Des. Devel.*, **11**, 321–331.
Steier, K., Hoffmann, N., and Schönert, K. (1974). 'Strukturänderungen von quarz und kalkstein infolge druckbeanspruchungen', *Chemie Ing. Tech.*, **46**, 739.
Steier, K., and Schönert, K. (1972). 'Verformung und bruchphänomene unter druckbeanspruchung von sehr kleinen körnern aus kalkstein, quarz und polystyrol', Dechema Monograph 69, 167–192.
Veldkamp, J. D. B. (1980). 'A description of jetmilling in terms of plastic deformation and fracture', *Proc. Europ. Symp. Part. Technol.*, Amsterdam, 13–15 June, 339–409.

Weichert, R., and Schönert, K. (1978). 'Heat generated at the tip of a moving crack', *J. Mech. Phys. Solids*, **26**, 151–161.
Wells, A. A. (1953). 'The mechanics of notch brittle fracture', *Weld Res.*, **7**, 34–56.
Westbrook, J. H. (1966). 'Some effects of adsorbed water on the plastic deformation of non-metallic solids'. In: *Environment Sensitive Mechanical Behaviour* (Westwood, A. R. C., and Stoloff, N. S., Eds), Gordon & Breach, New York, 247–266.

CHAPTER 4

Particle shape, size and surface

NOTATION FOR CHAPTER 4

a	Area
A	Amplitude in Fourier harmonic series
$b_{i,j}$	Fraction of material in size interval j which falls into size i after breakage

b^* Steepness constant in log–normal equation
B_n Cumulative fraction by mass falling below value z_n (equation (4.43))
$B(y,x)$ Cumulative fraction by mass falling below size y when size x is broken
$B(z)$ Cumulative fraction by mass less than relative size z
$\mathbf{B}$ Breakage distribution matrix
c' Parameter in log–normal equation
d Diameter
d_{80} Diameter corresponding to 80% passing
f Feed element
$\mathbf{f}$ Feed vector
f_{H} Heywood surface shape factor
$f(x)$ Density or frequency size distribution
Subscripts: L by length
M by mass
N by number
S by surface
$F(x)$ Cumulative size distribution
$g(x)$ Function of particle size
$g(\bar{x})$ Function of mean particle size
G Jaw crusher gap
$\mathbf{I}$ Unit matrix
k_{H} Heywood volume shape factor
K Constant
L Length
M Mass
Subscripts: a of tamped aggregate in angularity test
e of sieved fraction passing length gauge
g of sieved fraction passing special gauge
j on jth sieve
s of sieved size fraction
t of total sample or charge
w of water to fill cylinder in angularity test
n Exponent, number
n' Exponent in RRB equation; uniformity index
n_i Number of particles of size x_i
p Product element (suffix n for n stages of operation)
$\mathbf{p}$ Product vector
P Probability
$r(\theta)$ Radius at an angle θ
R Common ratio of sizes in geometric progression
s Specific surface area
Subscripts: m by mass
v by volume

S	Specific rate of breakage (selection) function		
$\mathbf{S}$	Specific rate of breakage (selection) matrix		
t	Time		
T	Thickness		
u	Parameter in equation (4.21)		
V	Volume		
W	Width		
x	Particle size		
	Subscripts:	A	Equivalent sphere diameter passing square aperture
		C	Equivalent circle diameter having same outline perimeter
		CH	Maximum chord diameter
		D	Drag diameter
		f	Fineness number
		F	Feret's diameter
		FF	Free-fall diameter
		g	Geometric mean of log–normal distribution
		H	Heywood diameter
		mo	Mode of distribution
		me	Median of distribution
		max	Maximum
		min	Minimum
		M	Martin's diameter
		P	Projected area diameter
		S	Surface diameter
		SH	Shear diameter
		ST	Stokes' diameter
		SV	Specific volume diameter
		V	Volume diameter
		50	Value corresponding to 50% ordinate
		84.1	Value corresponding to 84.1% ordinate
$\bar{x}$	Mean particle size		
	Subscripts:	a	Arithmetic mean
		c	Cubic mean
		g	Geometric mean
		h	Harmonic mean
		q	Quadratic mean
x'	Reference particle size		
y	Particle size (less than x)		
z	Relative particle size		
z_{max}	Maximum relative particle size		
z''	$z/(1-z)$		
$\alpha(\alpha')$	Exponent (factor) related to stress-activated flaws along an edge		
$\beta(\beta')$	Exponent (factor) related to stress-activated flaws in a surface		

$\gamma(\gamma')$	Exponent (factor) related to stress-activated flaws in a volume
$\delta\theta(L)$	Change of slope along a length L
Δx	Aperture of mesh $(i-1)$ minus aperture of mesh i
ε	Shape factor $= \tan^{-1}(W/T)$
ζ	Parameter in log–normal frequency distribution
η	Phase angle
θ	Angle
λ	Poisson distribution parameter
Λ	Standardized normal variate corresponding to the probability $(1-P')$
μ	Mean
$\mu_{\ln z''}$	$\ln[z_{50}(1-z_{50})]$
π	Ratio of circle circumference to diameter
ρ	Density
ρ_r	Relative density
σ	Standard deviation (84.1% size minus 50% size)
σ_g	Ratio of 84.1% size to 50% size
$\sigma_{\ln z''}$	$\ln[\langle z_{84.1}(1-z_{50})\rangle/\langle z_{50}(1-z_{84.1})\rangle]$
$\psi_{sv,x}$	Specific surface coefficient
ψ_{sx}	Surface coefficient
ψ_{vx}	Volume coefficient

4.1 INTRODUCTION

Because there exists such a vast bibliography on particle shape, size and surface, it is only possible in this handbook to outline the main features of the subject. In particular, the measurement of particle size and surface is regarded as a subject in its own right, and students are directed to the appropriate textbooks, which are indicated.

Characteristics of particle shape can be considered independently of size (though there are links in comminution) but size is strongly related to surface. Hence particle shape is first discussed, followed by size and surface.

In the last decade there has been an expansion of interest in particle shape with the advent of image analysers and associated computers. Thus an appropriate space has been allocated to the subject and a reference list is provided separately from that for the rest of the chapter.

4.2 PARTICLE SHAPE

Note: For references see §4.5.1.

4.2.1 The importance of particle shape

The following examples illustrate the importance of particle shape in industrial processes (Rose, 1961).

Inks, paints and cosmetics have better hiding power and optical properties if the particles contained therein are flaky rather than round. Rubber reinforcement, on the other hand, needs granular particles because if they are allowed to elongate, they tend to form a grain pattern which tears preferentially in one direction. However, elongated reinforcements such as fibres are better for plastics than granular shapes because plastics need high impact strength, not high tensile strength. Smooth spherical grains are called for in the explosives industry because irregular particles make the initiators chemically unstable, especially if they have sharp edges or corners. The abrasives industry, by contrast, welcomes highly angular grains, but with the proviso that they have a low aspect ratio, since flat particles tend to slide rather than roll and hence have a low probability of presenting their sharp features to the surface to be abraded (Giltrow and Groszek, 1969).

The size reduction conditions which will result in the production of these shapes must therefore be known to satisfy the needs of industry.

4.2.2 Qualitative definitions of particle shape

In British Standard 2955 (1958) an attempt has been made to define shapes qualitatively using the following ten terms, which certainly have general descriptive value; but for scientific shape assessment, mathematical definitions are also needed.

Acicular	Needle-shaped
Angular	Sharp-edged or having roughly polyhedral shape
Crystalline	Of geometric shape, freely developed in a fluid medium
Dendritic	Having a branched crystalline shape
Fibrous	Regularly or irregularly thread-like
Granular	Having approximately equidimensional but irregular shape
Irregular	Lacking any symmetry
Modular	Having a rounded irregular shape
Spherical	Globule shaped

4.2.3 Basic quantitative definitions of particle shape

4.2.3.1 Fundamental concepts

The definition of particle shape is complex, since the general term 'shape' implies two distinct characteristics, namely geometrical form and the relative proportions of length L, width W and thickness T (Heywood, 1937). For example, cylinders can be long and thin like dowel rods, or equidimensional when the length equals the diameter, or again flat and coin-like; and similarly for other geometrical forms like rectangular prisms, spheroids and octohedra. In principle, the concept can be extended to irregularly shaped particles.

By normalizing the relative proportions, for example into W/L and T/W, two parameters are obtained. These, together with the geometrical form, make three factors which are essential to define shape. These can sometimes be combined to a fewer number which has a statistical meaning, but this has to be demonstrated as being valid in specific circumstances.

4.2.3.2 Shape factors

Many different shape factors have been suggested as listed by Davies (1975), but the concept will be illustrated by perhaps the best known factor introduced by Heywood (1933, 1937, 1947). Extending the ideas of Martin (1923), Heywood conceived of spheres of equal volume and of equivalent surface to the irregularly shaped grain. Starting with a projected image of the particle resting in its stable position, the area of the image is assessed and the diameter x_H of the equivalent circle is calculated. The volume V of the particle is then defined by:

$$V = k_H x^3 \tag{4.1}$$

where k_H is the volume constant for that particle.

Similarly the surface area a of the particle is defined by:

$$a = f_H x^2 \tag{4.2}$$

where f_H is the surface constant.

The ratio f_H/k_H is used as a shape factor, combining the volume and surface constants, and some values are given for square-ended prisms (Heywood, 1933) in Table 4.1, where the length of the square side is L_1 and the length of the prism is L_2.

Similar results to those in Table 4.1 for prisms are obtained for cylinders, spheroids and octahedra of varying proportions and the following general deductions can be made:

(1) When the dimensions of a particle are approximately the same in three dimensions the shape factor has a minimum value.
(2) When the shape is either flattened or elongated, the shape factor increases.

Table 4.1. Shape factors for square ended prisms

Description	L_2/L_1	f_H	k_H	f_H/k_H
Flat tablet	0.1	1.885	0.0696	27.08
Flat tablet	0.4	2.827	0.2784	10.16
Cube	1.0	4.712	0.6958	6.77
Long prism	5.0	3.454	0.3112	11.10
Long prism	10.0	3.298	0.2202	14.98

Reprinted by permission of the Council of the Institution of Mechanical Engineers from Proc. Inst. Mech. Eng., *125, 390.*

The above is based on the Heywood diameter of the circle of equivalent area to that of the projection of the particle in a stable position, but others have used statistical diameters such as x_M proposed by Martin (1923) or x_F proposed by Feret or a sieve size as discussed by Harris (1960). For definitions of statistical diameters, see Table 4.4, and for the range of shape indices (ratios of characteristic diameters) see Tsubaki and Jimbo (1979b).

4.2.3.3 *Flakiness index*

Aggregates are classified as flaky when they have a thickness (smallest dimension) of less than 0.6 of their nominal size as judged by two limiting square sieve apertures. From a given sieved size fraction of mass M_s, the mass fraction M_g which passes through a slot sieve of special design is found. Then the flakiness index is $100M_g/M_s$. Full details are given in British Standard 812, Part 1 (1975).

The flakiness index is important in the choice of aggregates for roadmaking, and typically required values from British Standard 63, Part 2 (1971) are:

Nominal size (mm)	*Flakiness index*
50–40	40
28–10	10

4.2.3.4 *Elongation index*

Aggregates are classed as elongated when they have a length (greatest dimension) of more than 1.8 of their nominal size. From a given sized square mesh fraction of mass M_s, the mass proportion M_e is found for which the longest particle dimension passes between appropriate pegs in a specially designed apparatus. Then the elongation index is $100M_e/M_s$. Full details are given in British Standard 812, Part 1 (1975). Markwick (1941) has shown that a relation exists between the elongation and flakiness indices, and today the elongation index is rarely called for, but only the flakiness index is specified.

4.2.3.5 *Angularity number*

Angularity, the absence of rounding of particles in aggregate, affects the ease of handling mixtures with binders and is important in abrasive cutting. It is determined from the proportion of voids in a sample of the aggregate when subjected to a standard system of tamping in a cylinder of standard dimensions. See British Standard 812, Part 1 (1975) for details. Then the angularity number is:

$$67 - 100M_a/\rho_r M_w$$

where ρ_r is the relative density of the aggregate when determined by a standard procedure. Typical values of angularity number are zero for a rounded shape, up to 12 for an angular product, with 6 to 9 as the most common values.

4.2.4 Sieving methods of shape characterization

Bond (1954), Lees (1964) and more recently Nakajima *et al* (1978, 1979) all appreciated the value of sieving as that means of determining irregular particle shape characteristics which obviates the necessity of the time-consuming process of measurement of individual particles.

Following Nakajima *et al.* (1978), the maximum distance between two points on a particle is denoted by L (length) and the minimum by T (thickness). Thickness is associated with the dimensions of the gaps in a series of slot sieves. The dimension W (width) is associated with the size of the normal square mesh sieve opening. In both square and slot sieving, the dimension of length is largely lost, but it was found this is not a hindrance to realistic shape assessment. L, W and T are taken to lie in mutually perpendicular directions and the basic cross-sectional shape is taken to be an ellipse with W and T as major and minor axes. This is justified in that the outcome of the analysis was in agreement with experimental findings for the material which was used, namely hematite. About one hundred grains were sampled from the material on each of square mesh sieves sized 5.3, 8.3 and 11.3 mm, and L, W and T were measured by callipers. When T was plotted against W, distinctive boundaries were found between the sieve sizes as indicated in Figure 4.1. Similarly

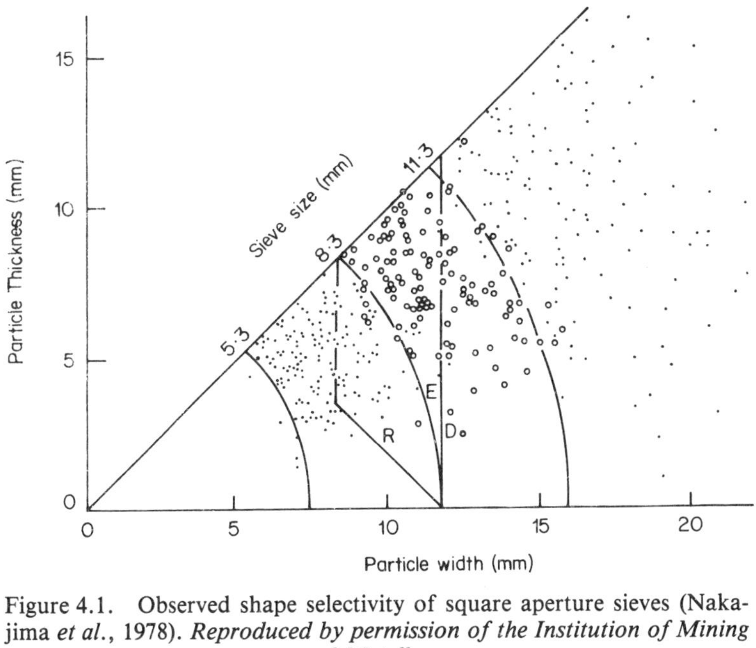

Figure 4.1. Observed shape selectivity of square aperture sieves (Nakajima *et al.*, 1978). *Reproduced by permission of the Institution of Mining and Metallurgy*

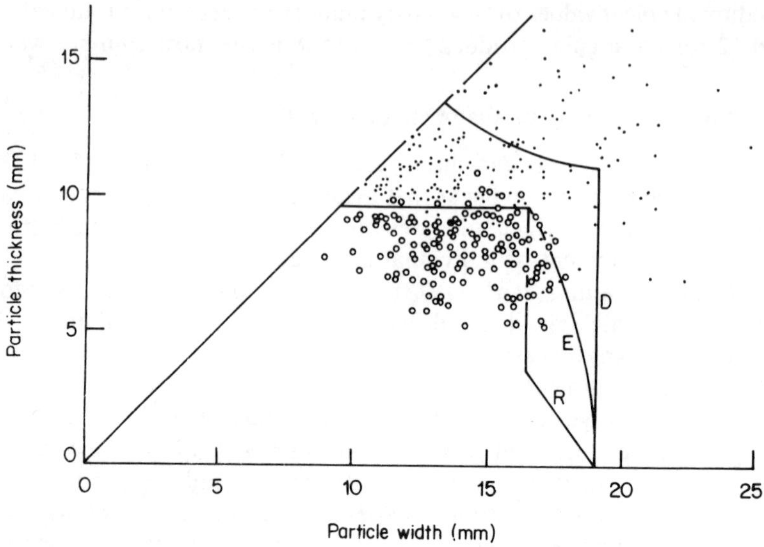

Figure 4.2. Observed shape selectivity of slot aperture sieves (Nakajima *et al.*, 1978) *Reproduced by permission of the Institution of Mining and Metallurgy*

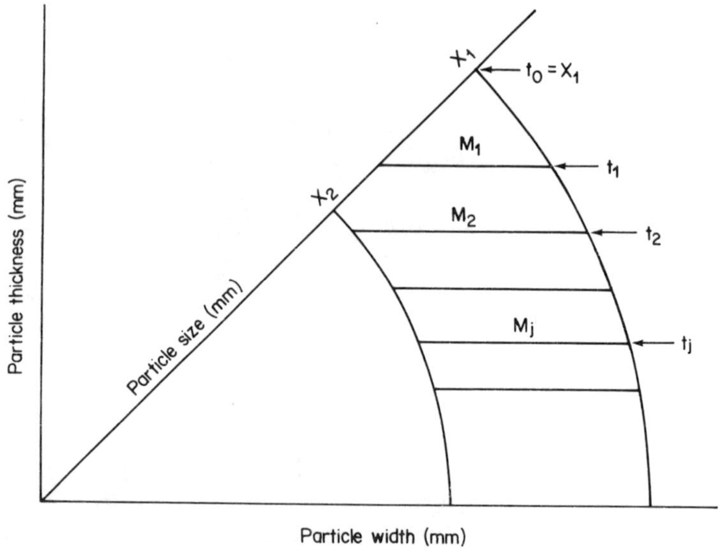

Figure 4.3. Sub-divided zones by square and slot aperture sieves (Nakajima *et al.*, 1978) *Reproduced by permission of the Institution of Mining and Metallurgy*

when the same particles were passed through a slot sieve sized 16.5×9.5 mm, distinctive boundaries appeared on the T versus W plot as seen in Figure 4.2. Combining the results allowed Figure 4.3 to be drawn, with distinctive subdivided zones, showing the masses M_j retained corresponding to the thickness T_j for the series $1,2,3,\ldots,j$.

A shape factor $\varepsilon = \tan^{-1}(W/T)$ was defined, assumed independent of size, and linear dimensions were converted to mass on the assumption (proven for hematite) that mass is proportional to the product of particle dimensions. An expression was derived for M_j in terms of three parameters, namely ε, the known size distribution and an assumed shape distribution. By a rapid

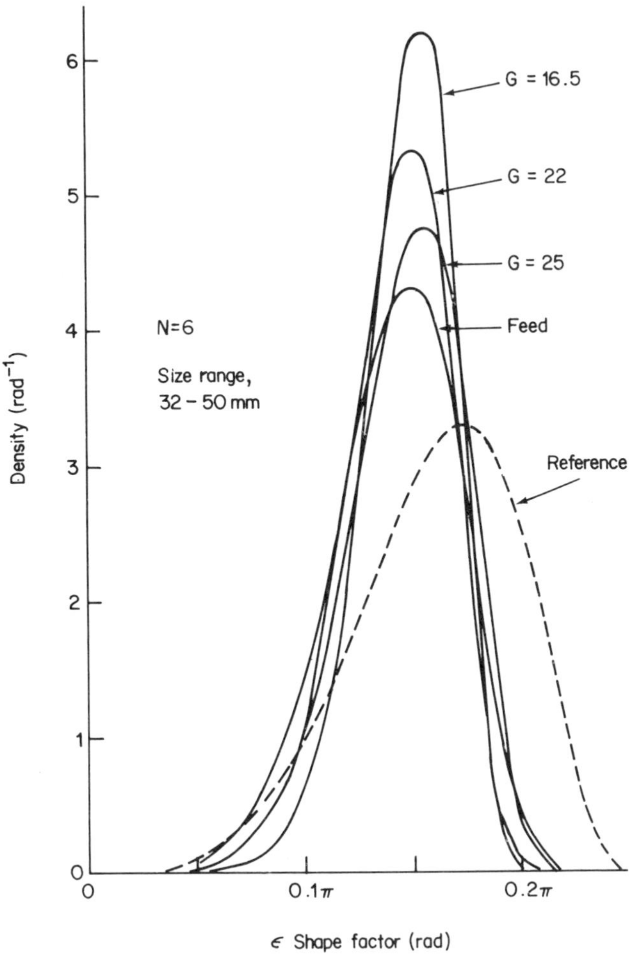

Figure 4.4. Shape distribution of one size range of the product from a cone crusher (Nakajima *et al.*, 1978) *Reproduced by permission of the Institution of Mining and Metallurgy*

iterative calculation, that shape distribution was determined which gave the nearest value to the known M_j.

A typical example of the frequency–shape distribution for a given particle size range in terms of the shape factor ε (measured in radians) is shown in Figure 4.4. The material was hematite reduced in a 2.1 m (7ft) cone crusher for which G is the gap setting (mm). N is the number of sub-divided zones in the sieve size range. An experimental check revealed an error of only 2%, which suggests that the equation derived for the assumed shape was sufficiently versatile to describe practical shape distributions.

Increase in particle size tends towards smaller shape factors and for the curve to become sharper. This change in shape of the coarser fractions was attributed to the failure of the crusher to discharge particles with larger shape factors through the slots between the crushing surfaces.

This work needs to be repeated with materials other than hematite to verify its wider applicability, but it appears that sieving methods of particle shape characterization have considerable potential.

4.2.5 Image analysis methods of shape characterization

4.2.5.1 Basic principles

The introduction of image analysers combined with computer techniques has opened the door to methods of particle shape measurement which are more flexible than formerly. However, the problem remains of the time required to treat a large number of particles, and if attempts are made to reduce this by sampling, there is always the difficulty of getting a truly representative sample (see §4.3.9.3). However, image analysers are being used increasingly and recent lines of exploration will be summarized.

The usual procedure is to fix an image of the particle photographically with a microscope and analyse the same with a stylus pen (graf pen) digester and a microcomputer. The image can thus be transformed into N (e.g. 150–200) points and digested in terms of either (x,y) or (r,θ) coordinates as required. The centre of gravity is normally taken as the origin for either system of coordinates. The N points can be used to reproduce the original particle shape quite closely.

4.2.5.2 Shape factor exploration

The immediate effect of the introduction of such a new powerful technique as the image analyser is to explore the many available shape factors to see which is most relevant to a given situation. Tsubaki and Jimbo (1979a) have tried out several shape indices, each of which is the ratio of characteristic diameters such as x_H/x_C or of statistical diameters such as x_F (for definitions see §4.3.1). They were able to recognize groups of similar shape indices by comparison of the indices with morphological features of grains, and to express the differences in triangular diagrams for a range of materials such as limestone, glass powder and atomized stainless steel powder. This work is valuable in providing a body of background information on particle shape which has not been available before.

4.2.5.3 Special representation of form and roundness

Davies (1975) has advanced a three-dimensional space representation of both 'form' and 'roundness' in which triaxial form factors are assessed in horizontal triangular diagrams while roundness is assessed in the vertical axis, as shown in Figure 4.5. The method has proved to be tedious even with current image analysers, and awaits the appearance of more advanced computer interfaced imaging systems to make it practically viable. However, the concept is quite basic, as it differentiates between geometrical shapes in moving vertically from the cube to the sphere, and the relative proportions of those basic shapes in the triangular horizontal diagrams.

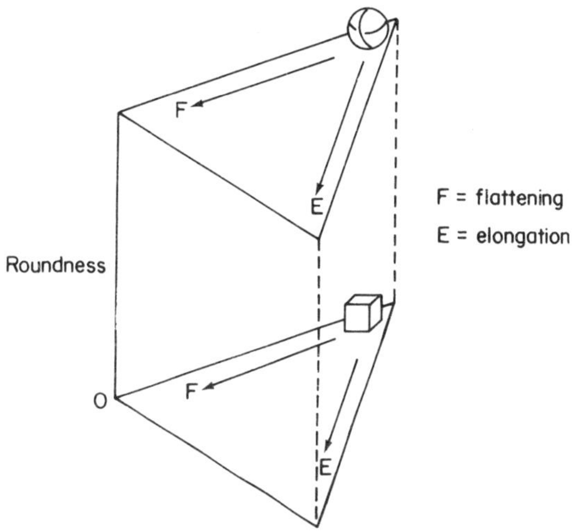

Figure 4.5. Simple space representation of form and roundness (Davies, 1975). *Reproduced by permission of Elsevier Sequoia*

4.2.5.4 Use of harmonic series

Several workers such as Gotoh and Finney (1975), Meloy (1977a, 1977b), Holt (1981a) and Luericens *et al.* (1982) have used harmonic series such as those of Walsh and Fourier to assist in shape characterization.

Concentrating on the truncated Fourier series, it takes the form:

$$r(\theta) = A_0 + \sum_{n=1}^{N} A_n \cos(n\theta - \eta_n) \qquad (4.3)$$

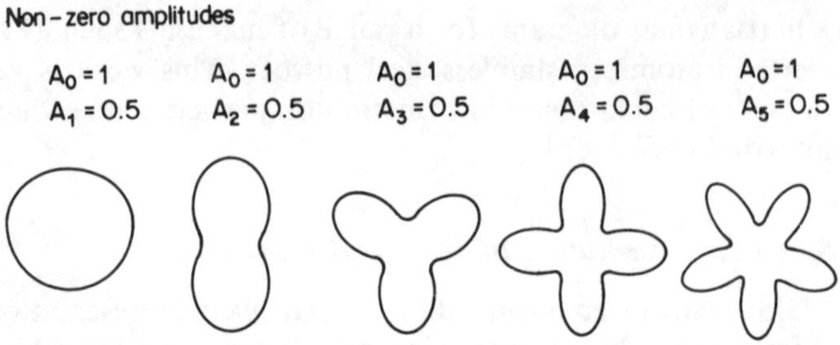

Figure 4.6. Simple shapes for harmonic analysis (Holt, 1981a). *Reproduced by permission of the Institution of Chemical Engineers*

where $r(\theta)$ is the radius at an angle θ, η_n is the nth phase angle, A_n is the nth amplitude, and N is the number of terms after which the series is truncated.

This method of describing particle shape was originally reported by Cheng and Sutton (1971) and developed by Beddow and Philips (1975). Because it runs into problems with re-entrant shapes, Fong *et al.* (1979) have advanced a procedure due to Zahn and Roskies (1972) which overcomes the difficulty by using the parameters arc length L and change of slope $\delta\theta(L)$ from a starting point on the perimeter of the image.

In equation (4.3), A_0 is usually set at unity so that all the coefficients are normalized.

The effect of simple combinations of A_0 and A_n (n from 1 to 5) are exemplified in Figure 4.6 taken from Holt (1981a), but for most irregular particle shapes, of course, a much larger number of combinations is needed.

Figure 4.7 (for $N = 10$) shows that agreement between actual and derived shapes can be quite close. The significance of the amplitudes in Figure 4.6 is

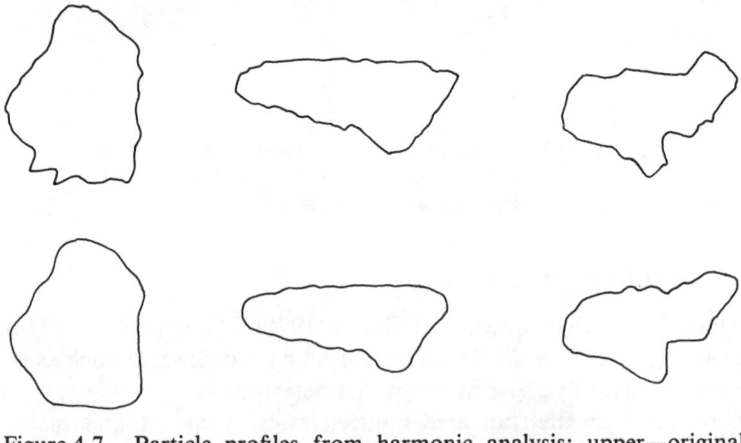

Figure 4.7. Particle profiles from harmonic analysis: upper—original shape, lower—regenerated shape (Holt, 1981a). *Reproduced by permission of the Institution of Chemical Engineers*

that, if the perimeter of the grain is highly convoluted, then a large number of amplitudes may be necessary to describe it, whereas if the perimeter is smooth, a smaller number will do. Further, the greater the magnitude of the amplitude, the further the departure from equidimensional shape.

4.2.5.5 Experimental results

Figure 4.8 supplied by Holt (1981a) shows the comparative magnitudes of the harmonic numbers from 1 to 10 for single particles in the size range 710–1000 microns. The particles had been treated as follows:

(a) crushed slowly in a hydraulic press (2×10^{-5} m/s);
(b) crushed between parallel plates (2 m/s);
(c) fired from a compressed air gun (100 m/s);
(d) reduced in size in a laboratory hammer mill over a range of conditions.

The greatest magnitudes were produced by low-speed crushing, showing that the particles were more elongated than those from high-speed impact. Charles (1956) noticed the same effect. The product of the hammer mill showed still lower amplitude, and hence the shapes from the process were more cubiform than those from the single particle crushing.

Figure 4.9 applies to a smaller size range (125–250 microns) compared with Figure 4.8, and although the general pattern is similar, the magnitudes of

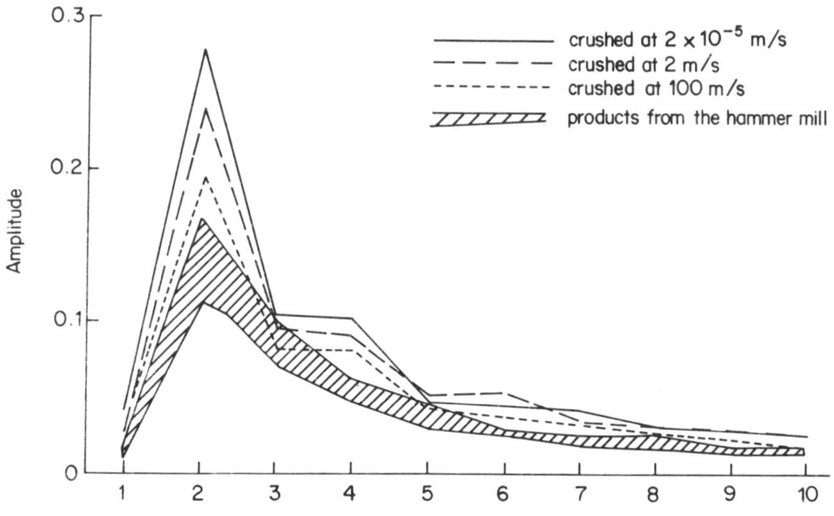

Figure 4.8. Harmonic amplitudes for glass particles of size 710–1000 microns produced by single particle crushing (Holt, 1981a). *Reproduced by permission of the Institution of Chemical Engineers*

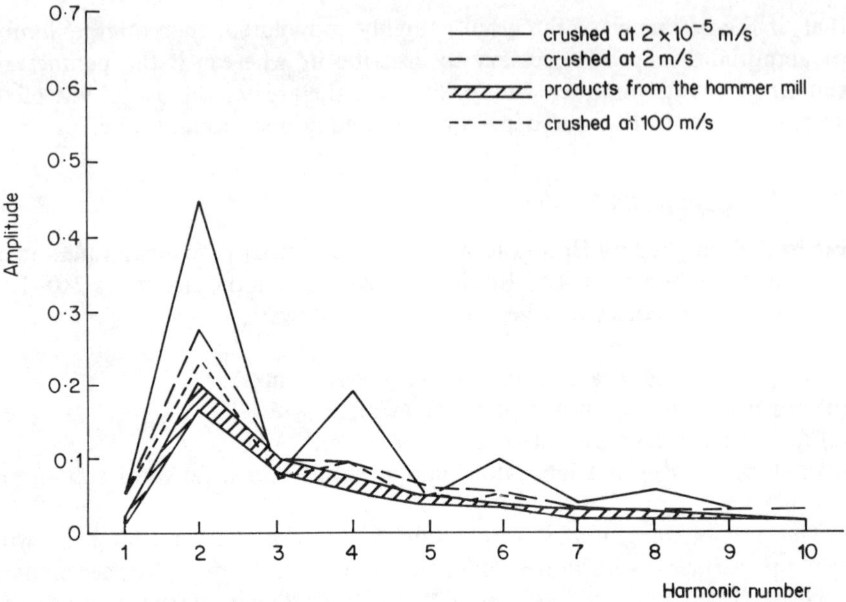

Figure 4.9. Harmonic amplitudes for glass particles of size 125–500 microns produced by single particle crushing (Holt, 1981a). *Reproduced by permission of the Institution of Chemical Engineers*

the amplitudes are greater than the corresponding ones in Figure 4.8. Thus the smaller particles were more elongated and generally less granular than the larger ones, though the corners were rounded by tumbling within the mill casing. Harris and Stambolitzis (1969) also found for primary breakage of glass spheres under slow compression that particles became more elongated with decreasing particle size.

This phenomenon of particle shape being a function of particle size was observed by Gaudin (1926) when grinding quartz in a ball mill with ball size relative to feed size greater than a certain critical ratio. In a rod mill it has long been known that small sized particles tend to be acicular and larger sizes granular.

More recently this has been borne out in a practical experience of aggregate production for road-making by Schmidt (1981), who finds that only particular size fractions from the products of stone crushing provide the cubiform grades necessary for a good quality surface. Schmidt therefore recommends the testing of samples from discrete size bands of crushed products and the use of only those bands which conform to the required shape standard.

Holt (1981b) attempts to explain the size dependence of shape in terms of the proposal that 'a particle with a large number of faces is more likely to be equidimensional on statistical grounds, whereas a particle with a small number of faces is probably elongated or non-granular'. The proposal is illustrated by computer generated random lines intersecting a plane with a given line density.

The 'particles' of an indicated size have an average of 5.5 faces in the first and 4.6 in the second. In the second also a series of smaller 'particles' have only 3.3 faces on the average.

This cannot be the whole story, however, because Schmidt (1981) shows that grades both smaller and greater have particles with less cubiformity than the optimum grade. Thus further investigation is necessary because both machine effects and material properties are likely to be involved, as is discussed more generally in the next section.

4.2.6 Effect of material and machine on particle shape

Holt (1981b) has pointed out conflicting conclusions on the relative effects of material and machine on the shape of product particles. Bone (1954) took the view that 'the character of the material being broken has more influence on the shape of the product particles than the type of reduction machine used or the nature of the reduction process'. Rose (1961) took the opposite point of view: 'whilst the properties of the material determine to some extent the particle shape after comminution, the type of mill has, almost certainly, an even more marked effect'. In neither case was evidence put forward and therefore there is no basis for analysis of the statements. Clearly, with materials like shale, with pronounced bedding planes, compression parallel to or perpendicular to the bedding plane will produce differently shaped particles, as has been demonstrated by Heywood and Pryor (1945/46). Though similar in plan view, the grains were thinner in elevation when compressed parallel to the bedding plane. Mica and schistose rock behave similarly.

However, for silicaceous rocks and limestone, Heywood and Pryor (1945/46) found that during grinding a state of equilibrium is gradually attained, such that a mean shape arises with length to breadth ratio in the range 1.2 to 1,5 and breadth to thickness ratio in the range 1.6 to 1.8. The type of size reduction machine was not stated but it is likely to have been a ball mill.

A study by Shergold (1959), however, states clearly the machines and materials which were used, and product shape comparison is made in terms of the flakiness index, all in the context of roadstone aggregates crushed from 3 inches to 0.75 inches over a range of reduction ratios. He comes to the conclusion that the dominant factor in deciding shape is not type of machine or type of material, but reduction ratio; the smaller the ratio, the more cubiform the particle. For a given reduction ratio, the order of merit of machines to produce cubiform products was as follows:

impact breaker with unrestricted outlet
impact breaker with restricted outlet
jaw crusher
crushing rolls
cone crusher

No precise relation was found between the type of rock and the shape of the product, but the range of variation between rocks tested was of the same order as that between machines, and in both cases less than the variation due to reduction ratio. The rocks tested were for the most part hard and close grained:

Andesite	Granite
Blast furnace slag	Hornfels
Diabase	Limestone
Electric furnace slag	Quartzite gravel
Flint gravel	Quartzite sandstone
	Trachite

The experiments of Shergold (1959) did not include electrohydraulic crushing, which Yigit *et al.* (1967) have demonstrated most convincingly to be superior to any mechanical crusher in respect of cubiform shape and fewness of fines. However, the economics of electrohydraulic crushing are as yet unfavourable in the context of large-scale aggregate production. (see §11.4.)

For the fine grinding of bituminous coal, Austin *et al.* (1963) report that shape factor is a function of coal rank, in common with other properties such as calorific value. In the range of dry mineral matter free volatile percentages from 15% to 25%, most ground products tend to granular shapes (shape factor 0.7 to 0.8), while lower or higher rank coals, i.e. below 15% or above 25% volatile, tend to be less cubical (shape factors 0.6 to 0.7). The standard Hardgrove machine and tumbling ball mills were used in the tests.

Apart from the above, no detailed work on shape factors in the context of fine grinding appears to have been attempted, but Woźniak (1980) has studied the control of shape in the context of particles to give the maximum abrasion. As indicated in §4.2.1, angular shapes and a granular form are sought rather than a flaky form. The merit of impact machines for this purpose was stressed, as long as reduction ratio was limited, which again turned out to be an important parameter. Rose (1961) found that rolls gave angular grains as long as the relative peripheral speed of the rollers was limited. Thus, as would be expected, all the preferred machines for angular shaped products have no bed of particles in which the rounding of corners occurs.

Overall, it would appear to be dangerous to generalize on the relative effects of material and machine as regards product shape. For materials with strong bedding planes, the product will be flaky whatever the machine used. For machines with beds of particles the product will have rounded particles, whatever the basic form. For angularity or cubiformity of product, apart from electrohydraulic crushing, impact machines with low reduction ratio are required. This much is clear, but no doubt there is scope for closer examination of the relative effects of machine and material in specific processes and for exploration of the effects in each of several size ranges of the product.

4.2.7 Review

It cannot be denied that the characterization of particle shape is difficult, which accounts for the paucity of research work until recent years when the availability of new equipment has catalysed new interest in the subject. Development is still in an exploratory stage, both of shape concepts and of practical methods of applying the concepts, and future work should prove to be both fascinating to students and valuable to industry.

4.3 PARTICLE SIZE

Note: For references see §4.5.2.

4.3.1 Definitions of individual particle size

There are basically three groups of definitions of irregularly shaped particle size (Svarovsky, 1977).

The first group links the diameter of an equivalent sphere that would have the same property as the particle itself, such as volume etc. These are listed in Table 4.2.

The second group of definitions uses the diameter of a circle that would have the same property as the projected outline of the particles, as listed in Table 4.3.

The third group of definitions is associated with a linear dimension, as measured by a microscope, parallel to a fixed direction and may be classified as a statistical diameter as given in Table 4.4.

The equivalent circle and the statistical diameters have been found to have relevance to particle shape characterization. (see §§4.2.3 and 4.2.5.2.)

It is important to select the diameter most relevant to the process being considered.

Table 4.2. Definitions of equivalent sphere diameters

Symbol	Diameter	Equivalent property of sphere
x_V	Volume	Volume
x_S	Surface	Surface
x_{SV}	Surface volume	Surface to volume ratio
x_D	Drag	Resistance to motion in the same fluid at the same velocity
x_{FF}	Free-fall	Free-falling speed in the same liquid with the same particle density
x_{ST}	Stokes	Free-falling speed, if Stokes' law obeyed
x_A	Sieve	Passing through the same square sieve aperture

Reproduced by permission of Butterworths, London.

Table 4.3. Definitions of equivalent circle diameter

Symbol	Diameter	Equivalent property of a circle
x_H	Heywood	Projected area when the particle rests in a stable position (Heywood, 1937)
x_P	Projected area	Projected area if the particle is randomly orientated
x_C	Projected	Perimeter of the outline

Reproduced by permission of Butterworths, London.

Table 4.4. Definitions of statistical diameters

Symbol	Diameter	Dimension measured
x_F	Feret's	Distance between two tangents on opposite sides of the particle
x_M	Martin's	Length of the line which bisects the image of the particle
x_{SH}	Shear	Particle width obtained with an image shearing microscope eyepiece
x_{CH}	Maximum chord	Maximum length of a line limited by the contour of the particle

Reproduced by permission of Butterworths, London.

For measurement of the appropriate particle diameters, see §§4.3.9.1 and 4.3.9.2.

4.3.2 Types of particle size distribution

In practice there are four types of size distribution:

by number	$f_N(x)$	frequency distribution
by length	$f_L(x)$	density distributions
by surface	$f_S(x)$	
by mass (or volume)	$f_M(x)$	

The term 'frequency' distribution is sometimes used in the literature where length, surface or mass is involved, but this handbook reserves it for number only.

The density distributions are related to the frequency distribution as follows:

$$f_L(x) = K_1 x f_N(x) \tag{4.5}$$

$$f_S(x) = K_2 x^2 f_N(x) \tag{4.6}$$

$$f_M(x) = K_3 x^3 f_N(x) \quad (4.7)$$

The constants K_1, K_2 and K_3 contain a shape factor which may be particle-size dependent, and hence an accurate conversion from one distribution to another is not possible without a quantitative knowledge of the shape factor's dependence on size (Svarovski, 1977). For some evidence of shape change with size reduction see Harris and Stambolitzis (1969) and §4.2.5.5.

Size distribution by mass is the most commonly used method, and Figure 4.10 shows a typical form of the distribution.

Density distributions may also be expressed in cumulative form $F(x)$ given by:

$$F(x) = \int f(x)\mathrm{d}x$$

or, otherwise stated, the density is the derivative of the cumulative.

Cumulative functions are given as either oversize or undersize, because

$$F(x)_{\text{oversize}} = 1 - F(x)_{\text{undersize}}$$

Figure 4.11 is illustrative of both density and cumulative plots for the same

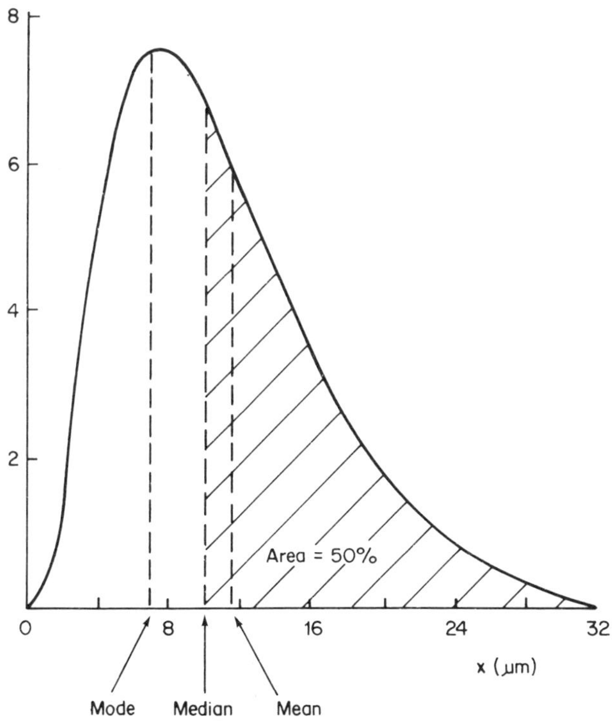

Figure 4.10. The mode, median and mean of a size distribution

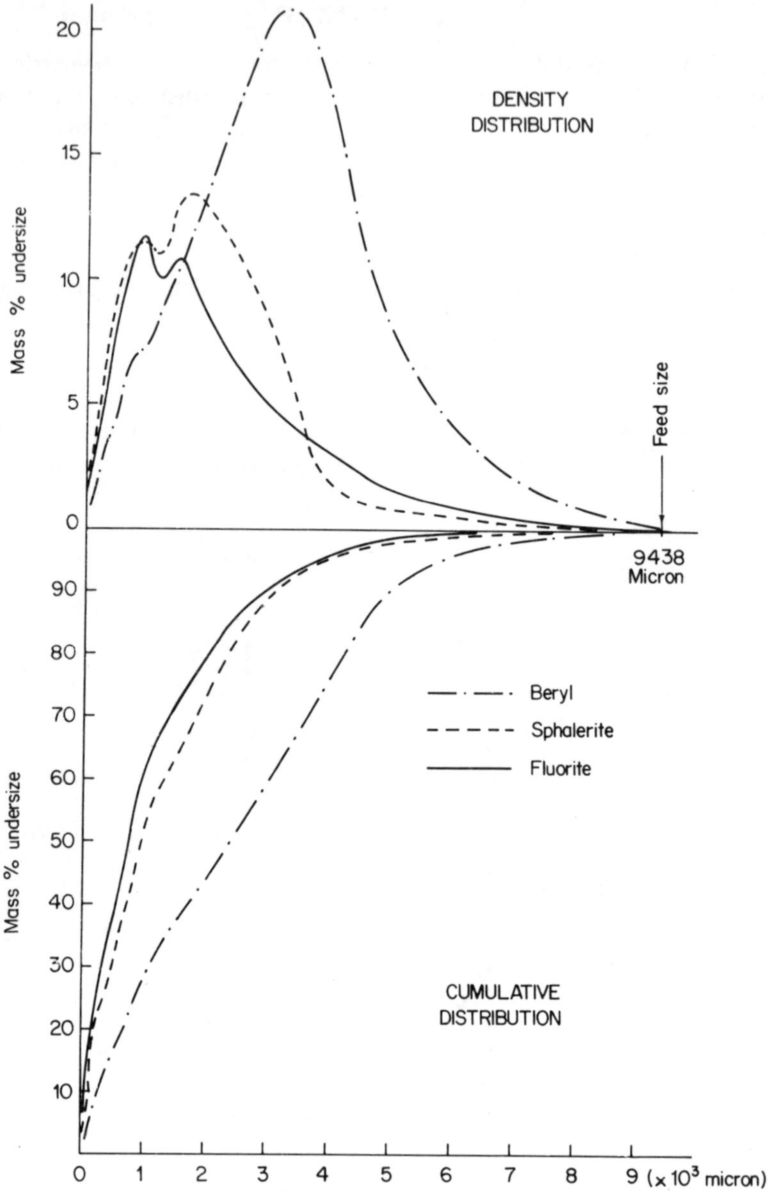

Figure 4.11. Size distribution of products of three minerals each having identical treatment in a pendulum crusher (Hukki, 1944)

size distribution in the cases of three minerals. The bimodal character of the sphalerite and fluorite density distributions is lost in their cumulative counterparts. This highlights the value of the density distribution to demonstrate unique features of a dispersion.

4.3.3 Average particle size in a distribution

There are many average sizes which can be defined for a size distribution, each designed to represent it by a single figure. The average, of course, gives no guide to the distribution spread.

The three most important averages are depicted in Figure 4.10. They are the mode, the median and the mean. The mode x_{mo} is the size corresponding with the peak of the distribution curve. The median x_{me} is the size which divides the distribution into two equal parts. There are several definitions of mean particle size and their general form may be written (Svarovski, 1977) as:

$$g(\bar{x}) = \int_0^\infty g(x)f(x)\,\mathrm{d}x \tag{4.8}$$

where $f(x)$ is the frequency distribution (by number) or density distribution by length, surface or mass, and $g(x)$ is a particular function of particle size as given in Table 4.5.

If no mathematical function for a distribution is known, then graphical means must be used to calculate the appropriate mean. However, it is very much easier to handle mean sizes from analytical functions using modern computer techniques, and hence the fitting of mathematical expressions to size distribution curves is becoming increasingly important. (see §4.3.8.)

Table 4.5. Types of mean diameter

Form of $g(x)$	Name of mean diameter	
x	Arithmetic mean	$\bar{x}_a$
x^2	Quadratic mean	$\bar{x}_q$
x^3	Cubic mean	$\bar{x}_c$
$\log x$	Geometric mean	$\bar{x}_g$
$1/x$	Harmonic mean	$\bar{x}_h$

Reproduced by permission of Butterworths, London.

4.3.4 Factors affecting particle size distribution

4.3.4.1 Statistical nature of breakage

As indicated in §3.5.3, if several particles of a given material, each having the same size and shape, are given the same comminution treatment—such as a single blow by falling from a given height on to a hard plate, as in a drop shatter test—the resulting particle size distributions will not be all identical. They will vary within certain limits. Hence to obtain realistic mean size distributions from such tests, it is important to arrange for a large number of particles to fall simultaneously and pool all the product particles before sizing by sieving.

Similarly for single-particle compression tests, successive tests on exactly similar particles must be performed, and the products pooled before the size distribution is assessed to obtain a valid mean.

Clearly, the product from a tumbling ball mill test would not need such a procedure, as a sufficient number of small breakages take place in the milling process for the resulting size distribution to be repeatable from a given sample, for a given time of grind under identical treatment.

4.3.4.2 Effect of mineral properties

Hukki (1944) reports the results of impact tests (pendulum crusher) on five minerals, namely beryl, quartz, sphalerite, fluorite and galena, each having identical treatment. The density distributions of three of the products are given in Figure 4.11, and as would be expected, they differ considerably. The bimodal character of the density distributions is apparent, even if for beryl only a point of inflexion occurs.

4.3.4.3 Effect of feed size and time of grind

For silicate sand ground in a laboratory ball mill, Jimbo (1973) demonstrated the effect of feed size on product size distribution for a given time of grind. Figure 4.12, a linear cumulative plot, shows that at relatively short grind times of the order of 4 minutes, the finer the feed the finer the product, as would be expected; but at longer times such as 64 minutes, the grades are almost

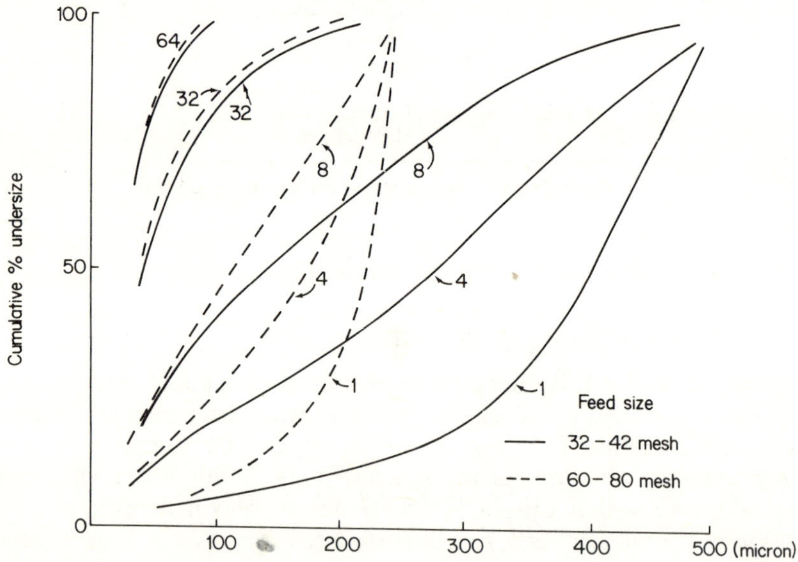

Figure 4.12. The effect of feed size on cumulative product size distribution for silicate sand ground in a laboratory ball mill. Numerals indicate time of grind in minutes (Jimbo, 1973). *Reproduced by permission of the author*

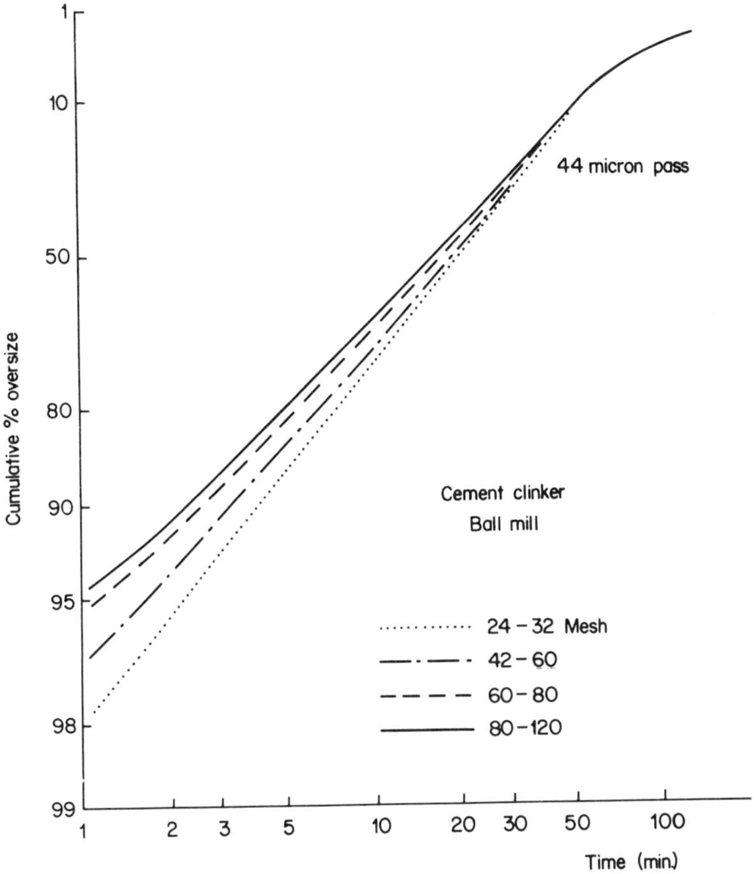

Figure 4.13. The effect of time of grind on product cumulative size distribution for a range of feed sizes of cement clinker (Jimbo, 1973). *Reproduced by permission of the author*

exactly similar. Figure 4.13 shows the same feature for the 44 micron pass of cement clinker. It appears that after a certain time the product size distributions tend to coincide, for a given material, whatever the initial size distribution. (see §4.3.5.4.)

Heywood (1957), in what is now regarded as a classical experiment, used density distributions which are even more illuminating than the cumulative plots by Jimbo (1973). A screened fraction of anthracite (600–1600 microns) was ground in a ball mill and the size distribution was checked at regular time intervals, depicted in a three-dimensional plot reproduced as Figure 4.14. After a short period of grinding the distribution is bimodal, but as comminution proceeds a change to a unimodal distribution takes place. Jimbo (1973) reports the same effect. This is further discussed in §§4.3.5.2. and 4.3.5.3.

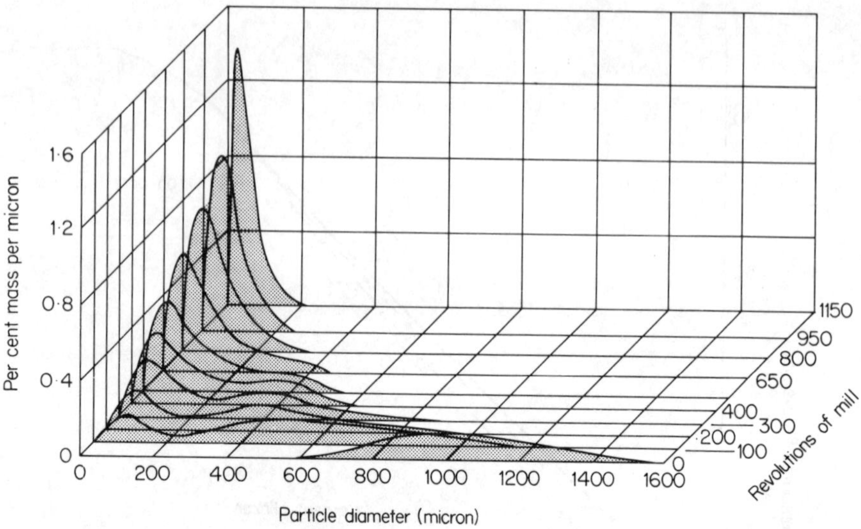

Figure 4.14. The effect of time of grind on product density size distribution for a monomodal feed size of anthracite (Heywood, 1957)

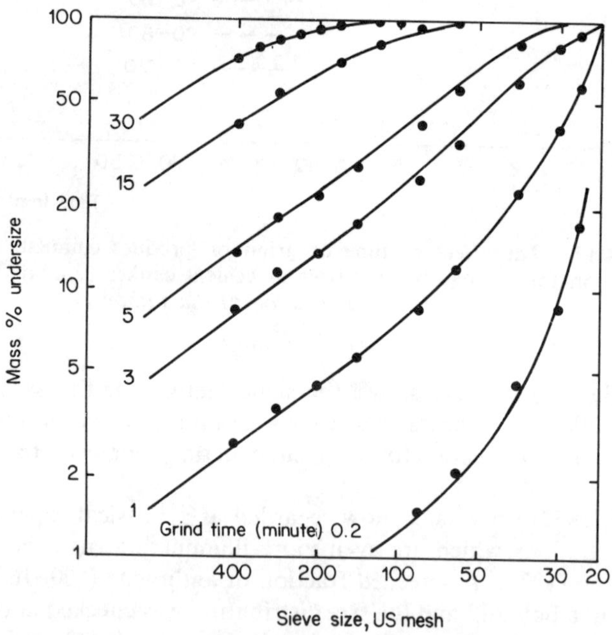

Figure 4.15. Cumulative log–log size distributions for cement clinker ground in a batch ball mill over a range of grinding times (Austin and Bhatia, 1971/72). *Reproduced by permission of Elsevier Sequoia*

This change from multimodal to unimodal distribution is masked in common forms of plotting, such as a log–log plot of mass percentage undersize versus grinding time (Figure 4.15), which are commonly used in grinding kinetic studies. In as much as mathematical equations can be applied successfully to fit such curves implies that the modal changes are accommodated thereby, despite the visual masking. This applies equally to a Rosin–Rammler–Bennett plot (see §4.3.8.2(a)).

4.3.4.4 Effect of method of comminution

Snow and Paulding (1973) have demonstrated for primary breakage almost identical dispersion arrays from impact and compression comminution of 0.75 inch diameter glass spheres, as seen in Figure 4.16. The bimodal pattern is noted. For tests on quartz, Rumpf (1965) has produced data, summarized in Figures 4.17 and 4.18 which shows an apparent difference between the size distributions for impact and compression; but reference back to Figure 4.15 reveals that the compression tests have just not proceeded as far as the impact tests in the progressive change of distribution.

Increasing the speed of a single impact has the same effect on the pattern of size distribution change as increasing the time of grind in a ball mill.

Thus it appears that all methods of comminution give product arrays which pass through the same cycle of changes as breakage proceeds.

4.3.4.5 Effect of impact energy versus velocity of impact

The effect of increased velocity of impact has been seen in the cumulative plot in Figure 4.17. Figure 4.19 is a density distribution from pendulum tests on galena by Hukki (1944). The two curves presented are associated with equal conditions, except that one is given twice the externally applied energy than the other, and inherently a greater velocity of impact. The patterns of the distribution arrays are very similar: peaks at A and B are retained at A′ and B′, but the peak C has become a point of inflexion C′. This appears to be the first stage of simplification from a trimodal to a bimodal distribution. What is not clear from this experiment is whether impact energy or velocity of hit is the more important parameter, but Snow and Paulding (1973) have elucidated this point, as indicated in Figure 4.16. Looking at the two curves for 62 in lb units of impact energy, that corresponding to the greater height of drop—and therefore to higher velocity of hit—produces only slightly increased fines over that from the lower height. Hence it is deduced that impact energy is the prevailing external parameter.

Guided by fundamental considerations of the stress pattern and the resulting energy density distribution within a loaded sphere, Snow and Paulding (1973) go further and conclude that it is the breaking strength of the particle up to the moment of fracture, rather than the magnitude of the externally applied impact energy, which really decides the resulting size distribution, and this is

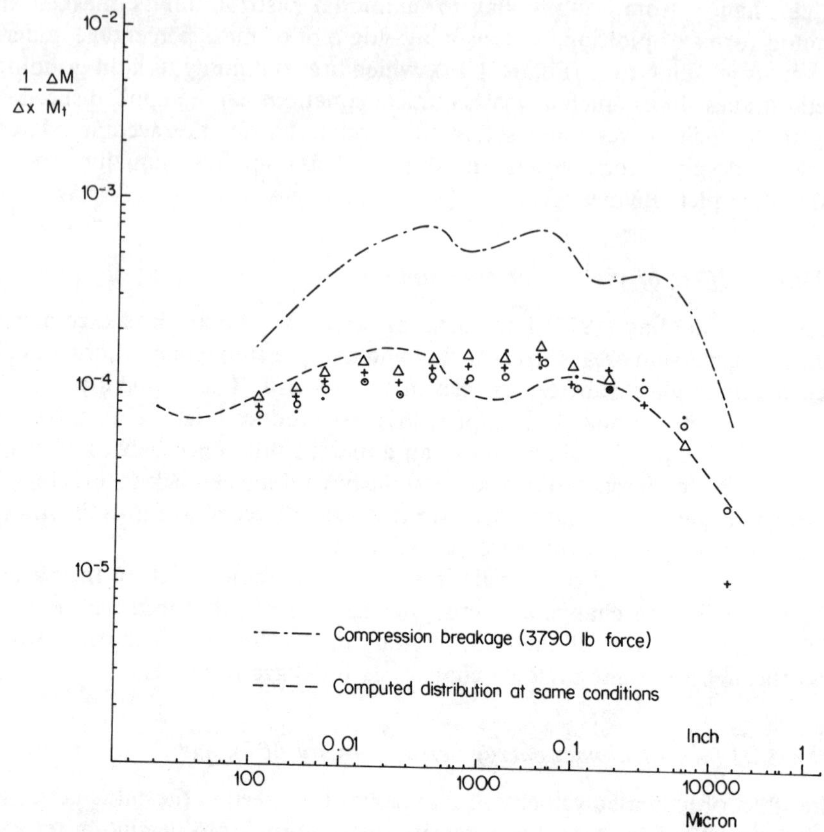

Figure 4.16. Comparison of density distribution from breakage of 0.75 inch diameter glass spheres by impact and compression and from computation based on energy density within a stressed particle (Snow and Paulding, 1973). *Reproduced by permission of the authors*

Test conditions

Symbol	Drop height (ft)	Impactor mass (640 g units)	Impactor energy (in-lb)
●	2	1	31
+	2	2	62
Δ	2	3	93
○	1	4	62

borne out by the correspondence they obtain between the distribution computed on the above basis and that obtained in practice, for both impact and compression in Figure 4.16. The externally applied energy is obviously necessary, but all of it is not used for fracture.

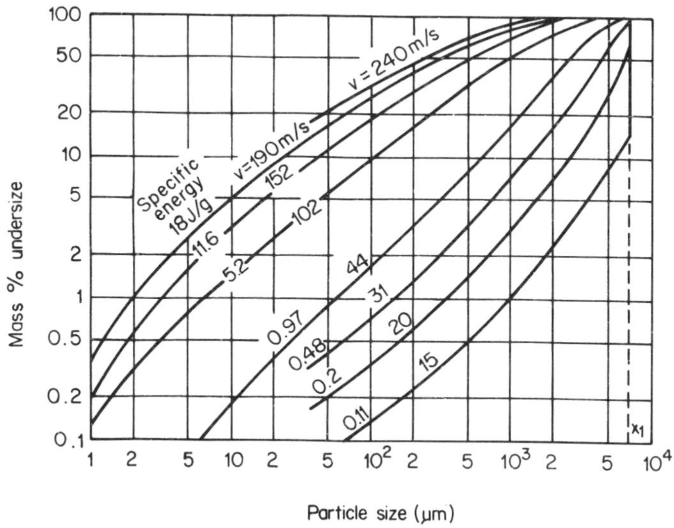

Figure 4.17. Comparison of mean cumulative distributions for impact breakage of quartz over a range of impact velocities. Each curve is the mean of at least 15 tests (Rumpf, 1965). *Reproduced by permission of VCH Vergalsgesellschaft*

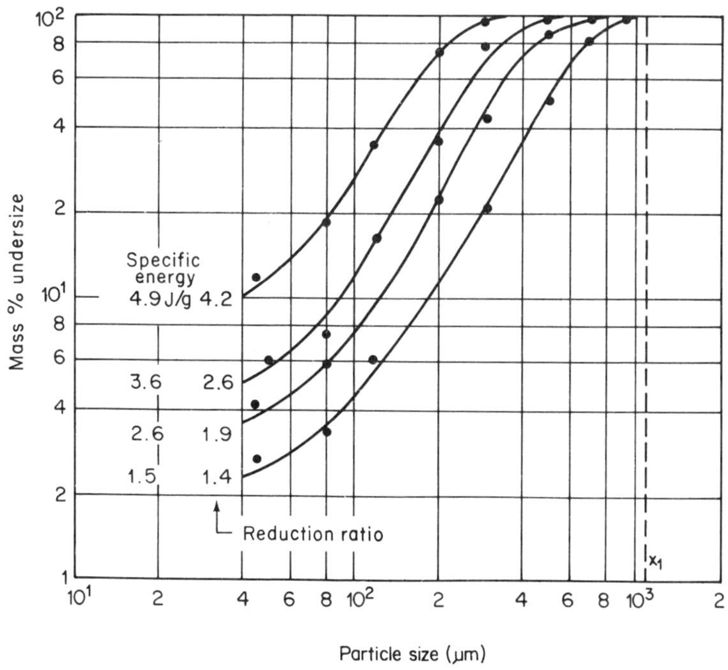

Figure 4.18. Comparison of mean cumulative distributions for compression breakage of quartz over a range of reduction ratios. Each curve is the mean of at least 15 tests (Rumpf, 1965). *Reproduced by permission of VCH Vergalsgesellschaft*

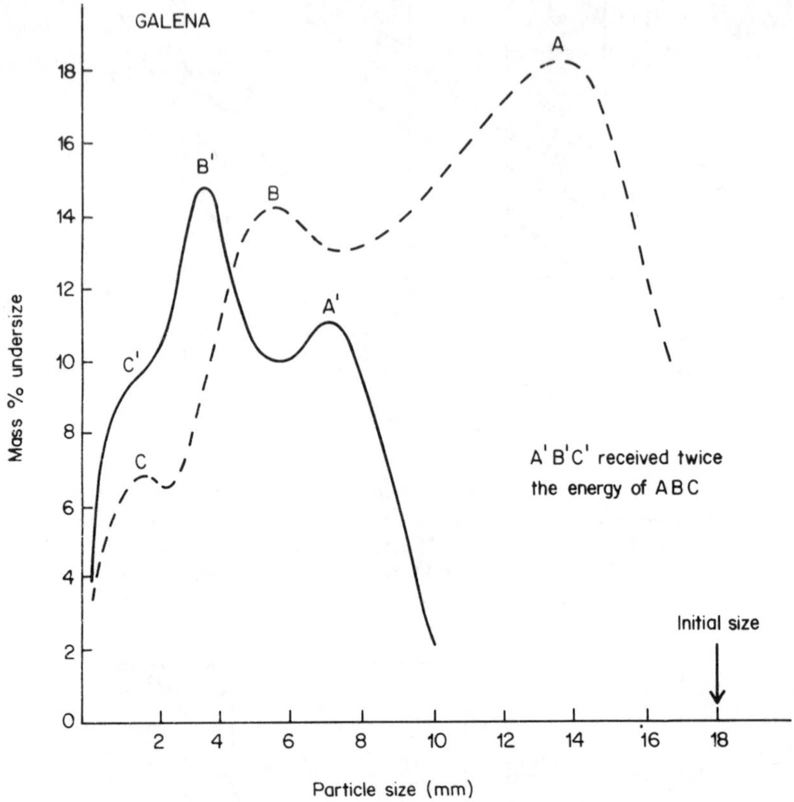

Figure 4.19. Comparison of density distributions for impact breakage of galena samples having the same mass but one having twice the externally applied energy of the other. A′B′C′ received twice the energy of ABC (Hukki, 1944)

4.3.4.6 Effect of single versus repeated fracture

Strictly, because of the conclusions of Snow and Paulding (1973), single and repeated fracture should be compared on the basis of the same breakage strength at the moment of fracture, but such data are obviously very hard to come by. Hukki (1944) has published results of pendulum impact tests on common salt with the same energy input but one a single impact and the other repeated smaller impacts. The resulting size distributions are plotted in Figure 4.20. Higher fineness is achieved with the repeated compared with the single fracture, and the trimodal distribution of the single fracture has become bimodal in the repeated fracture test. (see §4.3.5.5.)

4.3.4.7 Machine effects

Moving from laboratory equipment to commercial sized machines, a few examples are taken of methods of change of product size distribution.

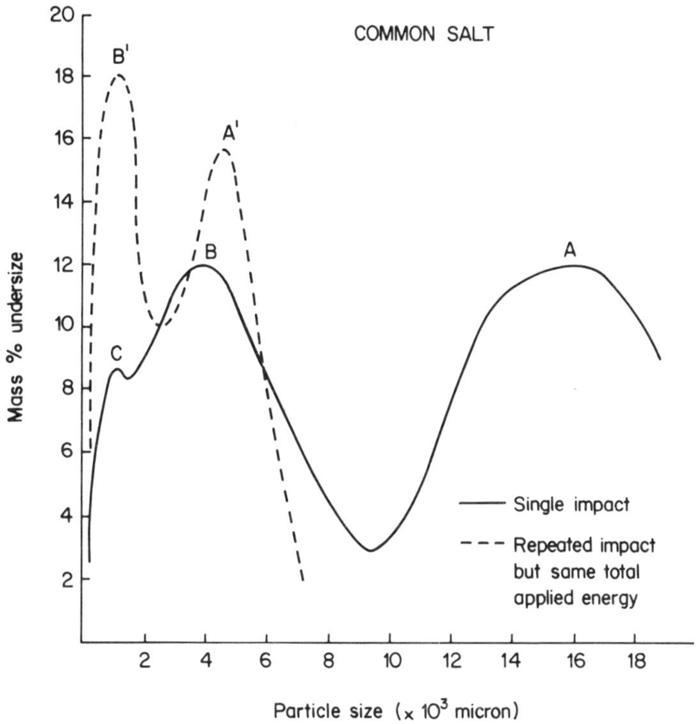

Figure 4.20. Comparison of density distribution for common salt with single and repeated impact but the same externally applied energy (Hukki, 1944)

For an open-circuit crusher, Heywood (1957) has shown the effect of reduced jaw gap on the size density curve, given in Figure 4.21. All the distributions are bimodal. With the reduction of jaw gap the coarser mode moves toward a smaller particle size but the finer mode moves only slightly towards a smaller particle and increases in intensity. These features correspond with the pattern presented in Figure 4.14 for increased time of grind.

For a closed-circuit machine, that is one having a classifier loop, it is possible to change the output size distribution by adjustment of the classifier setting, e.g. by change of rotational speed for a rotating blade classifier or by change of blade angle for a static centrifugal classifier. This, of course, changes the recirculation rate in the classifier loop for a constant feed rate. However, for constant classifier setting, reduction of feed rate alone will increase fineness of product at constant power utilization, and a capacity fineness factor has to be established for a given mill. Figure 4.22 is an example of such a factor for a coal pulveriser of the ring-ball type.

The density–size distributions for these closed-circuit machines are likely to have moved well into the monomodal phase.

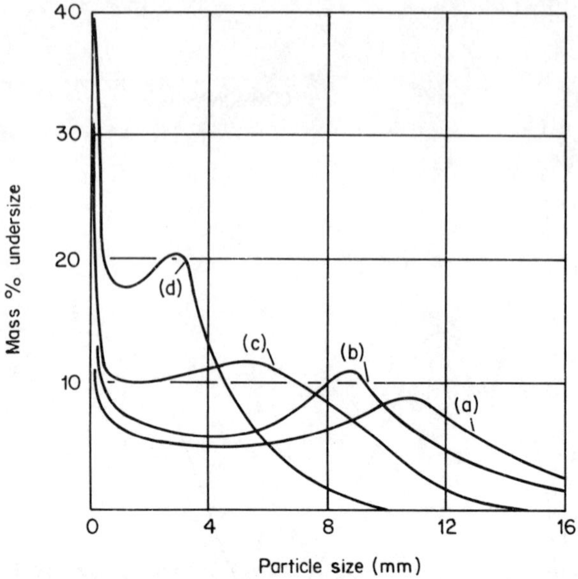

Figure 4.21. Density size distributions from a jaw crusher. Jaw gap: (a) 9.9 mm, (b) 7.4 mm, (c) 5.3 mm, (d) 2.8 mm (Heywood, 1957)

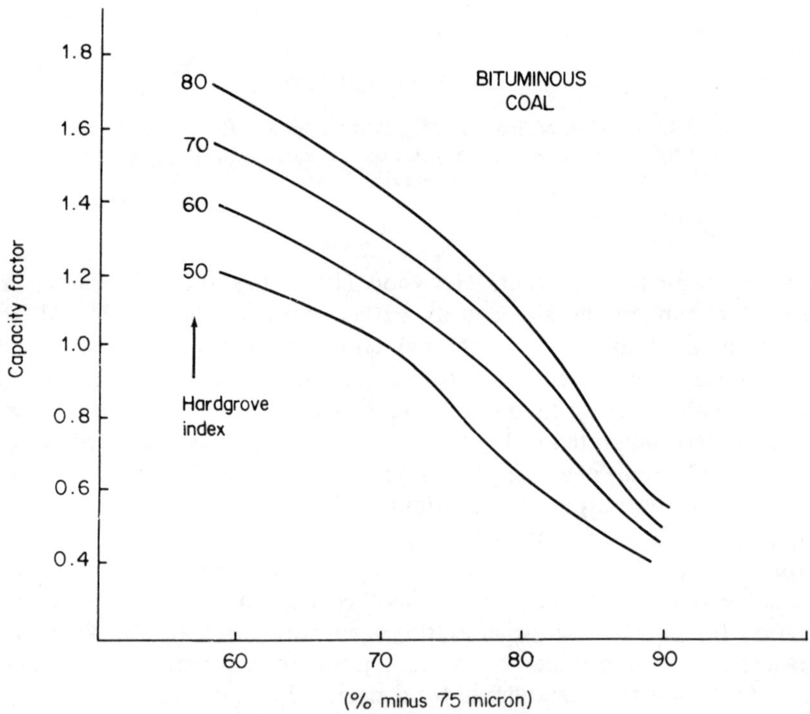

Figure 4.22. Constant power capacity factor change with product size distribution for a ring-ball mill grinding coal with the Hardgrove index as a parameter (Adapted from Babcock & Wilcox Co., New York, 1955)

4.3.5 Reasons for size distribution changes as grinding proceeds

4.3.5.1 Survey

Reasons are now sought for the observed patterns associated with comminution size distributions, summarized as follows:

(1) In primary breakage, the density distribution is multimodal.
(2) A monomodal pattern gradually emerges with grinding beyond primary breakage (see §4.3.4.3).
(3) For a given material, whatever the feed size, after continuous grinding the distributions tend to coincide (see §4.3.4.3).
(4) Repeated fracture produces more fines than single fracture for the same expenditure of applied energy (see §4.3.4.6).

These are now discussed in turn.

4.3.5.2 Multimodal distribution from primary breakage

In §3.2.3.2 it was noted that a brittle particle, stressed by an impressed force, develops a cone of high stress intensity beneath the force contact area, where the energy density is also highest. Hence particle degradation would be expected to be highest in this zone, which has been confirmed experimentally for primary breakage, as follows.

Arbiter *et al.* (1969) produced Figure 4.23 for 25 mm diameter glass spheres having undergone primary breakage by low-velocity double impact and by slow compression. Figure 4.23(a) gives the log–log plot as produced by Arbiter and co-workers which reveals two distinct slopes associated with the 'complement' and the 'residue'. The finest grade was found to come from the conical zones beneath the force contact areas. Between these cones, along the core of the sphere where breakage is by compression, came a relatively fine grade, though coarser than that from the cones. From the rest of the sphere, where breakage occurs by tension, came lune-shaped coarse fragments. In Figure 4.23(a) the fines from the cones and the central core are classed together as the 'complement' and the coarse as 'residue', each giving a distinct straight line in the cumulative log–log plot. Looking at the density–size plot in Figure 4.23(b) for the same data, it is evident that a very early stage of comminution progress has been achieved, since the coarse grade peak is high relative to the fine grade peak, which is just developing.

Rumpf (1965) has demonstrated three distinct size groups from impact breakage of 8 mm diameter glass spheres at a velocity of 30 m/s (see Figure 4.24). It is thus easy to understand the trimodal distributions found by Hukki (1944), as seen in Figure 4.19 for example.

Whiten *et al.* (1979) obtained very similar plots to those of Arbiter *et al.* (1969) when they crushed approximately equally sized irregular particles of hematite and of andesite between platens. Figure 4.25 for both materials

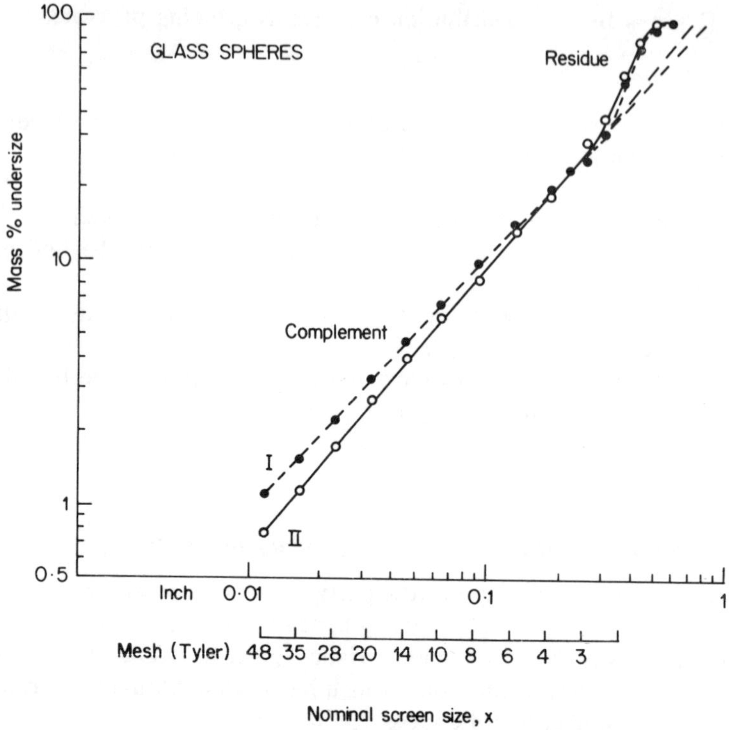

Figure 4.23(a). Cumulative size distribution from breakage of 1 inch diameter glass spheres by (I) slow compression, (II) double impact (Arbiter *et al.*, 1969). *Reproduced by permission of the American Institute of Mining Metallurgical and Petroleum Engineers*

shows the two distinct slopes of the curve on the cumulative log–log plot for the first major fracture, and less pronouncedly for subsequent fractures. Whiten and co-workers also traced the fines to the central core of particles and further found that the maximum fines proportion by mass of the whole was 0.2, and when only the first breakage was considered, the fines rarely exceeded 0.1.

The reasons for multimodal distributions from initial fracture are made clear from the above experiments.

4.3.5.3 Monomodal distribution from continued comminution

As breakage proceeds beyond the primary stage it would be expected that the larger particles would be preferentially broken for two reasons: firstly the smaller are stronger than the larger, and secondly the smaller will tend to be shielded in the interstices between the larger. Hence it could be forecast that with continued grinding the peak in the density–size array associated with the larger particles would disappear and that associated with the fines would mount higher and higher: and so it is, as seen in §4.3.4.3 and Figure 4.14.

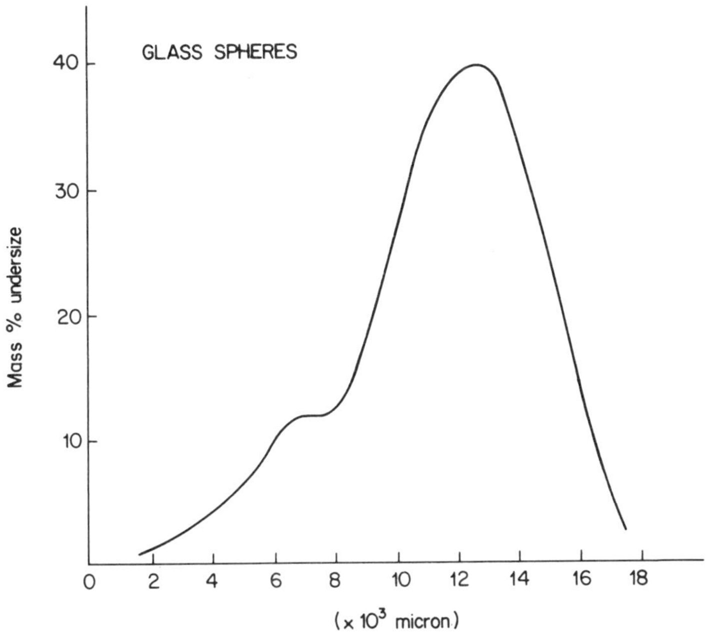

Figure 4.23(b). Density–size distribution from breakage of 1 inch diameter glass spheres by slow compression: same data and source as Figure 4.23(a).

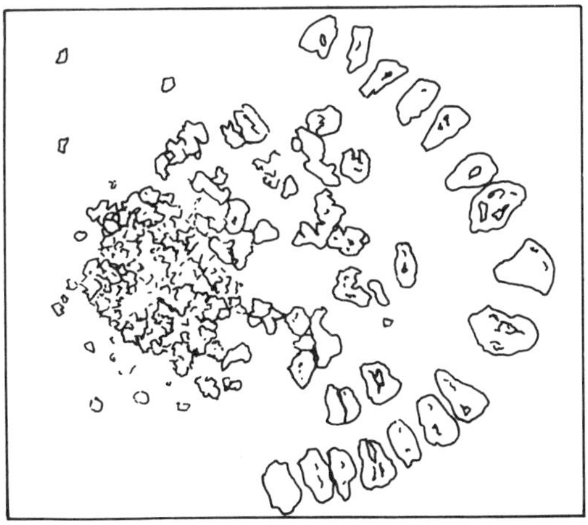

Figure 4.24. Three distinct groups of particles from impact breakage of an 8 mm diameter glass sphere at a velocity of 30 m/s (Rumpf, 1965). *Reproduced by permission of VCH Vergalsgesellschaft*

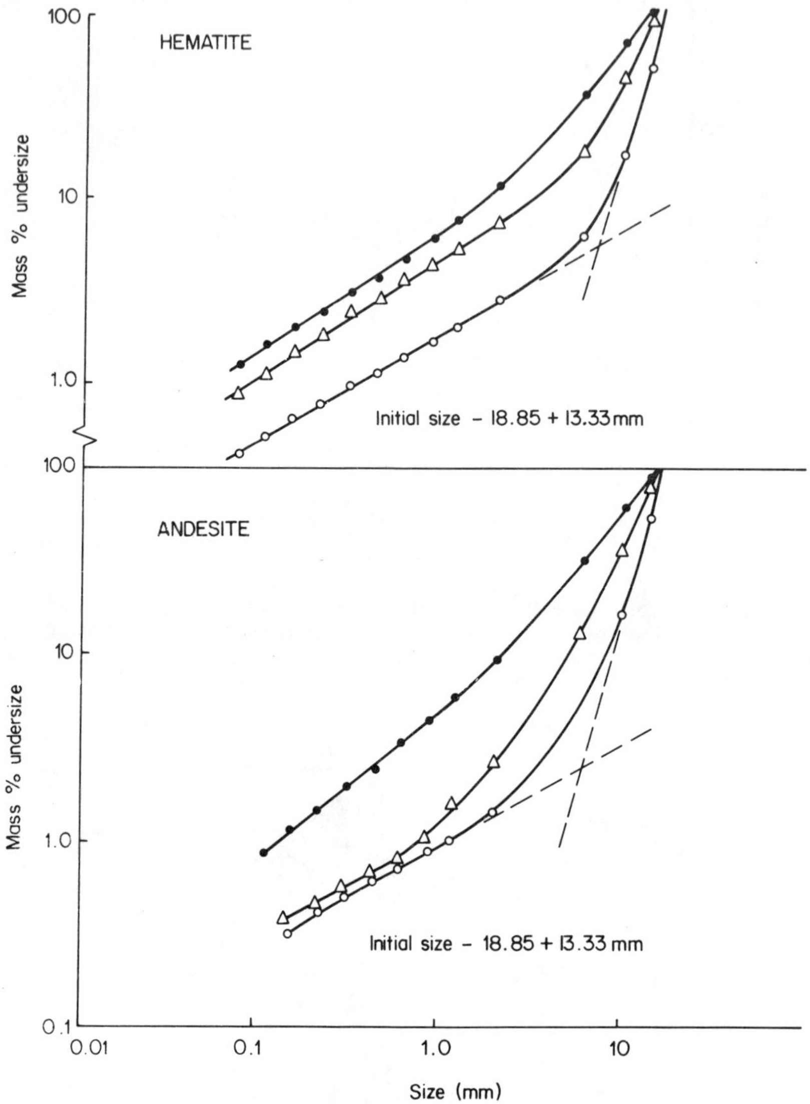

Figure 4.25. Cumulative size distributions from the crushing of approximately equal sized irregular particles of hematite and andesite between platens (Whiten *et al.*, 1979). *Reproduced by permission of Melbourne University*

○ One major fracture
△ Two major fractures
• Five major fractures

4.3.5.4 *Coincidence of product distributions from differing feed sizes*

Similarly for two feed distributions of a material undergoing comminution, one coarser than the other, the larger particles of the coarser will break more easily than the smaller ones of the finer. Hence both sized feed distributions would be expected to tend toward the same product distribution with continued grinding: and so it it, as shown in §§4.3.4.3 and Figure 4.12.

4.3.5.5 *Fines from repeated fracture versus a single blow*

If the same number of equisized particles is given equal amounts of externally applied energy, and for one set it is confined to a single blow but in a second set repeated fracture is applied, then it would be expected that more fines would result in the second set. This follows from the fact that at each particle impingement the cone beneath the contact area produces fines, whereas the rest of the particle splits into comparatively coarser pieces. In other words, the greater the number of contacts, the greater the fines for a given expenditure of energy, which is in accord with experience, demonstrated in §4.3.4.6 and Figure 4.20.

4.3.6 Effects associated with very fine grinding

4.3.6.1 *Survey*

When very fine grinding is undertaken, effects occur which are not present in coarse grinding. These include aggregation and agglomeration, the theme of §4.3.6.2, the influence of the gas surrounding a particle, discussed in §4.3.6.3 and plastic deformation, a factor in the grind limit considered in §4.3.7.

The forces and processes which lead to aggregation and agglomeration have been considered by Rumpf (1972). They include solid and liquid bridge formation between small particles, van de Waals forces and electrostatic forces. An effect associated with the last-named has been described by Owe Burg and Avis (1970–71) who explored increased humidity (and hence reduced electrostatic force) in the fine grinding of common salt. Further, Gregg (1968) has discussed the increased chemical activity of surface layers under fine grinding conditions. It is not always easy to be sure of the exact forces which produce a given effect in fine grinding, and the subject needs much more study. However, the effects themselves are now quite well documented.

4.3.6.2 *Aggregation and agglomeration*

Opoczky (1980) made carefully conducted experiments using a laboratory ball mill for the grinding of such materials as calcite, quartz and cement clinker for periods up to 80 hours. She assessed the products at various stages of size reduction by many means, including size distribution, surface formation, chemical analysis and observation with a scanning electron microscope. This work resulted in the need to distinguish between two phases of particle cohesion in fine grinding: aggregation and agglomeration.

Aggregation is defined as the weak reversible adherence of particles in a disperse system due to cohesion and van der Waals forces. It usually takes place at an earlier stage in a grinding process than agglomeration and is prevalent with soft, easily ground materials such as gypsum, burnt lime and some limestones.

Agglomeration, on the other hand, is a compacting and irreversible adherence of particles due mainly to chemical forces, which may cause some fundamental change in the ground product. It is a feature of materials which are difficult to grind, such as quartz and cement clinker, in which crystal structure changes have been observed.

As far as size distribution is concerned, both aggregation and agglomeration imply a coarsening of measured grades. An example of aggregation in the lower sieve range occurred in the grinding of limestone in a ring-ball mill, which gave very different analyses between dry and wet sieving. In the dry process, aggregates passed through a mesh as wholes, but when wet, they were broken up into their smaller components and therefore returned a finer distribution.

Figure 4.26 (Opoczky 1980) illustrates the above effects for calcite and quartz. Figure 4.26(a) shows that the calcite size distribution had flattened out

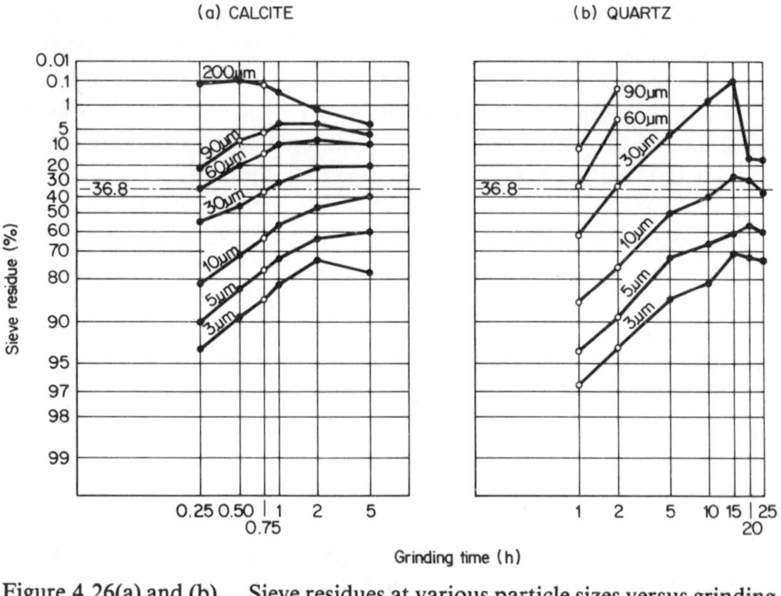

Figure 4.26(a) and (b). Sieve residues at various particle sizes versus grinding time for calcite and quartz (Opoczky, 1980)

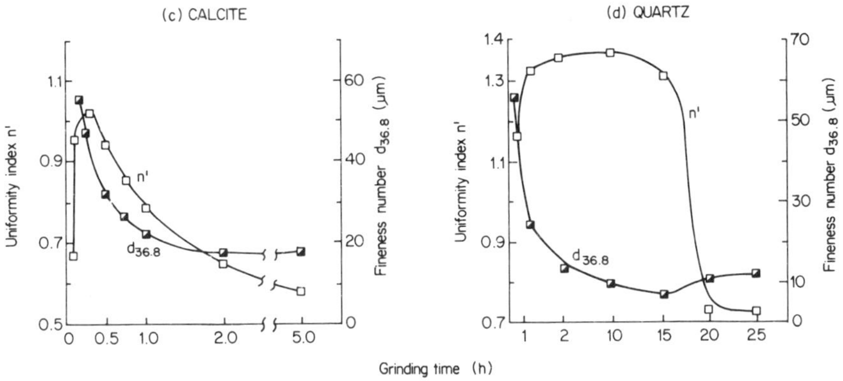

Figure 4.26(c) and (d). Variation of uniformity index n' and fineness number $d_{36.8}$ with grinding time for calcite and quartz (Opoczky, 1980). *Reproduced by permission of DECHEMA, Frankfurt*

after only 1 hour's grind, and after 2 hours had turned downward, demonstrating increased particle size with continued grinding. The same trend, shown in Figure 4.26(b) for quartz, took 15 hours. Viewing samples in the scanning electron microscope revealed two simultaneous processes: the breakage of grains together with the attachment of small ones to each other and to the surface of larger ones. Aggregation was in progress, and it was also revealed in Figures 4.26(c) and (d) by concurrent decrease of the uniformity index n' and the fineness number $d_{36.8}$ (see §4.3.8.2(a)). This net size reduction, even though aggregation was occurring, was confirmed by increase in total surface (gas adsorption).

With further grinding, however, a further change was noticed: $d_{36.8}$ reached a minimum and then began to increase, n' levelled off and total surface rapidly fell, as seen in Figure 4.38. Not even nitrogen could find many pores to fill after this final change which marked the onset of agglomeration, and coincided with observed crystal structure modifications. In confirmation of this, Steier and Schönert (1972) agreed that high compression of certain limestones, quartz and polystyrene could cause very dense agglomerates, and Gregg *et al* (1954) had earlier noticed lattice distortion of kaolin without chemical decomposition as the accompaniment of very fine grinding.

4.3.6.3 Effects of gas surrounding a particle

Careful work by Cottaar and Rietema (1984) has demonstrated the influence of interstitial gas in the context of the tumbling mill for particles in the range 100 to 10 microns. They found, for example, a general increase in specific rate of breakage with increased gas pressure, and hence of gas viscosity, because then the bulk of the powder remained in the toe of the mill and was acted on by falling grinding media. At low gas viscosity, the powder does not flow so readily and is carried up with the lifting balls and falls into the toe of the mill with them, making breakage correspondingly ineffective. This research work is being continued in the Netherlands.

4.3.7 The grind limit

4.3.7.1 Dynamic equilibria on continued grinding

Herbst and Sepulveda (1978) have suggested an *apparent* comminution limit for a given material in a particular type of mill, such as a tumbling ball mill, since the same material may be ground to a finer size distribution in an attrition mill, perhaps because of the greater shear forces which may be brought into play.

It is interesting to note that, for a pin mill, the grind limit was formerly attributed to the slowing down of the collision speed of small particles by the air currents produced by the rotating pins. However, Marshall (1967) demonstrated that, when the mill is operated *in vacuo*, no change in the grind limit is discernible. Hence he argues that another effect must come into operation to compensate for the handicap of reduced collision speed in air. This other effect he identifies as fluid energy grinding which invokes considerable shear in high-velocity gradients in the air.

However, for a given material ground in a given mill, say a tumbling ball mill, there appears to exist an equilibrium state of grinding, as originally conceived by Hüttig (1952), in which small particles are forming larger particles as fast as larger particles are being disintegrated. Opoczky and Farnady (1984) have taken the concept one stage further, by defining a temporary dynamic equilibrium associated with aggregation, and later, after further grinding time, a permanent dynamic equilibrium associated with agglomeration,the terms aggregation and agglomeration being defined as in §4.3.6.2. They demonstrate these distinct equilibria for each of calcite, quartz and cement clinker, combining the zones of increasing and decreasing fineness into a three-dimensional 'curve of state', having axes in three directions at right-angles: grinding time x, particle size y and grinding rate z.

Krycer and Hersey (1981) have determined the final agglomeration equilibrium for each of lactose, sucrose and paracetamol ground in a vibration mill for long periods, the finally produced granules having been formed by the agglomeration of finer particles.

4.3.7.2 Brittle/plastic transition considerations

Looking back to §3.2.6.4 it is seen that the transition to completely plastic behaviour of normally brittle materials occurs in the range 35 microns for coal down to about 0.5 microns for glass, judged by measurements made to date.

That materials other than those listed have a still lower brittle/plastic transition is suggested from the work of Kondo and Gaudin (1964) who, via surface area measurements and subsequent calculation, assessed particle size distributions in the range 1–0.03 microns for albite, a sodium felspar. Whatever the plastic/brittle transition size, it would be expected that this size would bear some relation to the grind limit, since further comminution action could only change its shape, but not its size. For albite, Figure 4.27 does show on a log–log plot of cumulative size distribution a concave downward turn to the curves as if they were becoming asymptotic to a size of about 0.01 micron.

By surface calculation based on a measured grade in the sieve range, Hukki and Venho (1972) and Hukki (1975) extended the work to include calcite and cement and obtained the curves shown in Figure 4.28. The calcite curves appear to approach asymptotically a size of about 0.1 microns and the cement curves a size in the range 0.5–1 micron. Now from §3.2.6.4 the brittle/plastic transition for one cement was assessed at 10–20 microns, which is at least one order of magnitude greater than the limit suggested by Figure 4.28.

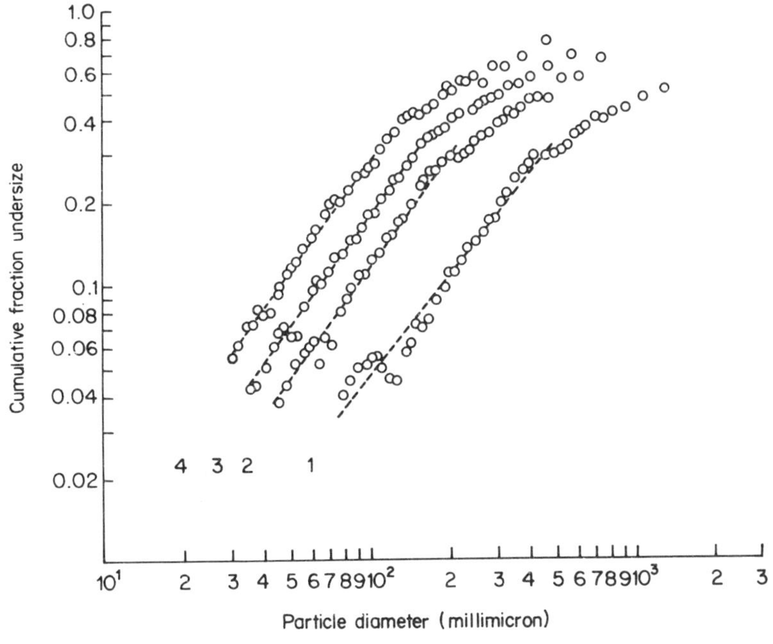

Figure 4.27. Cumulative size distributions of albite in the sub-micron range assessed via surface area measurement (Kondo and Gaudin, 1964.) *Reproduced by permission of the American Institute of Mining, Metallurgical and Petroleum Engineers*

Time of grind:
1. 5 hr, 2. 20 hr, 3. 40 hr, 4. 80 hr

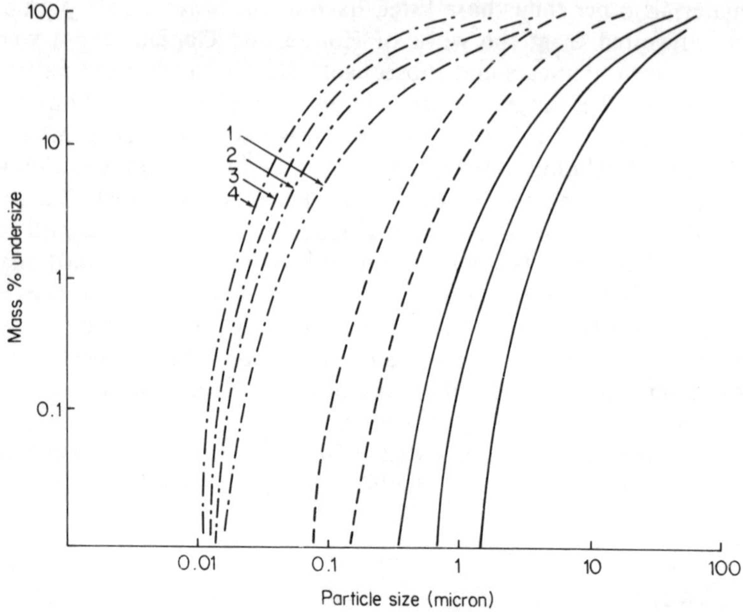

Figure 4.28. Cumulative size distributions for albite, calcite and cement assessed via surface area measurement and calculation. 1, 2, 3 and 4 are as in Figure 4.27. *Reproduced by permission of McGraw-Hill Inc.*
—— Cement (Hukki, 1975)
--- Calcite (Hukki, 1975)
-·-·- Albite (Kondo and Gaudin, 1964)

On the one hand, it must be emphasized that the Hukki proposed limits in Figure 4.28 are calculated values based on surface area and they are therefore very insecure. On the other hand, experimental work by Hess and Schönert (1981), resulting in the value d_c = 10–20 microns, was based on single-particle investigation and therefore the particles concerned were not subject to the varying forces which operate in a bed. Again, based on the work of Opoczky (1980), the small particles undergoing agglomeration could hardly be called cement clinker, although they were initially, because of chemical changes associated with agglomeration. Hence it is not surprising that exact correspondence is not reached between the single-particle brittle/plastic transition size and the grind limit for a bed of particles.

However, it would appear that the trend of the curves in Figure 4.28 is realistic for a particular mill grinding a given material, and that log–log cumulative size distributions become concave downwards as micron and submicron sizes are reached, as indicated in Figure 4.30; but at the present state of our knowledge, that size has not been fully explored for particle beds. The resolution of this problem awaits deeper investigation.

4.3.8 Mathematical presentation of size distribution data

4.3.8.1 Survey

There exists a wide range of mathematical functions which have been used to describe size distribution data.

Early equations took relatively simple forms for several reasons. Firstly, products from practical mills were considered as distinct from those from primary breakage, and for such mills it has been found in §4.3.5.3 that monomodal density distributions are likely. Secondly, one aim was to obtain a straight-line graphical correlation to assist in the establishment of 'laws'. However, the straight lines only ever applied over a limited size range. Endeavours to extend the range of applicability of formulae involved their increased complexity, as did the increased interest in primary breakage, with which is associated multimodal density distributions. Beside this, a search was initiated for parameters in the equations which had links with particle physical properties. Initial attempts were fraught with failure, but the search continues with some recent limited success (see §4.3.8.4)

The complexity of the size distribution problem calls for a system which is flexible enough to be tailored to any breakage pattern, and Broadbent and Callcott (1956a,b) found a powerful answer in vector and matrix algebra which has been extensively adopted in a modified form. Preliminary discussion of this concept, together with breakage functions, is commenced in this section, to be more fully considered in Chapters 8 and 9.

With the above in mind, the sub-divisions of this section are:

(1) commonly used functions in §4.3.8.2
(2) more complex functions in §4.3.8.3
(3) functions involving particle physical properties in §4.3.8.4
(4) matrix notation for size distributions in §4.3.8.5

4.3.8.2 Commonly used functions

(a) Rosin–Rammler–Bennett (RRB) References: Rosin and Rammler (1933), Bennett (1936), Harris (1968)

The equation is:

$$F(x) = 1 - e^{(-x/x')^{n'}} \tag{4.9}$$

Beke (1964) and Bailey (1969) have presented derivations of equation (4.9). See also §9.2.1.

When $x = x'$, $F(x) = 1 - 1/e = 63.2\%$ passing $= 36.8\%$ retained.

Equation (4.9) may be rearranged to give:

$$\log_e \log_e \left[\frac{1}{1 - F(x)}\right] = n'(\log_e x - \log_e x') \tag{4.10}$$

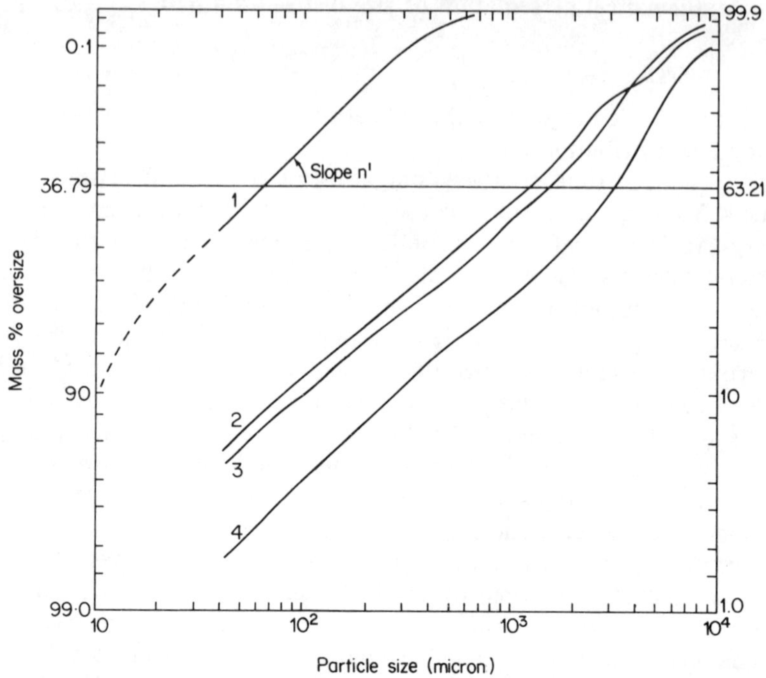

Figure 4.29. Rosin–Rammler–Bennett plot of cumulative size distributions
1 Pulverized fuel (bituminous coal)
2 Fluorite
3 Sphelerite } from Figure 4.11 (Hukki, 1944)
4 Beryl

Thus when the left-hand side of equation (4.10) is plotted against $\log_e x$ a straight line of slope n' results if the dispersion follows the RRB equation, and special graph paper is available to facilitate the plotting, as illustrated in Figure 4.29. For pulverized fuel (bituminous coal) the value of n' approximates to unity over the straight-line portion of the curve which is valid in the fine-to-medium size range, but not in the coarse or the very fine. For analysis of the RRB formula, see Beke (1946), who calls n' the uniformity index of the distribution pattern, because small values indicate a scattered distribution, whereas larger values imply a steeper curve and thus an increasingly uniform particle size structure. This is further reinforced by Beke (1964) in his equation (1.49a) which demonstrates that the uniformity index is a measure of the standard deviation on a logarithmic argument, the two being inversely proportional. This principle has been effectively used by Opoczky (1980) in aggregation studies. In this context, values of n' fell into the range 0.5–1.5 (see §4.3.6.2). She also used $d_{36.8}$ as a fineness number.

(b) Gaudin–Schuhmann References: Schuhmann (1940), Harris (1968)

The formula takes the form:

$$F(x) = \left(\frac{x}{x'}\right)^n \tag{4.11}$$

or

$$\log F(x) = n(\log x - \log x') \tag{4.12}$$

For a derivation of equation (4.11) see Beke (1964) and Bailey (1969). With logarithmic plotting, for homogeneous materials, a straight-line is usually given over a considerable portion of the fine range, but Heywood (1957) has shown mathematically that the slope of the graph must decrease from that of the straight-line section with increasing particle size. Fuerstenau (1961) indicates that for highly heterogeneous materials no straight-line portion exists. Figure 4.30 shows a typical plot for a homogeneous material. The slope of the straight-line section gives the index n and its intercept on the 100% axis gives x', which is called the modulus. In the micron and sub-micron range, the

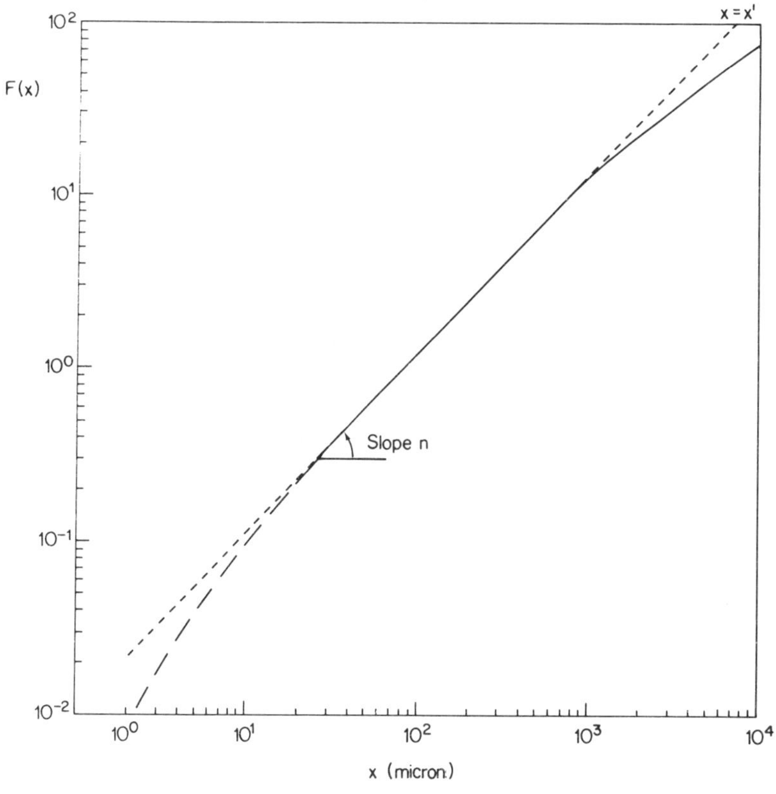

Figure 4.30. Schuhmann plot of cumulative size distribution (see equations (4.11) and (4.12)). (Hukki, 1944, Test 18)

straight line may not obtain and the curve is concave downwards as indicated in §4.3.7, at some small particle size. As for the RRB formula, Beke (1964) has presented a full analysis of the Schuhmann formula.

(c) Broadbent–Callcott References: Broadbent and Callcott (1956a,b)

The equation may be written:

$$F(x) = \frac{1 - e^{-x/x'}}{1 - e^{-1}} \tag{4.13}$$

and Bailey (1969) gives a derivation as a special case of a more general formula.

Equation (4.13) is similar to equation (4.9) except that x' corresponds to the 100% passing size instead of the 63.2% passing size, which is the case for the RRB equation. Further, when $x = x'$, $F(x) = 1$, and when $x = 0$, $F(x) = 0$; i.e. the function lies between zero and unity, which is often convenient in practical size analysis.

(d) Gaudin–Meloy References: Gaudin and Meloy (1962), Bergstrom (1966), Harris (1968)

The equation can be stated as:

$$F(x) = 1 - \left(1 - \frac{x}{x'}\right)^n \tag{4.14}$$

The advantage of the Gaudin–Meloy over the RRB equation is the expansion in the coarse region (though contracted in the fine) and the fact of a finite upper limit not possessed by the RRB.

4.3.8.3 *More complex functions*

Rumpf (1965) was quite definite in his comment on 150 single-particle tests performed at Karlsruhe, using both impact and compression breakage, that none of the commonly used two-parameter equations as considered in §4.3.8.2 could satisfy the size distributions produced. It is understandable, therefore, that more complex equations have been sought and examples are given below.

(a) Log–normal distribution Reference: Svarovsky (1977)

The normal probability graph is the well-known Gaussian frequency bell-shaped curve. It rarely applies to the products of comminution because most distributions are skewed to the right.

The formula for normal distribution is:

$$\frac{dF}{dx} = \frac{1}{\sigma\sqrt{(2\pi)}} \exp\left(-\frac{x - \bar{x}_a}{2\sigma^2}\right) \tag{4.15}$$

which is a two-parameter function. One parameter is the arithmetic mean $\bar{x}_a$ (which is also the median of the distribution) and corresponds to 50% size.

The other parameter is the standard deviation σ, which is the 84% size minus the 50% size.

The log–normal distribution is more widely used than the normal distribution. It is skewed to the right and gives equal probability of ratios of sizes rather than differences of sizes. It may be obtained from the normal distribution (equation (4.15)) by substitution of $\ln x$ for x, $\ln x_g$ for x_a and $\ln \sigma_g$ for σ, giving

$$\frac{dF}{d(\ln x)} = x\frac{dF}{dx} = \frac{1}{\ln \sigma_g \sqrt{(2\pi)}} \exp\left[-\frac{(\ln x - \ln x_g)^2}{2\ln^2 \sigma_g}\right] \tag{4.16}$$

where x_g is the geometric mean of the distribution dF/dx. Svarovsky (1977) shows that the equation is more usually expressed in a form in which the variable x occurs only once and which is mathematically more convenient:

$$\frac{dF}{dx} = \frac{1}{x_m \ln \sigma_g \sqrt{(2\pi)}} \exp\left(-\ln^2 \frac{\sigma_g}{2}\right) \exp\left[-\frac{(\ln x - \ln x_m)^2}{2\ln^2 \sigma_g}\right] \tag{4.17}$$

where x_m represents the mode of the distribution dF/dx, and σ_g is the ratio of the 84% size to the 50% size. Equation (4.17) can be written more simply as:

$$dF(x) = \zeta \exp[-b^* \ln^2(x/x_m)]\, dx \tag{4.18}$$

where b^* is the steepness constant $1/(2\ln^2\sigma_g)$, and:

$$\zeta = (b^*/\pi)^{1/2} \exp\left(-\frac{1/4b^*}{x_m}\right) \tag{4.19}$$

Further

$$x_m = c' x_g = c' x_{50} \tag{4.20}$$

where $c' = \exp(-1/2b^*)$.

Based on equation (4.18), probability graph paper is available on which $F(x)$ is plotted against $\ln x$ and distributions which follow the log–normal law give straight lines. Two extra scales are provided for b^* and c' so that ζ and x_m may be easily evaluated from equations (4.19) and 4.20) respectively.

The log–normal equation has been widely used for the analysis of comminution functions, not only in relation to size distribution but also with respect to energy. In fact, the central limit theorem of statistics, which states that a sum of n independently occurring quantities are normally distributed for n approaching infinity, can also be stated as: 'a product of n independent quantities are log–normally distributed as long as the scatter of one quantity does not dominate'. Thus, since for both size distribution and applied energy distribution a log–normal law is found to apply, it is deduced that for each, several independent factors are operating, no one of which dominates.

(b) Truncated log–normal distribution (TLND) References: Baumgardt *et al.* (1975), Klotz and Schubert (1982), Barnard (1983)

A major disadvantage of the log–normal equation is the fact that the abscissa values pass to $\pm\infty$. By contrast, the truncated log–normal cuts the abscissa in definite points, and may be represented as:

$$F(z) = \int_{-\infty}^{u} \frac{1}{\sqrt{(2\pi)}} e^{-u^2/2}\, du \tag{4.21}$$

where

$$z = \text{relative particle size}$$

$$z'' = z/(1-z)$$

$$u = (\ln z'' - \mu_{\ln z''})/\sigma_{\ln z''}$$

$$\mu_{\ln z''} = \ln\left(\frac{z_{50}}{1-z_{50}}\right)$$

$$\sigma_{\ln z''} = \ln\left[\frac{z_{84.1}(1-z_{50})}{z_{50}(1-z_{84.1})}\right]$$

If experimental data can be represented by the TLND, a plot of $F(z)$ against $z/(1-z)$ on log-probability paper will yield a straight line.

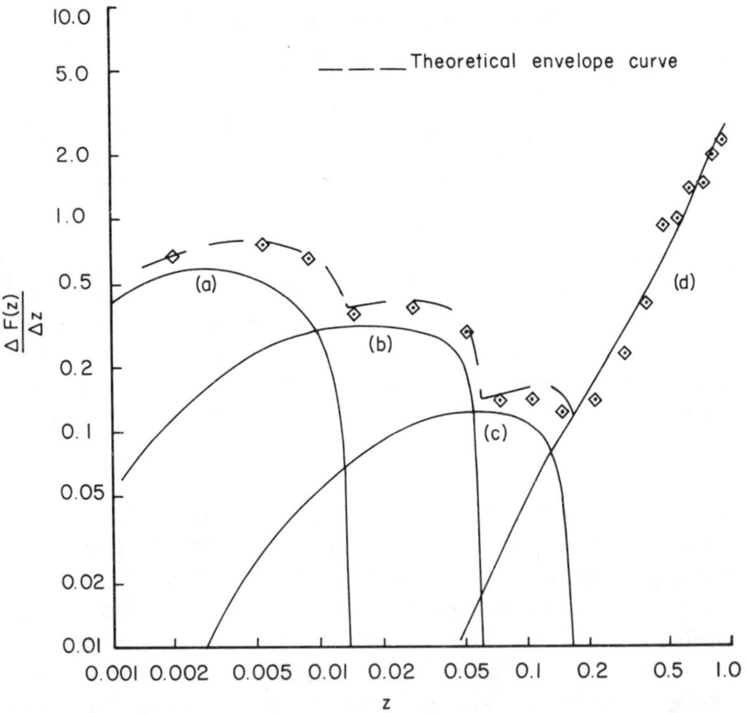

Figure 4.31. The fit of four truncated log–normal distributions to the density data for 36 mm brown coal char particles dropped from 2 m (Barnard, 1983). *Reproduced by permission of the author*

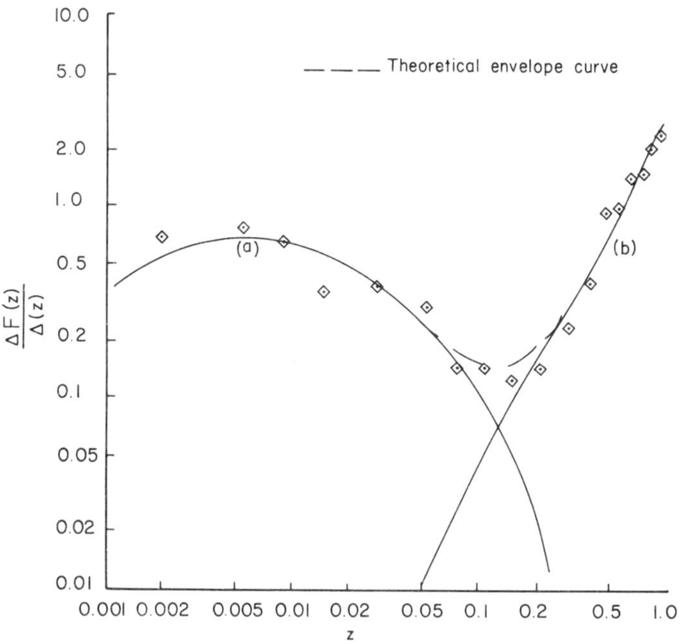

Figure 4.32. The fit of two truncated log–normal distributions to the density data for 36 mm brown coal char particles dropped from 2 m (Barnard, 1983). *Reproduced by permission of the author*

The TLND was used effectively by Baumgardt *et al.* (1975) to represent the cumulative array of fragments produced in both slow compression and double-impact experiments on glass spheres and in the compression of irregularly shaped particles between hard platens. Barnard (1983) also found excellent fits to cumulative coarse fragment data when using equation (4.21) in drop-shatter tests both for bituminous coal (Figure 4.34(d)) and brown coal char (Figure 4.35(d))

For density distributions, $\Delta F(z)/\Delta z$ is plotted against z, as has been tried by Klotz and Schubert (1982) to describe the distribution resulting from the crushing of quartzite. Good fits of multimodal experimental points were obtained by using four overlapping truncated log–normal curves. Barnard (1983) similarly found a good fit to the density data for the breakage of brown coal char particles by using four overlapping curves, as demonstrated in Figure 4.31, but an almost equally good fit using two overlapping curves, shown in Figure 4.32. The interesting feature of Figure 4.32 is that the fines from areas under particle contact surfaces could be related to one of the curves and the coarse from the rest of the particle to the other curve. This is in keeping with the findings of others demonstrated in §4.3.5.2.

(c) Snow (1973) three-parameter equation Another density distribution equa-

tion is that of Snow (1973), which may be written:

$$\frac{\Delta M_i}{M_t}\frac{1}{\Delta x}=\frac{1.533}{x_{mo}}\left(\frac{x}{x_{mo}}\right)^{0.46}\exp\left[-\left(\frac{x}{x_{mo}}\right)^2\right] \tag{4.22}$$

where

$$\frac{\Delta M_i}{M_t}\frac{1}{\Delta x}=\frac{\text{mass on sieve } i}{\text{total sample mass}}\;\frac{1}{\begin{bmatrix}\text{aperture of}\\ \text{mesh}(i-1)\end{bmatrix}-\begin{bmatrix}\text{aperture of}\\ \text{mesh } i\end{bmatrix}}$$

x_{mo} = distribution mode

This form of equation was adopted by Snow and Paulding (1973) in the analysis of tests involving regular and irregular shapes of a range of materials in which they found a similar pattern emerged, but the distributions were displaced left or right according to the material being ground and to the energy imparted to it. Hence they adopted an equation with the mode as the reference size, to facilitate comparison of distributions.

A typical result is seen in Figure 4.16 for 0.75 inch diameter glass spheres for various energy inputs (experimental points). It is clear that equation (4.22) follows the pattern of the experimental data and also equally fits:

(1) the experimentally found distribution from the crushing of irregularly shaped particles of several ores;
(2) the calculated size distribution based on energy density distribution within a stressed particle.

The general agreement between theory and experiment in this difficult area of size distribution is most heartening feature of the work of Snow and Paulding (1973). If the truncated log–normal equation had been used, it could well have improved the fit to experimental data.

(d) Three-parameter equations of Harris and Bergstrom References: Harris (1966), Harris (1968), Bergstrom (1966)

Clearly better curve fitting is possible by three-parameter equations than with two parameters. Examples are:

Harris (1966)

$$F(x)=1-\left[1-\left(\frac{x}{x_{max}}\right)^{n_1}\right]^{n_2} \tag{4.23}$$

Bergstrom (1966)

$$F(x)=\left[1-\left(\frac{x}{x_{max}}\right)^{n_1}\right]^{n_2} \tag{4.24}$$

Later Harris (1968) demonstrated that both equations (4.23) and (4.24) are members of the same family of expressions. Malghan and Mular (1982) similarly point out that the normal distribution, the log–normal distribution, the RRB, Gaudin–Schuhmann as well as Harris' and Bergstrom's formulae can all be shown to be special cases of the same general function. However, although these equations are statistical in origin, the parameters have not been shown to relate to physical properties of the particles.

4.3.8.4 Functions incorporating physical properties of particles

(a) Edge, surface and volume flaws Gilvarry (1961), by applying the Poisson law of probability mathematics (Fry, 1928) to random fracture within a particle, produced an equation of the form:

$$B(y,x) = 1 - \exp\left[-\left(\frac{y}{\alpha'}\right) - \left(\frac{y}{\beta'}\right)^2 - \left(\frac{y}{\gamma'}\right)^3\right] \tag{4.25}$$

where

$B(y,x)$ = cumulative fraction by mass of particles which fall below size y when particles of size x are broken

α', β', γ' = constants associated with fracture initiating flaw density respectively in flaw edge length, flaw area and flaw volume

In similar vein, Klimpel and Austin (1965) gave:

$$B(y,x) = 1 - \left[1 - \left(\frac{y}{x}\right)\right]^{\alpha}\left[1 - \left(\frac{y}{x}\right)^2\right]^{\beta}\left[1 - \left(\frac{y}{x}\right)^3\right]^{\gamma} \tag{4.26}$$

where

α = number of fracture planes originating from edge flaws per L_0, where L_0 is the edge length of the original particle

β = number of fracture planes originating from flaws in the particle surface per a_0, where a_0 is the original particle surface

γ = number of fracture planes originating from flaws within the particle volume

Equations (4.25) and (4.26) are interesting because they were the first attempts to link some physical property of the particle with the equation representing the particle sizes produced from breakage.

(b) Volume flaws only considered It was later discovered by Austin and Klimpel (1968) that both equations were incorrect because of invalid statistical reasoning; in fact, the equations applied only to cumulative fractional number and not to mass. Hence, in reality, they were only empirical relationships with no physical counterpart. Further it was found not possible to correct them into relationships involving mass distribution as explained by Barnard and Bull (1979).

Assuming that volume flaws only are important, Austin and Klimpel (1968) developed a valid equation involving mass distribution, and putting $z = y/x$ it could be written:

$$B(z) = 1 - (1 + \gamma z^3)\,(1 - z^3)^\gamma \tag{4.27}$$

No experimental verification of equation (4.27) seems to have been attempted until Barnard (1983) carefully conducted a series of drop shatter tests on both brown coal char (easy to shatter) and bituminous coal (more difficult to shatter). In these tests it was noticed that the coarse fragments of a particle due to tension breakage accounted for about 90% of the volume of the particle. The remaining 10% were fines from comparatively small conical volumes under the force contact areas, and were due to compressive crushing. This was in agreement with the Whiten *et al.* (1979) findings that, for the first breakage event in crushing a variety of rock particles between hard platens, the fines rarely exceeded 10% of the total mass. (see §4.3.5.2).

Since, in the context of a shatter index, Barnard was chiefly interested in the larger fragments, by a histogram analysis he was able to separate the coarse distribution from the fines distribution. It was to this coarse distribution that mathematical relations were particularly applied (See Figure 4.33.)

Now, Klimpel and Austin (1965) had assumed that particles of equal size would contain the same number of stress-activated volume flaws γ, but they later (1968) conceded this to be unlikely. Barnard (1983) made the more plausible assumption that the number of volume flaws would be Poisson-distributed. This discontinuous array possesses the characteristic that the variate runs from zero to infinity through integral values only and it expresses the probability of the occurrence of stress-activated flaws in terms of a single parameter λ such that:

$$P(\gamma/\lambda) = e^{-\lambda}\,\frac{\lambda^\gamma}{\gamma!} \tag{4.28}$$

The mean of this distribution is λ.

For this to be a valid distribution, the sum of its terms must be unity, or:

$$\sum_{\gamma=0}^{\infty}\left(e^{-\lambda}\,\frac{\lambda^\gamma}{\gamma!}\right) = 1$$

However, in the case under consideration, $\gamma = 0$ is untenable, since it implies a particle has no flaws and hence will resist breakage. Therefore the variate γ had to be limited to the range 1 to ∞ by truncation at the lower end. This resulted in:

$$P(\gamma/\lambda) = \frac{e^{-\lambda}\lambda^\gamma}{\gamma!(1 - e^{-\lambda})}, \qquad \gamma = 1, 2, 3 \ldots \tag{4.29}$$

The mean value μ_γ of this distribution becomes $(\lambda/1 - e^{-\lambda})$. Combining equations (4.27) and (4.29) gives:

$$B(z) = \sum_{\gamma=1}^{\infty}\left[\frac{\lambda^\gamma e^{-\lambda}}{\gamma!(1 - e^{-\lambda})}\left\{1 - (1 + \gamma z^3)(1 - z^3)^\gamma\right\}\right] \tag{4.30}$$

and is called the truncated Poisson–Austin–Klimpel (TPAK) equation.

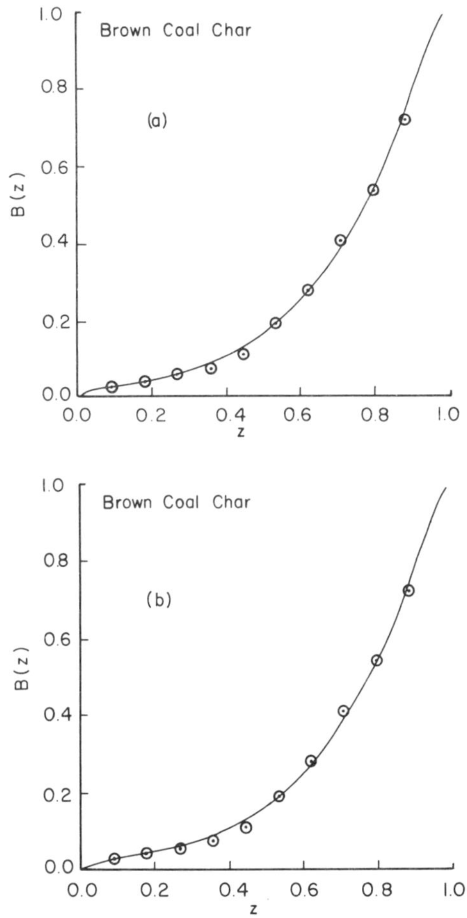

Figure 4.33. The fits of (a) two and (b) four truncated log–normal distributions to the breakage function data for 36 mm brown coal char particles dropped from 2 m (Barnard, 1983). *Reproduced by permission of the author*

In applying equation (4.30) Barnard (1983) found that for values of $\gamma > 30$, only negligibly small values of $B(z)$ resulted, and hence summation was validly made over the range of γ from 1 to 30.

In the context of the consideration of volume flaws only, Barnard then tested equations (4.27) and (4.30) which are both essentially single-parameter expressions, since γ and λ are related. For comparison with two parameter expressions, Barnard chose the Harris (1966) equation (4.23) modified to:

$$B(z) = 1 - (1 - z^{n_1})^{n_2} \tag{4.31}$$

and also the truncated log–normal distribution (4.21).

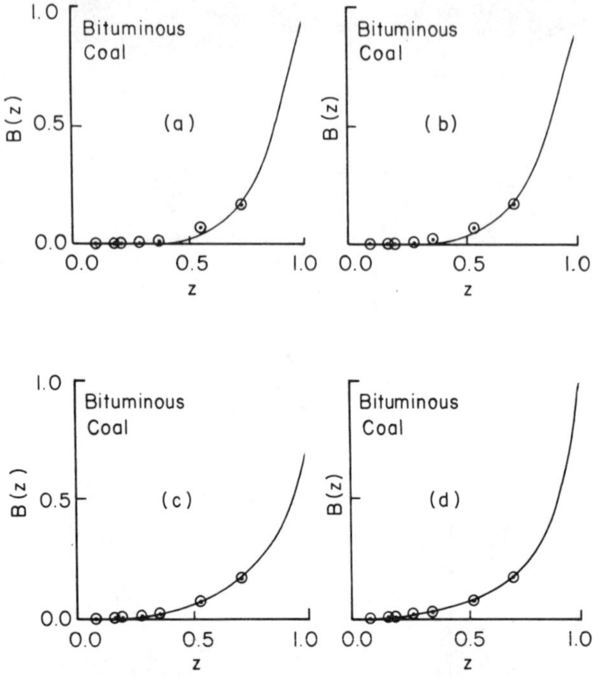

Figure 4.34. The fits of four equations to the cumulative coarse-fragment distribution data for 9 mm bituminous coal particles dropped from 5 m (Barnard, 1983). *Reproduced by permission of the author*
(a) Austin and Klimpel (1968), equation (4.27)
(b) Truncated Poisson–Austin–Klimpel, equation (4.30)
(c) Harris (1966), equation (4.31)
(d) Truncated log–normal, equation (4.21)

Results for these four equations are given in Figure 4.34 for bituminous coal and Figure 4.35 for brown coal char, and the conclusions are based on a statistical index of determination based on a least-squares principle:

(1) the two-parameter equations (4.31) and (4.21) gave slightly better fits than the single-parameter equations, but the parameters had no physical significance.
(2) of the single-parameter equations, the TPAK equation (4.30) gave a better fit to the experimental data than the Austin–Klimpel equation (4.27), but in each case the parameter γ did have physical significance, as follows in sub-section (c).

(c) Physical significance of γ It was noted in paragraph (b) that the mean

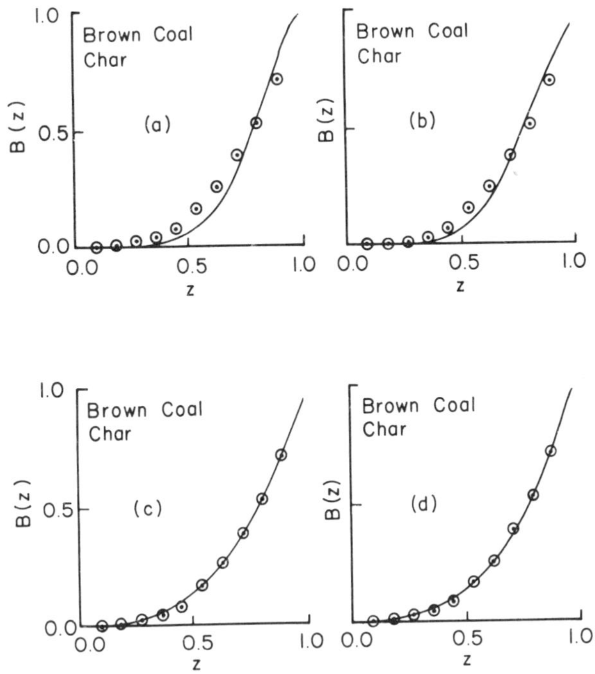

Figure 4.35. The fits of four equations to the cumulative coarse-fragment distribution data for 36 mm brown coal char particles dropped from 2 m. (a), (b), (c) and (d) are as in Figure 4.34 (Barnard, 1983). *Reproduced by permission of the author*

value μ_γ of the Poisson distribution is given by:

$$\mu_\gamma = \frac{\lambda}{1 - e^{-\lambda}} \tag{4.32}$$

where μ_γ is the number of stress activated flaws per particle.

Barnard (1983) was able to discover that μ_γ fell into the range 1–5 for the coarse distributions he studied. Now if $\gamma = 1$, the particle breaks into two pieces; if $\gamma = 2$, into three pieces, and hence in general $(\gamma + 1)$ progeny particles are formed from γ volume flaws.

By comparing the surviving mono-sized particles that were drop-shattered with the total number dropped it was possible to assess directly the probability P' of survival $(0 \leqslant P' \leqslant 1)$. The probability of breakage is clearly $(1 - P')$. When $(1 - P')$ was plotted against $\ln \mu_\gamma$ on log-probability paper, a linear relation was established that was statistically significant for both brown coal char and bituminous coal. Hence

$$\Lambda(1 - P') = K_4 \ln \mu_\gamma + K_5 \tag{4.33}$$

where K_4 and K_5 are constants that depend on the natural shape of the particles.

Equation (4.33) is probably the most important outcome of the work of Barnard, as it links the probability of breakage with the particle physical parameter γ, the number of stress activated volume flaws. For the present it must be regarded as no more than a hypothesis until it has been explored for other brittle materials and over a greater range of particle sizes and impact energies; but it could well come to be regarded as a breakthrough in the advance of comminution science.

4.3.8.5 *Matrix notation for size distributions*

(a) Vector concept for feed and product distribution Comminution starts with a feed of a certain size distribution and ends with a product of different distribution, each of which is generally sized by sieving, the apertures sizes of which normally form a geometric progression. Thus the pertinent information can be listed in three columns, typically as shown in Table 4.6.

Broadbent and Callcott (1956a,b) conceived the powerful idea of regarding these lists as column vectors, in the mathematical sense. The feed and product vectors are denoted by $\boldsymbol{f}$ and $\boldsymbol{p}$ respectively.

Table 4.6. Vector concept for feed and product distribution

Particle size range		Proportion by mass of particles			
(Screen apertures in mm)		Feed		Product	
1	38.1–33.9	f_1	0.513	p_1	0.118
2	33.9–30.2	f_2	0.256	p_2	0.222
3	30.2–26.9	f_3	0.085	p_3	0.116
·	26.9–24.0	·	0.029	·	0.090
·	24.0–21.4	·	0.013	·	0.056
$n-1$	21.4–19.1	f_{n-1}	0.012	p_{n-1}	0.054
n	< 19.1	f_n	0.092	p_n	0.314

(b) Matrix concept to describe breakage distribution behaviour In changing feed to product a breakage process occurs. Broadbent and Callcott's second idea was that this process could be described by a matrix by which the feed vector is multiplied to produce the product vector. This is now illustrated following Jowett. (1971).

During comminution, particles of all sizes are broken, products falling generally into lower size ranges, though some may remain in the same size range if, say, abrasion only should occur. Writing a mass balance on the operation in which $\boldsymbol{f}$ is changed to $\boldsymbol{p}$ gives the following:

$$
\begin{aligned}
f_1 &= \sum[p_{1,1} \quad p_{1,2} \quad p_{1,3}\ldots p_{1,n}] \\
f_2 &= \sum[0\cdot \quad p_{2,1} \quad p_{2,2}\ldots p_{2,n-1}] \\
f_3 &= \sum[0\cdot \quad 0\cdot \quad p_{3,1}\ldots p_{3,n-2}] \\
f_n &= \sum[0\cdot \quad 0\cdot \quad 0 \quad \ldots p_{n,1}]
\end{aligned}
$$

where the sequence of figures in the subscripts to each element in the product is as follows: the first figure refers to the size fraction in the feed from which the particle element originates; and the second figure refers to the number of the size range into which the particle element falls, below the original size from which the element came.

Now the above array has a matrix form denoted by $n \times n$. In fact it is an $(n-1) \times (n-1)$ lower triangular matrix. It could represent one stage of comminution, for which the general term may be written:

$$f_i = \sum_{j=1}^{n-i+1} p_{i,j} \tag{4.34}$$

If $b_{i,j}$ is defined as the proportion of f_i falling into the element earmarked $p_{i,j}$, then equation (4.34) can be written:

$$f_i = \sum_{j=1}^{n-i+1} b_{i,j}\, f_i, \quad \text{with} \quad \sum_{j=1}^{n-i+1} b_{i,j} = 1 \tag{4.35}$$

$b_{i,j}$ is called the 'breakage distribution function' (originally just called the 'breakage function') because it describes the size distribution change for particles which are actually broken in a stage of comminution.

Otherwise:

$$\begin{bmatrix} b_1 & 0 & 0 & \cdots & 0 \\ b_2 & b_1 & 0 & \cdots & 0 \\ b_3 & b_2 & b_1 & \cdots & 0 \\ \vdots & \vdots & \vdots & & \vdots \\ b_{n-1} & b_{n-2} & b_{n-3} & \cdots & b_1 \end{bmatrix} \begin{bmatrix} f_1 \\ f_2 \\ f_3 \\ \vdots \\ f_{n-1} \end{bmatrix} =$$

$$\begin{bmatrix} & & & & b_1 f_1 \\ & & & b_2 f_1 & b_1 f_2 \\ & & b_3 f_1 & b_2 f_2 & b_1 f_3 \\ & & \vdots & \vdots & \vdots \\ b_{n-1} f_1 & \cdots & b_3 f_{n-3} & b_2 f_{n-2} & b_1 f_{n-1} \end{bmatrix} = \begin{bmatrix} p_1 \\ p_2 \\ p_3 \\ \vdots \\ p_{n-1} \end{bmatrix}$$

or in algebraic matrix symbols,

$$\mathbf{B} f = p \tag{4.36}$$

(c) Matrix concept to describe breakage rate Equation (4.36) only tells part of the comminution story. It would apply to a drop shatter test in which only primary breakage is arranged for, but it takes no account of breakage rates which change from stage to stage of further breakage. To take account of this, Broadbent and Callcott (1956a,b) introduced a specific rate of breakage function S, which they called a selection function. Hence $b_1 f_1$ in equation (4.36)

is replaced by $S_1 b_1 f_1$ etc. to obtain the actual quantity of material of size f_1, which is broken. Then in matrix notation, the left-hand side of equation (4.36) may be written:

$$\begin{bmatrix} S_1 & 0 & 0 & \dots & 0 \\ 0 & S_2 & 0 & \dots & 0 \\ 0 & 0 & S_3 & \dots & 0 \\ \vdots & \vdots & \vdots & & \vdots \\ 0 & 0 & 0 & \dots & S_{n-1} \end{bmatrix} \begin{bmatrix} b_1 & 0 & 0 & \dots & 0 \\ b_2 & b_1 & 0 & \dots & 0 \\ b_3 & b_2 & b_3 & \dots & 0 \\ \vdots & \vdots & & & \vdots \\ b_{n-1} & b_{n-2} & b_{n-3} & \dots & b_1 \end{bmatrix} \begin{bmatrix} f_1 \\ f_2 \\ f_3 \\ \vdots \\ f_{n-1} \end{bmatrix}$$

or $\mathbf{S} \quad \mathbf{B} \quad f$

which is the matrix array corresponding to particles which are broken. Now for the general stage of broken particles $S_i f_i$, the residue $(1 - S_i) f_i$ passes unbroken, and hence the matrix array for unbroken particles can be written:

$$\left\{ \begin{bmatrix} 1 & 0 & 0 & \dots & 0 \\ 0 & 1 & 0 & \dots & 0 \\ 0 & 0 & 1 & \dots & 0 \\ \vdots & \vdots & \vdots & & \vdots \\ 0 & 0 & 0 & \dots & 1 \end{bmatrix} \begin{bmatrix} S_1 & 0 & 0 & \dots & 0 \\ 0 & S_2 & 0 & \dots & 0 \\ 0 & 0 & S_3 & \dots & 0 \\ \vdots & \vdots & \vdots & & \vdots \\ 0 & 0 & 0 & \dots & S_{n-1} \end{bmatrix} \right\} \begin{bmatrix} f_1 \\ f_2 \\ f_3 \\ \vdots \\ f_{n-1} \end{bmatrix}$$

or simply $(\mathbf{I} - \mathbf{S}) f$

The overall product is the sum of the breakage and residue matrices:

$$\begin{aligned} p &= \mathbf{SB}f + (\mathbf{I} - \mathbf{S})f) \\ &= f(\mathbf{BS} + \mathbf{I} - \mathbf{S}) \end{aligned} \tag{4.37}$$

where **I** is the unit matrix, **B** is the breakage distribution matrix, and **S** is the specific rate of breakage matrix.

For n stages of operation, equation (4.37) becomes:

$$p_n = f[\mathbf{BS} + \mathbf{I} - \mathbf{S}]^n \tag{4.37a}$$

General term notation, as equation (4.35), is to be preferred over matrix notation, and is used generally in this handbook.

4.3.9 Measurement of particle size

4.3.9.1 Summary of available methods

The detailed description of available methods is outside the scope of this handbook and reference should be made to bibliographies such as Allen (1972) and standard textbooks such as Allen (1975). Other more recent references are listed below for the convenience of students. An asterisk indicates that online

Table 4.7. Methods of measurement of particle size (based on Malghan and Mular, 1982)

Technique	Example	Approximate size range (micron)	Diameter measured (See §4.3.1)
Field scanning	Microscope counting	1–150	Length, area
	Electron microscope	0.001–5	Statistical
Stream scanning	Coulter counter	1–500	Volume
	Light scattering photometer*	0.5–50	Volume
	Ultrasonic attenuation*	70–2000	Volume
Classification	Sieve: Wire-mesh	40–12500	Sieve
	Electro-formed	5–40	Sieve
	Elutriation	2–50	Stokes
Sedimentation	Pipette	1–50	Stokes
	Balances	1–60	Stokes
	Photosedimentometer	1–60	Stokes
Centrifugal action	Bahco	3–100	Stokes

Reproduced by permission of the American Institute of Mining, Metallurgical and Petroleum Engineers.

methods for wet processes are available. (Online systems for dry processes are less well developed.)

Dias (1980)*; Scarlett (1980); Sresty and Venkateswar (1980); Johnston and Swanson (1982); Malghan and Mular (1982)*

From the above literature it is possible to list the main available types of particle size measurement, as in Table 4.7, though other methods are being developed, including laser techniques.

4.3.9.2 *Comparisons of size distributions as measured by different methods*

It is common practice to use wire mesh sieves, as described in British Standards 410 (1976) and 1796 (1976), down to about 40 microns, depending on the particular series adopted, such as Tyler, etc. (see Table 12.2). To obtain the distribution in the immediately lower sizes, the following are examples of what is available:

(1) electroformed sieves (wet) down to 5 μm (Hidaka and Miwa 1979);
(2) Bahco centrifugal separator (dry) down to 3 μm (ANSI PTC 28 (1973), Graham and Hanna (1962);
(3) wide-angle photosedimentometer (wet) down to 1 μm (Allen, 1968/69);
(4) Coulter counter (Allen, 1975).

Since each device measures its own distinctive diameter, as indicated in Table 4.7, a hiatus would be expected in the region of 40 μm where the

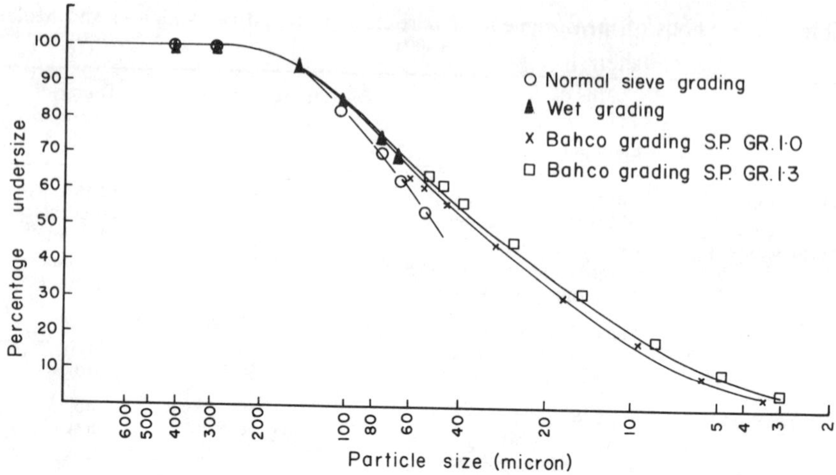

Figure 4.36. Cumulative size distributions for a bituminous coal sample by sieving and by Bahco centrifugal separator (Babcock Power Ltd., Research Report 5/57/1, 1957). *Reproduced by permission of Babcock Power Ltd.*

transition from wire sieves occurs. Figure 4.36 shows this transition for bituminous coal from wire sieve to the Bahco (Babcock Power, 1957). The graph reveals for this particular case that aggregation of grains was occurring during dry sieving. The aggregates were dispersed by wet sieving and the agreement between wet sieving and the Bahco results was very good.

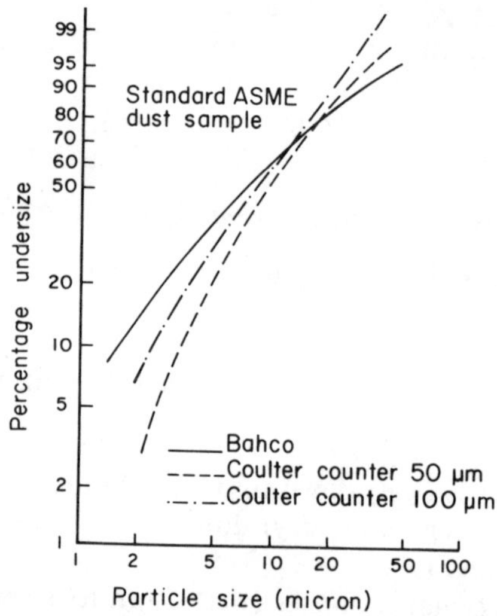

Figure 4.37. Comparison of cumulative size distributions from Bahco centrifugal separator and the Coulter counter (Babcock Power Ltd.)

The wide-angle photosedimentometer (Allen, 1968–69) is a versatile instrument, which, when linked with appropriate auxiliary equipment, will automatically print out both size distribution and specific surface (see §4.4.3.3). Bucsky (1975) commends the instrument because of relative freedom from human error, good reproducibility, ease of statistical analysis of output and relative speed, which is of the order of hours, however, compared with modern laser granulometer techniques which may take only minutes (Opoczky, 1981). When sieving is performed down to 40 μm, and a photosedimentometer is used for smaller sizes, a change of slope is usually obtained at 40 μm, when each is plotted, for example, on the RRB grid.

Comparisons between distributions by using the Coulter counter and the Bahco machine on samples from the same batch as a standard ASME dust are given in Figure 4.37. It is clear that the volume diameter from the Coulter counter is different from the Stokes diameter from the Bahco machine, but they appear to have common values in the region of 15 microns.

British Standard 3406 (Part 3) (1963) gives approximate values as follows for conversion between sieve, projected and Stokes diameters, with the warning that caution is needed in their application because of variations arising from shape factors and the size range of interest.

	Multiply by
Sieve to projected	1.40
Sieve to Stokes	0.94
Projected to Stokes	0.67

Johnston and Swanson (1982) give comparisons of distributions over a wider range of instruments based on standard samples of dust as used by the AC Spark Plug Division of General Motors. The instruments are:

A Coulter counter measuring volume diameter;
B HIAC and Royco measuring the area of a shadow cast on a window;
C πMC which views particles deposited on a substrate surface, giving the longest end-to-end particle distance;
D Quantimet, which gives the horizontal width whatever the orientation;
E Micrometrics X-ray sedimentometer which gives mass distribution directly;
F Microtrac by Leeds & Northrup, which uses a light scattering technique.

Some of the findings for this particular powder are:

(1) There is agreement between A, B and C over the range 4–40 microns, but B instruments do not see so many small particles as A.
(2) C sees more small particles than B, but not so many as A.
(3) D sees more small particles than C.
(4) F sees fewer small particles but more large particles than E.

This illustrates the need to select the instrument suited to a particular requirement both of diameter and size range which are appropriate to the task in hand.

4.3.9.3 Sampling

Any accurate size distribution implies accurate sampling, which is not always given the attention it deserves.

For sampling from dry bulk powders, Allen (1975) gives four main methods: cone and quartering, the scoop system, chute splitting and the spinning riffler, which last is specially recommended—see British Standard 1017 (Part 1) (1977).

When a sample is to be taken of a powder being carried by a gas in a duct, then an isokinetic method is highly desirable; that is, the speed of the gas entry to the probe is arranged to be that of the gas approaching the probe—see Hawkesley *et al.* (1977) and British Standard 893 (1978). A recently developed probe in this category by Leschonski and Rothele (1978) has proved itself to be accurate for on-stream measurements of high powder concentrations in streams moving at up to 50 m/s.

If only a very small sub-sample is needed, a portion of that by any of the above methods is spatulated on to a watchglass with a suitable dispersing agent to produce a viscous paste. This is thoroughly blended and a further portion is diluted with a suitable electrolyte ready for analysis. This electrolyte may be a simple sodium chloride or phosphate solution, with perhaps glycerol added to increase viscosity (Malghan and Mular, 1982).

4.4 PARTICLE SURFACE

Note: For references see §4.5.2.

4.4.1 Problems in the definition of surface area

The surface of any solid is not continuous. Even that of a defect-free crystal, when viewed through an electron microscope, is 'rough' although it appears flat and smooth when seen directly. At the macroscopic level, most solids have pores, cracks or fissures which make assessment of surface area from size analysis very difficult. A further factor is that particle shapes can vary with particle size range, even when they are the product of a single comminution process. (See §4.2.5.5).

Elongated particles will have proportionately greater specific surface than the cubiform particles that pass through the same square mesh sieve. Hence a shape factor must be introduced to accommodate the trend.

Direct measurement of surface area is very method-dependent. For example, in air permeability systems, little contribution to the assessed pressure loss will be made by cul-de-sac pores, which will strongly affect gas

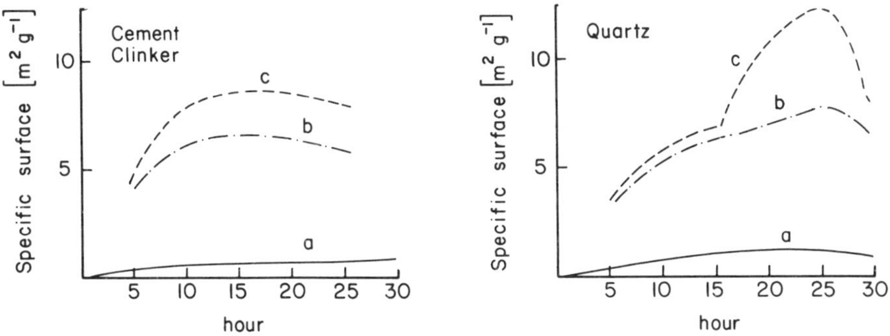

Figure 4.38. Comparison of specific surface by air permeability and by gas adsorption methods for cement clinker and quartz (Based on Opoczky, 1977). *Reproduced by permission of Elsevier Sequoia*

a 'External' surface by air permeability method (Blaine)
b 'Total' surface by BET method—nitrogen adsorption
c 'Total' surface by BET method—water vapour adsorption

absorption systems. Hence a much higher specific surface will be obtained from a gas adsorption than from an air permeability method (see Figure 4.38). In fact Opoczky (1980) has called the former the 'total' and the latter the 'external' surface.

Thus no one definition can be given of the specific surface of a powder: it depends entirely on how it is assessed, and any measure of its value must be accompanied by a statement of the method of assessment. Within these limitations, however, surface is perhaps the best way of expressing fineness as a single measure, as it does take account of the whole of a distribution, whereas d_{80}, for example, just takes account of its coarser end.

4.4.2 Calculation of specific surface by mass

4.4.2.1 Uniformly sized particles of constant shape

At the macroscopic level, the area of a sphere of diameter d is πd^2 and its mass is $\rho\pi d^3/6$. Thus its specific surface by mass $s_m = 6/\rho d$, where ρ is the material density.

Similarly for other geometric shapes. For example a rectangular parallelepiped of size $L \times 2L \times 4L$ has a surface of $28L^3$, a mass of $8\rho L^3$ and a specific surface by mass of $3.5/\rho L$.

Hence for uniformly sized particles of constant shape but not of simple geometric form (British Standard 4359 (Part 3) 1970):

$$\text{surface area} = \psi_{sx} x^2$$

$$\text{mass} = \psi_{vx} \rho x^3$$

$$\text{specific surface by mass } s_m = \frac{\psi_{sx}}{\psi_{vx}} \frac{1}{\rho x} = \psi_{sv,x} \frac{1}{\rho x} \tag{4.38}$$

where x is a characteristic linear dimension, ψ_{sx} is a surface coefficient, ψ_{vx} is a volume coefficient, and $\psi_{sv,x}$ is a specific surface coefficient. These coefficients depend on particle shape and roughness and must be separately assessed.

In all the above there is a consistency of pattern; the specific surface is inversely proportional to a linear dimension related to the particle. Hence the smaller the particles, the greater is their specific surface.

4.4.2.2 Non-uniformly sized particles of constant shape

By similar reasoning to the above, the specific surface by mass for non-uniformly sized particles of constant shape is given by:

$$s_m = \psi_{sv,x}\sum(n_i x_i^2)/\rho\sum(n_i x_i^3) \tag{4.39}$$

where n_i is the number of particles of size x_i.

In practice, class i will contain a finite, though small, range of particles and the range is normally decided by sieves in $2^{1/2}$ or $2^{1/4}$ size series—see British Standard 410 (1976).

4.4.2.3 Non-uniformly sized particles of variable shape

(a) General equation If the distribution of particle shape is the same for all classes i, then the coefficient $\psi_{sv,x}$ can be retained outside the summation brackets. If, as is often the case, particle shape is a function of particle size, then the expression for specific surface by mass is given by:

$$s_m = \sum(\psi_{sv,x_i} n_i x_i^2)/\rho\sum(n_i x_i^3) \tag{4.40}$$

where ψ_{sv,x_i} is the specific surface coefficient pertaining to size x_i—see British Standard 4359 (Part 3) 1970.

It is instructive to state the specific surface in the form of equation (4.40) because it underlines the considerable problems which are confronted in any accurate assessment of surface:

(1) ψ_{sv,x_i} must be found for each value of x_i, which is not easy.
(2) Counting n_i is tedious and almost impossible for the smaller sizes.
(3) The characteristic dimension x must be appropriate to the process concerned (consult Table 4.2).

Only to the extent that these actions are fulfilled will an accurate value of surface be obtained. However, calculation of specific surface avoiding counting of particles can be performed as shown below.

(b) Specific mathematical equations (spherical particles) Since the density distribution by mass is usually available, it is useful to express specific surface

in terms of $f_M(x)$. From equation (4.7):

$$f_M(x) = K_3 x^3 f_N(x)$$
$$= [K_2 x^2 f_n(x)]\,[K_6 x]$$

Hence, using equation (4.6):

$$f_S(x) = f_M(x)/K_6 x$$

or

$$S_V = \int_{x_{min}}^{x_{max}} f_S(x)\,dx$$
$$= K_7 \int_{x_{min}}^{x_{max}} \frac{f_M(x)}{x}\,dx \qquad (4.41)$$

K_7 generally contains a shape factor which must be separately assessed; but if particles are assumed to be spheres, $K_7 = 6$.

Making this assumption Barnard (1983) has shown that if $f_M(x)$ is taken to be a truncated log–normal distribution, then:

$$S_V = \frac{6}{x_{max}}\left[\left(\frac{z_{max}}{z_{50}} - 1\right)\exp\left(\frac{1}{2}\,\sigma_{\ln z''}\right) + 1\right] \qquad (4.42)$$

Barnard (1983), in analysing the bimodal distributions from shatter tests, required two or four superimposed truncated log–normal distributions to make valid representation, as seen §4.3.8.3(b), and gives corresponding expressions to equation (4.42) for these cases also.

Bull (1955) developed an approximate but easily applied expression for specific surface of a powder in which the finest fraction of the cumulative distribution, i.e. that below the lowest sieve size used, is represented by the graph of a cubic function and the coarser fractions by a series of chords drawn between the experimentally determined points, corresponding to sieve sizes in geometric progression of common ratio R. The cubic equation for the fine end of the spectrum is chosen to pass through the origin with zero gradient and through the point corresponding to the lowest sieve size with the same gradient as the chord between the points corresponding to the finest and second finest sieve sizes used. By this method was developed the equation:

$$S_V = \frac{1}{x_{max}}\left[\frac{18B_1}{Z_1} - \frac{3(B_2 - B_1)}{Z_2 - Z_1}\right]$$
$$+ \frac{6\ln R}{x_{max}Z_1}\left[\frac{-B_1}{R-1} + \frac{B_2}{R} + \frac{B_3}{R^2} + \ldots + \frac{B_{n-1}}{R^{n-2}} + \frac{B_n}{R^{n-2}(R-1)}\right] \qquad (4.43)$$

where n is the number of experimental points, Z_1 is the finest relative size, Z_2 is the second finest relative size, and B_n values are the cumulative fractions by mass of broken material finer than the corresponding Z_n values.

The area of greatest uncertainty is that below the finest sieve size used. To

Table 4.8. Comparison of calculated specific surface areas by volume (m^2/m^3) for bituminous coal broken in a shatter test (Barnard, 1983, Table 9.1.1)

1 Run number	2 Bull method	3 Bull/Bahco method	4 Four superimposed truncated log–normal distributions	5 Two superimposed truncated log–normal distributions
19A	502	853	–	498
25A	734	849	–	686
3A	789	870	–	695
6A	934	–	836	828
9A	1306	1465	–	1094
12A	1388	–	1392	1138
32A	1447	–	1442	1255

Reproduced by permission of the author.

obtain some concept of the degree of this uncertainty, Barnard (1983) used the Bahco instrument to assess the sub-sieve size distribution and compared the resulting specific surface with those using the straight Bull method.

Table 4.8 displays the values determined by the four methods indicated in the column headings, and is of interest to show the degree of variation in any one horizontal line for a particular sample. By comparing columns 2 and 3, for example, it is evident that the sub-sieve information by the Bahco instrument consistently gives a higher surface than without it. Similarly, by comparing columns 4 and 5 it is seen that the four superimposed truncated log–normal distributions consistently return higher s_v values than if only two superimpositions are made. Against this must be weighed the extra time and effort involved in column 3 over column 2 and in column 4 over column 5, and Barnard concluded that for his purpose, when reviewing all the relevant data, the Bull method gave the most consistent and yet the most easily obtained values.

4.4.3 Practical measurement of specific surface

4.4.3.1 Survey

In view of the severe problems encountered in the calculation of specific surface from size distribution, direct measurements have been explored, of which those of greatest interest are:

(1) air permeability in §4.4.3.2
(2) gas adsorption in §4.4.3.3
(3) photosedimentometry in §4.4.3.4
(4) laser granulometry in §4.4.3.5

Others of lesser interest include gas diffusion and hindered settling (Allen, 1975; Malghan and Mular, 1982).

4.4.3.2 Air permeability method

Air permeability systems involve the assessment of pressure loss across a bed of powder to be tested, under carefully controlled flow conditions—see British Standard 4359 (Part 2) 1982.

Constant pressure systems are typified in the Fisher sub-sieve sizer (Gooden and Smith, 1940) and constant volume types include the Blaine instrument (Rigden, 1947; ANSI/ASTM C204-78a). The Blaine specific surface has become meaningful in the cement industry as one condition for satisfactory concrete manufacture, if not less than 225 m^2/kg for ordinary Portland cement (British Standard 12, 1978). According to Malghan and Mular (1982), air permeability methods are reproducible when bed porosity lies in the range 0.4–0.8, and they are not sensitive to cul-de-sac pores which may exist in the particles.

4.4.3.3 Gas adsorption methods

Gas adsorption methods depend on the multilayer adsorption of a gas on the surface of a solid as described by Brunauer *et al.* (1938)—see British Standard 4359 (Part 1) 1969.

Commercial instruments are available over a range of types from the single-point method of Strolein Areameter (Dusseldorf), to the automatic instruments giving a direct readout of surface area (Micromeritics, Norcross, USA). It is seen from Figure 4.38 that specific surface from gas adsorption instruments is normally much greater than that from air permeability instruments, though it is rather an extreme case there depicted.

4.4.3.4 Photosedimentometry

Of the photosedimentometer methods, those having wide-angle features are to be preferred, particularly those which also automatically give a printout of both surface area by mass and cumulative size distribution for sub-sieve fine powders (Allen, 1968/69). The instrument is well suited to the characterization of classifiers because it measures the equivalent Stokes diameter, and has been advocated for routine grade efficiency testing of hydrocyclones of small diameter, because of savings in time and effort over other methods (Svarovsky and Svarovska, 1984).

It is noted that, for energy utilization experiments at the Karlsruhe University, in West Germany, where many single-particle experiments have been performed, all product surface was measured by a combination of methods (Schönert, 1972). Fragments larger than 30 microns were sieved, and their specific surface calculated for the size distribution. For the fragments smaller than 30 microns, a suspension was made and its light extinction measured, from which the specific surface was calculated. The two values were combined to obtain the specific surface of all the fragments.

4.4.3.5 Laser granulometry

Felton and Watson (1984) have outlined the theory for a laser granulometer which employs a diffraction technique. A parallel beam of laser light is passed through the particle field which is to be assessed, and is brought to a focus by a lens. The resulting diffraction pattern is viewed by a detector lying in the focal plane of the lens. The detector consists of multiple (e.g. 30) semi-annular photodetectors around a central alignment diode. The signals from the detector are transmitted via a multiplexor and an analogue-to-digital converter to a microcomputer, which gives a direct printout of the required distribution.

The sample of dry powder under investigation is either added to water to make an opaque field, or it can be suspended dry by means of air jets. In either case, precautions must be taken against the largest particles falling out of suspension during the short time of the exposure to the laser beam, otherwise a skewing of the distribution occurs.

For non-spherical particles there is a broadening of the distribution compared with that from spheres. However, the ability of the device to measure particle concentration simultaneously with particle size enables a shape factor to be determined for non-spherical particles of constant density, which may be used as a corrective.

The laser granulometer is comparatively expensive, but its speed and accuracy largely compensate for this.

Some manufacturers are Malvern Instruments in the UK, and CILAS in France and Belgium.

4.4.4 Specific surface and comminuted product size

It would not be expected that a high degree of correlation would obtain between specific surface and a point fineness such as d_{80} or a percentage passing a given size, such as 75 microns. This can be seen from Figure 4.30, for it will be conceded that a succession of products from a grinding process, each of which passes through d_{80} (10^4 microns in this case) are likely to have different distributions in the micron and sub-micron ranges. Since it is these small particles which contribute so massively to specific surface, even small changes in distribution in this range can effect large changes in surface.

For narrow size ranges, specific surface varies approximately inversely as mean particle size. This would be expected from equation (4.38). However, over a complete size distribution, it would not hold. Attempts in the cement industry to correlate the Blaine index with percentage passing 75 microns have met with very little success. However, for several materials including cement clinker, Kihlstedt (1962) has claimed a relation in the form:

$$s_v = K_8/(d_{80})^n \tag{4.41}$$

with n approximating to 0.5 for some materials; but there were wide discrepancies, particularly with variation of feed distribution. It is to be noted that the experiments were conducted over very limited values of d_{80}. Such empirical relations as equation (4.41) are to be viewed with great reserve because other workers have found very different results under different conditions and over different size ranges (Harris, 1966). This would be expected from equation (4.40).

4.4.5 Review of surface versus size measurement

In former years when Rittinger's law (see §6.5.1), which relates comminution energy to new surface formed, was of great importance in the minds of experimenters, the assessment of particle specific surface was pursued vigorously (Beke, 1964). Today, in a climate of mathematical simulation of breakage, which inherently involves size distribution rather than surface, interest in specific surface has waned accordingly. There are still contexts in which it is important, such as the cement industry, as noted in §4.4.3.2; but even here improvement in the understanding of the breakage process has been in terms of size distribution rather than surface, as indicated by Austin *et al.* (1980) for the airswept ball mill and by Austin *et al.* (1975) for the non-airswept case. In fact, Austin *et al.* (1976) have demonstrated for the non-airswept mill that a narrow distribution of certain sizes of cement is a valuable addition criterion to the Blaine index for the most acceptable concrete specification. Opoczky (1981) has urged the maximization of the 3–30 micron band for this purpose, and Frigione and Marra (1976) have related cement strength to the uniformity index n' of the RRB distribution, among other factors. (see §4.3.8.2(a)).

Another area in which surface area of sub-sieve sized particles is important is the paint industry which uses the BET gas absorption system (British Standard 4359 (Part I) 1969). But Coulter counter and other 'diameter' methods are becoming increasingly accepted rather than surface methods, and this appears to be the recent trend.

However, specific surface measurement will continue to be a valuable laboratory aid. Figure 4.38 (Opoczky, 1977), for example, illustrates for cement clinker and quartz that after a certain time of grinding the 'total' surface begins to fall, coinciding with the onset of aggregation, confirmed by electron microscope examination. The surface phenomena are thus in agreement with the fine grinding particle size observations of §4.3.6.2.

4.5 REFERENCES

4.5.1 Applicable to §4.2 — Particle shape

Beddow, J. K., and Philip, G. C. (1975), 'On the use of a Fourier analysis technique for describing the shape of individual particles', *Planseeberichte fur Pulvermetal.*, **23**, 3–14.

Bond, F. C. (1954). 'Control particle shape and size', *Chem. Eng.*, **61**, 195–198.

British Standard 63, 'Specification for single-sized roadstone and chippings', Part 2 (1971), Metric units.

British Standard 812, 'Methods for sampling and testing of mineral aggregates, sands and fillers', Part 1 (1975), 'Sampling, size, shape and classification'.

British Standard 2955 (1958), 'Glossary of terms relating to powders'.

Charles, R. J. (1956). 'High velocity impact in comminution', *Min. Eng. Trans. AIME*, **8**, (10), 1028–1032.

Cheng, D. C. H., and Sutton, H. M. (1971). 'Absolute determination of particle size and shape', *Nature* (*Phys. Sci.*), **232**, 30 Aug., 192.

Davies, R. (1975). 'A simple feature-space representation of particle shape', *Powder Technol.*, **12**, 111–124.

Fong, S. T., Beddow, J. K., and Vetter, A. F. (1979). 'A refined method of particle shape representation', *Powder Technol.*, **22**, 17–21.

Gaudin, A. M. (1926). 'An investigation of crushing phenomena', *Trans. AIME*, **73**, 253–316.

Giltrow, J. P., and Groszek, A. J. (1969). 'Effect of particle shape on the sliding abrasiveness of lamellar solids', *Wear*, **13**, 317–329.

Gotoh, K., and Finney, J. L. (1975). 'Representation of the size and shape of a single particle', *Powder Technol.*, **12**, 125–130.

Harris, C. C. (1960). 'A method for the routine measurement of particle shape factors in the sieve range', *Nature*, **187**, (4735), 401–403.

Harris, C. C., and Stambolitzis, G. (1969). 'Relationship between particle size distribution by number and weight', *Powder Technol.*, **2**, 58–60.

Heywood, H. (1933). 'Calculation of the specific surface of a powder', *Proc. Inst. Mech. Eng.*, **125**, 383–416.

Heywood, H. (1937). 'Numerical definitions of particle size and shape', *Chem. Ind.*, **56**, (17), 149–154.

Heywood, H. (1947). 'The scope of particle size analysis and standardisation', *Proc. Symp. Particle Size Anal.*, Inst. Chem. Eng. Suppl. 25, 14–24.

Heywood, H., and Pryor, E. J. (1945/46). 'Application of sizing analysis to mill practice', *Trans. IMM* (*London*), **55**, 405.

Holt, C. B. (1981a). 'Characterisation of the shape of particles produced by crushing using harmonic series analysis', *Inst. Chem. Eng. Symp. Series* 63, Birmingham, UK, D2/H/1–13.

Holt, C. B. (1981b). 'The shape of particles produced by comminution—a review', *Powder Technol.*, **28**, 59–63.

Lees, G. (1964). 'The measurement of particle elongation and flakiness: a critical discussion of British Standard and other methods', *Mag. Concrete Res.*, **16**, (4), 225–230.

Leuricens, D. W., Beddow, J. K., and Vetter, A. F. (1982). 'Morphological Fourier descriptors', *Powder Technol.*, **31**, 209–215.

Markwick, A. H. D. (1941). 'The grading and shape of commercial sizes of aggregates', *J. Soc. Chem. Ind.*, **61**, 85–91.

Martin, G. (1923). 'Researches on the theory of fine grinding', *Trans. Ceram. Soc.*, **23**, 61.

Meloy, T. P. (1977a). 'Fast Fourier transforms applied to shape analysis of particle silhouettes to obtain morphological data', *Powder Technol.*, **17**, 27–35.

Meloy, T. P. (1977b). 'A hypothesis for morphological characterisation of particle shape and physiochemical properties', *Powder Technol.*, **16**, 233–253.

Nakajima, Y., Whiten, W. J., and White, M. E. (1978). 'Method for measurement of particle shape distribution by sieves', *Trans. IMM* (*London*), **88**, C194–C203.

Nakajima, Y., Whiten, W. J., and White, M. E. (1979). 'Behaviour of non-spherical particles on screening', *Trans. IMM (London)*, **88**, C88–C92.

Rose, H. E. (1961). 'Particle shape, size and surface area', *Proc. Powders in Industry Symp.*, London, UK, 29–30 Sept. 1960, Soc. Chem. Ind., 130–149.
Schmidt, H. (1981). 'Impact reduction: Part III, Problem of the cubiformity of crushed products', *Aufbereit Tech.*, **10**, 523–532.
Shergold, F. A. (1959). 'A study of the granulators used in the production of road making aggregates', DSIR Road Research Technical Paper 44.
Tsubaki, J., and Jimbo, G. (1979a). 'A proposed new characterisation of particle shape and its application', *Powder Technol.*, **22**, 161–169.
Tsubaki, J., and Jimbo, G. (1979b). 'The identification of particles using diagrams and distributions of shape indices', *Powder Technol.*, **22**, 171–178.
Woźniak, K. (1980). 'Einfluss der technischen bedingungen der erzeugung von schliefkorn auf seine form', *Proc. Europ. Symp. Particle Technol.*, Amsterdam, 3–5 June, 214–229.
Yigit, E., Johnston, H. A., and Maroudas, N. G. (1967). 'Shape and size distribution of electrohydraulically crushed aggregate', *Quarry Managers J.*, Dec., 467–472.
Zahn, C. T., and Roskies, R. Z. (1972). 'Fourier descriptors for plane closed curves', *IEEE Trans. Comput.*, **C21**, 269–281.

4.5.2 Applicable to §§4.3 and 4.4—Particle size and surface

Allen, T. (1968–69). 'Determination of size distribution and specific surface of fine powders by photoextinction methods', *Powder Technol.*, **2**, 133–140, 141–153.
Allen, T. (1972). 'Particle size measurement and surface area determination—bibliography', University of Bradford.
Allen, T. (1975). *Particle Size Measurement*, Chapman & Hall, London.
ANSI/ASTM C204—78a, 'Standard test method for fineness of Portland cement by air permeability apparatus'.
ANSI, Powder Test Code No. 28 (1973), 'Determining the properties of fine particle matter'.
Arbiter, N., Harris, C. C., and Stambolitzis, G. A. (1969). 'Single fracture of brittle spheres', *Trans. SME/AIME*, **244**, 118–133.
Austin, L. G., and Bhatia, V. K. (1971–72). 'Experimental methods for grinding studies in laboratory mills', *Powder Technol.*, **5**, 261–266.
Austin, L. G., Gardner, R. P., and Walker, P. L. (1963). 'The shape factors of coal in a standard Hardgrove mill', *Fuel*, **42**, 319–323.
Austin, L. G., and Klimpel, R. R. (1968). 'Statistics of random fracture', *Trans. SME/AIME*, **4**, 219–224.
Austin, L. G., Luckie, P. T., and Seebach, H. M. (1976). 'Optimization of a cement milling circuit with respect to particulate size distribution and strength development by simulation models', Dechema Monograph 79, 519–537.
Austin, L. G., Luckie, P. T., and Wightman, D. (1975). 'Steady state simulation of a cement milling circuit', *Int. J. Miner. Process.*, **2**, 127–150.
Austin, L. G., Weymont, N. P., and Knobloch, O. (1980). 'The simulation of air-swept cement mills, Part I: The simulation model', *Proc. Europ. Symp. Particle Technol.*, Amsterdam, 640–655.
Babcock & Wilcox Co. (1955). 'Steam—its generation and use', page 17-17.
Babcock Power Ltd (1957). 'Determination of particulate size in the sub-sieve range', Research Dept. Report 5/57/1.
Bailey, P. (1969). 'A mathematical model for vertical spindle pulverising mills', Central Electricity Generating Board, Harrogate, UK, Report SSD/NE/R168.
Barnard, W. J. (1983). 'Primary breakage of brittle particles using drop shatter test apparatus', D.Ph. thesis, University of Melbourne, Australia.
Barnard, W. J., and Bull, F. A. (1979). 'Primary breakage of brittle particles', *Proc. 4th Tewksbury Symp. Fracture*, Melbourne, Australia, Feb. 1979, 20: 1–5.

Baumgardt, S., Buss, B., May, F., and Schubert, H. (1975). 'On the comparison of results of single grain crushing under different kinds of loads', *Proc. XIth Miner. Process Congr.*, Cagliari, Sardinia, 1–32.

Beke, B. (1964). *Principles of Comminution*, Akadémiai Kiadó, Budapest.

Bennett, J. G. (1936). 'Broken coal', *J. Inst. Fuel*, **10**, 22–39.

Bergstrom, B. H. (1966). 'Empirical modification of the Gaudin-Meloy equation', *Trans. AIME*, **235**, 45.

British Standard 12 (1978), 'Ordinary and rapid hardening Portland cement'.

British Standard 410 (1976), 'Specification of test sieves'.

British Standard 893 (1978), 'Methods for the measurement of the concentration of particulate materials in ducts carrying gases'.

British Standard 1017, 'Methods for sampling of coal and coke. Part 1 (1977) Sampling of coal'.

British Standard 4306, 'Methods for the determination of particle size of powders. Part 3 (1963) Air elutriation methods'.

British Standard 4359, 'Methods for the determination of specific surface of powders: Part 1 (1969) Nitrogen adsorption (BET method); Part 2 (1982) Recommended air permeability methods; Part 3 (1970) Calculation from particle size distribution'.

Broadbent, S. R., and Callcott, T. G. (1956a). 'A matrix analysis of processes involving particle assemblies', *Phil. Trans. R. Soc.*, **249**, 99–123.

Broadbent, S. R., and Callcott, T. G. (1956b). 'I: A new analysis of coal breakage processes', *J. Inst. Fuel*, **29**, 524–528; 'II: A matrix representation of breakage', *Ibid.*, **29**, 528–539.

Brunauer, S., Emmett, P. H., and Teller, E. (1938). 'Adsorption of gases in multimolecular layers', *Am. Chem. Soc. J.*, **60**, 309–319.

Bucsky, G. (1975). 'Accounting for the difference between particle sizing techniques', *Process Eng.*, July, 52–54.

Bull, F. A. (1955). 'The designation of the fineness of pulverised brown coal', *J. Inst. Fuel*, **28**, 163–170.

Cottaar, W., and Rietema, K. (1984). 'The effect of interstitial gas on milling', *Powder Technol.*, **38**, 183–194.

Dias, L. S. (1980). 'Continuous particle size measurement—a status report', *Instrum. Min. Metall. Ind.*, **7**, 111–124.

Felton, P. G., and Watson, D. J. (1984). 'Measurement of solid particle size distribution and concentration', *Proc. 9th Powder & Bulk Solids Handling & Processing Conf.*, Rosemont, Illinois, USA, 15–17 May, 294–303.

Frigione, G., and Marra, S. (1976). 'Relationship between particle size distribution and compressive strengths of Portland cement', *Cem. Concrete Res.*, **6**, 113–128.

Fry, T. C. (1928). *Probability and its Engineering Uses*, Van Nostrand, New York, 214.

Fuerstenau, D. W. (1961). 'Size distributions resulting from the comminution of heterogeneous minerals', *Trans. AIME*, **220**, 395–396.

Gaudin, A. M., and Meloy, T. P. (1962). 'Model and a comminution distribution equation for single fracture', *Trans. SME/AIME*, **223**, 40–43.

Gilvarry, J. J. (1961). 'Fracture of brittle solids. Part I, Distribution function for fragment size in single fracture (theoretical)', *J. App. Phys.*, **32**, 391–399.

Gooden, E. L., and Smith, C. M. (1940). 'Measuring average particle diameter of powders', *Ind. Eng. Chemi. Anal.*, **12**, (8), 479–482.

Graham, A. L., and Hanna, T. H. (1962). 'The micro-particle classifier', *Ceramic Age*, Sept., 38–40.

Gregg, S. J. (1968). 'Surface chemistry study of comminuted and compacted solids', *Chem. Ind.*, 11 May, 611–617.

Gregg, S. J., Hill, K. J., and Parker, T. W. (1954). 'The grinding of kaolinite, Part I: A preliminary study', *J. App. Chem.*, **4**, 631–632.

Harris, C. C. (1966). 'On the role of energy in comminution: a review of physical and mathematical principles', *Trans. IMM (London)*, **75**, C37–C56

Harris, C. C. (1968). 'The application of size distribution equations to multi-event comminution processes', *Trans. SME/AIME*, **241**, 343–358.

Harris, C. C., and Stambolitzis, G. (1969). 'Relationship between particle size distribution by number and weight', *Powder Technol.*, **2**, 58–60.

Hawksley, P. G. W., Badzioch, S., and Blackett, J. H. (1977). '*Measurement of Solids in Flue Gases*', Inst. Fuel, London.

Herbst, J. A., and Sepulveda, J. L. (1978). 'Fundamentals of fine and ultrafine grinding in a stirred ball mill', *Proc. Tech. Program Int. Powder Bulk Handling Process.*, Rosemont, Illinois, Ind. Sci. Conf. Management, Chicago, 452–470.

Hess, W., and Schönert, K. (1981). 'Brittle-plastic transition in small particles', *Inst. Chem. Eng. Symp. Series No. 63*, Birmingham, UK, D2/J/9.

Heywood, H. (1957). 'Principles of crushing and grinding in chemical engineering practice'. In: Cremer, H. W., and Davis, T. (Eds.), Vol. 3, Butterworths, 1–23.

Hidaka, J., and Miwa, S. (1979). 'Fractionation and particle size analysis of fine powders by microsieve', *Powder Technol.*, **24**, 159–166.

Hukki, R. T. (1944). 'An experimental study of the principles of comminution', D.Sc. thesis, Massachusetts Institute of Technology.

Hukki, R. T., and Venho, J. E. (1972). 'Estimation of sub-sieve size distribution of ground products by computer', Dechema Monograph 69, 557–574.

Hukki, R. T. (1975). 'The principles of comminution: an analytical survey', *Eng. Min. J.*, **176**, 106–110.

Hüttig, G. F. (1952). 'Zermahlungsvorgähge als chemisches problem', Dechema Monograph, Band 245–268, 96–115.

Jimbo, G. (1973). 'Bi-modal size distribution model in size reduction process', *Proc. Harold Heywood Memorial Symp.*, 17–18 Sept., Loughborough University of Technology.

Johnston, P. R., and Swanson, R. (1982). 'A correlation between the results of different instruments used to determine the particle size distribution in AC fine test dust', *Powder Technol.*, **32**, 119–124.

Jowett, A. (1971). 'An introduction to the assessment of energy requirements and product size in comminution', *Miner. Sci. Eng.*, **3**, (4), 33–34.

Kihlstedt, P. G., (1962). 'The relationship between particle size distribution and specific surface in comminution, *Proc. 1st Europ. Symp. Zerkleinern*, 10–13 Apl. Frankfurt, VDI Verlag, Dusseldorf, 205–216.

Klimpel, R. R., and Austin, L. G. (1965). 'The statistical theory of primary breakage distribution for brittle materials', *Trans. SME/AIME*, **232**, 88–94.

Klotz, K., and Schubert, H. (1982). 'Crushing of single irregularly shaped particles by compression: Size distribution of progeny particles', *Powder Technol.*, **32**, 129–137.

Kondo, Y., and Gaudin, A. M. (1964). 'Sizing comminution products in the extremely fine range', *Trans. SME/AIME*, **229**, (3), 292–299.

Krycer, I., and Hersey, J. A. (1981). 'Grinding and granulation in a vibratory ball mill', *Powder Technol.*, **28**, 91–95.

Leschonski, K. (1979). 'Sieve analysis, the Cinderella of particle size analysis methods?', *Powder Technol.*, **24**, 115–124.

Leschonski, K., and Rothele, S. (1978). 'A new probe for highly concentrated gas-solid flow systems', *Proc. Int. Symp. In-stream Measurements of Particle Solid Properties*, Bergen, Vol. II, 1–18.

Malghan, S. G., and Mular, A. L. (1982). 'Measurement of size distribution and surface area of granular materials'. In: *Design and Installation of Comminuted Circuits*, (Mular, A. L., and Jergensen, G. V., Eds.), SME/AIME, New York, 48–70.

Marshall, V. C. (1967). 'The mechanism of comminution in pin mills', Dechema Monograph 57, 729–739.

Marshall, V. C. (Ed.) (1974). 'Comminution: a report by Inst. Chem. Eng. Working Party', Inst. Chem. Eng., London, 8–9.

Opoczky, L. (1977). 'Fine grinding and the agglomeration of silicates', *Powder Technol.*, **17**, 1–7.

Opoczky, L. (1980). 'Fine grinding and the interaction of particles', *Proc. Europ. Symp. Particle Technol.*, Amsterdam, 3–5 June, 302–305.

Opoczky, L. (1981). 'Ways of influencing the particle size distribution of ground products in dry ball mills', *Interceram.*, (4), 387–389, 403.

Opoczky, L., and Farnady, F. (1984). 'Fine grinding and states of equilibrium', *Powder Technol.*, **39**, 107–115.

Owe Burg, T. G., and Avis, L. E. (1970–71). 'Exploratory experiments in kinetics of comminution', *Powder Technol.*, **4**, 27–31.

Rigden, P. J. (1947). 'The specific surface of powders: a modification of the theory of the air permeability method', *J. Soc. Chem. Ind.*, **66**, 130–136.

Rosin, P., and Rammler, E. (1933). 'The laws governing the fineness of pulverised coal', *J. Inst. Fuel*, **7**, 29–36.

Rumpf, H. (1965). 'Die enzelkornzerkleinerung als grundlage einer technischen zerkleinerungswissenschaft', *Chemie. Ing. Tech.*, **37**, 187–202.

Rumpf, H. (1972). 'Adhesion of solid particles to solid surfaces', *J. Res. Assoc. Powder Technol., Japan,* **9**, 3–15.

Scarlett, B. (1980). 'The measurement of particle size', *Proc. Europ. Symp. Particle Technol.*, Amsterdam, 3–5 June, 1180–1199.

Schönert, K. (1972). 'Role of fraction physics in understanding comminution phenomena', *Trans. SME/AIME*, **252**, 21–26.

Schuhmann, R., (1940). 'Principles of comminution. Part I, Size distribution and surface calculations', *Trans. AIME*, Publn. 1189.

Snow, R. H. (1973). 'Grinding mill simulation and scale-up of ball mills', *Proc. Int. Conf. Particle Technol.*, IITRI, Chicago, 28–38.

Snow, R. H., and Paulding, B. W. (1973). 'Production of fragment size distribution from calculations of internal stress in spheres and discs', *Proc. Harold Heywood Memorial Symp.*, 17–18 Sept., Loughborough University of Technology.

Sresty, G. C., and Venkateswar, R. (1980). 'Particle size analysis', *Proc. Int. Symp. Fine Particle Process.*, Las Vegas, Nevada, Feb. 1980, AIME, New York, Vol. 1, 329–365.

Steier, K., and Schönert, K. (1972). 'Verformung und bruchphänomene unter druckbeansprungung von sehr kleinen körnern aus kalkstein, quarz und polystyrol', Dechema Monograph 69, 167–192.

Svarovsky, L. (1977). *Solid–Liquid Separation*, Butterworth, London, 6–29.

Svarovsky, L., and Svarovska, J. (1984). 'Application of WASP photosedimentometer to grade efficiency testing', *Proc. 3rd Europ. Symp. Particle Characterisation*, (PARTEC), Nuremburg, 9–11 May.

Wadell, H. (1932). 'Volume, shape and roundness of rock particles', *J. Geol.,* **40**, 443–451.

Whiten, W. J., Walter, G. W., and White, M. E. (1979). 'A breakage function suitable for crusher models', *Proc. 4th Tewksbury Symp.*, Melbourne, Australia, **19**, 1–32.

CHAPTER 5

Classification

NOTATION FOR CHAPTER 5

a'	Proportion of classified stream reporting to coarse stream independently of size
A	Area
b'	Proportion of classified stream reporting to fine stream independently of size
$B[r(x)]$	Binomial function
$c(x)$	Classification function

C	Solids concentration in gas or liquid Subscripts: c coarse stream f fine stream o feed
C'	Recirculation ratio = M_c/M_f
d	Cut size of particles associated with classification (Tromp) curve Subscripts: a analytical p cumulative 25 corresponding to 25% feed of coarse product (and similarly for other numerals)
d'	Cut size of particles associated with selectivity curve: subscripts as for d
D	Diameter Subscripts: i classifier inlet at blade pivot pcd o classifier outlet (id)
D^*	Disc diameter ($\bar{D}^*$ is mean value)
E	Empirical capacity factor at base conditions in screen sizing
$f(x)$	Density–size distribution function Subscripts as for C
$\bar{f}(x)$	Mean value of $f(x)$
$f_o'(x)$	Density size distribution function of reconstituted feed
$F(x)$	Cumulative size distribution function Subscripts as for C
$F_o'(x)$	Cumulative size distribution function of reconstituted feed
g	Acceleration due to gravity
g^*	Centrifugal acceleration Subscripts: x at radius x R at radius R
H	Duct height
H_i	Cone classifier blade height
K	Constant
L	Length of screen
M	Mass flow rate Subscripts as for C
n	Number or exponent
Δp	Pressure difference
P	Probability
P_n	Probability of particle passing through a screen after n presentations
Q	Correction factors in screen sizing
r	Number
R	Radius
R_e	Radius at entry of classifier
$s(x)$	Selectivity function
$\hat{s}(x)$	Estimated selectivity function

t	Time
T	Thickness of material of square mesh
T^*	Particle thickness ($\bar{T}^*$ is mean value)
v	Velocity
	Subscripts: i at classifier inlet (blade pivot pcd)
	M maximum
	R tangential at radius R
	x tangential at radius x
V	Volumetric flow rate
	Subscripts as for C
W	Width of square mesh aperture
W^*	Particle width
x	Particle size
	Subscripts: cm coarse fraction minimum
	cM coarse fraction maximum
	fm fine fraction minimum
	fM fine fraction maximum
	m minimum
	M maximum
	om feed minimum
	oM feed maximum
	t associated with terminal velocity
X	Particular function of x, e.g. log x
y	Independent variable
Y	Inverse of classification function $c(x)$
z	Independent variable
Z	Factor embodying cyclone geometry in equation (5.45)
α	Size modulus parameter
β	Distribution modulus parameter
γ	$[f_o(x_i) - f_f(x_i)]$ in equation (5.28)
ε	Error associated with density distribution
E	Error associated with cumulative distribution
ζ	$[f_c(x_i) - f_o(x_i)]$ in equation (5.28)
η	Viscosity of air
θ	Shape factor = $\tan^{-1}(W^*/T^*)$
λ	Likelihood factor
Λ	Exponent inversely proportional to the slope of $c(x)$ at d_{50}
Λ'	Exponent associated with d_{25} and d_{75}
μ	Suffix indicating mean value of partial derivatives
ξ	Screen load factor
ρ	Density
	Subscripts: l liquid
	s solid
σ	Standard deviation
ϕ	Function

φ	Angle with vertical of classifier cone
Ψ	Yield = M_c/M_o
Ψ'	Net yield = $(\Psi - a')/(1 - a')$
ψ	Ratio of tangential to radial component of inlet velocity
ω	Angular velocity
	Subscripts: R at radius R
	x at radius x

5.1 INTRODUCTION

Classification is a process of particle stream separation, ideally at some particular size, called the cut size. In practice this ideal is never attained, so that some particles below cut size join the coarse fraction and vice versa, as seen in Figure 5.1, which is based on cumulative size distributions.

The most important industrial application of classifiers is to reduce overgrinding in a mill by separating the grinding zone output into coarse and fine fractions, so that fines can be removed as they are produced and the coarse returned for further size reduction. Many consider that the most fruitful area for improvement of a mill-classifier system is the reduction of the overlaps across the cut size in the classifier.

Methods for the measurement of the effectiveness of separation are discussed, as are means to attain greater effectiveness. In both, mathematical simulation of classification is being used increasingly in association with computerized statistical analysis of errors, which are inevitably associated with all classifier assessments.

The many types of classifiers are surveyed, with emphasis on their method of operation. Dry screening is first discussed as the most direct method because no carrying fluid is necessary and particle density is not important. However, continuous screening is difficult in small particle sizes and most often industrial classification uses air or water as the carrying medium. Thus carrying medium

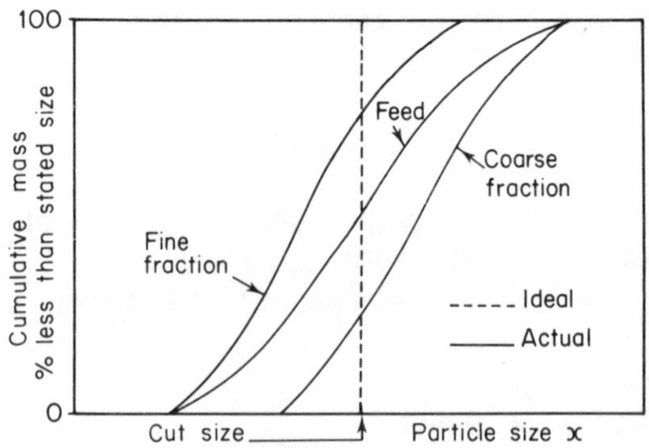

Figure 5.1. Cumulative size distributions of feed and of coarse and fine streams after classification

properties must be considered as well as those of the carried particles. Hence after screening there follows consideration of the many separation processes involving a fluid medium.

In this chapter the emphasis is on the classifier *per se*: the study of mill/classifier systems is the subject of Chapter 10.

5.2 REPRESENTATION OF CLASSIFIER PERFORMANCE

5.2.1 Survey

Four levels of representation of classifier performance are recognized, in which:

(1) Only the total mass of the coarse (or the fine) stream is considered relative to the mass of the feed, resulting in the term 'yield'. This is the crudest form of assessment.
(2) Size distributions of feed, coarse and fine streams are included, leading to the term 'recovery'.
(3) Measurement is made of the extent to which coarse particles appear in the fine stream and vice versa. This is the most meaningful representation of classifier performance, resulting in the selectivity function and the classification function, which are defined below. Two important concepts are involved here, both of which need to be assessed, namely (a) the size modulus or cut size at which separation is occurring, and (b) the distribution modulus, or sharpness of cut, which is a measure of the spread about the cut size.
(4) Provision is made for stochastic errors, which are inherent in practical measurements.

Each of these four areas is discussed in turn.

The word 'efficiency' has been omitted deliberately from definitions in this chapter because of its many conflicting connotations in the literature.

5.2.2 Yield

Yield Ψ is simply the ratio of the mass M_c of the coarse fraction to the mass M_o of the feed, or:

$$\Psi = \frac{M_c}{M_o} = 1 - \frac{M_f}{M_o} \tag{5.1}$$

Some writers call this 'total efficiency'.

Be definition (see §4.3.2) the density distribution, the derivative of the cumulative, gives the fraction of particles of size x. Hence the mass of particles of size x respectively in the feed, coarse and fine fractions may be represented as:

$$M_o \frac{d}{dx} F_o(x), \quad M_c \frac{d}{dx} F_c(x) \quad \text{and} \quad M_f \frac{d}{dx} F_f(x)$$

By mass balance we have:

$$M_o \frac{d}{dx} F_o(x) = M_c \frac{d}{dx} F_c(x) + M_f \frac{d}{dx} F_f(x) \tag{5.2}$$

Applying equation (5.1) to (5.2) we have:

$$\frac{d}{dx} F_o(x) = \Psi \frac{d}{dx} F_c(x) + (1 - \Psi) \frac{d}{dx} F_f(x)$$

Integration gives:

$$F_o(x) = \Psi F_c(x) + (1 - \Psi) F_f(x) \tag{5.3}$$

After rearrangement we have:

$$\Psi = \frac{F_o(x) - F_f(x)}{F_c(x) - F_f(x)} \tag{5.4}$$

Equation (5.4) offers a useful method of obtaining a mean value of Ψ from the cumulative size distributions of feed, coarse fraction and fine fraction, as shown in Figure 5.2, where $[F_o(x) - F_f(x)]$ is plotted against $[F_c(x) - F_f(x)]$. A mean line is drawn through the scatter of points, the slope of which determines Ψ (Svarovsky, 1977).

When solids are suspended in liquids as in hydrocyclones it is not easy to measure directly the solids mass flow rates, and then it is more practically convenient to measure volumetric flow rates V_o, V_c and V_f together with concentrations C_o, C_c and C_f, where the suffixes o, c and f denote feed, coarse stream

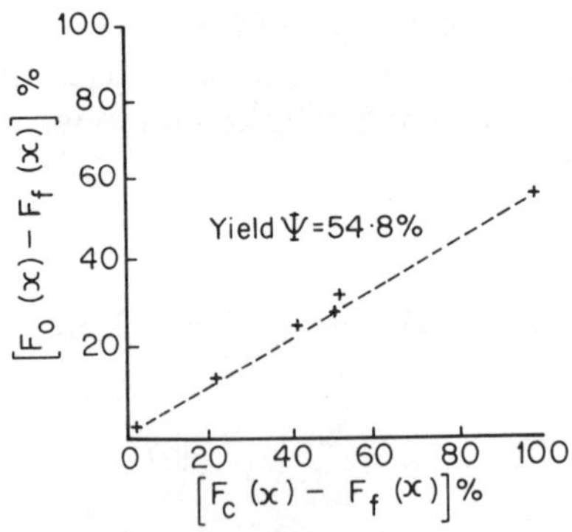

Figure 5.2. Experimental value of yield Ψ for a hydrocyclone using equation (5.4) (Svarovsky, 1977). *Reproduced by permission of Butterworths, London*

and fine stream respectively. Then the yield Ψ is given by:

$$\Psi = \frac{C_c V_c}{C_o V_o} = 1 - \frac{C_f V_f}{C_o V_o} = \frac{1}{1 + (C_c V_c / C_f V_f)} \tag{5.5}$$

5.2.3 Recovery

Recovery takes some account of size distribution, to the extent that it is defined as:

$$\frac{\text{Mass of fine stream finer than } x}{\text{Mass of feed finer than } x} = \frac{(1 - \Psi)F_f(x)}{F_o(x)}$$

$$\frac{\text{Mass of coarse stream coarser than } x}{\text{Mass of feed coarser than } x} = \frac{\Psi[1 - F_c(x)]}{1 - F_o(x)}$$

whichever best characterizes the separation process in hand. Other more complex expressions have been derived and used extensively in the past, as summarized by Leschonski (1981) for example, but the present trend is toward classification and selection functions as better measures of performance than recovery concepts.

5.2.4 Classification (Tromp) curve

Equation (5.2) may be written for density distribution as:

$$M_o f_o(x) = M_c f_c(x) + M_f f_f(x)$$

Dividing through by M_o, we have:

$$f_o(x) = \Psi f_c(x) + (1 - \Psi) f_f(x) \tag{5.6}$$

The three terms of equation (5.6) are called 'weighted density distribution functions' and are plotted against x in Figure 5.3.

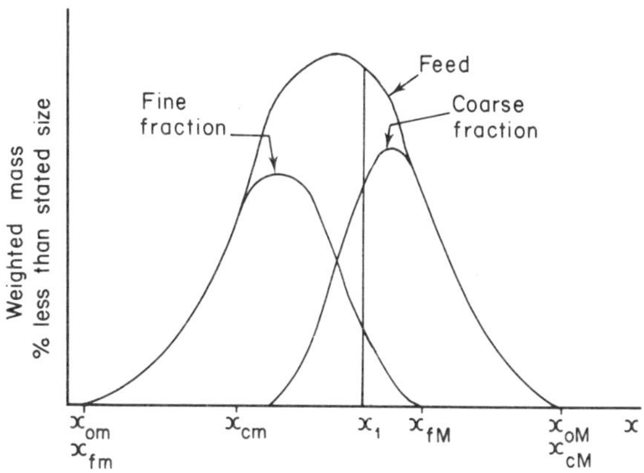

Figure 5.3. Weighted density distributions of feed and of coarse and fine streams after classification

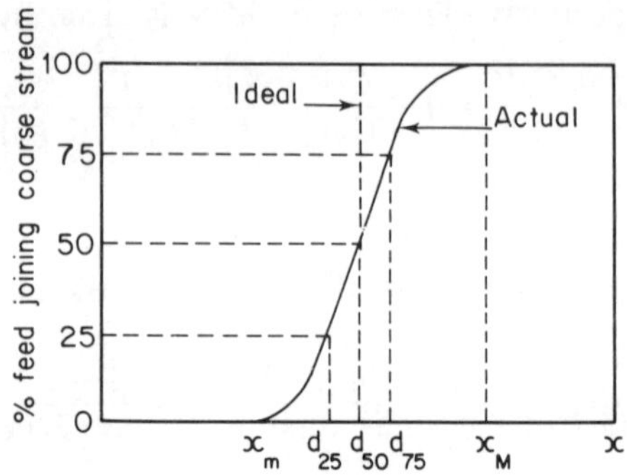

Figure 5.4. Classification (Tromp) curve

For any particular particle size such as x_1, the shaded ordinate gives the amount of that size in the coarse stream, i.e. $\Psi f_c(x_1)$. Similarly the full ordinate at x_1 gives the amount of that size in the feed, namely $f_o(x_1)$. The ratio $\Psi f_c(x_1)/f_o(x_1)$ gives the fraction of feed of size x_1 which appears in the coarse stream. When this ratio is obtained for a range of values of x and plotted against x, there results the classification curve shown in Figure 5.4. This curve is often called the grade efficiency curve because it is the summary of the partition performance of each size grading. It is also called the Tromp diagram after the name of the man to whom it is attributed (Tromp, 1937).

The mathematical representation of the classification curve is the classification function $c(x)$. Hence from the above definition in terms of density distribution:

$$c(x) = \Psi f_c(x)/f_o(x) \tag{5.7}$$

or in terms of cumulative distribution:

$$c(x) = \Psi \frac{d}{dx} F_c(x) \Big/ \frac{d}{dx} F_o(x)$$

$$= \Psi \frac{dF_c}{dF_o} \tag{5.8}$$

Several mathematical forms have been proposed to fit the classification curve as summarized by Luckie and Austin (1974). Among these forms, two have been explored by Luckie and Austin (1974) in practical examples and found to be equally valid. They are an RRB type suggested by Plitt (1971) and Reid (1971) and also a log–normal distribution form proposed by Aso (1957).

5.2.5 Features of the classification curve

5.2.5.1 Survey

The features of the classification curve which are of interest are:

(1) limits of separation;
(2) size modulus or cut size;
(3) distribution modulus or sharpness of cut.

5.2.5.2 *Limits of separation*

In this chapter, the suffix m denotes minimum and M denotes maximum.

From the classification curve in Figure 5.4 it is noted that, above x_M, $c(x)$ continues to be unity and below x_m, $c(x)$ continues to be zero.

Similarly for the density distributions, Figure 5.3 indicates the limits of separation for the coarse fraction from x_{cm} to x_{cM} and for the fine fraction from x_{fm} to x_{fM}.

5.2.5.3 *Size modulus or cut size*

In principle, any value of x in the overlap region between coarse and fine separation, that is between x_{cm} and x_{fM} in Figure 5.3, could be used as the cut size. However, three definitions are commonly used: (1) the equiprobability partition size, d_{50}; (ii) the analytical cut size from density distribution, d_a; and (iii) a cut size from cumulative distribution considerations, d_p.

(i) The cut size d_{50} corresponds to 50% of the feed passing to the coarse stream as seen in Figure 5.4. It is therefore that size which has equal probability of passing to either coarse or fine streams.

(ii) The analytical cut size d_a is such that the coarse and fine streams contain equal quantities of misplaced material, indicated by areas I and II in Figure 5.5. When no errors exist in the distributions, the following analysis based on Leschonski (1981) holds good for d_a:

$$\text{Area I} = \Psi \int_{x_{cm}}^{d_a} f_c(x)\,dx = \Psi F_c(d_a) \tag{5.9}$$

$$\text{Area II} = \int_{d_a}^{x_{fM}} [(1-\Psi)f_f(x)]\,dx$$

By using equation (5.6), this may be written as:

$$\text{Area II} = \int_{d_a}^{x_{fM}} [f_o(x) - \Psi f_c(x)]\,dx \tag{5.10}$$

which is representative of the shaded area III. Hence, if area I = area II, each also equals area III.

Integrating (5.10) and equating to area I from (5.9), we obtain:

$$\Psi F_c(d_a) = F_o(x_{fM}) - F_o(d_a) - \Psi F_c(x_{fM}) + \Psi F_c(d_a)$$

or

$$F_o(d_a) = F_o(x_{fM}) - \Psi F_c(x_{fM})$$

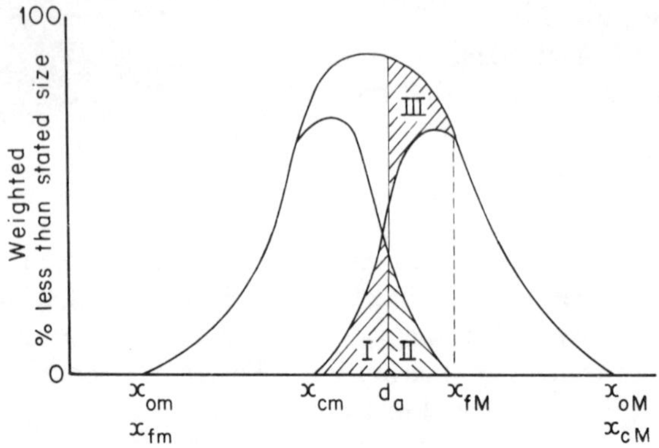

Figure 5.5. Weighted density distributions of feed and of coarse and fine streams, illustrating analytical cut size d_a (Leschonski, 1981). *Reproduced by permission of the author*

Applying equation (5.3), this becomes:

$$F_o(d_a) = (1 - \Psi)F_f(x_{fM})$$

or, since $F_f(x_{fM})$ is unity:

$$F_o(d_a) = (1 - \Psi) \tag{5.11}$$

Hence as Ψ is easily found, d_a is equally easily found as that abscissa on the cumulative feed distribution corresponding to the ordinate $(1 - \Psi)$.

(iii) The cut size d_p is that for which the cumulative fraction of the coarse stream is equal to the fraction oversize of the fine stream, as indicated in Figure 5.6. d_p is even more sensitive to feed size distribution than d_a, but both may be derived from size distributions without first having to obtain the classification curve; wherein lies their practical value.

d_a and d_p are close to d_{50} for small difference between areas I and II, but as the difference increases, so d_a and d_p may become further removed from d_{50}.

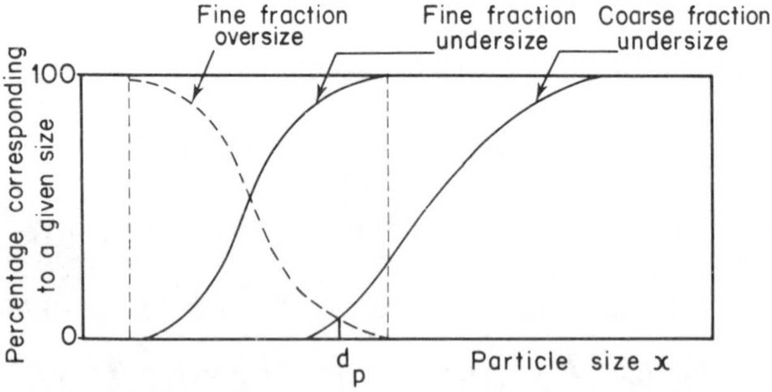

Figure 5.6. Cut size d_p from cumulative size distributions of coarse and fine streams after classification (Leschonski, 1981). *Reproduced by permission of the author*

5.2.5.4 *Distribution modulus or sharpness of cut*

In order to specify sharpness of cut, a measure is required which applies to the overlap size range between x_m and x_M in Figure 5.4.

Many such measures have been adopted as summarized by Leschonski (1981), but perhaps the most common is d_{25}/d_{75}, or if a wider spread is to be accommodated, d_{10}/d_{90}. In either case, the nearer the ratio approaches unity, the sharper the separation. For steep classification curves $(d_{90} - d_{10})/d_{50}$ may be appropriate.

In mathematical modelling of classifiers, Austin and Klimpel (1981) have preferred the following:

$$1/c(x) = 1 + (x/d_{50})^{-\Lambda} \tag{5.12}$$

where Λ is the slope of the classification curve at d_{50}. Alternatively (Klimpel 1982), the exponent may take the form:

$$\Lambda = K_1/\ln(d_{25}/d_{75}) \tag{5.12a}$$

where K_1 is a constant.

Equation (5.12) conveniently embodies both size and distribution moduli.

5.2.6 Selectivity curve

5.2.6.1 *Survey*

The classification curve depicted in Figure 5.4 is asymptotic to the abscissae at ordinate values of zero and unity. This is not often the case in practice, and four types of deviation from the pattern of Figure 5.4 are considered:

(1) an apparent bypass effect, and its variation with classifier load;
(2) the effect of comminution within the classifier;
(3) a shortfall of data in smaller values of x;
(4) a fish-hook effect.

Graphs embodying such deviations are called selectivity curves to distinguish them from the classification curve. Some writers call the selectivity curve the 'practical grade efficiency curve' and the classification curve the 'reduced grade efficiency curve'. The mathematical representation of the selectivity curve is the selectivity function $s(x)$ and practical ways of accommodating its idiosyncrasies are discussed.

5.2.6.2 *Apparent bypass effect*

Plitt (1971), in the context of hydrocyclone operation, discusses the liquid that is discharged with the coarse product which fills the interstices between the particles. In this liquid is a proportionate share of all sizes of particles, and hence

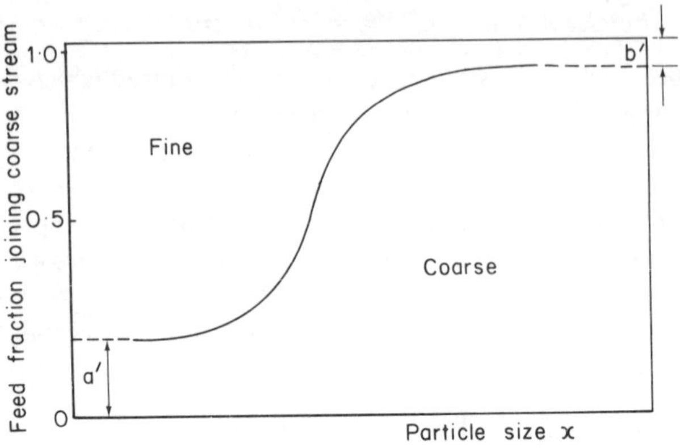

Figure 5.7. Selectivity curve showing bypass effect (Luckie and Austin, 1974). *Reproduced by permission of the Institution of Mining and Metallurgy*

a portion of each size fraction virtually bypasses the classifying action. Other writers refer to this bypass as 'dead flux'. This has the effect on the selectivity curve of causing the lower asymptote to occur at the ordinate value a', as shown in Figure 5.7. Luckie and Austin (1974) have expressed the effect more generally by thinking of the parameter a' as representing the arrival of particles in the coarse stream independently of their size, as may be seen from Figure 5.7 viewed as a partition curve.

To take account of the bypass effect on yield, there has been defined a net yield Ψ' (also called a reduced total efficiency) and expressed as $\Psi' = (\Psi - a')/(1 - a')$.

Kelsall (1963) and Schönert (1969), among others, have shown that the bypass parameter a' varies with classifier feed rate, and hence it is difficult to describe a single selectivity function which is representative of the classifier. However, Yoshiota and Hotta (1955) furnish evidence that when selectivity functions corresponding to different operating conditions are plotted against the normalized parameter x/d_{50}, the problem is overcome, because all curves map on to a common functional form, within the experimental scatter.

5.2.6.3 The effect of comminution within the classifier

Leschonski (1981) notes the effect of any comminution which may take place within the classifier. It clearly means that the coarse stream is finer than it would be without this size reduction, and hence in the larger of the sizes separated there is a lower proportion than otherwise. This causes the selectivity curve to be asymptotic to an ordinate $(1 - b')$ as shown in Figure 5.7. Luckie and Austin (1974) have expressed the effect more generally by thinking of the parameter b' as representing the arrival of particles in the fine stream, independently of their size, as may be seen by viewing Figure 5.7 as a partition curve.

5.2.6.4 *A shortfall of data in smaller values of x*

A further difficulty is illustrated in Figure 5.8, where the selectivity curve does not reach far enough into the smaller particle sizes for the value of a' to be determined. Attempts to extrapolate the selectivity curve have not met with success, and therefore Austin and Klimpel (1981) propose a trial and error system of arriving at the parameter a' by minimizing the difference between observed and calculated values of the known part of the selectivity curve on the basis of theoretical work by Klimpel (1980) along the lines of §5.2.6.7. These papers should be consulted for computational analysis details.

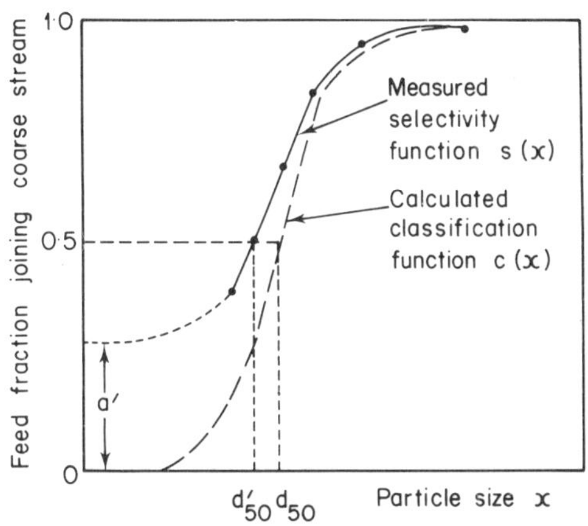

Figure 5.8. Example of measured selectivity curve which falls short of the area which decides the value of the intercept a'. (Austin and Klimpel, 1981). *Reproduced by permission of Elsevier Sequoia*

5.2.6.5 *Fish-hook effect*

It is a feature of certain mechanical air separators possessing rotating vanes (see §5.3.8(b)) that the fine particle stream from a primary classification is met by a recirculating air stream containing still finer particles under conditions which allow recoating of the separated particles by the still finer particles. This secondary action leads to a fish-hook pattern of selectivity curve shown in Figure 5.9 (solid line) after Luckie and Austin (1975).

Leschonski (1981) shows a similar effect from classifiers in which there is incomplete feed dispersion at the classifier entry, or even within the classification zone, whereby aggregates of fine particles may be classified as coarse particles and thus pass to the coarse stream. If, during the analysis, the aggregates are dispersed, the analysis detects more fines than the classifier did. Clearly the cure is better dispersion of feed prior to classification; but without such a cure, Luckie and Austin (1975) have proposed a model with two

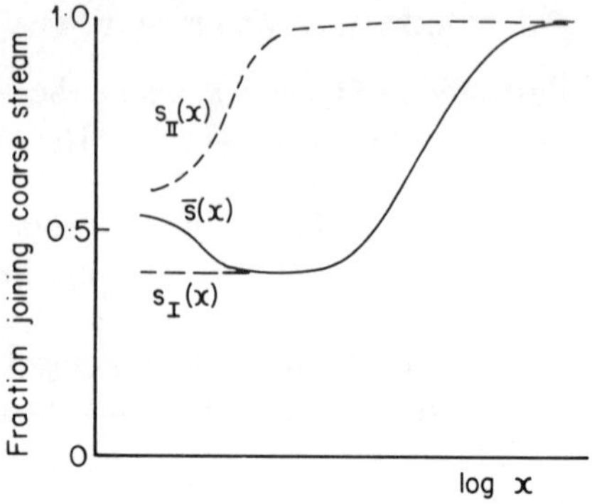

Figure 5.9. Fish-hook type of selectivity curve from a mechanical air separator (Luckie and Austin, 1975). *Reproduced by permission of the Institution of Mining and Metallurgy*

classifiers in series, including recirculation of fines, to handle the analysis, having selectivity functions $s_I(x)$ for the primary action and $s_{II}(x)$ for the secondary action. Then the overall selectivity function $\bar{s}(x)$ for the system is given by:

$$\bar{s}(x) = \frac{s_I(x)}{\{1 - [1 - s_I(x)]\,[1 - s_{II}(x)]\}} \tag{5.13}$$

Functions $s_I(x)$ and $s_{II}(x)$ are shown in Figure 5.9 by broken lines. The statistical mathematical treatment of these functions is an extension of the Austin method given in §5.2.6.7 and the relevant papers should be consulted for further details.

Finch (1983) has offered an alternative to the two-classifiers-in-series model by the introduction of an entrainment component to the recovery to the coarse stream which is particle size dependent, i.e. $s(x) = c(x) + a'(x)$, where $a'(x)$ is assumed to be a linear function of x. This concept is logical, in that it is well known that entrainment increases with reduced particle size, but further work may reveal that the linear function is but an approximation to the true relation between entrainment and particle size for hydrocylones.

5.2.6.6 Representation of selectivity function

In order to use known mathematical expressions for the classification function $c(x)$, a relation is required between $s(x)$ and $c(x)$. Following Luckie and Austin (1974), this is given by:

$$s(x) = (1 - a' - b')\,c(x) + a' \tag{5.14}$$

Svarovsky (1977) has suggested a graphical means of finding $s(x)$ from the cumulative size distributions of, say, feed and coarse streams. Equation (5.8)

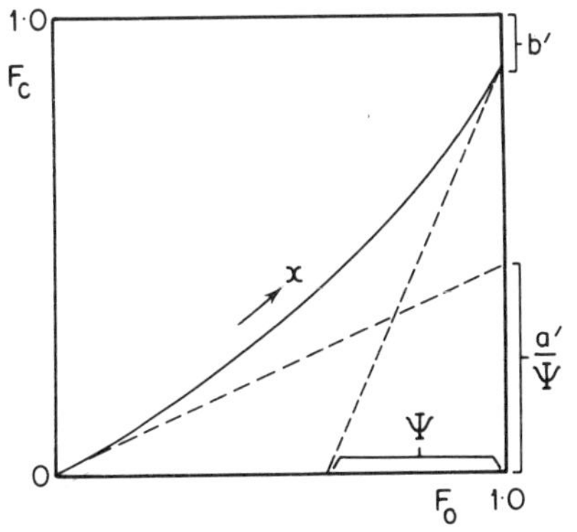

Figure 5.10 Graphical determination of selectivity function $s(x)$ from feed and coarse stream size distributions (Svarovsky, 1977). *Reproduced by permission of Butterworths, London*

applies, giving:

$$s(x) = \Psi dF_c/dF_o$$

Now the maximum value of $s(x)$ from Figure 5.7 is $(1 - b')$, corresponding to $x \to \infty$, giving:

$$dF_c/dF_o = (1 - b')/\Psi$$

Similarly the minimum value of $s(x)$ is a', corresponding to $x \to 0$, giving:

$$dF_c/dF_o = a'/\Psi$$

Hence, it F_c is plotted against F_o over a range of values of x, the result is Figure 5.10. The limiting values of the slope of the curve can be immediately inserted as asympototes as indicated, simplifying the location of the curve through a set of points which often show considerable experimental scatter.

5.2.6.7 *Austin method of analysis of classifier data*

In order to accommodate the problems summarized in §5.2.6.1, Luckie and Austin (1974) have proposed the following method of analysis, geared particularly to the hydrocyclone:

(1) Feed, coarse and fine streams are simultaneously sampled for size distribution and solids concentration.
(2) Because of measurement errors, discussed further in §5.2.7.2, and to effect a mass balance over the system for each size fraction, a reconstituted feed size distribution $F_o'(x_i)$ is obtained by using equation (5.24), with $E_i = 0$,

to give:

$$F'_o(x_i) = \frac{C'}{1+C'} F_c(x_i) + \frac{1}{1+C'} F_f(x_i) \tag{5.15}$$

with $n > i > 1$, where there are n size divisions in, say, $1/\sqrt{2}$ intervals, and where C' is the circulation ratio, i.e. the ratio of mass flow rates of fine, coarse and feed $= 1{:}C'{:}(1 + C')$.

(3) Using equation (5.15) with now zero error, and summing over the range of i values, C' can be calculated from:

$$C' = \frac{\sum F_f(x_i) - \sum F'_o(x_i)}{\sum F'_o(x_i) - \sum F_c(x_i)} \tag{5.16}$$

(4) Estimated selectivity values $\hat{s}(x)$ are obtained from

$$\hat{s}(x_i) = \frac{C'}{1+C'} \frac{f_c(x_i)}{f'_o(x_i)} \tag{5.17}$$

where $f_o(x_i)$ is the mass fraction of the coarse stream less than x_i, and $f'_o(x_i)$ is the mass fraction of the reconstituted feed less than x_i.

At this stage, from (1)–(4) it is possible to tabulate data for the following estimated parameters over a range of particle sizes: $F_o(x)$, $F'_o(x)$, $F_c(x)$, $F_f(x)$ and $\hat{s}(x)$.

(5) Assuming all particles have the same specific gravity, and that the classifier is designed to minimize mutual interference of coarse and fine streams, if the probability of a particle appearing in the coarse stream for a given set of conditions is P, then the probability of a particle taken at random from the total population of particles appearing in the coarse stream is $P(1-P)$. Further, the probability of finding r of them in the coarse stream if there are n of them in the feed is given by the binomial distribution in the form:

$$B\left[r(x)\right] = \left[\frac{n(x)}{r(x)}\right] P(x)^{r(x)} \left\{1 - P(x)\right\}^{n(x)-r(x)} \tag{5.18}$$

where $P(x)$ is the probability of particles of size x reporting to the coarse stream for a given set of conditions, and corresponds to $s(x)$. Hence the selectivity function can be estimated by maximizing the likelihood function of the binomial distribution, defined as:

$$\lambda = \sum \ln B[r(x)]$$

Maximization is given by equating to zero the partial derivatives of this function with respect to the parameters to be maximized, i.e.:

$$\frac{\partial \lambda}{\partial a'} = \sum n \left\{ \frac{\hat{s}(x) - s(x)}{s(x)[1 - s(x)]} \right\} \frac{\partial s(x)}{\partial a'} = 0 \tag{5.19}$$

where $\hat{s}(x)$ is the estimated selectivity value from equation (5.17).

Expansion by the Taylor Maclaurin thorem is necessary to obtain a set of soluble simultaneous linear equations.

(6) This set of simultaneous equations is combined with an equation of the form of equation (5.12), embodying both size and distribution moduli, but here expressed as:

$$Y = \alpha + \beta X \tag{5.20}$$

where Y is the inverse of the classification function $= 1/c(x)$, α is the size modulus parameter, and β is the slope of the straight line expressed by equation (5.20) and gives the distribution modulus.

(7) The relation between $s(x)$ and $c(x)$ given in equation (5.14) is then introduced, involving the parameters a' and b', for which estimated values must be initially assumed.

(8) Finally, a mathematical expression for $c(x)$ is adopted as indicated in §5.2.4, such as an RRB or a log–normal distribution function.

(9) All this results in a set of simultaneous equations which contain all the variables. These are continuously solved by computer means until the differences between two successive sets of parameters are less than some prescribed amount, that is, until convergence to unique values of a', b', α and β is attained. Then valid values of $s(x)$ and $c(x)$ can be assessed for each size range of x.

Examples are given in Luckie and Austin (1974) which should be consulted for details.

5.2.6.8 *Alternatives to the Austin method*

It has been attempted to estimate the parameters a' and b' by independent means. For example, in a hydrocyclone context, Lynch (1970) assumed that a', as a limiting value, is defined by the percentage feed water in the coarse stream (underflow). However, Austin and Klimpel (1981) and Klimpel (1982) have demonstrated that water split without the additive XFS 4274 is at a value greater than the true value of a', and with the additive at a value less than a'. The reason advanced is that, with slurries, a relatively high solids content creates a higher concentration at the wall than at the centre of the hydrocyclone, resulting in a hindered movement condition. It appears that some fraction of all sizes of particles becomes locked in the layer and leaves in the underflow. Thus there is no reason why this fraction should equal the water split, but in some circumstances it provides a reasonable approximation to it (Rogers, 1982).

Another assumption has been that $c(x)$ values for fine sizes approximate to zero and then the parameter a' is estimated as $s(x)$. However, Luckie and Austin (1974) quote examples to show that this is not always true.

Hence the method of §5.2.6.7 must be generally preferred, in which all the variables are included in all the equations and a statistical computer exercise is applied for the estimation of these variables. This method, though complex, is clearly more accurate than others proposed so far.

5.2.7 Stochastic error assessment

5.2.7.1 Survey

Four approaches to the analysis of errors in classification are recognized:

(1) mass balance for each size in the distributions leading to reconstitution of feed;
(2) analysis assuming normal distribution of errors;
(3) analysis involving other error structures, such as rectangular, or double exponential form;
(4) identification of stream combinations which are likely to produce least errors.

5.2.7.2 Size analysis errors

Assuming that no errors exist, the mass balance expressed in equation (5.2) may be divided throughout by M_f, to give:

$$\frac{M_o}{M_f}\frac{d}{dx}F_o(x)=\frac{M_c}{M_f}\frac{d}{dx}F_c(x)+\frac{d}{dx}F_f(x) \tag{5.21}$$

Defining $C' = M_c/M_f$, noting $(1 + C') = M_o/M_f$ and integrating, equation (5.21) becomes:

$$(1+C')F_o(x)=C'F_c(x)+F_f(x) \tag{5.22}$$

Similarly it may be shown that:

$$(1+C')f_o(x)=C'f_c(x)+f_f(x) \tag{5.23}$$

These equations will not be true balances in practice because of errors, and hence, considering each size fraction, there must be written:

$$(1+C')F_o(x_i)=C'F_c(x_i)+F_f(x_i)+\mathrm{E}_i \tag{5.24}$$

and

$$(1+C')f_o(x_i)=C'f_c(x_i)+f_f(x_i)+\varepsilon_i \tag{5.25}$$

where ε_i for one size fraction is independent of that for another size fraction but will all be of similar magnitude. E_i, however, will be the sum of a series of ε values and hence will not possess equal variance or be statistically independent, which is important to notice.

One method of handling the problem is to assume that the coarse and fine streams are more accurately sized than the feed, that is that ε resides chiefly in the feed distribution, and use equation (5.24) to obtain a reconstituted feed distribution $F'_o(x_i)$, as exemplified in equation (5.15) of the Austin method of analysis of classifier data. (see §5.2.6.7.)

Collecting terms in C' in equation (5.25) results in the following expression:

$$f_o(x_i) - f_f(x_i) = C' \, [f_c(x_i) - f_o(x_i)] + \varepsilon_i \tag{5.26}$$

ε_i can be reduced to zero by replacing $f_o(x_i)$ by $f'_o(x_i)$.

5.2.7.3 *Assumption of normal distribution of errors*

The standard approach to the handling of variability in data is to formulate the estimation by the statistical least squares regression method (Draper and Smith, 1981). This implies the basic assumption that each component of error is statistically independent of the ε of other components and is normally distributed according to equation (4.15), with a zero mean and constant variance σ_ε^2. Mass ratio parameters are then chosen to fulfil the condition of:

$$\text{minimized} \sum_{i=1}^{n} \varepsilon_i^2 \tag{5.27}$$

Klimpel (1980) shows that this can be done, for example, in respect of the circulation ratio C' by rearranging equation (5.26) into a form for which condition (5.27) is satisfied, namely:

$$\begin{aligned} C' &= \sum_{i=1}^{n} \Big[f_o(x_i) - f_f(x_i) \Big] \Big[f_c(x_i) - f_o(x_i) \Big] \Big/ \sum_{i=1}^{n} \Big[f_c(x_i) - f_o(x_i) \Big]^2 \\ &= \sum_{i=1}^{n} \Big[\gamma \Big] \Big[\zeta \Big] \Big/ \sum_{i=1}^{n} \Big[\zeta \Big]^2 \quad \text{for convenience} \end{aligned} \tag{5.28}$$

A hidden assumption in equation (5.28) is that the experimental error is associated with the γ component while ζ is error-free. This is never achieved, of course, in practice but it does imply that, since $f_o(x_i)$ is common to both γ and ζ, $f_c(x_i)$ is more accurately assessed than $f_f(x_i)$, which may or may not be the case. An approximate indication of the sensitivity of the calculated mass ratios can be obtained, however, from the formula:

$$\text{Standard error } (C') = \frac{\sigma_\gamma}{[\sum(\zeta - \bar{\zeta})^2]^{1/2}} \tag{5.29}$$

where $\bar{\zeta}$ indicates the mean value over the n components and σ_γ is an independent measure of the variance associated with the experimental technique.

In view of the indication in §5.2.7.2 that E_i values for cumulative distributions are not independent measures of variance, they should not be used without suitable, but complex, corrections. It is much simpler to work with density distributions for this reason.

5.2.7.4 *Error structures other than those with normal distribution*

Klimpel (1980) has found that the assumption of normal distribution of errors is frequently not justified, and that one of two other error structures may prevail, having either a double exponential or a rectangular form as shown in Figure 5.11. These allow for a broader spread of errors than the normal curve.

Using the statistical concept of maximum likelihood (Draper and Smith, 1981) it can be shown that the most efficient and least-biased criterion of mass ratio estimation is as follows:

Method	*Criterion*
Double exponential	Least absolute sum, i.e. minimize $\sum_{i=1}^{n} \lvert \varepsilon_i \rvert$
Rectangular	Minimax, i.e. minimize $[\text{maximum} \lvert \varepsilon_i \rvert]$

Klimpel (1973) illustrates the use of these criteria in general-purpose curve fitting. One advantage of these two criteria over the least-squares criterion is that they are not as strongly influenced by the presence of errors in the ζ component which may be regarded as the independent variable in equation (5.28). The choice of which criterion to adopt depends on the underlying error structure in any particular situation. Klimpel (1980) proposes that this is determined by the following method. Let r replicated samples be taken of each stream into and from a classifier in as short a time as possible. Then, after the components of each replicated sample of each stream are analysed, several estimates of the components $f_o(x_i)$, for example, are available and average $f_o(x_i)$ values can be calculated. At this point, a series of plots (one for each component i) can be made of the r residuals, $\varepsilon_i = [f_o(x_i) - \bar{f}_o(x_i)]$ against frequency of occurrence. The shape of these curves matched against the curves of Figure 5.11 will indicate the criterion to be chosen.

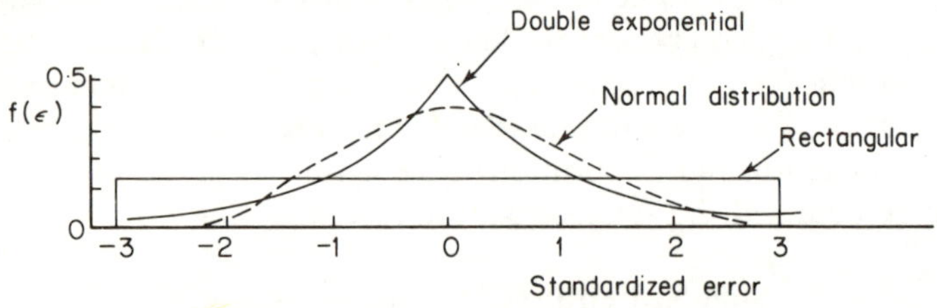

Figure 5.11. Illustration of error distribution functions (Klimpel, 1980). *Reproduced by permission of the American Institute of Mining, Metallurgical and Petroleum Engineers*

5.2.7.5 *Stream selection considerations*

Assuming that instrument calibration is not at fault, i.e. that systematic errors can be ignored, there still remain random errors in all classification measurements, whether of yield or of the classification and selectivity functions as illustrated by Leschonski (1981). These errors can be estimated theoretically for a function ϕ from the standard deviations of the independent variables y and z determined from repeated assessments, using the equation (Valikanov, 1965):

$$\sigma_\phi^2 = \left(\frac{\partial \phi}{\partial y}\right)_\mu^2 \sigma_y^2 + \left(\frac{\partial \phi}{\partial z}\right)^2 \sigma_z^2 \tag{5.30}$$

where the suffix μ indicates that the partial derivatives are evaluated close to the mean value of ϕ.

Svarovsky (1977) applies equation (5.30) to the case of the yield from a hydrocyclone, for which equation (5.5) is applicable. Taking each form of equation (5.5) in turn, the following expressions are obtained for the standard deviation of Ψ:

For coarse and feed

$$\sigma_\Psi = \bar{\Psi}\left[\left(\frac{\sigma_{C_c}}{C_c}\right)^2 + \left(\frac{\sigma_{V_c}}{V_c}\right)^2 + \left(\frac{\sigma_{C_o}}{C_o}\right)^2 + \left(\frac{\sigma_{V_o}}{V_o}\right)^2\right]^{1/2} \tag{5.31}$$

For fine and feed

$$\sigma_\Psi = (1 - \bar{\Psi})\left[\left(\frac{\sigma_{C_f}}{C_f}\right)^2 + \left(\frac{\sigma_{V_f}}{V_f}\right)^2 + \left(\frac{\sigma_{C_o}}{C_o}\right)^2 + \left(\frac{\sigma_{V_o}}{V_o}\right)^2\right]^{1/2} \tag{5.32}$$

For coarse and fine

$$\sigma_\Psi = \bar{\Psi}(1 - \bar{\Psi})\left[\left(\frac{\sigma_{C_c}}{C_c}\right)^2 + \left(\frac{\sigma_{V_c}}{V_c}\right)^2 + \left(\frac{\sigma_{C_f}}{C_f}\right)^2 + \left(\frac{\sigma_{V_f}}{V_f}\right)^2\right]^{1/2} \tag{5.33}$$

Svarovsky (1977) points out that, assuming the same value K inside the square roots, i.e. the same accuracy of the measured variables, the standard deviation of Ψ will be different in each case, as seen in Figure 5.12.

It is deduced that the errors from using fine plus coarse are the least errors over the whole range of possible values of $\bar{\Psi}$, with the advantage being most

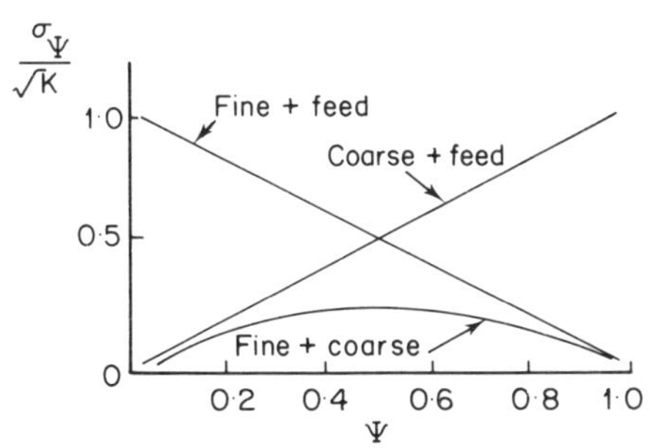

Figure 5.12. Random errors in the determination of yield (Svarovsky, 1977). *Reproduced by permission of Butterworths, London*

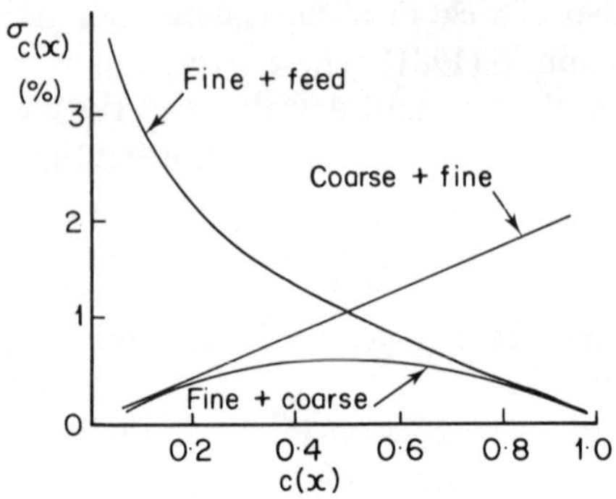

Figure 5.13. Example of random errors in the classification function $c(x)$ determination (Svarovsky, 1977). *Reproduced by permission of Butterworths, London*

marked for $\overline{\Psi} = 0.5$. For highly efficient separation, fine plus feed streams give comparable errors to those from fine plus coarse.

Svarovsky (1977) obtains a very similar graph to Figure 5.11 for the standard deviation of the classification function $c(x)$ if the new value of K' (inside the square root) is independent of particle size. However, in practice, particle size measurement becomes more difficult with decreasing size, which skews the curve, as shown in Figure 5.13. This does not, however, affect the deductions at the end of the previous paragraph, which therefore remain true for the classification function $c(x)$ as for Ψ.

5.3 CLASSIFICATION METHODS

5.3.1 Preview

The main division of classification methods is by operating principles, with subsidiary division into wet or dry processes. Indications are also given as appropriate whether a device is suitable for laboratory use or for full-scale commercial operation.

Sieves and screens are first considered because they need not involve a particle carrying medium, as do the remaining systems. Hence the chapter is arranged as follows:

(1) Sieves and screens in §5.3.2
(2) Crossflow system in §5.3.3
(3) Counterflow gravity systems (elutriation) in §5.3.4
(4) Inertia systemss in §5.3.5
(5) Centrifugal systems without moving parts in §5.3.6
(6) Centrifugal systems with rotating walls in §5.3.7
(7) Mechanical rotor systems in §5.3.8
(8) Combined systems in §5.3.9

In all of the above, because of the multitude of variations in each category, typical examples only are taken which reveal the process functions. Further, more emphasis is placed on examples associated with comminution than with other unit operations such as gas cleaning for dry systems or leaching and dewatering for wet systems.

5.3.2 Sieves and screens

5.3.2.1 Basic principles

(i) Generally, sieves relate to laboratory batch processes whereas screens belong to industrial continuous processes.

In each, mechanical partition of particles is effected by means of a surface containing uniformly sized and spaced apertures, round, square or rectangular in shape. Sufficiently small particles, at least in two dimensions, fall through the openings, but larger ones are retained on the surface. Because the bed of particles to be separated has a finite thickness, means (usually shaking or vibration) must be found to assist the movement of the particles to the surface to have an opportunity of acceptance or rejection. Thus the mathematics of probability is relevant to sieving and screening.

(ii) What appears to be so simple from a practical standpoint is extremely complex from a process standpoint. An attempt is made to highlight the main problems and present some solutions. For example, the effect of time on the size distribution which passes through a sieve or a screen is shown in Figure 5.14 for a single size fraction (Whitby, 1959) and in Figure 5.15 for a complete distribution (Heywood, 1945). Thus there can never be finality in the production of a definite distribution, but only an approximation to it. Hence also a sufficient time must be allowed to obtain a close enough approximation. A good practical end-point is when the residue per unit time does not change by more than 0.1% per minute (Leschonski, 1979).

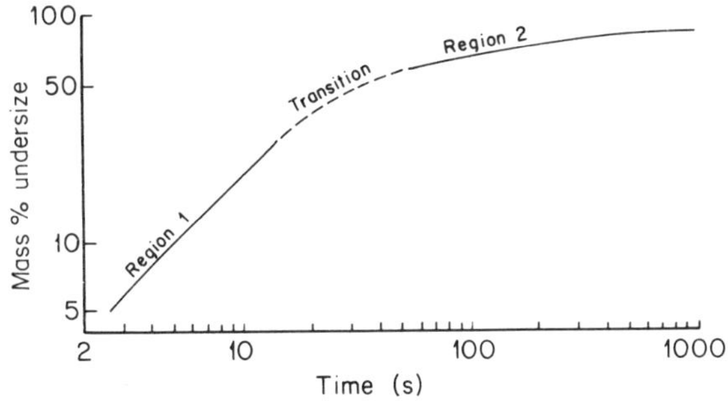

Figure 5.14. Mass percentage passing a given sieve as a function of time (Whitby, 1959). *Reprinted, with permission. Copyright ASTM*

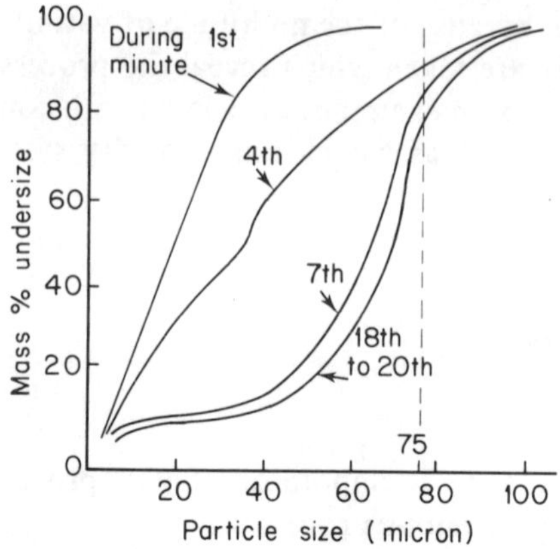

Figure 5.15. Variation of the cumulative size distribution with sieving time (Heywood, 1945). *Reproduced by permission of the Institution of Mining and Metallurgy*

(iii) Typical size distributions of feed, fine and coarse fractions from a screening process are shown in Figure 5.16 for the cumulative mode and Figure 5.17 for the weighted density mode. Clearly, because of the screen barrier, no misplaced coarse particles should find their way into the fine stream, but misplaced fines are easily carried into the coarse stream. However, Heywood (1945) has shown that the maximum particle size x_{fM} that just passes a square mesh of side W is about $1.4W$, exemplified in a particle of relatively small thickness and having a width of the order of the diagonal of the square mesh. As noted in §4.2.4, the dimension of length is largely lost in sieving.

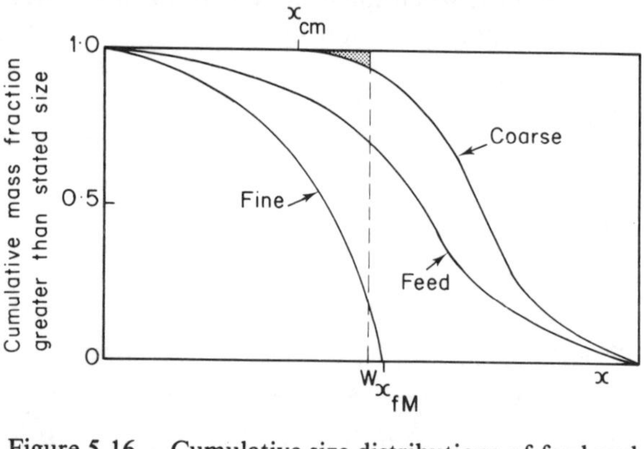

Figure 5.16. Cumulative size distributions of feed and of coarse and fine fractions after passing through a square mesh of screen aperture W. Shaded area represents fines remaining in coarse fraction

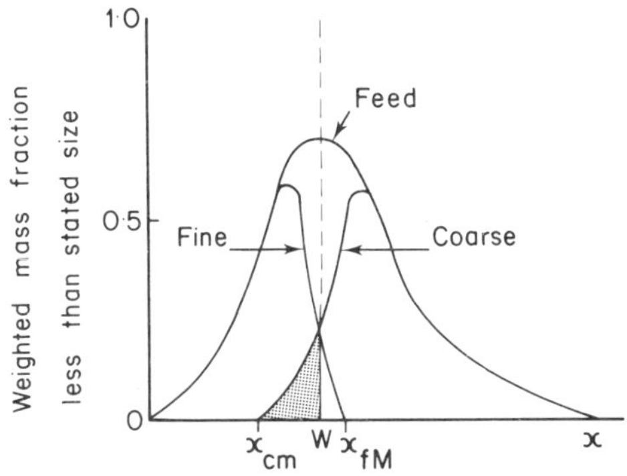

Figure 5.17. Weighted density distributions of feed and of coarse and fine fractions after passing through a square mesh screen of aperture W. Shaded area represents fines remaining in coarse fraction

(iv) A look at Figure 5.17 shows that a classification (Tromp) curve can be drawn for the sieving or screening process and that the d_{50} and d_a concepts apply as for other partition methods. (See §5.3.2.3.) The slope of the classification curve at d_{50} is usually steeper than for other partition methods because of the relatively high separation performance of a screen.

(v) Since coarse particles may become coated with fines under certain conditions of sieving, a selection curve with a fish-hook pattern is possible, as discussed in §5.3.2.6.

5.3.2.2 Empirical design procedure

Manufacturers have designed screens on an experience-determined base capacity per unit screen area for each particular type of equipment, adjusted by a range of factors, which may include bulk density of feed, free area of mesh, percentage greater than mesh size in the feed, aperture shape, slope of screen, moisture in feed, and so on (Gluck, 1965). Wilkinson (1971) exemplifies the appropriate factors for equipment operating on sinter, and Colman and Tyler (1978) for crushed ore using a vibratory screen. Matthews (1972) provides design data adapted from Hewitt Robins Inc., and Foreman (1979) has useful information relative to the screening of coal. Schmidt (1981) gives tabulated data for a range of materials including coal and coke. Again, Nichols (1982) has discussed the Allis Chalmers calculation of screen area A, given by:

$$\mathrm{A} = M_o/EQ_1Q_2Q_3$$

where M_o is the required capacity, E is the empirically found capacity per unit area of screen for base conditions, which are defined, Q_1 is an oversize correction factor (graph given), Q_2 is a half-size correction factor (graph given), and Q_3 is the product of all other factors (tabulated).

It might be thought that screen design would be easy with such a breadth of information available; but it must be used with great caution because considerable judgement based on experience of particular types of plant needs to be exercised in applying the factors to a specific problem.

Hence the screening process is best simulated by mathematical models discussed in §§5.3.2.5, 5.3.2.6 and 5.3.2.7, which give an insight into what is really happening in any one particular case.

5.3.2.3 Definition of separation size

Most often the separation size is assumed to be the mesh opening W.

Brereton (1975b) has indicated that d_{50} is more representative of actual screening performance than W.

Leschonski (1979) has for twenty years used an analytical cut size d_a proposed by Andreasen (1927) and by Andersen (1931). When sieving has proceeded to the agreed end-point—for example, when the increase in residue is less than 0.1% per minute—each sieve in a stack is carefully removed. Gentle tapping of the frame removes the particles that are held in the mesh, which could have fallen through the mesh if sieving had continued. These particles therefore characterize the cut size at the time of cessation of sieving. The particles thus collected fall into a very narrow size range and the average (say by average equivalent volume diameter) is taken to be the cut size.

The relative positions of W, d_{50} and d_a vary in practice, though they are often quite close to each other.

5.3.2.4 Criteria of performance

The usually accepted definition of screen performance based on the concept of recovery is:

$$\frac{\text{Mass of fines below the separation size}}{\text{Mass of feed below the separation size}}$$

The value obtained by this definition clearly depends on which separation size is adopted and on the accuracy with which the denominator is determined—for example, by careful sieving of a sample of the feed through a mesh of the same size as the screen.

More complex expressions for screen performance have been proffered, as summarized by Leonard (1974); but Brereton (1975a) has demonstrated the superiority of systems based on the probability of a particle passing through a screen aperture, leading to the establishment of classification functions, as discussed in §5.3.2.5. Then sharpness of cut, related to the slope of the classification curve at d_{50}, is the criterion of performance, as for other separation methods.

5.3.2.5 *Probability considerations*

(a) Ideal conditions For unrestricted passage of a sphere of diameter x ($x \leqslant W$) through a square aperture of side W formed by material of thickness T (see Figure 5.18), the probability P of passage through the screen at one presentation is given by Gaudin (1939) as:

$$P = (W - x)^2/(W + T)^2 \tag{5.34}$$

Harris (1961) reasons that, because the probability of failure at one attempt is $(1 - P)$, the probability of failure after n attempts is $(1 - P)^n$, and hence the probability of success after n attempts is given by:

$$P_n = (1 - P)^n$$

or

$$P_n = 1 - \left[1 - \left(\frac{1 - x/W}{1 + T/W}\right)^2\right]^n \tag{5.35}$$

Though based on ideal assumptions, equation (5.35) is valuable for the assessment of the minimum number of presentations necessary to reach a probability of, say, 95% success. Taking $T/W = 0.2$ and $x/W = 0.9$ as typical ratios, $P_n = 1 - (0.8227)^n$. Plotting P_n against n, as in Figure 5.18, it is clear that at least 16 unrestricted presentations would be necessary.

This calculation underlines the need for two practical screen features:

(1) Bed agitation to bring successive layers of material to the surface. In fact, it is good practice to effect agitation in such a way that stratification occurs, with smaller particles lying nearer the surface than larger—see Schmidt (1981), Foreman (1979) and Matthews (1972).
(2) Sufficient screen length must be allowed to provide ample opportunity for particle presentation to the surface.

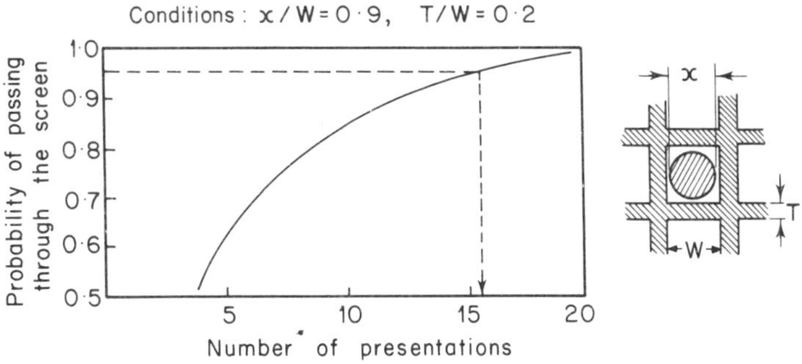

Figure 5.18. Probability of a particle passing a square mesh, based on equation (5.35)

(b) Practical conditions In practice, particles are not spheres nor is their passage through a mesh unhindered, and hence equation (5.35), as it stands, is inadequate for assessment of screen performance of design. Ferrara and Preti (1975) and Brereton and Dymott (1974) use equation (5.35) to derive kinetic equations for two distinct regimes: the sparse regime in which no particle interaction occurs, and the crowded regime in which there are interactions, but a fresh supply of undersize particles is always available at the screen surface. Hence the presentation of particles to the surface is taken to be constant for the crowded regime, but it is assumed to decrease with each vibration for the sparse regime. Hence different equations apply in each regime. These kinetic models have the following shortcomings, however:

(1) It is difficult to measure screen performance along the length of the screen— a necessary step in the development of the kinetic data.
(2) Active screening processes include the diffusion of particles through the bed to the surface, for which factors are not included by either Ferrara and Preti or by Brereton and Dymott.

Whiten (1972), in a crusher context, also starts with equation (5.35) and takes the number of presentations n at the surface to be a function of screen length L and screen load, for which a factor ξ is derived. Then:

$$n = K_2^2 L \xi \tag{5.36}$$

In addition, he includes a fines factor K_3, which allows for the bypassing of some of the material at the bottom end of the feed size distribution directly to oversize. A single factor K_3 is inadequate because adherence of fine particles to the oversize produces a fish-hook type of selection curve as shown in Figure 5.9 and this requires special treatment, as discussed in §5.3.2.6.

Attempts have been made to include factors for the angle of inclination of the screen and the corresponding collision of particles with screen wires (Brereton, 1975a). English (1974) has gone further by including in the probability equation a term for the blinding of the screen, discussed at length by Rose (1976). However, in practice, blinding is largely prevented, either by heated decks or by a deck of bouncing rubber balls housed below the screen, for which there is an optimum load as given by Stairmand (1973), or by rotating brushes (Schmidt, 1981) or by discreet selection of the screen vibration mode (Blachetta, 1984). Blinding, the gradual build-up of layers of fines to block the screen openings, must be differentiated from pegging, which is blockage by larger pieces of material becoming wedged in a mesh opening.

Batterham *et al.* (1980) have objected to equation (5.35) on the grounds that the probability approaches very abruptly to the upper limit set by $x = W$ and have preferred an equation developed by Mular and Herbst (1978) from probability considerations of a particle appearing above or below a screen surface, to produce a classification function given by:

$$1/c(x) = 1 + \exp[(d^3{}_{50} - x^3)/\Lambda] \tag{5.37}$$

where Λ is a slope parameter. As derived, equation (5.37) applies to a narrow size range and it was therefore necessary to integrate the classification function over the applicable range of x. To assist in this integration, Batterham found that the following close approximation to equation (5.37) was useful:

$$1/c(x) = 1 + \exp[(d_{50} - x)/\Lambda] \tag{5.38}$$

with Λ inversely related to the slope of the classification curve at d_{50} and independent of d_{50}.

The integration between x_1 and x_2 gave:

$$c(x) = 1 + \frac{\Lambda}{x_2 - x_1} \ln\left\{\frac{1 + \exp[(d_{50} - x_2)/\Lambda]}{1 + \exp[(d_{50} - x_1)/\Lambda]}\right\} \tag{5.39}$$

The effect of particle shape on the probability of a particle passing through a screen has been studied by Nakajima *et al.* (1979). They concluded that the probability could be safely taken as independent of particle length. A shape factor $\theta = \tan^{-1}(W^*/T^*)$ was used to incorporate particle width and thickness, and a term in the form of $\cos^4(\theta - \pi/8)$ was introduced into the probability function to account for the slower movement down the screen of flatter particles compared with those of more granular shape. The screen performance calculation based on the shape-adjusted probability function gave improved accuracy of representation over that without a shape function.

5.3.2.6 Selectivity function for dry screening with small additions of moisture

Batterham *et al.* (1980) have developed a model for the analysis of dry screening in the context of iron ore crushing in which limited moisture (up to 3%

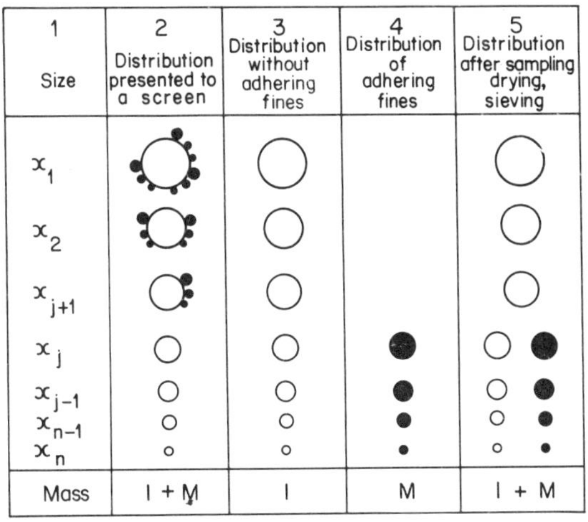

Figure 5.19. Size distributions as presented to a screen and after sampling, drying and sieving in the laboratory (Batterham *et al.*, 1980). *Reproduced by permission of CSIRO, Australia*

of feed) is added to prevent electrostatic effects, but which also causes larger particles to pass through the screen with adhering fines. Batterham found that by sieving a sample of a stream, a distribution was obtained as in column 2 of Figure 5.19. If another sample of the same stream was first thoroughly dried, detaching adhering particles, the distribution was as column 5. By mass balance, the distribution of the adhering fines was obtained as in column 4 and that for the distribution without them as in column 3. Hence the classification curve could be established and d_{50} and Λ found for insertion into equation (5.39). The method has proved satisfactory in plant operation over several years.

5.3.2.7 Selectivity function for wet screening

Rogers (1982) has found that the value of $c(x)$ given by equation (5.37) is not appropriate to the screening of coal slurries containing a high percentage of solids (>60%) but favoured the following:

$$1/c(x) = 1 + (d_{50}/x)\exp\left\{\Lambda[1 - (x/d_{50})^3]\right\} \tag{5.40}$$

This value of $c(x)$ was inserted in equation (5.14) with $b' = 0$, giving

$$s(x) = (1 - a')c(x) + a' \tag{5.41}$$

A good fit was demonstrated of actual and calculated values of $s(x)$ plotted against x for coal slurries passing through Derrick high frequency vibrating screens with woven wire cloth, and Rogers intends to continue the work by trying models proposed by Batterham *et al.* (1980), Brereton and Dymott (1974) and Molerus and Hoffmann (1969).

5.3.2.8 Types of sieving and screening equipment

Only brief references and descriptions are given as follows.

(a) Sieving equipment for laboratory use

British Standard 410 (1976), 'Specification for test sieves'
British Standard 1796 (1976), 'Methods for test sieving'
ISO/R565, 'Woven wire cloth and perforated plate in test sieves'
Orr *et al.* (1979): Automatic wet sieving

(b) Microsieves Hidaka and Miwa (1979) give the conditions for successful microsieving:

(1) The velocity of the dispersing medium through the sieve must be high.
(2) The sieve surface must be immersed in the dispersing medium to reduce surface tension effects.
(3) The particles must be fluidized by vibration to reduce resistance to passage through the sieve.

When these conditions are satisfied, microsieves give results which compare favourably with photosedimentation and the Andreasen pipette methods.

(c) Static screens The grizzly consists of parallel bars or rods with uniformly clear openings, often tapered from feed to discharge to provide self-cleaning characteristics. The bars may lie in a horizontal plane over a bin, or be inclined for use ahead of crushers, and are usually used dry.

By contrast, the sieve bend, shown in Figure 5.20, is employed for wet separation of small particles. It is constructed of horizontal wedged bars 1–5 ft in length, arranged in either a curved or a flat plane, having the narrow wedged sections outwardly placed. The gaps between the wide wedged sections are in the range 0.08–0.005 in (2–0.125 mm), through which liquid and fines may pass, leaving coarser material in the overflow. The advantages of the sieve bend are that it contains no moving parts, has low operating costs and has the ability to separate at large capacity without blinding. The addition of sieve bend screens to cyclone plant has approximately halved the vibrating screen area required in a coal preparation plant (Sokaski *et al.*, 1979).

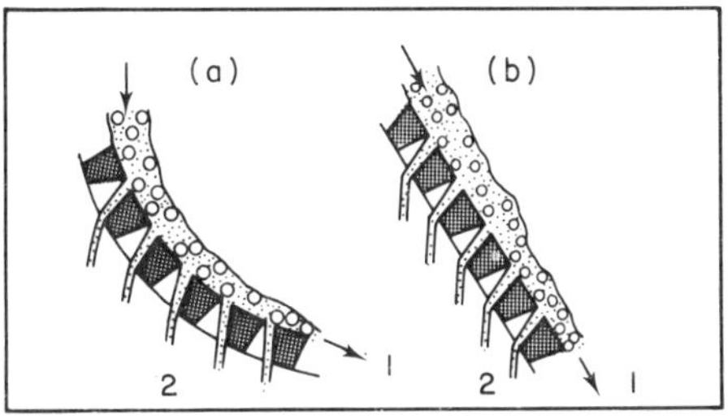

Figure 5.20. Cross-section of sieve bend: (a) curved, (b) straight (Schubert, 1980). *Reproduced by permission of DECHEMA, Frankfurt*
1 = underflow, 2 = overflow

(d) Dynamic screens Table 5.1, based largely on Matthews (1972), summarizes the main types of dynamic screens. Notice that there is a general trend for the higher frequencies of screen movement to be associated with the finer separation sizes.

Other references of interest are: Colman and Tyler (1978), Foreman (1979) —which contains a glossary of screening terms—Sokaski *et al.* (1979), and Pritchard (1980).

(e) Special features of screens Woven wire cloth is acceptable for comparatively unabrasive materials like coal and bauxite, but for ores, rubber and polyurethane panels are giving long life, but cost more initially. Hence an economic balance must be struck to decide on their use in a particular plant.

Table 5.1. Types of dynamic screens

Screen Type	Feature	Stroke	Frequency (min^{-1})	Separation size range	Wet or dry	Material type
Vibration	High-speed	Low-amplitude	3000	Ultra-fine	D	Wide range
	Inclined screen	Circular	*c*1000	Medium-fine	W or D	
	Horizontal screen	Pulsating	*c*1000	Close sizing of medium range	W or D	
Sifter	According	Circular		4–300 mesh	D	Light and bulky materials
	to	Gyratory	500–600	4–300 mesh	D	in chemical pigment
	stroke	Spiral		4–325 mesh	W or D	& food industries
Oscillating	Horizontal screen	Linear	100–400	200 mesh	D	Light and free-flowing
Reciprocating (shaking)	Horizontal screen	Substantially linear	200–300		D	Good for accurate sizing of large lumps
Centrifugal	Vertical cylindrical screen	Gyratory (1000 mm^{-1})	70	35 mesh	W or D	Coal, sand or gravel dewatering
Revolving (Trommel)	Nearly horizontal cylindrical screen	—	15–20		W or D	Sand, gravel sizing. Washing sticky materials
Vibrating grizzly	Nearly horizontal spaced bars	Linear	Low	Very coarse	D	Large rock lumps

Reproduced by permission of the Institution of Chemical Engineers.

Heated decks are employed, not to dry the material being screened, but to keep the wires sufficiently dry to reduce surface tension to a level at which blinding is reduced (Pritchard, 1980).

5.3.3 Crossflow systems

This is perhaps the oldest known method of separation, as practised by the ancients on threshing floors, where the wind carried off the chaff, leaving the grain on the floor. The large density difference between grain and chaff made this a practical success, but attempts to winnow out fines from a falling curtain of material of constant density by a cross current of air, as seen in the Szego classifier (Figure 5.21), although demonstrated in model size, proved difficult in large sizes. One reason lay in the large diameter needed to obtain a thin enough falling curtain for the winnowing to be efficient, and another in the fear of blockage of narrow ducts in practical conditions of moisture-laden materials such as coal.

A more successful application has been the classification of thin flakes, which is demanded by the fact that optimal reinforcing of plastics by mica requires flakes with an aspect ratio (equivalent diameter/average thickness) greater than 30. This has been successfully investigated by Etkin *et al.* (1980) using the Toronto Infrasizer Mark III (Etkin and Haasz, 1980). In a rectangular duct of height H, feed is injected from the top into a smoothly flowing air current. The ideal particle trajectory is a straight line of slope H/x, where x is the distance downstream of the vertical through the point of injection at

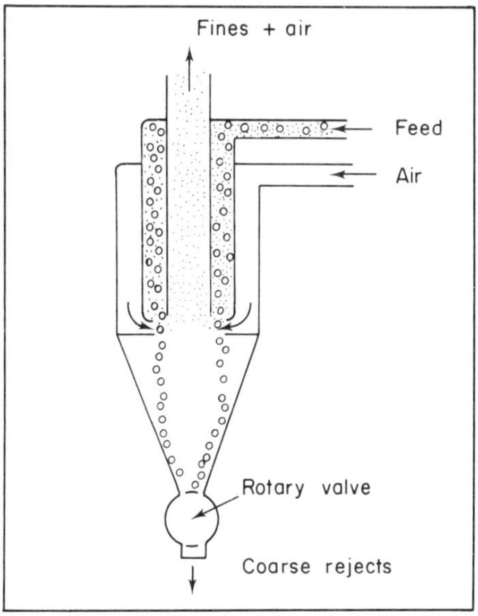

Figure 5.21. Principle of Szego classifier

which the particle lands. The floor of the duct is covered with a succession of adjacent narrow trays, each having a length equal to the duct width. Then if $\bar{D}^*$ is the mean value of the diameter and $\bar{T}^*$ the mean value of the thickness of mica circular discs that fall into one tray, from theoretical considerations of the probabilities of angles of orientation of discs, Etkin *et al.* (1980) expected the following relation:

$$\bar{D}^*\bar{T}^* = (x/H)^{-1} \tag{5.42}$$

Although randomly shaped mica flakes fell upstream of their expected location, the relationship of equation (5.42) held remarkably well. Assuming constant x for any one tray, the product $\bar{D}^*\bar{T}^*$ was therefore approximately constant. Now by sieving the contents of that tray, $\bar{D}^*$ could be constant between two given sieve sizes, and hence also $\bar{T}^*$, if their product is constant. Thus the aspect ratio $\bar{D}^*/\bar{T}^*$ is also constant, and by suitable selection of parameters those particles having an aspect ratio greater than 30 can be isolated.

Crossflow is also a useful means of feed dispersion preparatory to further classification, as exemplified in the Hukki (Larox) device discussed in §5.3.6.1(iii).

High-velocity crossflow systems as designed by Rumpf and Leschonski (1967) have been applied to cement industry practice (Quittkat, 1973).

5.3.4 Counterflow gravity systems (elutriation)

Leschonski (1981) describes several counterflow gravity classifiers, including the batch Gonell elutriator and two continuous systems: the Analysette 8, which allows exact predetermination of cut size, and the zig-zag classifier. In the latter, vortex formation leads to the acceleration of the main flow owing to reduction of effective tube cross-section. Then fines follow the main stream and coarse particles glide to the wall, falling back against the stream. Although sharpness of cut at each stage is low, by virtue of a succession of stages, a required cut size can usually be attained, even at high velocity, with its attendant turbulence. Leschonski (1981) gives a selection of references for further reading.

Elutriation classification is also inherent in the airflow pattern of some vertical spindle mills, as indicated in Figure 5.22. Just above the grinding zone there is a reduction of plug flow velocity to about 6 ft/s owing to increased cross-sectional area to airflow. Single-particle terminal velocity considerations would suggest that the maximum size x_t of coal particle of specific gravity of 1.5 that could be lifted to the classifier entry would be about 500 microns. In fact, because of particle-to-particle interaction, wall effects, departure from plug flow conditions and differing particle shapes, a whole range of sizes is lifted, as seen typically for coal in Figure 5.23. A separation does take place, however, and a coarser grade is returned to the grinding zone over the top of

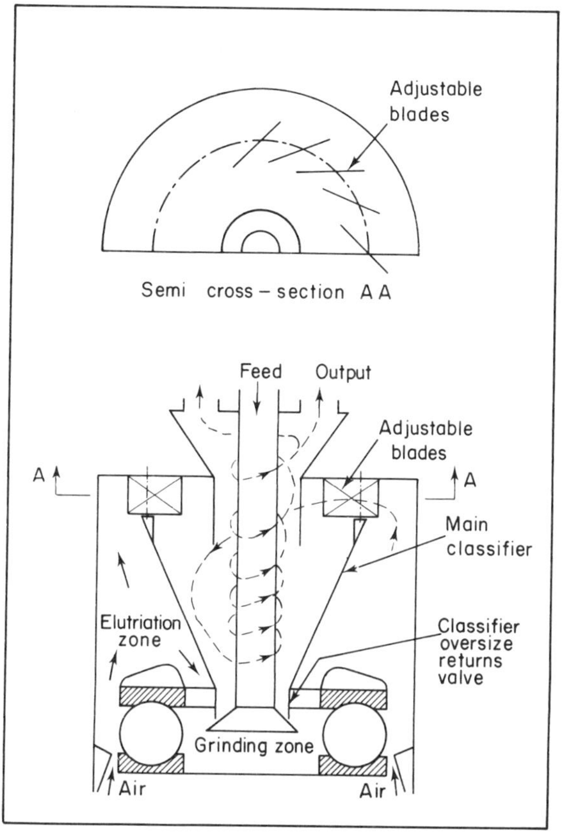

Figure 5.22. Vertical spindle mill cross-sections, showing elutriation zone in which an initial classification occurs prior to the main classification

the upper grinding ring than is lifted to the classifier, where the main partition occurs.

5.3.5 Inertia systems

The inertia system is well-exemplified in the type of classifier often fitted to hammer mills for brown coal pulverization, as illustrated in Figure 5.24(a). The particle-laden gas stream is made to turn through 180° in the region marked B. Coarse particles, by their inertia, fall into the region C, but fines are able to follow the gas to the exit D. Rastyapin *et al.* (1973) considerably improved the separation performance of the classifier by adopting the multiple louvre pattern shown in Figure 5.24(b).

In a type of inertia system seen in Figure 5.25, developed by Müller and Wessel (1963), the solids are introduced countercurrent to the airflow, and

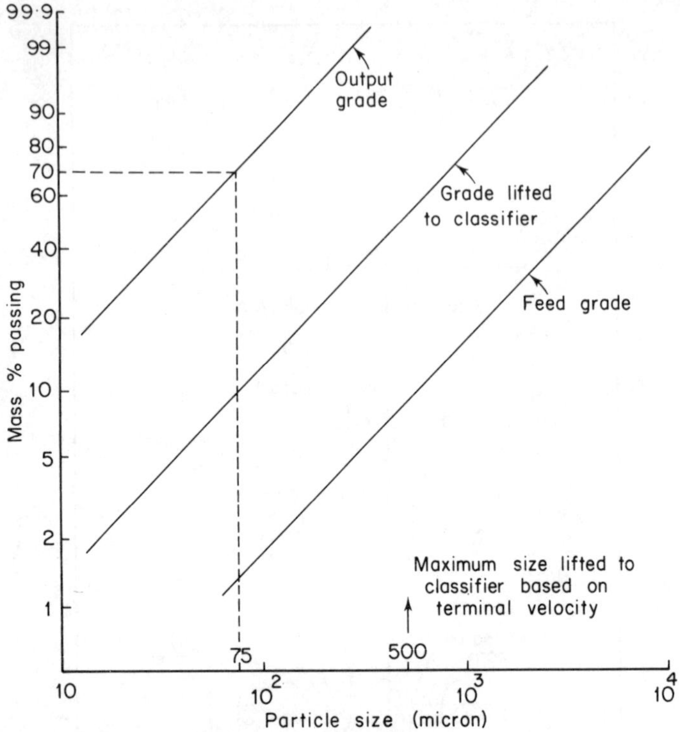

Figure 5.23. Typical RRB size distributions for coal ground in a vertical spindle mill as shown in Figure 5.22

follow a trajectory according to their mass and density and the fluid velocity. Wessel also devised an analogue computer method of calculating the trajectories followed by the particles, which is of great assistance in assessment of cut size.

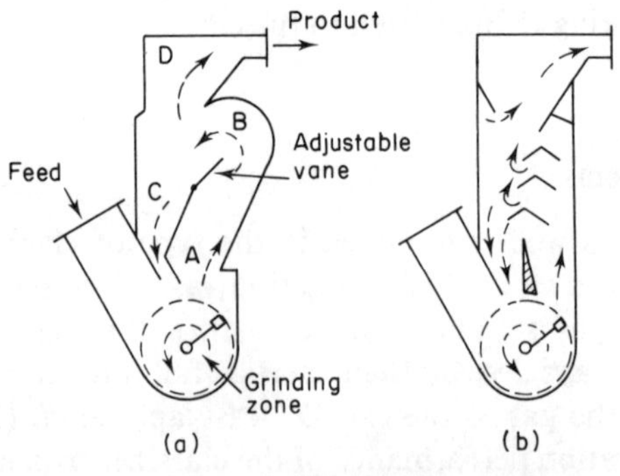

Figure 5.24. Inertia classifier as fitted to a hammer mill for brown coal grinding (Rastyapin *et al.*, 1973)

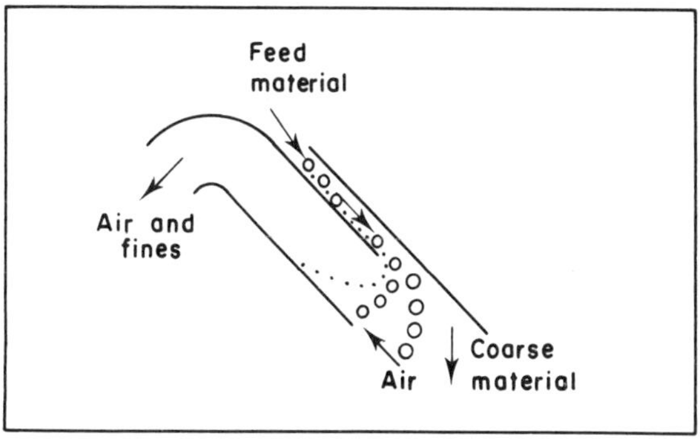

Figure 5.25. Principle of the counterflow inertia classifier (Leschonski, 1981). *Reproduced by permission of the author*

Jones *et al.* (1971) have produced a model for inertia classifiers, but no wall or flow separation effects are taken into account.

5.3.6 Centrifugal systems without moving parts (free vortex)

5.3.6.1 Air separators

(i) In the context of centrifugal systems, a distinction must be made between gas cleaning equipment, in which the aim is the removal of all solids, and classifiers in which a partition is sought at a specific cut size. Only the latter are discussed in this section.

(ii) A common high-capacity air separator in this category is the cone classifier, for which a typical cross-section is shown in Figure 5.22. To the feed is imparted a tangential component of velocity by means of blades set at an angle to the radial direction. Under the influence of centrifugal force, coarse particles are thrown to the cone and pass through the return valve back to the grinding zone, whereas particles less than cut size are carried up in the central helical spiral to the product outlet. The diameter D_o of the vortex finder is critical in the determination of the cut size and it may be assessed from equation (5.43), on the assumption that it approximates to the diameter D_c of the orbit of the cut size particle. (The principles of §5.3.6.2(ii) apply.) By equating centrifugal to drag forces on a cut size particle and linking with the geometry of the cone and blade, we obtain:

$$\frac{D_c}{D_i} = \frac{1 + 2H_i \tan \varphi / D_i}{1 + \dfrac{18 \eta H_i \tan \varphi}{\rho_s v_i \psi^2 d_{80}{}^2}} \tag{5.43}$$

The factor ψ must be determined empirically for a given classifier geometry and blade angle, but it will also change with classifier size for a given geometry and blade angle.

Equation (5.43) is, of course, the result of a very simplistic approach, as it does not take account of eddy currents which cause particle re-entrainment, or of wall effects or of particle-to-particle interactions. However, it has value in scaling from one size to another for classifiers of the same geometry as long as the variation of ψ with size has been established.

Operating on coal, with a value of $d_{80} = 100$ microns, recovery of a cone classifier may be of the order of 50% at full load, but this may rise to 80% at reduced load.

(iii) Another high-capacity classifier used in the cement industry is the Hukki (Larox) illustrated in Figure 5.26. Good dispersion, prior to main classification in a horizontal cyclone, is achieved by the feed falling across the inlet gas, allowing the coarsest particles to be removed immediately. Three means of adjustment, A, B and C, are provided, and optimal positions for a given duty are found pragmatically. In both the cone and Hukki (Larox) classifiers, sharpness of cut can be improved by a system of winnowing out fines from the coarse stream.

(iv) Lower capacity machines which offer improved sharpness of cut over the cone and Hukki (Larox) types, employ a spiral airflow in a horizontal plane, as shown in Figure 5.27, and a practical example is the Walther classifier. Leschonski (1981) gives a theoretical derivation of design formulae but demonstrates the problems involved, which are such that accurate assessment in advance cannot yet be made; only an experiment on the full-sized

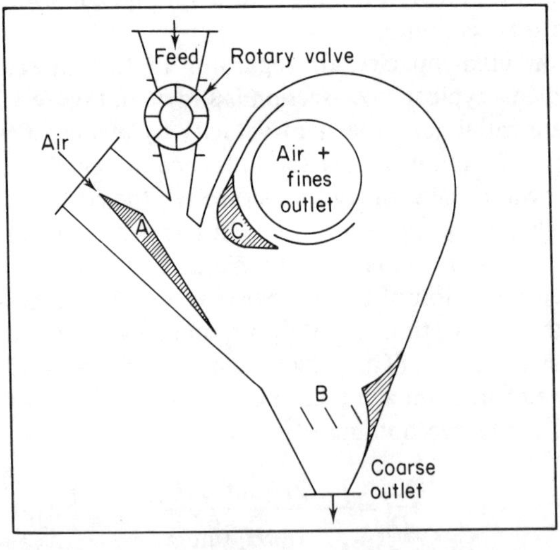

Figure 5.26. Hukki (Larox) classifier as used in the cement industry (Hukki, 1976). *Reproduced by permission of DECHEMA, Frankfurt*

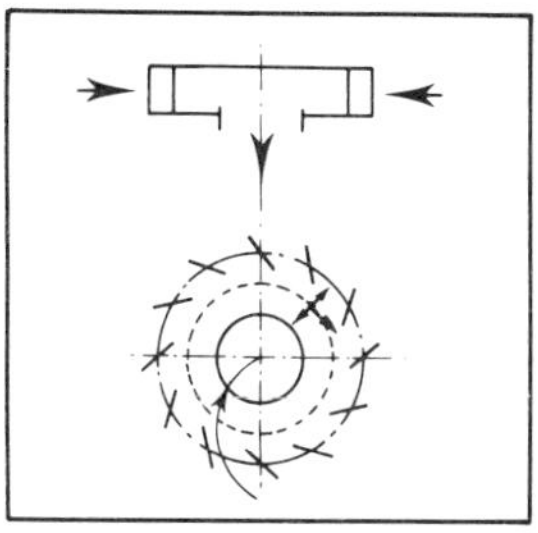

Figure 5.27. Horizontal spiral airflow classifier (Leschonski, 1981). *Reproduced by permission of the author*

equipment with a given material to be classified can safely decide cut size and sharpness of cut, because of unknowns associated with particle interaction and undesirable boundary-layer effects. However, computer programs have been developed at Clausthal University, West Germany, for particle trajectory determination with spiral flow, which is a first step towards design of classifiers of this type.

5.3.6.2 Hydrocyclones

(i) Hydrocyclones are often used as classifiers in wet systems (Lynch, 1965). Figure 5.28(a) gives a typical cross-section (Treasure, 1965) and Figure 5.28(b) the internal flow path, showing primary downward vortex and the inner secondary upward vortex. Under the influence of the primary vortex, larger particles are thrown to the cyclone wall and are carried out in the underflow. Fine particles are carried up in the secondary vortex to the overflow outlet.

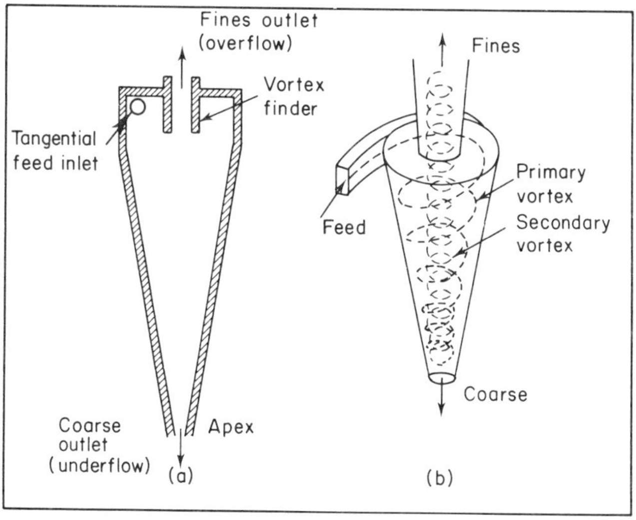

Figure 5.28. The hydrocyclone: (a) typical proportions, (b) internal vortex flow paths. *Reproduced by permission of McGraw-Hill Inc.*

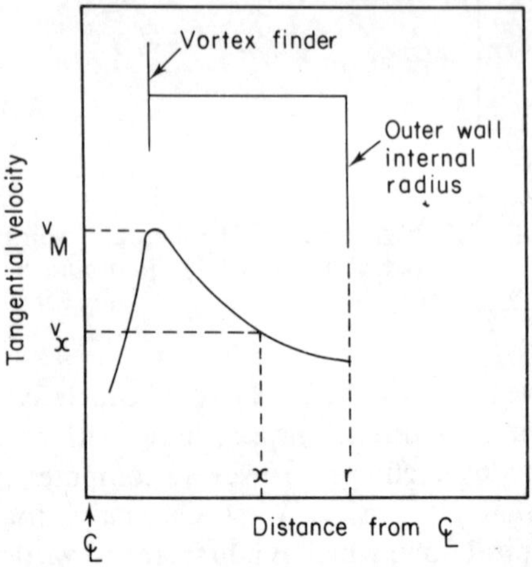

Figure 5.29. Tangential velocity pattern along a radius in a hydrocyclone (Trawinski, 1976). *Reproduced by permission of McGraw-Hill Inc.*

(ii) Figure 5.29 plots the variation of tangential velocity along a radius within the hydrocyclone. For the primary vortex the velocity is lowest at the wall and highest at the radius of the vortex finder. In the secondary vortex the tangential speed reduces as the centre-line is approached.

Taking friction forces into account, then for a free vortex:

$$v_x x^n = \text{constant}$$

where v_x is the tangential velocity at a radius x. If $n = 0.5$ (Krijgsman, 1949) for average solids content, and R is the internal cyclone wall radius, then:

$$v_x = v_R \left(\frac{R}{x}\right)^{0.5}$$

$$\omega_x = \omega_R \left(\frac{R}{x}\right)^{1.5}$$

The centrifugal accelerations at radii x and r respectively are:

$$g_x^* = x\omega_x^2$$

$$g_R^* = R\omega_R^2$$

Hence:

$$g_x^* = x\omega_R^2 \left(\frac{R}{x}\right)^3$$

$$= x\,\frac{g_R^*}{R}\left(\frac{R}{x}\right)^3$$

$$= g_R^* \left(\frac{R}{x}\right)^2 \qquad (5.44)$$

Equation (5.44) indicates that the secondary vortex has the highest centrifugal acceleration and therefore determines the cut size (Trawinski, 1976)

(iii) By equating centrifugal and drag forces at cut size orbit radius, and taking account of geometrical relationships, Trawinski (1976) developed the following formula for cut size d':

$$d' \triangleq \left[\frac{\eta}{g(\rho_5 - \rho_e)}\right]^{0.5} (Z)^{0.5} \frac{R^{0.5}}{\Delta p^{0.25}} \tag{5.45}$$

The first term collects together the characteristic data of the suspension and the second those of the particular geometrical ratios chosen in physical design. The third term is all-important because it shows that, for a given expenditure of energy (Δp limit), the cut size is a function of the cyclone diameter. Hence fine separation demands small diameters, and to obtain the required capacity, multiple cyclones in parallel are necessary. Similarly for large cut size, large-diameter cyclones are required. The third term also has application to the modelling of cyclones. For constant suspension conditions, geometrical ratios of cyclones and particle cut size, then $R \propto \sqrt{\Delta p}$. Hence, when operating a cyclone of half the full-scale diameter, only one quarter of the full-scale pressure loss should be used to maintain similarity.

(iv) Two of the problems associated with hydrocyclones have been discussed already: in §5.2.6.2 the bypass effect, and in §5.2.6.8 the attendant inaccuracy of the water split in the estimation of a'. A third problem is the design of the apex discharge which has a considerable effect on the selectivity curve.

Too much throttling of the underflow results in a rope discharge and a selectivity curve as in Figure 5.30(a). An umbrella discharge, caused by dilution of the underflow, however, gives a selectivity curve as in Figure 5.30(b). To control these features, specially designed and exchangeable apex 'stoppers' are available (Wills, 1980).

There also exist within the cyclone certain flows which are not indicated in Figure 5.28(b). One of these carries unclassified material across the top of the cyclone interior to the overflow and the other carries coarse material from the outer walls to the underflow (Plitt, 1976). Clearly these features reduce separation effectiveness. Kelsall and Holmes (1960) demonstrated that tangential

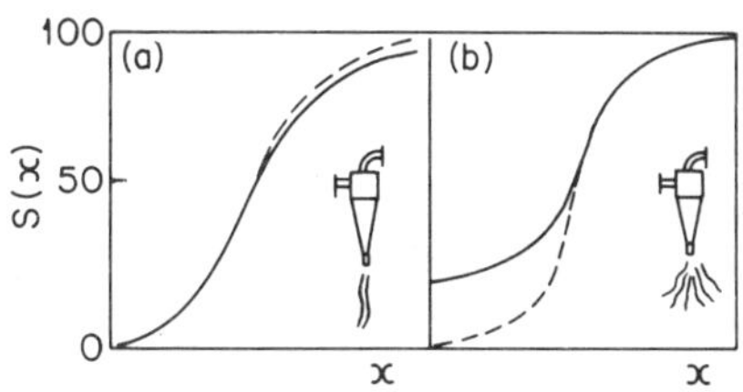

Figure 5.30. Influence of underflow nozzle throttling on the selectivity curve for a hydrocyclone: (a) rope discharge, (b) umbrella discharge (Trawinski, 1976). *Reproduced by permission of McGraw-Hill Inc.*

water injection in the apex area proved advantageous in the prevention of fines in the spaces between coarse particles discharged in the underflow.

Thus, although the hydrocyclone is a simple piece of equipment to construct, the internal flow patterns are complex and careful design is essential to attain a desired performance. Arterburn (1982) gives one manufacturer's method of design. Generally, a most exacting method of analysis of performance, as outlined in §5.2.6.7, is essential to obtain the classification function and its derivatives in the form of cut size and sharpness of cut.

5.3.6.3 Common features of free vortex systems

Whether used under wet or dry conditions, all free vortex systems are susceptible to reduced separation performance by increased solids loading or by a coarsening of the feed size distribution, because the spinning velocity is thereby reduced by momentum exchange. This is less marked in forced vortex classifiers because of the extra spin energy supplied by the rotor. (See §5.3.8(c).)

5.3.7 Centrifugal systems with rotating walls

5.3.7.1 Air separators

To remove the undesirable boundary-layer effects of spinning fluids at stationary walls, classifiers with rotating walls have been developed, exemplified in (i) the Bahco laboratory apparatus for size distribution measurement, and (ii) the Alpine Mikroplex spiral classifier for separation in the superfine regime.

(i) A cross-section of the Bahco separator is shown in Figure 5.31. The American version of this well-known device is described by Graham and Hanna (1962). Air passes through a control valve which is adjustable by an

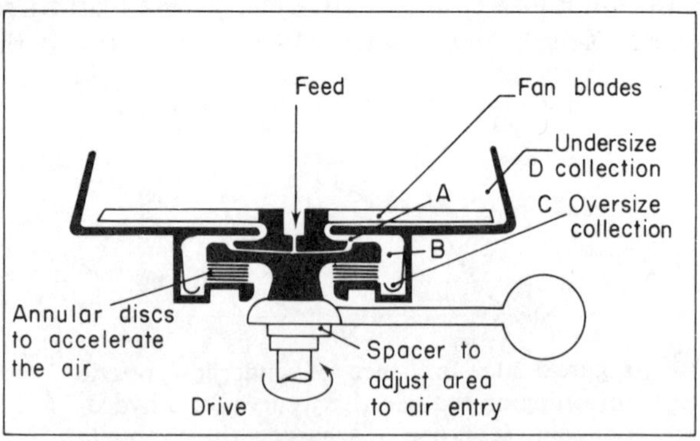

Figure 5.31. Cross-section of a Bahco separator

accurately sized spacer to give set radial and tangential components of velocity, at the entry annular space A, to the inwardly spiralling air in compartment B. A known mass of the powder to be separated is fed centrally and is introduced into the spiralling air at A. Particles coarser than cut size are thrown outward by centrifugal force and are collected in container C. Particles smaller than cut size are carried by the air up through the fan blades and are collected at D. The coarse collection is removed and weighed.

In operation, the largest spacer corresponding to the smallest cut size (about 3 microns) is first used. The coarse powder from that operation becomes the feed for the next stage with the next largest spacer, and so on. By this means the mass fractions of feed corresponding to each of the cut sizes are obtained and may be plotted as in Figure 4.36, after the instrument has been calibrated using standard powder.

(ii) The Alpine Mikroplex classifier is seen in cross-section in Figure 5.32. Air is sucked into the classifying zone *b* through a spiral casing with adjustable guide vanes *c*. The feed enters at *a*. Fines plus air leave through the central outlet *g*. Suction is applied by the fan *h* having radial, wear-resistant blades. Coarse particles pass along the inner side of the guide vanes, being constantly washed by the incoming air to remove fines and are separated from the flow by the blade *d* and the screw feeder *e*. The side walls forming the classifying zone *b* are driven at the same speed as the fan *h*.

Typical performance curves for this and other Alpine classifiers have been reported by Lauer (1965) and Leschonski (1981). In common with most classifiers, separation is poorer with increased solids loading, as shown in Figure 5.33, and the classification (Tromp) curve moves toward the coarser particle sizes, i.e. increased d_{50}.

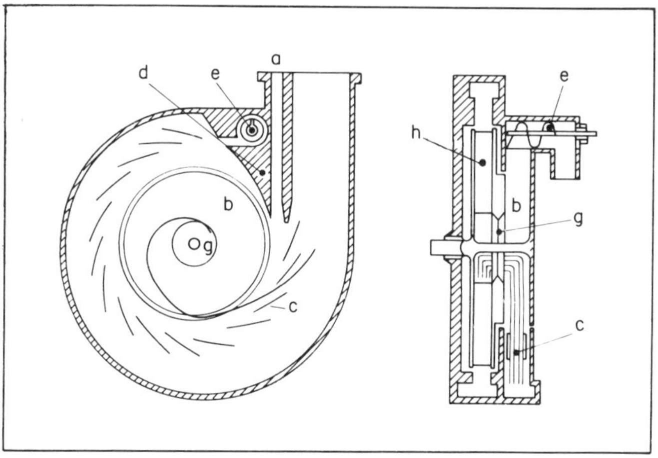

Figure 5.32. Alpine Mikroplex classifier with rotating walls Leschonski (1981)

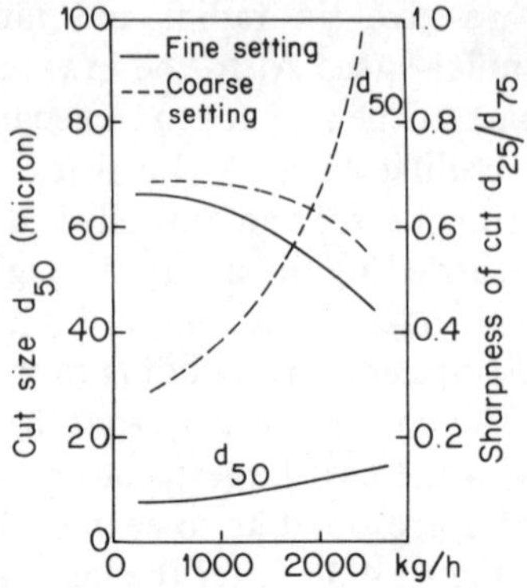

Figure 5.33. Variation of cut size and sharpness of cut with solids concentration for the Alpine Mikroplex classifier (Leschonski, 1981). *Reproduced by permission of the author*

5.3.7.2 Wet separators

These are exemplified in the solid bowl centrifuge given in cross-section in Figure 5.34. The bowl consists of a conical frustum which rotates about a horizontal axis. Within the bowl a screw conveyor rotates at a slightly greater speed than the bowl and in the same direction. Feed enters through the hollow shaft. Coarse particles are thrown to the wall of the bowl and are carried by the screw to the narrow end of the cone for discharge. Fines move with the liquid to the wide end of the shell, where they are carried out.

Treasure (1965) derives a formula for cut size d_{50} for a given slurry:

$$d_{50} = K_4\sqrt{V}/\omega \tag{5.46}$$

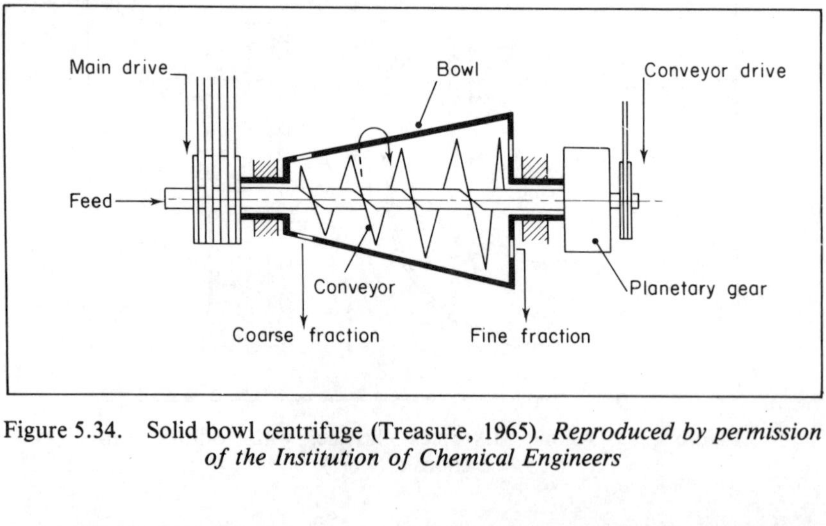

Figure 5.34. Solid bowl centrifuge (Treasure, 1965). *Reproduced by permission of the Institution of Chemical Engineers*

where V is the volume flow rate and ω is the angular velocity of the bowl. Hence fineness of product can be increased at the expense of throughput by increased speed of rotation.

The machine is capable of a cut size down to about 5 microns, but the separation is not sharp. (see §5.4.2.3.)

5.3.8 Mechanical rotor systems

Many classifiers, supplied by a large number of manufacturers, incorporate a mechanical rotor. Three main functions of the rotor may be discerned, dividing this group into three sub-groups as follows:

(a) Particle deflection Large particles are deflected and fall back into the grinding zone, as exemplified in the whizzer classifier shown in Figure 5.35. Fines pass through the blades to pass out with the rising air stream. Cut size is reduced by increase of rotor speed, but sharpness of cut is low.

(b) Air recirculation Air is recirculated within a mechanical air separator (depicted in Figure 5.36) which is constructed of an inner and an outer shell. The feed falls to a rotating distributor within the inner shell which causes particles to be dispersed in the annulus between the distributor periphery and the inner shell wall. Air, pulled in from the outer shell, passes upward through the descending curtain of particles, elutriating immediately the intermediate

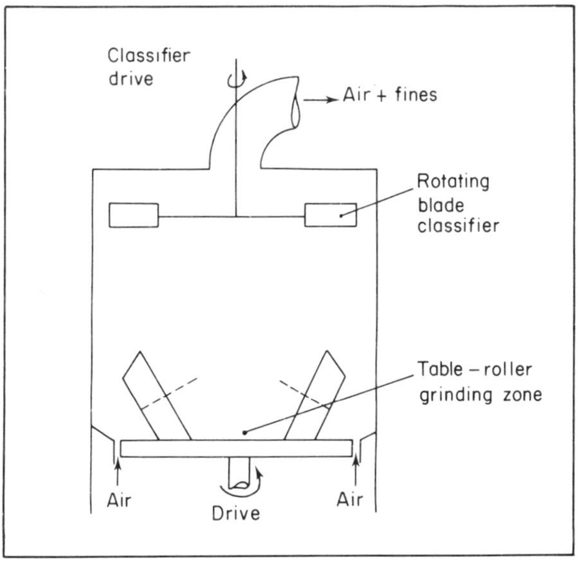

Figure 5.35. Diagrammatic cross-section of vertical spindle mill with rotating blade (whizzer) classifier

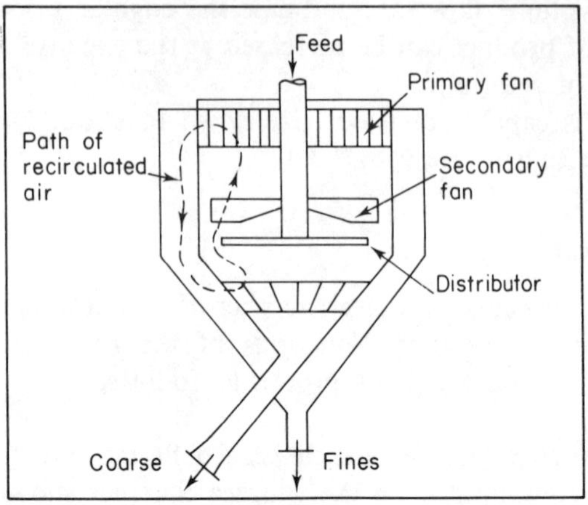

Figure 5.36. Mechanical air separator showing recirculating air pattern (Luckie and Austin, 1975). *Reproduced by permission of the Institution of Mining and Metallurgy*

and fine material, leaving the coarse to pass to the appropriate outlet. The elutriated particles enter a section where the secondary fan blades separate the intermediate material from the fines. The intermediate falls to join the coarse stream, but the fines pass with the recirculated air through the blades of the primary fan, which maintains the recirculation. The swirling of fines and air in the outer shell is supposed to produce a separation between them, allowing the fines to reach the appropriate outlet, and the air to be again recirculated. In fact, the recirculated air contains particles which are deposited on to coarse particles, producing a fish-hook selection curve as discussed in §5.2.6.5 (Luckie and Austin, 1975). The separation is relatively poor as indicated in the selectivity curves of Figure 5.37.

(c) Vortex production The forced vortex keeps large particles at its periphery but fines move along spiral paths to the centre with the outgoing air. An example is the Donaldson Acucut B18 classifier shown in Figure 5.38 in its original form (Leschonski, 1981). The feed was injected at one point in the circumference of the rotor housing and coarse particles were ejected from another. Figure 5.39 shows an improved version which provided feed dispersion and hence better yield and sharpness of cut, but the claim that the classifier will separate down to 1 micron appears to be only valid if the feed is so well dispersed that no aggregates exist larger than cut size, which is hard to achieve. Cut size decreases with increased rotor speed, as indicated in Figure 5.40.

There is evidence (Leschonski, 1981) that these classifiers in which a forced vortex is developed are less susceptible to reduction in separation performance

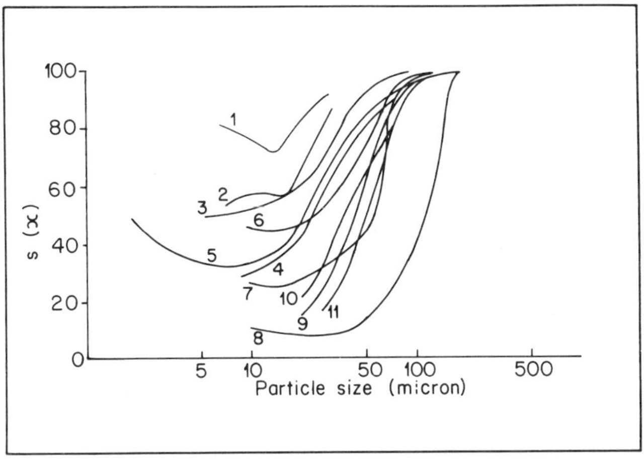

Figure 5.37. Selectivity curves for mechanical air separators: 1 and 2 WEDAG ZUB45; 3 and 4 Heyd 4.2 m diam.; 5 Streukegelsichter Polysius 3.3 m diam.; 6, 7, 8 and 9 Heyd 5 m diam.; 10 WEDAG ZUB45; 11 WEDAG ZUB25 (Leschonski, 1981). *Reproduced by permission of the author*

with increased solids loading or coarsened feed size distribution than their free vortex counterparts. However, it is not easy to produce a sufficiently controllable flow pattern in the classifying zone to attain a clean cut over a range of cut sizes; but this appears to have been successfully attained in the O'Sepa (Onuma and Furukawa, 1984) so much so that even dust-laden gases can be used as the carrying medium. The O'Sepa classifier is described in §5.4.2 as an example of good design.

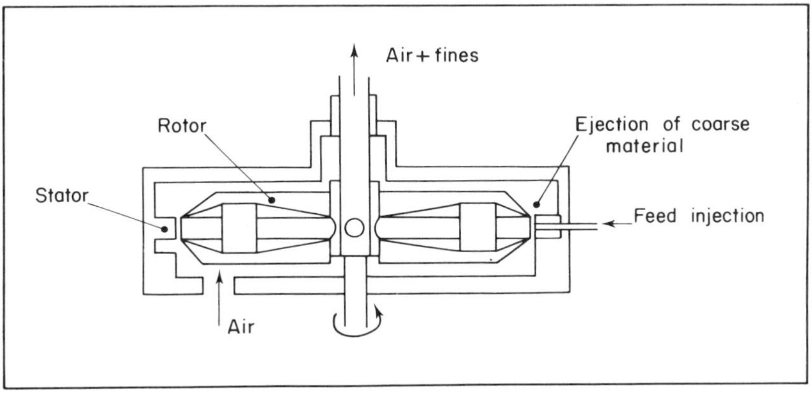

Figure 5.38. Diagrammatic cross-section of Donaldson Acucut B18 classifier as originally produced (Leschonski, 1981). *Reproduced by permission of the author*

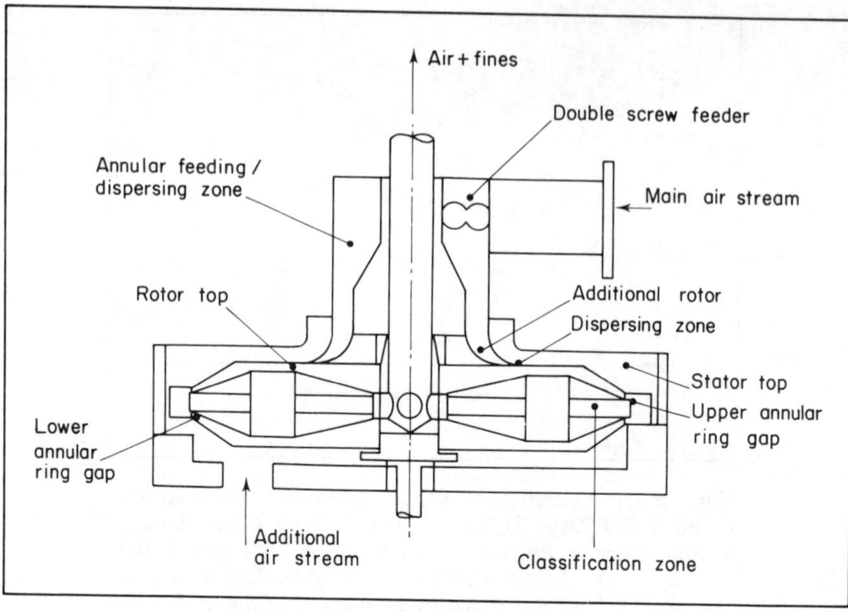

Figure 5.39. Diagrammatic cross-section of modified Donaldson Acucut B18 classifier to improve feed dispersion (Leschonski, 1981). *Reproduced by permission of the author*

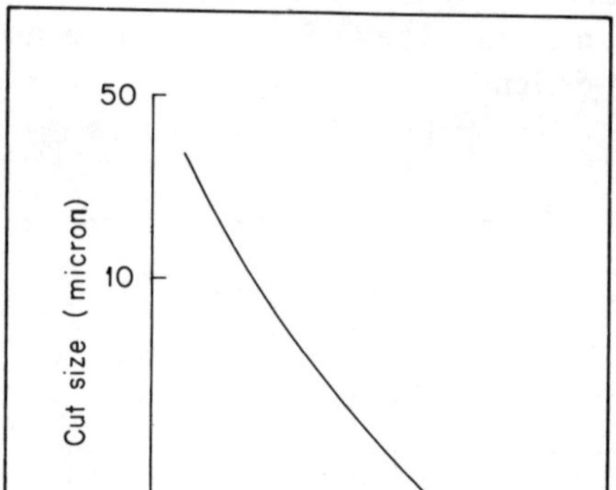

Figure 5.40. Cut size as a function of rotor speed for the Acucut series of classifiers (Adapted from Leschonski, 1981). *Reproduced by permission of the author*

5.3.9 Combined systems

5.3.9.1 Preview

In order to produce a low cut size and high sharpness of cut, classifiers have been designed which combine the principles outlined in §§5.3.3–5.3.8, and two examples are given here. The first is a laboratory instrument comparable to, but better than, the Bahco. The second is a piece of industrial equipment.

5.3.9.2 Inertia and centrifugal action

Leschonski and Rumpf (1968/69) devised what they called the counterflow equilibrium classifier shown diagrammatically in Figure 5.41. The feed sample is dispersed and accelerated in tube 1 and enters the flat classification chamber tangentially at 2 on the circumference. The air leaves the chamber together with the fines through the orifice 3 at the centre, whilst the coarse gathers at the circumference 4, spinning along the peripheral wall of the chamber. This rotating stream crosses the air entry jet 2 at a very acute angle and is therefore distributed fanwise under the influence of drag and inertia forces. The coarse particles reach the circumference at different points according to their size, but all join in the circumferential stream, which is removed intermittently through a special outlet 5 (that is kept closed during the actual classification process) to a filter 6. The fines follow the air to the centre, where they leave the chamber. Such a procedure allows very frequent repetition of the crossflow

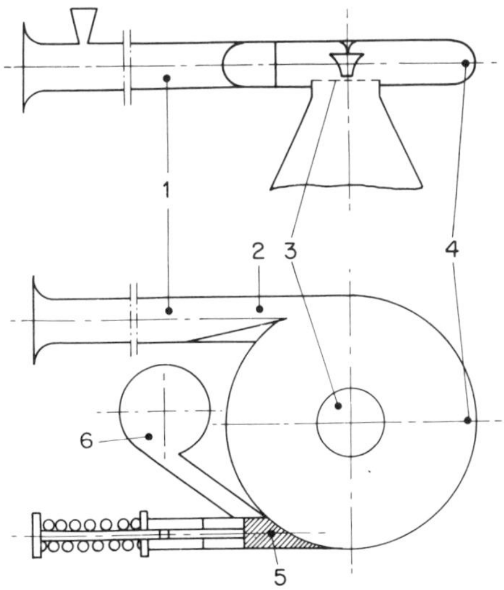

Figure 5.41. Counterflow equilibrium classifier cross-sections (Leschonski, 1981). *Reproduced by permission of the author*

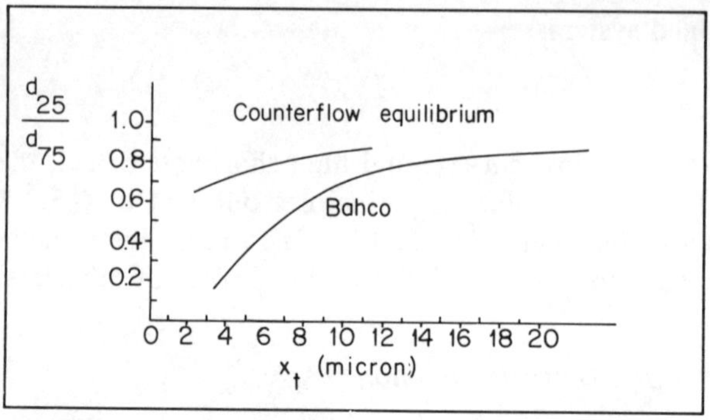

Figure 5.42. Comparison of sharpness of cut of the counterflow equilibrium classifier with the Bahco instrument (Leschonski, 1981). *Reproduced by permission of the author*

classification which can be of the order of 2000 times per minute. This rapid repetition, together with dispersion of aggregates by friction forces, leads to excellent sharpness of cut sizes, better than the Bahco in the 2–12 micron range, as seen in Figure 5.42.

5.3.9.3 Crossflow and centrifugal action

Crossflow and centrifugal action are combined in the crossflow elbow classifier developed in Karlsruhe (Maly and Löeffler, 1979). The principles of operation are indicated in Figure 5.43. Consider firstly the airflow. Its main entry stream

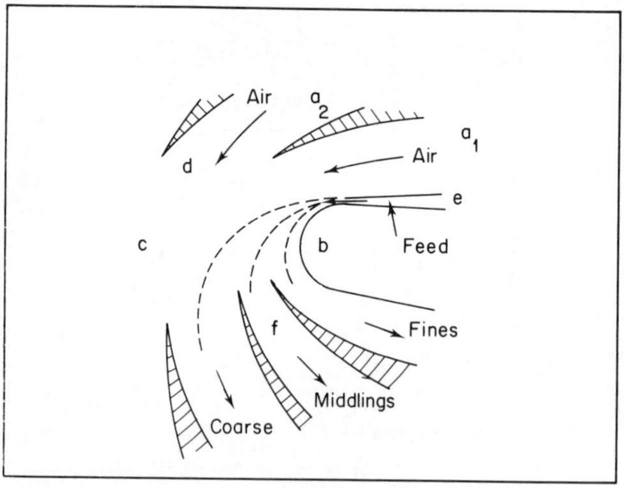

Figure 5.43. Principle of operation of the crossflow elbow classifier (Leschonski, 1981). *Reproduced by permission of the author*

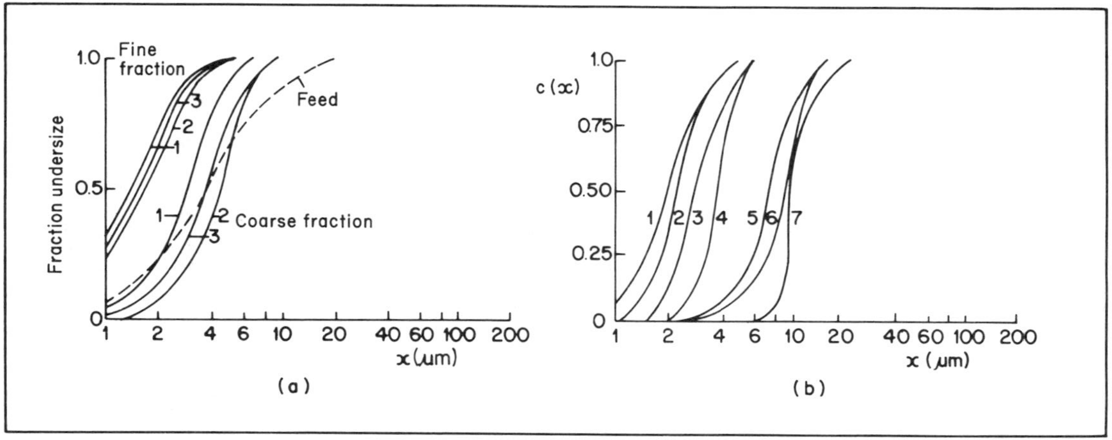

Figure 5.44. Crossflow elbow classifier using limestone: (a) cumulative size distributions, (b) classification (Tromp) curves (Leschonski, 1981). *Reproduced by permission of the author*

(a): (1) 15 kg/cm h (2) 30 kg/cm h (3) 150 kg/cm h

(b):

	capacity (kg/cm h)	d_{50}(μm)
(1)	130	1.95
(2)	20	2.30
(3)	20	3.00
(4)	80	3.80
(5)	200	7.50
(6)	80	9.50
(7)	20	10.50

is a_1, with a_2 as a secondary entry, arranged to bend round a solid wall b of specially designed curvature such that Coanda-effect forces (together with suction through the surface after 90° and up to 135° of turn) cause a vortex-free flow field to exist. Into this field the well-dispersed feed is introduced close to the wall b from an entry e. The particles follow trajectories according to their size, and are collected via precisely located curved blades, into coarse, medium and fine fractions as required.

The Japanese firm of Matsuzaka manufacture the classifier based on these principles in sizes up to 2 tonne/h. Figure 5.44(a) gives typical cumulative size distributions of limestone feed, coarse and fine fractions, and Figure 5.44(b) the classification (Tromp) curves for a range of conditions with d_{50} between 1.95 and 10.50 microns and capacity given in terms of kg/h per cm width of classifying zone.

5.4 CLASSIFIER SELECTION CRITERIA

5.4.1 Principles of good design

In the comminution world there is increasing awareness of the advantages to be gained from improved sharpness of cut. For example, with closed-circuit grinding in the cement industry, Hukki (1976) has demonstrated that if sharpness of cut was raised from the then norm of 50% to 80% there could be a simultaneous increase in plant capacity of 20% and a reduction of energy of 15%. Similar gains are probable in other industries.

The practical design features which contribute to a sharper cut include the following:

(1) good feed dispersion according to particle size prior to the application of separating forces;
(2) clear identification of an effective classifying zone in which a stable flow pattern is maintained;
(3) non-interaction of coarse and fine streams after separation;
(4) application of strong shear forces, where necessary, to break up aggregates;
(5) after primary separation, provision of secondary winnowing of fines from the coarse stream and removal of grits from the fine stream;
(6) where a coarse stream non-return valve is required to seal against airflow, it must be able to function efficiently at all loads.

5.4.2 Examples of bad and good design

5.4.2.1 Preamble

Selected examples are confined to industrial separators associated with comminution, rather than with laboratory equipment, and dry systems are first discussed, then wet. Each example is considered against the backcloth of the six desirable features listed in §5.4.1.

5.4.2.2 *Dry Systems*

(i) The widely used cone classifier shown in Figure 5.22 is an example of the feed being carried with the entering gas, in which the 180° turn through the blades throws the coarser particles to the cone wall, down which they fall to the oversize outlet. Thus feed dispersal assists the centrifugal action of the downward vortex. The Hukki (Larox) classifier, depicted in Figure 5.26, is a good example of feed dispersal for the case where the solids enter separately from the gas, and are thrown across the gas, again allowing coarse particles to move towards the coarse outlet.

(ii) In all free vortex systems, such as the cone and the Hukki (Larox) classifiers, it is difficult to obtain a clearly defined classifying zone and more difficult still to eliminate currents which re-mix separated streams. Hence sharpness of cut is not a good feature of free vortex systems. Hukki (1980, 1979) has advocated two-stage classification, which does help, but capital and running costs are doubled. Hukki and Airaksinen (1980) have given full details of classifier capacity, specific energy and output size distribution. It is clearly more advantageous to obtain the desired end in a single piece of equipment, and this has been achieved for the cement industry in the O'Sepa classifier (Onuma and Furukawa, 1984) as described in (iii).

(iii) The designers of the O'Sepa classifier (Figure 5.45) have chosen the forced vortex system as that which offers the best means of satisfying the requirements of §5.4.1. Particles to be separated enter via ducts 1 and fall on

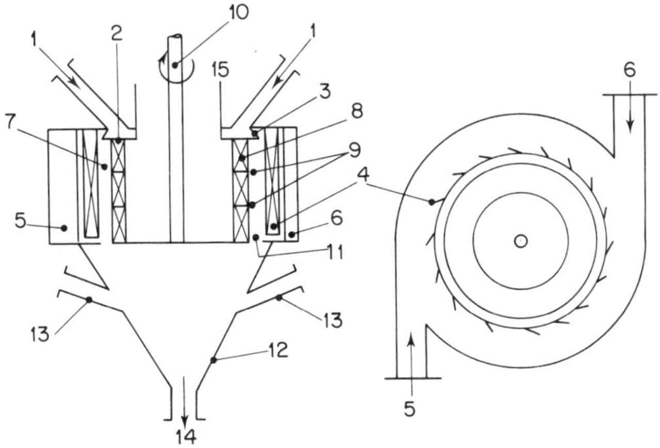

Figure 5.45. O'Sepa classifier cross-sections (Onuma and Furukawa, 1984). *Reproduced by permission of World Cement*

to the rotating plate 2 which disperses the feed according to size. The buffer 3 directs the larger particles, thrown out by centrifugal force, down the inside of the air guide vanes 4, the air having entered through tangentially placed ducts 5 and 6. The main classifying zone is 7 in which a precise horizontal vortex is maintained, partly by the guide vanes and partly by the rotating cage, of which the dispersion plate 2 is the top-most member. The cage contains vertically orientated flow-adjusting blades 8 and horizontal partition plates 9 which ensure that the vortex is even over the whole length of the classifying zone. The cage is driven by means of the shaft 10 at a speed rather greater than the rotational speed of the air in the classifying zone.

Coarse particles fall down through the annular slot 11 into a hopper 12 where they are washed by clean air entering through ducts 13, before leaving at exit 14. Fines are carried with the exiting air through the rotating cage and leave via the duct 15.

Because particles in the classifying zone follow spiral paths, changes in the speed of the rotor allow control of these paths, and hence cut size is adjusted by change of rotor speed.

Factors contributing to the sharp cut are:

(1) good feed dispersal;
(2) relatively low rotor speed, together with a long classifying rotor cage covering the whole length of the classifying zone, allowing sufficient

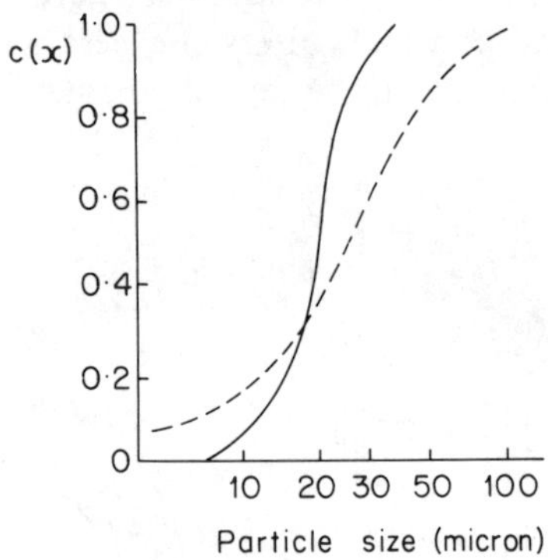

Curve	Separator	Solids concentration	Blaine cm^2/g	% > 30 micron
---	Cyclone	2 · 14	3960	5·5
——	O'Sepa	1 · 45	3000	23·5

Figure 5.46. Comparison of classification (Tromp) curves for cyclone and the O'Sepa separators (Onuma and Furukawa, 1984). *Reproduced by permission of World Cement*

residence time for particles to come under the influence of separating forces;

(3) maintenance of a stable separation vortex, eliminating currents which would tend to cause re-mixing of separated streams.

To give long life to the equipment, alumina ceramic coatings are given to the guide vanes, rotor blades and separation plates and other internal parts subject to wear.

Figure 5.46 shows a typical comparison of classification (Tromp) curves for the cyclone and O'Sepa separators.

Other plant operational advantages claimed by O'Sepa are typically:

(a) 19% production rate increase associated with a decrease of 8.4 kWh/tonne;
(b) specific surface area of cement decrease by 160 cm^2/g without any change in cement mortar strengths;
(c) lowering of cement temperature by 40°C without the installation of a cooler, by virtue of the fact that the O'Sepa classifier can accept a large volume of fresh air. For details of plant arrangement see Onuma and Furukawa (1984).

5.4.2.3 Wet systems

(i) In the bowl centrifuge (Figure 5.33) the surface current that carries the fines to the effluent port will be of variable depth in practice, and the starting radius of any particle will therefore vary considerably. Thus the classification zone is ill-defined and the separation size is the mean of a range of sizes. Further, the screw conveyor movement, designed to carry back the coarse particles thrown to the bowl periphery, will also cause re-mixing of separated streams (Treasure, 1965). This complex nature of flows within a spiral classifier has been confirmed, for a static wall model, by Stewart and Restarick (1967) operating with a laboratory model.

(ii) Similarly, the hydrocyclone has been found to suffer from complex flow problems, as indicated in §5.4.2.2(ii) for dry systems.

(iii) The above are only two examples from a wide range of available wet classifiers, as summarized by Stewart (1957). However, the indications are that, although cut sizes down to a few microns are possible, there is no hydraulic classifier with the sharpness of cut of the O'Sepa for dry systems. This represents a challenge to some enterprising organization.

5.4.3 Test of a classifier for a given duty

The following are required to be known in respect of a classifier for a given duty:

(1) solids output rate at a given size distribution for a feed of given size distribution;

(2) solids concentration in the carrying medium at inlet;
(3) cut size and sharpness of cut for the above conditions;
(4) pressure loss;
(5) wear life of parts subject to erosion;
(6) effectiveness of gas seal at solids oversize outlet, where applicable.

If several classifiers meet the specification, then that with the sharpest cut should be chosen, particularly if the classifier is part of a mill closed circuit, because this will imply a reduced recirculating load. However, a greater specific energy usually accompanies improved cut sharpness and therefore a compromise must be made. The overall system minimum cost will usually decide the optimum classifier.

5.5 REFERENCES

Andersen, J. (1931). 'Uber maschenweiten und korngrössen', *Zement*, **20**, 224–226, 242–245.

Andreasen, A. H. M. (1927). 'Einige betrachtungen und beobachtungen über Wirkungsweise des schüttelsiebes', *Sprechsaal*, **60**, 515–517.

Arterburn, R. A. (1982). 'The sizing and selection of hydrocyclones'. In: *Design and Installation of Comminution Circuits* (Mular, A. L., and Jergensen, G. V., Eds.), SME/AIME, New York, 592–607.

Aso, K. (1957). 'On the theory of the partition curve and its application to coal preparation or mineral dressing', *Mem. Fac. Eng. Kyushu University*, **17**, 18–83.

Austin, L. G., and Klimpel, R. R. (1981). 'An improved method for analysing classifier data', *Powder Technol.*, **29**, 277–281.

Batterham, R. J., Weller, K. R., Norgate, T. E., and Birkett, C. J. (1980). 'Screen performance and modelling with special reference to iron ore crushing plants', *Proc. Europ. Symp. Particle Technol.*, Amsterdam, 842–857.

Blachetta, W. (1984). 'New developments in the design of high efficiency screening machines', *Proc. 9th Powder Bulk Solids Handlg. Processg. Conf.*, Rosemount, Ill, 15–17 May, 245–271.

Brereton, T. (1975a). 'Probability screening and the effect of major operating variables', *Filtration Separation*, Nov/Dec, 692–696.

Brereton, T. (1975b). 'The determination of overflow recovery, size and efficiency of separation of a screening process', *Proc. Inst. Mech. Eng. Screen Conf.*, Paper C45/75.

Brereton, T., and Dymott, K. R. (1974). 'Some factors which influence screen performance', *Proc 10th Int. Miner. Min. Cong.*, April 1973, Publ. IMM (London), 181–194.

Colman, K. G., and Tyler, W. S. (1978). 'Selection guidelines for size and type of vibrating screens in ore crushing plants'. In: *Mineral Process Plant Design* (Mular, A. L., and Bhappu, R. B., Eds.), SME/AIME, New York, 341–361.

Draper, N. R., and Smith, H. (1981). *Applied Regression Analysis*, John Wiley, New York.

English, J. E. (1974). 'A new approach to the theoretical treatment of the mechanics of sieving and screening', *Filtration Separation*, March/April, 195–203.

Etkin, B., and Haasz, A. A. (1980). 'Performance characteristics of the Toronto infrasizer Mark III', *Proc. Int. Symp. Fine Particle Processg.*, Las Vegas, AIME, New York, 209–231.

Etkin, B., Haasz, A. A., Raimondo, S., and D'Eleuterio, G. M. T. (1980). 'Air classification of thin flakes', *Inst. Chem. Eng. Symp. Series* 59, Dublin, 5:3/1–5:3/23.

Ferrara, G., and Preti, U. (1975). 'A contribution to screening kinetics', *Proc. 11th Int. Miner. Process. Congr.*, Cagliari, 183–217.

Finch, J. A. (1983). 'Modelling a fish-hook in hydrocyclone selectivity curves', *Powder Technol.*, **36**, 127–129.

Foreman, W. E. (1979). 'Screening'. In: *Coal Preparation*, (Leonard, J. W., Ed.), SME/AIME, 8/3–47.

Gaudin, A. M. (1939). *Principles of Mineral Dressing*, McGraw Hill, New York, 145.

Gluck, S. E. (1965). 'Vibrating screens: surface selection and capacity calculations', *Chem. Eng.*, 15 March, 179–184.

Graham, A. L., and Hanna, T. H. (1962). 'The micro-particle classifier', *Ceramic Age*, Sept., 38–40.

Harris, C. C. (1961). 'Some aspects of screen sizing', *Columbia Eng. Q.*, **15**, (1), 18–23, 48.

Heywood, H. (1945). 'A study of sizing analysis by sieving', *Trans. IMM (London)*, **54**, 373–390.

Hidaka, J., and Miwa, S. (1979). 'Fractionation and particle size analysis of fine powders by microsieve', *Powder Technol.*, **24**, 159–166.

Hukki, R. T. (1976). 'About the ways and means to improve the performance of the closed grinding circuit', Dechema Monograph 79, 319–330.

Hukki, R. T. (1979). 'Fundamentals of the closed grinding circuit', *Eng. Min. J.*, April, 102–108.

Hukki, R. T., and Airaksinen, T. (1980). 'A study of the effect of improved pneumatic classification on production of micropowders', *Proc. Int. Symp. Fine Particle Process.*, Las Vegas, Feb., AIME, New York, 181–302.

Jones, G. J., Mobbs, F. R., and Cole, B. N. (1971). 'Development of a theoretical model for an inertia filter', *Pneumotransport*, **1**, 6–8 Sept., Paper B1.

Kelsall, D. F. (1963). 'Some applications of hydraulic cyclones in hydrometallurgical processes', *Trans. SME/AIME*, **220**, 225–231.

Kelsall, D. F., and Holmes, J. A. (1960). 'Improvement of classification efficiency in hydraulic cyclones by water injection', *Proc. Int. Miner. Process. Congr.*, IMM (London), 159–170.

Klimpel, R. R. (1973). 'The use of mathematical modelling to evaluate operating alternatives in an industrial comminution facility', *Proc. 1st Int. Conf. Particle Technol.*, IITRI, Chicago, 84–91.

Klimpel, R. R. (1980). 'Estimation of weight ratios given component make-up analysis of streams', *Trans. SME/AIME*, **266**, 1882–1886.

Klimpel, R. R. (1982). 'The influence of a chemical dispersant on the sizing performance of a 24 inch hydrocyclone', *Powder Technol.*, **31**, 255–262.

Krijgsman, C. (1949). 'De toepassing van de centrifugaalkracht in moderne kolenwasserijen', *Woordrachten Koninklijk*, Inst. van Ing., Nederland, **1**, (5), 691.

Lauer, O. (1965). 'Trenngrenze and trennschärfe des spiralwindsichter mit umlaufenden sichtraumwänden', *Aufbereit Tech.*, **6**, (4), 213–222.

Leonard, J. W. (1974). 'Determination of industrial screening efficiency', *Trans. SME/AIME*, **256**, 185–187.

Leschonski, K. (1979). 'Sieve analysis: the Cinderella of particle size analysis methods?', *Powder Technol.*, **24**, 115–124.

Leschonski, K. (1981). 'Classification of particles in gases', Int. Fine Particle Res. Inst. Report (May).

Leschonski, K., and Rumpf, H. (1968/69). 'Principle and construction of two new air classifiers for particle size analysis', *Powder Technol.*, **2**, 175–185.

Luckie, P. T., and Austin, L. G. (1974). 'Technique for the derivation of selectivity functions from experimental data', *Proc. 10th Int. Miner. Process Congr.*, (Jones, M. J., Ed.), London, April, IMM (London), 773–790

Luckie, P. T., and Austin, L. G. (1975). 'Mathematical analysis of mechanical air separator selectivity curves', *Trans. IMM (London)*, **84**, C253–C255.

Lynch, A. J. (1965). 'The characteristics of hydrocyclones and their application as control units in comminution circuits', Progress Report 6, Dept. Min. Metall., University of Queensland, Brisbane, Australia.

Lynch, A. J. (1970). 'Lecture notes on comminution and classification', University of Queensland, Brisbane, Australia.

Maly, K., and Löeffler, F. (1979). 'Die strahlumlenkwindsichtung, ein neues trennverfahren zur wirtschaftlichen herstellung sehr feinkörniger produkte', *Acta Pharmaceutica Techn.*, APV Informationsdienst Supplement 7, 27–32.

Matthews, C. W. (1972). 'Trends in size reduction of solids-screening', *Chem. Eng.*, 10 July, 76–83.

Molerus, O., and Hoffmann, H. (1969). 'Darstellung von windsichterfrennkurven durch stochastisches modell', *Chemie. Ing. Techn.*, **41**, 340–344.

Mular, A., and Herbst, J. A. (1978). 'Digital simulation: an aid for mineral processing plant design'. In: *Mineral Processing Plant Design*, (Mular, A. L., and Bhappu, R. D., Eds.), SME/AIME, New York, 306–338.

Müller, K., and Wessel, J. (1963). 'Berechnung vone teilchenbahnen im windsichter mit dem electronischen analogrechner', *Aufbereit Techn.*, **4**, (6), 235–242.

Nakajima, Y., Whiten, W. J., and White, M. E. (1979). 'Behaviour of non-spherical particles in screening', *Trans. IMM (London)*, **88**, C88–C92.

Nichols, J. P. (1982). 'Selection and sizing of screens'. In: *Design and Installation of Comminution Circuits* (Mular, A. L., and Jergensen, G. V., Eds.), SME/AIME, New York, 509–522.

Onuma, E., and Furukawa, T. (1984). 'O'Sepa: a new high performance air classifier', *World Cement*, Jan/Feb, 13–24.

Orr, C., Davis, D. K., and Camp, R. W. (1979). 'Particle size analysis by automatic siever', *Powder Technol.*, **24**, 143–149.

Plitt, L. R. (1971). 'The analysis of solid–solid separations in classifiers', *CIM Bull.*, **64**, 42–47.

Plitt, L. R. (1976). 'A mathematical model of the hydrocyclone classifier', *CIM Bull.*, **69**, (776), 114–123.

Pritchard, A. N. (1980). 'Vibrating screens in the mining industry', *Mine Quarry*, **9**, Oct., 46–50.

Quittkat, W. (1973). 'Der strahlwindsichter ein leistungs fähiger, raumsparender sichter für feinsttrennung', *ZKG*, **26**, 326–330.

Rastyapin, V. I., Marshak, Y. L., and Osokin, V. P. (1973). 'Comparative effectiveness of inertia classifiers', *Teploenergetica*, **20**, (7), 65–68.

Reid, K. J. (1971). 'Derivation of an equation for classifier reduced performance curves', *Canadian Metall. Q.*, **10**, 253–254.

Rogers, R. S. C. (1982). 'A classifier function for vibrating screens', *Powder Technol.*, **31**, 135–137.

Rose, H. E. (1976). 'The derivation of an equation for the performance of a screen', Dechema Monograph 79, 351–369.

Rumpf, H., and Leschonski, K. (1967). 'Prinzipien und neuere verfahren der windsichtung', *Chemie. Ing. Techn.*, **39**, (21), 1231–1241.

Schmidt, D. (1981). 'Screening as a production process', *Ind. Chem. Eng. Symp. Series* 63, Birmingham, D2/N/1–11.

Schönert, K. (1969). 'The influence of classifier characteristics on the stability of closed circuit grinding'. In: *A Decade of Digital Computing in the Mineral Industry*, (Weiss, A., Ed.), SME/AIME, New York, 677–694.

Schubert, H. (1980). 'Nassklassieren feier teilchen', *Proc. Europ. Symp. Technol.*, Amsterdam, 705–729.
Sokaski, M., Geer, M. R., and McMorris, W. L. (1979). 'Wet concentration of fine coal: Part I, Dense medium separation'. In: *Coal Preparation*, (Leonard, J. W., Ed.), SME/AIME, 10.3–10.21.
Stairmand, C. J. (1973). 'Sieves and sieving', *Proc. Harold Heywood Memorial Symp.*, Loughborough University of Technology.
Stewart, P. S. B., and Restarick, C. I. (1967). 'Dynamic flow characteristics of a small spiral classifier', *Trans. IMM (London)*, **76**, C225–C230.
Stewart, R. F. (1957). 'Wet classification'. In: *Chemical Engineering Practice* (Cremer, H. W., and Davies, T., Eds.), Butterworth, Vol. 3, 287–311.
Svarovsky, L. (1977). *Solid–Liquid Separations*, Butterworth, Chap. 3.
Trawinski, H. (1976). 'Theory, applications and practical operation of hydrocyclones', *Eng. Min. J.*, Sept., 115–127.
Treasure, C. R. G. (1965). 'Fine particle size classification', *Trans. Inst. Chem. Eng.*, **43**, T199–T205.
Tromp, K. F. (1937). 'Neue wege für die beurteilung der aufbereitung von steinkohlen', *Glückauf*, **11**, 125–131, 151–156.
Valikanov, M. A. (1965). *Measurement Errors and Empirical Relations*, Oldbourne Press, London.
Whitby, K. T. (1959). 'The mechanics of fine sieving', *Proc. Symp. Particle Size Measurement*, Boston, June 1958, ASTM Special Publn. 234, 3–25.
Whiten, W. J. (1972). 'The simulation of crushing plants with models developed using multiple spline regression', *J. South African Inst. Min. Metall.*, **72**, (10), 257–264.
Wilkinson, H. N. (1971). 'An investigation into the performance of vibrating screens for sizing cold sinter', *J. Iron Steel Inst.*, **209**, March, 178–189.
Wills, B. A. (1980). 'Factors affecting hydrocyclone performance', *Min. Mag.*, Feb., 142–146.
Yoshiota, N., and Hotta, Y. (1955). 'Liquid cyclone as a hydraulic classifier', *Chem. Eng. Japan*, **19**, (12). 632–640.

CHAPTER 6

Energy for size reduction

NOTATION FOR CHAPTER 6

2a	Crack length Subscripts: c critical i initial
A	Factor depending on material strength
D	Mill diameter inside liners for ball or rod mills
E	Mass-related specific comminution energy Subscripts: m minimum M maximum
$\bar{E}$	Statistical average mass-related specific comminution energy
E_A	Mass-related specific applied energy (net)
$E_A(x)$	E_A for a particle size of x
$\bar{E}_A$	Statistical average mass-related specific energy (net)
$\bar{E}_M$	Maximum value of $\bar{E}_A$
E_V	Volume-related specific applied energy (net)
$\bar{E}_V$	Statistical average volume related specific applied energy (net)
f	Function
f_1 etc.	Factor
F	Externally applied force
g_b	Average of last three ore grindabilities in Bond test
G	Specific crack extension energy, crack-surface related
G_c	Critical value of G
H	Drop height
HI	Hardgrove index
K	Constant
L_R	Rod length in rod mill
m	Mass
n	Exponent in equation (6.18)
$P(\bar{E}_A, x)$	Comminution energy probability distribution function for an applied energy E_A and an initial particle size of x
r	Exponent
R	Reduction ratio
R_o	Optimum reduction ratio
Δs	Increase in mass-related specific surface
Δs_V	Increase in volume-related specific surface
s_E	Utilization of energy per unit volume = $\Delta s_V/E_V$

U_f	Fracture energy
v	Velocity
V	Volume
x	Particle size
	Subscripts: o optimum 1 feed 2 product t sieve size for Bond work index test
x'	Size modulus
y	Distance
W_i	Bond work index
β	Fracture energy per unit area of crack
β_{max}	Maximum value of β

6.1 FUNDAMENTALS

6.1.1 Measures of size reduction

Any energy expenditure causing breakage must be related to the initial and final particle size conditions.

For single-particle tests, the initial conditions may be the diameter of the sphere used (cubes or cylinders are not valid—see §3.2.7) or for irregularly shaped particles, two narrowly spaced sieve sizes.

For particle bed tests, the initial condition may be an assemblage of spheres of stated diameter(s), or irregularly shaped particles between two sieve sizes (scalped feed), or a full distribution of sizes. In all cases, the final condition after breakage will be a complete size distribution.

Where there is no material change in the pattern of the size distribution between feed and product, the ratio of x_{80}, say, for feed and product has been used as the reduction ratio. This is open to the objection that, since so much more energy is used to break small than large particles, changes in the fine end of the size spectra need not be reflected in the x_{80} change, and hence a false size measure could be associated with a given energy expenditure. This disadvantage clearly applies to the size modulus x' but there is even more uncertainty with this measure of size reduction because of different patterns of size distributions at the coarse end of the size spectrum, that can be associated with a given modulus. (see Figure 4.30.)

To attempt to overcome these objections, change in specific surface between feed and product is widely used as a measure of size reduction. Again there are problems, because each method of surface assessment gives a different answer for given product and feed distributions (see §4.4.1), and hence strict comparison is only possible where the same method of surface measurement is uniformly applied. Further, one specific surface value can be obtained from a variety of size distributions.

The only really acceptable measure of size reduction is the combined statements of initial and final size distributions, and today's computers can easily handle them.

6.1.2 Definitions of energy for breakage

All the energy supplied to a particle is not used for breakage (see §4.3.4.5). Hence it is necessary to distinguish between

(1) the comminution energy per unit mass, E, which is that actually used in size reduction;
(2) the externally applied energy per unit mass, E_A.

If several particles of the same material, size and geometry are stressed in turn under similar conditions of stress application, the amount of comminution energy E is found not to be a unique quantity, but varies with the flaw pattern of the individual particle. Hence to obtain a unique measurement of comminution energy, a sufficient number (Beavan, 1980) of such particles must be similarly stressed to obtain a statistical average energy $\bar{E}$.

Equally, the externally applied energy E_A is not a unique quantity and a statistical average must be obtained from a sufficient number of exactly similar tests to obtain $\bar{E}_A$.

The externally applied energy is relatively easy to assess but the comminution energy is much more difficult to arrive at. One successful system is single-particle slow compression up to the point of the first breakage during which stress and strain are measured. The area under the stress–strain curve is proportional to the energy $\bar{E}$ absorbed. Another system which has proved to be applicable to spheres, but not to irregularly shaped particles, is that of breakage probability as discussed in §6.1.6. It is to be noted that both systems involve single-particle tests individually performed.

For single-particle breakage, in the context of the same material, particle size and geometry, $\bar{E}_A$ is always greater than $\bar{E}$; but in a multi-layer bed it is very much greater, because many particles will be stressed to levels below that critical for fracture and that associated energy is lost as heat. Further, frictional and other losses are considerable, as summarized in Figure 6.26.

Note that E_A and $\bar{E}_A$ are net applied energies: they do not include mechanical and electrical drive losses.

6.1.3 Grinding efficiency

One measure of grinding efficiency may be expressed by the ratio $\bar{E}/\bar{E}_A$. Since $\bar{E}$ may be obtained only from single-particle tests, it is a concept removed from common practice, but data are available in §6.4.2.

The older literature contains references to 'theoretical' efficiency, which was based on specific free surface energy. For quartz, for example, this has been quoted as 920 ergs/cm^2 (Edser, 1922), and in these terms grinding efficiencies have been rated in the range 0.1% to 1% (e.g. Hukki and Reddy, 1967). However, it has been shown in §3.3 that fracture energy for most materials is several orders of magnitude greater than that of specific free surface energy, which is therefore an inappropriate basis for efficiency rating.

In the past, too, attempts have been made to assess the heat loss from a size reduction process and so obtain comminution energy by difference from externally applied energy, as discussed in §6.4.3.

6.1.4 Utilization of energy

A more useful concept than efficiency is that of utilization of energy, often defined as new surface produced per unit of externally applied energy for unit mass of material, or $\Delta s/\bar{E}_A$.

For fundamental studies it is often more convenient to discuss utilization of applied energy in terms of unit volume. Then $s_E = \Delta s_V/\bar{E}_V$.

In special circumstances only it is possible to consider utilization of comminution energy $\Delta s/\bar{E}$.

6.1.5 Energy investment

Referring to §3.3, where G is the specific crack extension energy proportioned to U_f and β is the fracture energy, each expressed per unit area of crack, and combining equation (3.4) with equation (3.6), the condition for crack propagation may be stated as:

$$G \geqslant 2\beta \tag{6.1}$$

Following Rumpf (1973), G increases with crack length a. Hence β rises with G until, at the critical crack length a_c, the critical crack extension energy G_c and the maximum energy absorption β_{max} have been reached. Above G_c, G continues to rise with increasing value of a, while β remains sensibly constant at β_{max}. The unabsorbed energy $(G - G_c)$ is transformed into kinetic energy and the fracture accelerates to the maximum speed of crack propagation. Thus β_{max} is an important material factor in the breakage process, as already noted in §3.3.

Rumpf (1973) has taken the theory further by consideration of similarity laws for physically identical materials on the following ideal assumptions:

(1) particles are geometrically similar as well as physically identical;
(2) deformations and stress states are similar when the particles are stressed;
(3) deformations are purely elastic.

For such conditions, it follows that the critical crack extension energy G_c at which a crack starts to propagate is reached in a critical length a_c/x, which itself will remain constant for similarity to obtain.

If E_V is the energy supplied per unit volume of particle, dimensional analysis is shown by Rumpf (1973) to give:

$$G = (E_V x) f(a/x) \tag{6.2}$$

At the critical condition:

$$G_c = (E_V x) f(a_c/x) = 2\beta_{max} = \text{constant} \tag{6.3}$$

Hence $(E_V x)$ is also constant and has been called by Rumpf the 'energy investment'.

Also from similarity conditions, for the same material and the same flaw distribution $\Delta s_V x$ is constant. Hence:

$$s_E = \frac{\Delta s_V}{E_V} = \frac{\Delta s_V x}{E_V x} = \text{constant}$$

In other words, the utilization of energy is also constant.

Now by dimensional analysis for the case of non-similar flaw distribution, Rumpf (1973) proved that the following relations hold as long as the velocity of displacement is small relative to the velocity of elastic waves in the material, which is normally the case:

$$\Delta s_V x = f_1\left(\frac{E_V x}{\beta_{max}}, \frac{a_i}{x}\right) \tag{6.4}$$

$$s_E \beta_{max} = f_2\left(\frac{E_V x}{\beta_{max}}, \frac{a_i}{x}\right) \tag{6.5}$$

where a_i/x represents the initial crack length.

If $(E_V x)$ as well as β_{max} are constant for the conditions considered, then:

$$\Delta s_V x = f_1(a_i/x) \tag{6.6}$$

$$s_E \beta_{max} = f_2(a_i/x) \tag{6.7}$$

Hence, in order to obtain a similar fracture pattern, not only must $(E_V x)$ be constant, but also (a_i/x). This latter condition would not be expected to be satisfied with particles of the same material but of different size, since initial flaws in 50 mm diameter glass spheres are not likely to be 10 times larger than those in spheres of 5 mm diameter. Hence the practical conditions to test equations (6.6) and (6.7) are difficult to achieve. However, the $(E_V x)$ factor can be investigated in a plot of energy utilization against particle size, because the slope of a line then gives $\Delta s/(E_V x)$ which is parallel to the x-axis for constant $(E_V x)$.

It may be argued that irregularly shaped particles are never geometrically similar, nor are they similarly deformed, nor are they physically identical. In other words, the similarity conditions used by Rumpf may be thought to be too far from reality to be of value. However, in the context of statistical average energy $\bar{E}_V$ and newly created surface, the relation between the two has been found to have significance within certain confidence levels, and therefore such data are included in the test survey in §6.2. Further, the above concepts have proved of assistance in analysis of test data and give reason to separate slow compression from impact.

6.1.6 Probability applied to comminution energy

When particles of a given material, geometry and size are individually stressed to a predetermined maximum level under predetermined conditions, some break and some do not. For a sufficient number of such tests (Beaven, 1980), the ratio of those that break to those that are stressed is the breakage probability P. For a glass sphere, there is a definite stress at which breakage occurs; but for an irregularly shaped particle, because some small projection might break away before the main body of the particle fails, it has become the practice to agree a particle is considered to be crushed when the mass of the largest fragment is at least 10% less than that of the original particle, and breakage probability is assessed on this basis.

Now the breakage probability has been found to be a measure of the strength of the material for that particle geometry and those loading conditions, and hence also it would be expected that the breakage probability would be related to the energy required for actual breakage, i.e. to the comminution energy.

Baumgardt *et al* (1976) explored this concept further. For a given applied energy E_A, they argued that there would be a range of mass-related specific comminution energies from the minimum E_m up to the maximum E_M, for a given particle size. If this energy range E_m to E_M is divided into small intervals, then for a sufficient number of tests in each, there can be determined the ratio

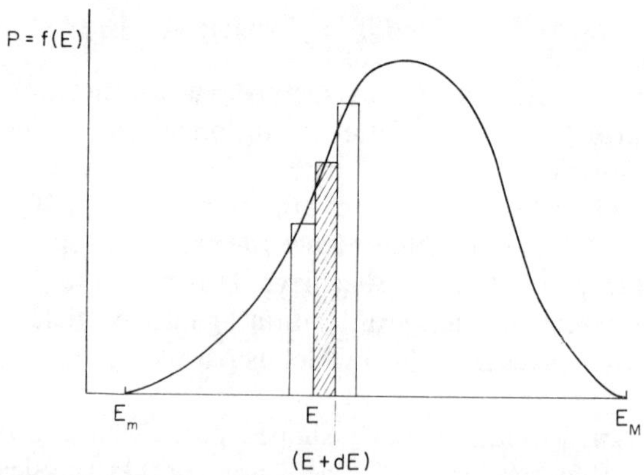

Figure 6.1. Probability density distribution of comminution energy of particles of a given material, size and geometry broken by a predetermined stress associated with a maximum comminution energy level E_M

of the number of particles that break to those that are stressed within each interval, i.e. a breakage probability for each interval. When these interval probabilities are plotted against the interval comminution energies, a distribution is obtained as in Figure 6.1. Summation of the interval probabilities over the range E_m to E_M permits the evaluation of the overall distribution probability of comminution energy expressed mathematically as

$$P(E_A, x) = \int_{E_m}^{E_M} f(E)\, dE \tag{6.8}$$

where $P(E_A, x)$ is the comminution energy distribution probability associated with a given applied energy E_A and an initial particle size x.

When this process is repeated over a range of values of E_A and of x, it is possible to plot $P(E_A, x)$ against E_A with x as a parameter, and a graph results as in Figure 6.9(a) for slow compression of glass spheres and as in Figure 6.15 for impact breakage of glass spheres. By suitable transforms, Baumgardt *et al.* (1976) demonstrated that truncated log–normal equations fitted these curves, as seen by the straight lines obtained in Figure 6.9(b).

6.2 SPECIALIZED ENERGY TESTS

6.2.1 Preview

(i) The following summarizes the tests reviewed:

(1) single-particle compression in §6.2.2.1;
(2) single-particle impact in §6.2.2.2;
(3) confined particle bed tests in §6.2.3.

(ii) To facilitate the further sub-division of single-particle compression and impact tests, a code system of letters has been devised, relatively positioned as shown in Table 6.1. Thus ISMF denotes an impact test with single stressing of a monolayer of equisized separated particles falling freely on to a hard plate. Note that the term 'single particle' is used to include both O and M tests, as the M tests may be thought of as a short-cut way of performing a succession of O tests, although the analysis of each proves to be different.

(iii) The methods of assessment of the externally applied energy and their approximate time scales are given in Table 6.2, based on Peters Rit *et al.* (1983).

(iv) Two types of tests on beds of particles must be distinguished: those in which the powder is completely confined, as in a cylindrical vessel surmounted by a piston, and those in which a roller passes over a bed which is not totally confined, e.g. Sligar (1979). In the unconfined case, the bed ahead of the roller is stressed in a way which may cause breakage of larger and hence weaker particles, in advance of the roller (Peters Rit and van Brakel, 1982). In this sense the tests can hardly be classed as fully controlled. On the other hand, the pre-roller breakage must occur in industrial mills of this type and hence

Table 6.1. Categories of single-particle tests

1	2	3	4*
C Slow compression by a press	S Single stress 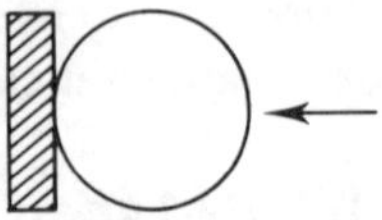	O one particle only	F Free-fall particle(s)
		M Monolayer of equisized separated particles	T Particle shot at target plate K Compressed air device
I Impact	D Double stress 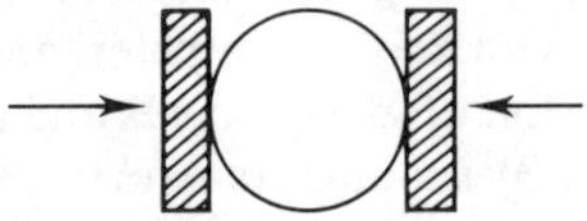		P Pendulum apparatus W Vertically dropped weight

* Applies only to impact tests

Table 6.2. Assessment of externally applied energy

Method	Time scale (s)	Measure of applied energy
Compression	1–10	$\int F\,dy$
Impact (drop weight)	10^{-4}–10^{-2}	mgH (*in vacuo*) or $\frac{1}{2}\ mv^2$
Impact (high-speed)	10^{-6}–10^{-4}	$\frac{1}{2}mv^2$

Reproduced by permission of the Institution of Chemical Engineers.

the tests are realistic. However, no energy data of significance were found for roller tests except that from Schwendig (1966); but because no surface was measured, it is difficult to relate to tests by other workers. Hence breakage by a roller passing over a bed is not further considered in this chapter.

6.2.2 Single-particle tests

6.2.2.1 Single-particle slow compression tests

(a) Significant test reports Significant single-particle slow compression test reports are summarized in date order in Table 6.3.

The problems of comparing the results of the various investigators is immediately apparent from Table 6.3 because of different materials, measures of size reduction and types of test. Despite this, analysis of tests that are comparable reveals the features set out below.

(b) Relation between specific energy and size modulus Bergstrom *et al.* (1961) and Bergstrom (1965) related energy for crushing to the modulus of size

Table 6.3. Significant single-particle slow compression test reports

Reference	Material	Particle shape	Type of test*	Energy measured	Measure of size reduction
Axelson and Piret (1950)	Quartz	Single crystal	CDO	Applied	Surface by GA†
Carey and Stairmand (1953)	Coal, quartz	Irregular	CDM	Applied	Size modulus
Bergstrom *et al.* (1961)	Glass, quartz, sapphire	Sphere	CDO	Applied	Size modulus
Bergstrom (1965)	Glass, quartz, sapphire, flint	Sphere	CDO	Applied	Size modulus
Rumpf (1965)	Glass, quartz, limestone, cement clinker	Sphere, irregular	CDO,CDM	Applied	Surface‡
Bown (1966)	Coal	Irregular	CDM	Applied	Size modulus
Schönert (1972)	Limestone, quartz	Irregular	CDM	Applied	Surface‡
Baumgardt *et al.* (1973)	Glass, cement clinker	Sphere	CDO	Applied, commi-nution	Surface calc. from size distribution§
Rumpf (1973)	Glass, quartz, limestone, sugar, cement clinker	Sphere, irregular	CDO,CDM	Applied	Surface‡
Baumgardt *et al.* (1976)	Glass	Sphere	CDO	Applied, commi-nution	Surface calc. from size distribution§

* For meaning of symbols see §6.2.1
† Gas adsorption
‡ Calculation from size distribution >40 microns and photosedimentation <40 microns
§ Via truncated log–normal equation for size distribution

distribution in slow compression tests using spheres of glass, quartz and sapphire and flint pebbles and obtained a relationship in the form:

$$E_A(x) \propto \frac{1}{x'} \tag{6.9}$$

where $E_A(x)$ is the applied energy per unit mass for a given initial size x obtained by assessment of the area under the stress–strain curve during compression, and x' is the product distribution size modulus (Figure 4.30).

Bergstrom found the same relationship as expressed in equation (6.9) applied to impact as well as slow compression crushing and to irregularly shaped particles as well as spheres by analysis of the work of Hukki (1944) and Charles (1957) for a wide range of materials. The equation expresses the well-known fact that greater specific energy is required to break small than large particles.

(c) Kinetic energy of particles from primary breakage In their slow compression tests on glass spheres, Bergstrom *et al.* (1961) obtained a smaller size modulus when the sphere was surrounded by a steel ring than by a gelatine

ring. In neither case did the ring touch the sphere initially. The smaller modulus was clearly due to secondary breakage of flying fragments in contact with the steel ring, whereas they were embedded in the gelatine ring without breakage. This demonstrates the considerable kinetic energy of fragments from primary breakage.

The same process occurs in primary impact breakage of single particles; and for both slow compression and impact tests, if the true size distribution or surface is required, the secondary breakage due to the kinetic energy of flying fragments must be prevented.

(d) Separated particle monolayer (CDM) tests Bown (1966) demonstrated that the Carey and Stairmand (1953) concept of 'associated energy' related to a given particle size suffered from the use of size modulus as the measure of size reduction, because the size distributions proved not to be straight lines on Gaudin-Schuhmann plots, as claimed by Carey and Stairmand. Hence it was difficult to be specific about the size modulus.

Rumpf (1965) reported continued CDM tests on quartz and limestone, the results of which are given in Figure 6.2. Both curves showed a succession of peaks and troughs corresponding to particle fractures, but those for the harder quartz had higher peaks and lower troughs than those for the softer limestone. Further, the mean slopes of the curves were different. For quartz, at each breakage, fragments were widely scattered owing to their kinetic energy and lay separated ready for further breakage. By contrast, the limestone particles

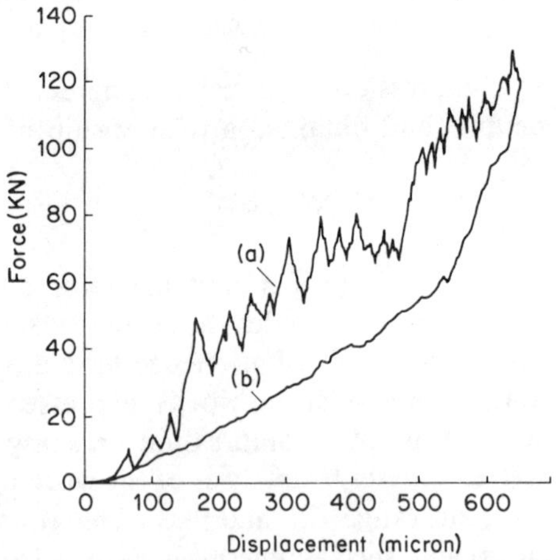

Figure 6.2. Separated particle monolayer (CDM) test data for 50 particles (1.1 mm diameter) broken by slow compression: (a) quartz, (b) limestone (Rumpf, 1965). *Reproduced by permission of VCH Vergalsgesellschaft*

tended to lie in decrepitated piles, and the steeper part of the curve above 500 microns displacement was due to briquetting of those piles.

Successive individual particle (CDO) tests have proved to be more valuable than CDM tests, because statistical averaging at a known initial particle size is then possible.

(e) Utilization of applied energy Axelson and Piret (1950) used the concept of energy utilization, but they did not perform sufficient tests at each particle size to obtain a statistical average applied energy for that size, and hence their results show a considerable scatter, as would be expected (see Figure 6.23) (Hukki and Reddy, 1967).

However, when averaged data are used, the energy utilization for slow compression is found to be a function of (i) the material undergoing breakage; (ii) initial particle size; (iii) energy investment; (iv) specific energy, and (v) shear.

(i) Change of material affects energy utilization as shown in Figure 6.3.
(ii) Increase of initial particle size is associated with:

(1) relatively constant energy utilization at constant energy investment for hard materials like quartz and cement clinker, in keeping with the forecast of Rumpf (1973) (see §6.1.5) and as seen in the solid lines in Figures 6.4 and 6.5;
(2) increasing energy utilization at apparently constant energy investment for softer materials like limestone and sugar, as given by the solid lines in Figures 6.6 and 6.7. The difference from hard materials is attributed to aggregation, a feature of these soft materials.

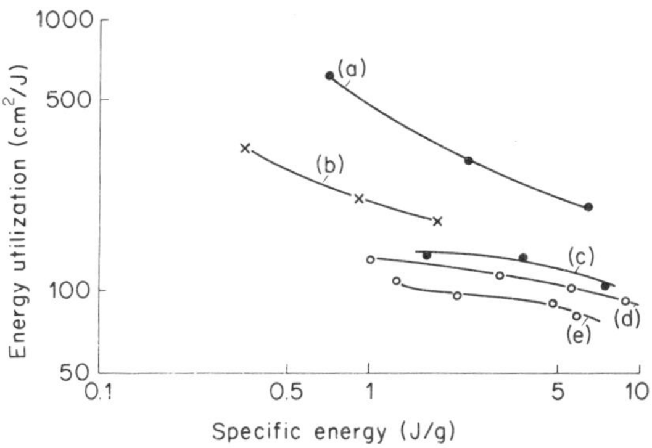

Figure 6.3. Energy utilization versus specific energy for a range of materials crushed by single-particle slow compression: (a) limestone 0.42 mm; (b) limestone 9.00 mm; (c) quartz 0.33 mm; (d) cement clinker 0.32 mm; (e) cement clinker 1.00 mm (Rumpf, 1965). *Reproduced by permission of VCH Vergalsgesellschaft*

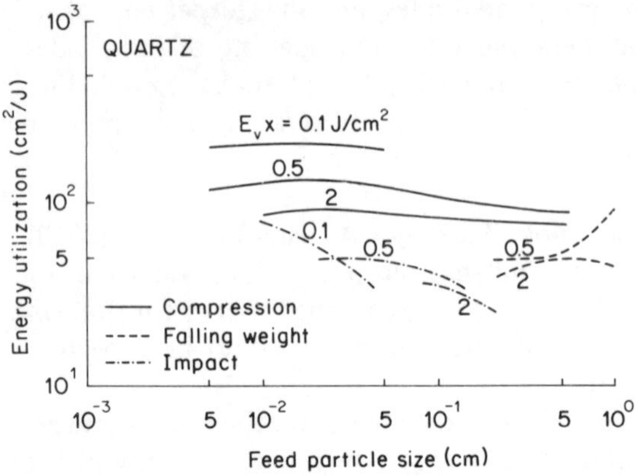

Figure 6.4. Energy utilization versus feed particle size for single particles of quartz crushed by slow compression, by falling weight and by impact (Rumpf, 1973). *Reproduced by permission of Elsevier Sequoia*

(iii) Increase in energy investment $\bar{E}_V x$ results in a decrease in energy utilization for all materials shown in Figures 6.4, 6.5, 6.6 and 6.7 owing to increasing frictional losses as fragments continue to be further stressed after the initial breakage.

(iv) The relation between energy utilization and energy per unit mass was forecast by Rose (1972) for glass from theoretical considerations, as depicted in Figure 6.8. The general pattern of the curves was substantiated by Hukki

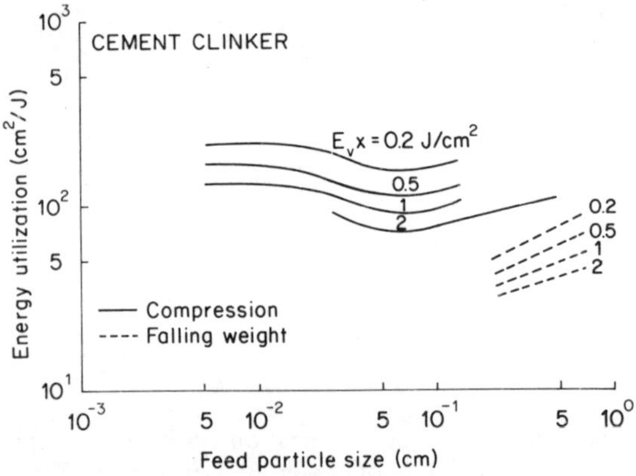

Figure 6.5. Energy utilization versus feed particle size for single particles of cement clinker crushed by slow compression and by falling weight (Rumpf, 1973). *Reproduced by permission of Elsevier Sequoia*

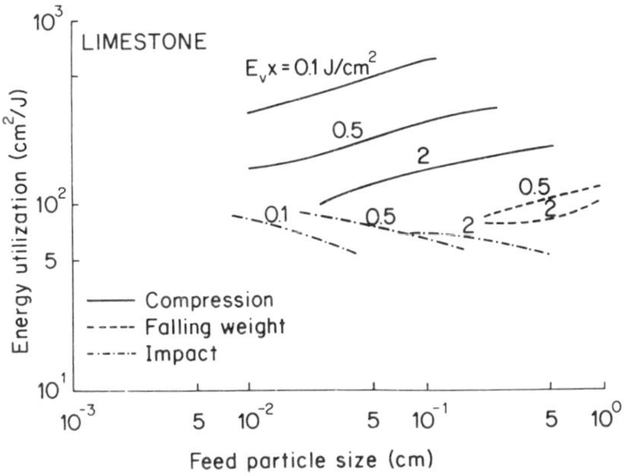

Figure 6.6. Energy utilization versus feed particle size for single particles of limestone crushed by slow compression, by falling weight and by impact (Rumpf, 1973). *Reproduced by permission of Elsevier Sequoia*

and Reddy (1967) over a wide range of materials and conditions (Figure 6.23). The single-particle slow compression results also follow the same general trend (Figure 6.3), but the peak appears not to be reached. The maximum points shown are likely to correspond to the first breakage. The reduction of energy utilization with increased specific energy is due to particle friction in the later stages of breakage.

(v) When shear is incorporated into slow compression, Hess and Schönert

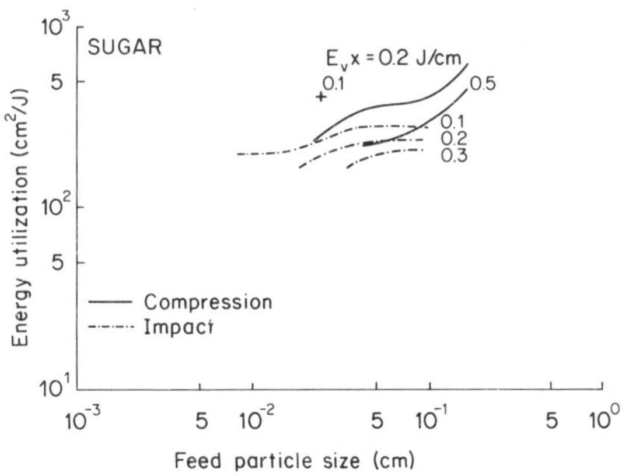

Figure 6.7. Energy utilization versus feed particle size for single particles of sugar crushed by compression and by impact (Rumpf, 1973). *Reproduced by permission of Elsevier Sequoia*

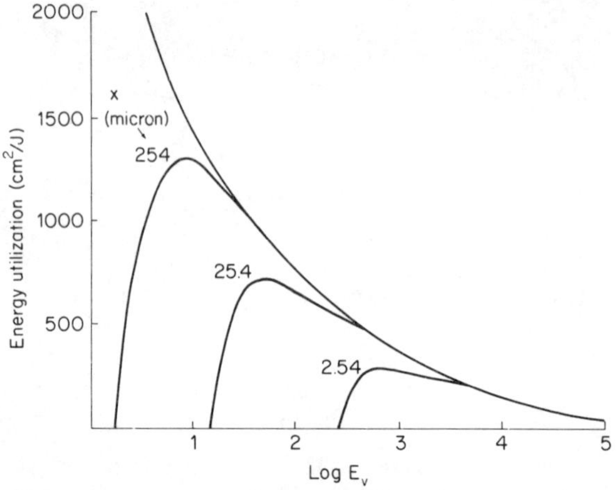

Figure 6.8. Theoretically deduced curves for energy utilization versus energy density for glass (Adapted from Rose, 1972). *Reproduced by permission of DECHEMA, Frankfurt*

(1976) show that the energy utilization is decreased when the ratio of tangential to radial forces is greater than unity for both spheres and irregularly shaped particles of limestone.

(f) Utilization of comminution energy Applying the probability method of §6.1.6, Baumgardt *et al.* (1976) were able to plot the comminution energy probability function $P(\bar{E}_A, x)$ against $\bar{E}_A$ with x as a parameter, resulting in Figure 6.9(a) for glass spheres. By using suitable transforms, indicated by equation (4.21), they also showed that a three-parameter truncated log–normal distribution applied as indicated by the straight lines obtained on the log-probability grid in Figure 6.9(b). Thus a linear relationship was established between the comminution energy probability function and a function of the applied specific energy. From this it was then possible to determine the increase in surface area Δs associated with each value of comminution energy $\bar{E}$ over a range of values of x, resulting in the straight line in Figure 6.10(a). The utilization of comminution energy could then be plotted against specific comminution energy for a given value of x as seen in Figure 6.10(b), giving a straight line parallel to the specific energy axis. Thus, for these conditions utilization of comminution energy is independent of the level of specific comminution energy, which is certainly not true of the utilization of applied energy—witness Figure 6.3.

It must be noted that while the above has been accomplished for glass spheres it is found not to be possible for irregularly shaped particles, since the loading geometry is then a matter of pure chance and the slow compression comminution energy cannot be deduced from the applied energy on prob-

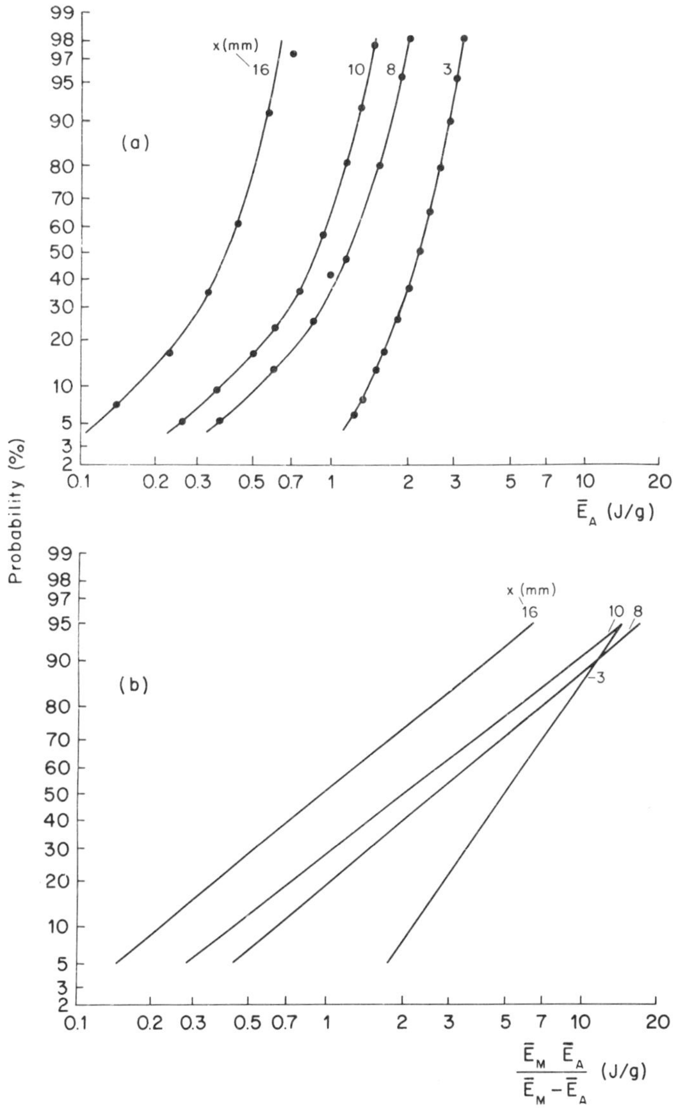

Figure 6.9. Breakage probability $P(\bar{E}_A, x)$ of glass spheres under single-particle slow compression for $x = 3$, 8, 10 and 16 mm diameter (3 mm diameter spheres had different loading velocity from 8, 10 and 16 mm diameter): (a) experimental points, (b) transformed according to equation (4.21) (Baumgardt *et al.*, 1976).
Reproduced by permission of Bergakademie, Freiberg

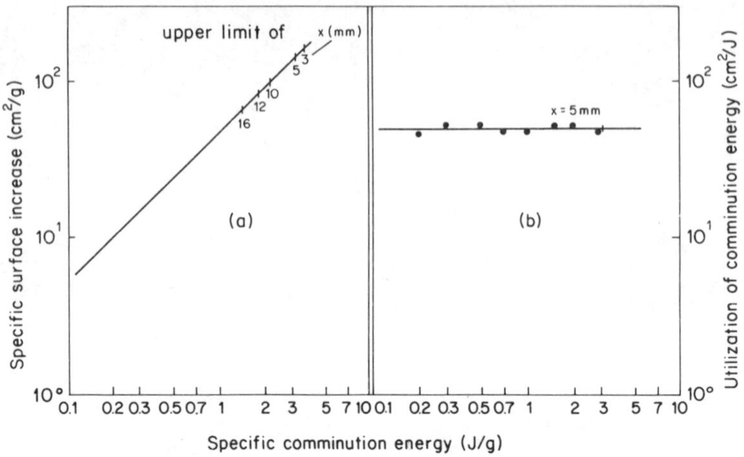

Figure 6.10. For single-particle slow compression breakage of glass spheres, specific comminution energy plotted against (a) specific surface increase, (b) utilization of comminution energy (Baumgardt, *et al.*, 1976). *Reproduced by permission of Bergakademie, Freiberg*

ability considerations. It is still possible, however, with irregularly shaped particles and using slow compression to assess the comminution energy via the stress–strain curve prior to the first breakage. This is clearly unattainable, however, in the context of single-particle impact size reduction, as the time scale is too short.

6.2.2.2 *Single-particle impact tests*

(a) Significant test reports Significant single-particle impact test reports on non-plastic materials are summarized in Table 6.4 in date order. (For impact tests on plastic materials such as nylon, PMMA, etc., see Okuda and Choi, 1980).

(b) Utilization of applied energy From the test data summarized in Table 6.4 it is demonstrated below that for single-particle impact breakage the utilization of applied energy is a function of (i) the material undergoing breakage; (ii) initial particle size; (iii) energy investment; (iv) specific energy; (v) impact velocity; and (vi) angle of impingement.

(i) Change of material alters the energy utilization for otherwise similar conditions, evidenced by a comparison of the chain-dotted lines in Figures 6.4 and 6.6 for quartz and limestone respectively. In this context, Rose (1974) demonstrated a break in the plot of energy utilization against specific energy

Table 6.4. Significant single-particle impact test reports for commonly ground materials

Reference	Material	Particle shape	Type of test*	Energy measured	Measure of size reduction
Rumpf (1965) (includes Priemer, etc.)	Glass, quartz, limestone, cement clinker	Sphere, irregular	IDOW etc.	Applied	Surface†
Behrens (1965)	Glass, pyrites, limestone, quartz	Sphere, irregular	ISOW etc.	Applied	Surface†
Hildinger (1969)	Glass, quartz, limestone, cement clinker	Sphere, irregular	IDOW	Applied	Suface†
Tschorbadjski (1969)	Brown coal	Irregular	ISOT	Applied	Surface
Lenkewitz (1970)	Coal	Irregular	ISOT	Applied	Surface
Schönert (1972)	Glass, quartz, limestone, etc.	Sphere, irregular	IDOK IDOW	Applied	Surface†
Baumgardt *et al.* (1973)	Glass, cement clinker	Sphere	IDOK IDOW ISOF	Applied, commin.	Surface calc. from size distribution‡
Rumpf (1973)	Glass, quartz, limestone, cement clinker	Sphere, irregular	IDOW ISOF	Applied	Surface†
Rose (1974)	Cement clinker	Sphere	IDOW	Applied	Surface
Baumgardt *et al.* (1976)	Glass, cement clinker	Sphere	IDOK IDOW ISOF	Applied, commin.	Surface calc. from size distribution‡
Smith (1979)	Coal, cement clinker	Irregular	ISOF	Applied	Surface calc. from size distribution§
Kerlin (1979)	Coal	Irregular	ISOT	Applied	Integration of size distrib. function
Barnard (1983)	Brown coal char, bituminous coal	Irregular	ISMF	Applied	Surface calc. from size distribution‖

*For meaning of symbols, see §6.2.1
†Calculation from size distribution > 40 microns and photosedimentation < 40 microns
‡Via truncated log–normal equation for size distribution
§Assuming spherical particles
‖By method of Bull (1955)

when crushing heterogeneous minerals, attributed to a change from severing of the matrix only to that of embedded particles and the matrix.

(ii) Barnard (1983) discovered an increase of energy utilization with increase of particle size at constant specific energy in the low-velocity regime, evidenced in Figure 6.11, for both brown coal char and bituminous coal.

By contrast, Kerlin (1980), also using coal, confirmed the earlier work of Priemer (1965) using limestone, that at constant impact velocity, in the high-velocity regime, increased initial particle size reduced the utilization of energy (see Figures 6.13(a) and (b)). The reconciliation of these apparently conflicting

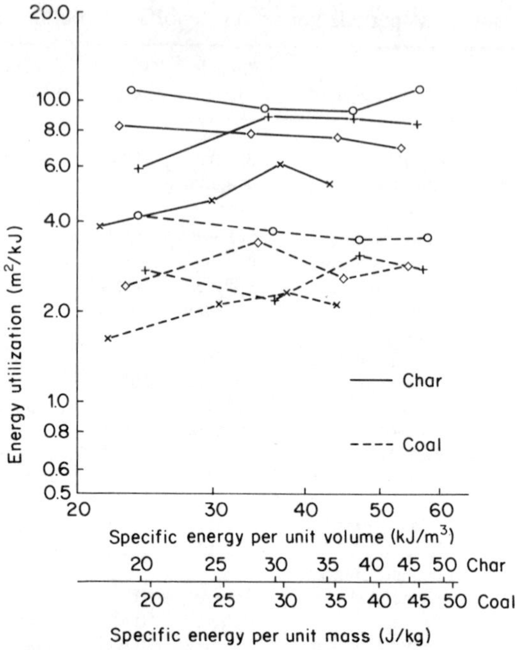

Figure 6.11. Utilization of energy versus specific energy per unit volume, with particle size as a parameter, for shatter tests on brown coal char and bituminous coal (○36 mm +27 mm ◇16 mm ×9 mm) (Barnard, 1983). *Reproduced by permission of the author*

trends is found in Figure 6.20, based on Schönert (1972), where it is clear that at high levels of specific energy the utilization decreases with increasing particle size, whereas at low specific energy the reverse is the case, for impact breakage.

However, Rumpf (1973), introducing the concept of energy investment, did not obtain such a clear trend, as indicated by the dashed and chain-dotted lines in Figures 6.4 for quartz, 6.5 for cement clinker and 6.6 for limestone. Interrelation of the variables is such that the trend of energy utilization with increasing initial particle size may either increase or decrease according to the regime of specific energy for a given material.

(iii) Increasing energy investment tends to reduce energy utilization as found in Figures 6.4, 6.5, 6.6 and 6.7 for impact tests, including falling weight tests.

(iv) Figure 6.12 demonstrates, for both glass spheres and limestone particles, an optimum specific energy for maximum utilization of energy, which maximum is dependent on the initial particle size. Reduction of utilization from the maximum with increase of specific energy beyond the optimum is due to particle-to-particle friction as breakage proceeds during the impacting process.

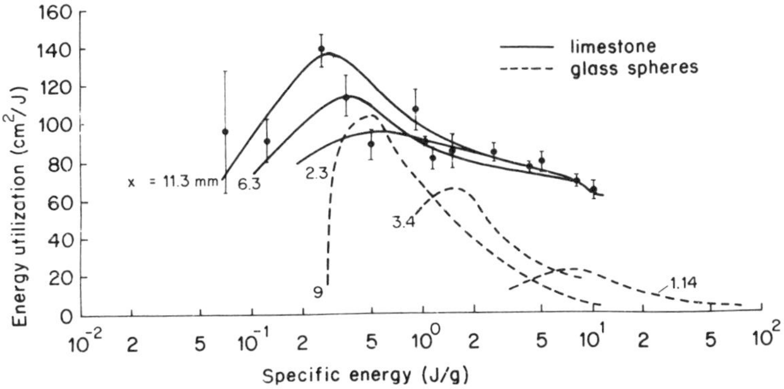

Figure 6.12. Energy utilization versus specific energy for single-particle impact tests on glass spheres and limestone particles (Hildinger, 1969). *Reproduced by permission of VCH Vergalsgesellschaft*

(v) Similarly an optimum velocity exists for maximum energy utilization as given in Figure 6.13, and is likely to be a reflection of the optimum specific energy, as was seen in the context of particle size distribution in §4.3.4.5.

(vi) From Priemer (1965), as particle size is reduced, a higher impact velocity is required to cause breakage.

(vii) The angle of impact has been studied by Tschorbadjski (1969) during tests on moist brown coal. In Figure 6.14 the inverse of energy utilization is plotted against angle of impact, 90° being normal to the surface, and initial particle size is a parameter. As the angle is reduced from 90° the normal component of applied stress is reduced and the shear component is increased. At 60° it appears that the optimum combined effect of the two components occurs.

For a general discussion of the theory of impact breakage, see Rumpf (1959)

(c) Utilization of comminution energy Applying the concepts of §6.1.6 to double-impact crushing of glass spheres, Baumgardt *et al.* (1976) obtained a plot of $P(\bar{E}_A, x)$ against $\bar{E}_A$ with x as a parameter as given in Figure 6.15. Comparison with Figure 6.9 for slow compression reveals a change of shape, attributed by Baumgardt to energy absorbed by the anvil, since the use of harder steel with a higher modulus of elasticity brought the shape of the curves nearer to that from compression tests. (Bergstrom *et al.* (1961) found 89% of input energy went to the particle and 11% to the anvil, for their conditions of test.) However, by the use of a four-parameter log–normal distribution, these curves could be brought to straight lines when plotted on log-probability paper, and thus a relation between breakage probability and comminution energy for glass spheres was obtained in the context of impact (whereas only three parameters were needed in the context of slow compression).

It was then possible to plot specific comminution energy against newly created surface and against utilization of comminution energy for glass

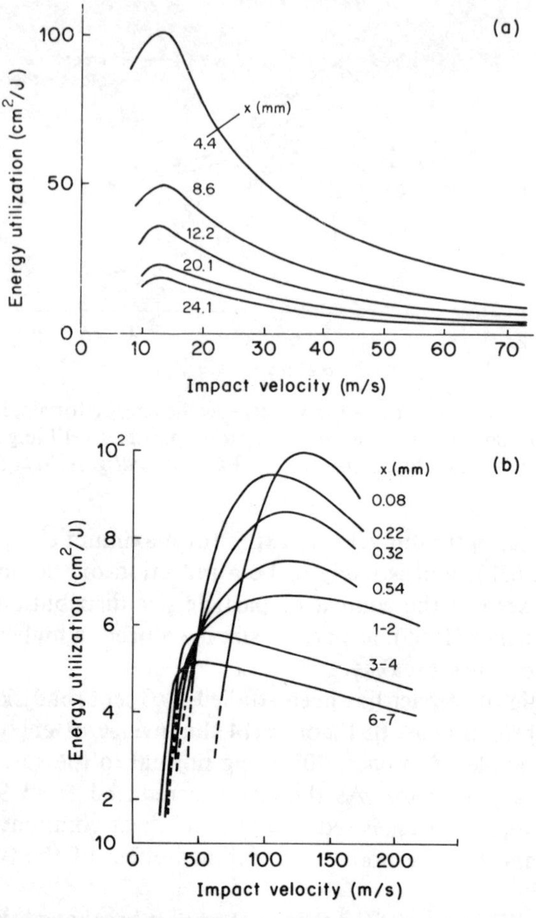

Figure 6.13. Energy utilization versus impact velocity, with particle size as a parameter, for (a) coal (Kerlin, 1979). (b) limestone (Rumpf, 1965). *Reproduced by permission of VCH Vergalsgesellschaft*

spheres, shown in Figure 6.16. As for slow compression, utilization of comminution energy was found to be independent of specific comminution energy for a given particle size, as indicated by the line parallel to the specific energy axis in Figure 6.16(b).

Also on this plot is the graph for utilization of applied energy for glass spheres, which gives the characteristic shape found by other workers in Figure 6.23.

In the context of impact tests, attempts by Baumgardt *et al* (1976) to obtain comminution energy for irregularly shaped particles, as distinct from spheres, by the breakage probability concept did not prove possible.

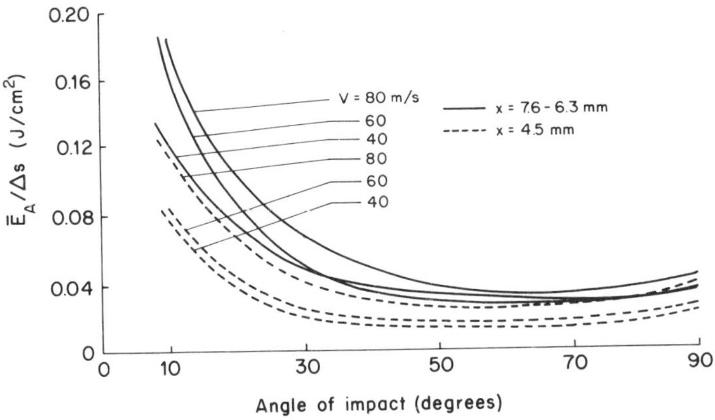

Figure 6.14. Inverse of energy utilization versus angle of impact, with impact velocity and particle size as parameters, for the breakage of brown coal (Tschorbadjski, 1969). *Reproduced by permission of Verlag für Aufbereitung*

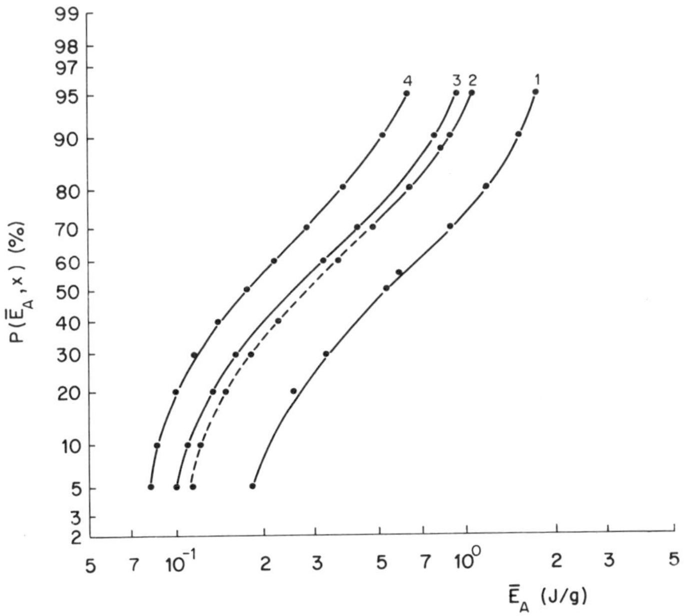

Figure 6.15. Breakage probability $P(\bar{E}_A, x)$ versus $\bar{E}_A$ with particle size as a parameter for glass spheres broken by double impact at $\upsilon = 3.8$ m/s: 1 $x = 5$ mm; 2 $x = 8$ mm; 3 $x = 10$ mm; 4 $x = 16$ mm (Baumgardt *et al.*, 1976). *Reproduced by permission of Bergakademie, Freiberg*

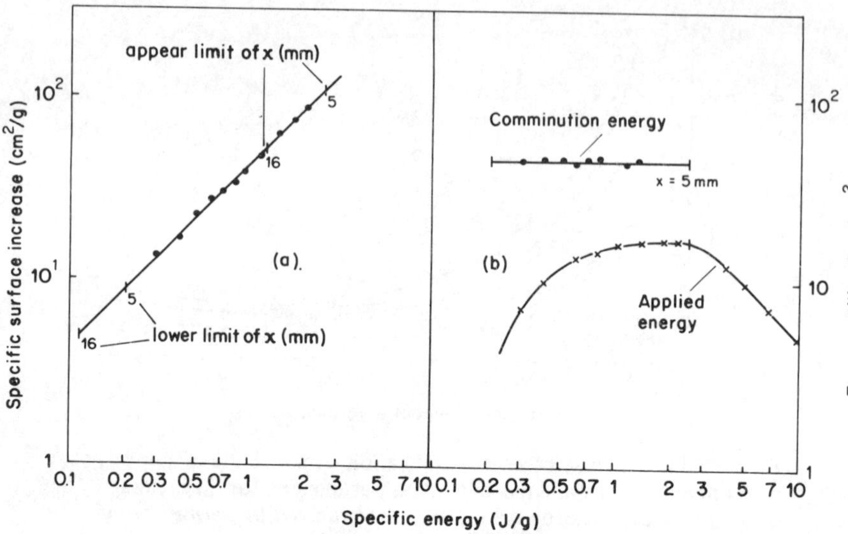

Figure 6.16. For single-particle double-impact breakage of glass spheres, specific energy plotted against (a) specific surface increase, (b) utilization of specific energy (comminution and applied) (Baumgardt *et al.*, 1976). *Reproduced by permission of Bergakademie. Freiberg*

6.2.3 Confined particle bed tests

6.2.3.1 Significant test reports

Significant tests of breakage in confined beds of particles are summarized in Table 6.5 in date order.

By comparison with Tables 6.3 and 6.4 it is evident from Table 6.5 that fewer tests in beds have been conducted than with single particles, which represents a research gap, since most industrial comminution involves beds, and single-particle tests cannot adequately represent the interaction of particles of different size, shape and orientation (Broadbent and Callcott, 1956). There is a particular dearth of slow compression bed tests in which energy has been measured, which could yield valuable insight into the operation of vertical spindle mills, for example.

6.2.3.2 Utilization of applied energy

For the tests in Table 6.5 involving the measurements of product surface, the utilization of applied energy appears to be a function of (i) the material undergoing breakage; (ii) initial particle size; (iii) specific energy.

(i) The effect of change of material on energy utilization is seen in Figure 6.17, which summarizes the slow compression and impact tests of Johnson

Table 6.5. Significant test reports of breakage in confined particle beds

Reference	Material	Method of stressing	Method of energy measurement	Measure of size reduction
Gross and Zimmerley (1930)	Quartz	Impact	mgH	Surface*
Johnson *et al.* (1949)	Quartz	Impact Compression	mgH Stress–strain curve	Surface by GA, PE†
Axelson *et al.* (1951)	Quartz, calcite, fluorite, topaz, sphalerite, salt, apatite, glass	Impact	mgH	Surface by GA, PE†
Batra and Biswas (1971)	Quartz	Impact	mgH	Distribution modulus
Rose (1974)	Cement clinker spheres	Impact	mgH	Surface
Buss *et al.* (1982)	Quartz	Compression	Stress–strain curve	Calculation from size distribution

*By initial rate of dissolution of quartz in hydrofluoric acid
†GA = gas adsorption, PE = permeability

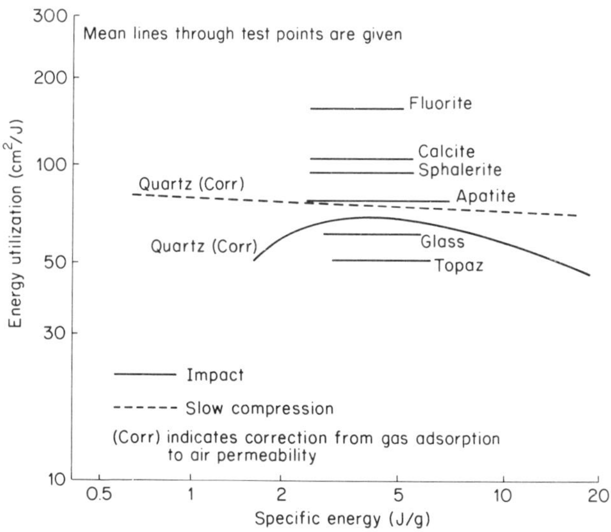

Figure 6.17. Correlation of energy utilization data for a range of minerals crushed in a confined bed. Surface measured by air permeability (Deduced from tabulated data in Axelson *et al.*, 1951; correction from Johnson *et al.*, 1949). *Reproduced by permission of the American Institute of Mining, Metallurgical and Petroleum Engineers*

et al. (1949) and Axelson *et al.* (1951). In this work sufficient tests were done using both gas adsorption and air permeability methods of product surface measurements for a correction from one to the other to be made for quartz. This has allowed the quartz data to be included with that for other minerals, all on the basis of air permeability.

(ii) Figure 6.18 demonstrates the effect of initial particle size on energy utilization in drop weight tests using a bed of cement clinker. The special features of this test must be noted, since the bed surrounded four hard steel spheres lying touching each other on the hard base of the containing cylinder, and a fifth similar sphere attached to a large mass was allowed to fall centrally to the four spheres in the bed, forcing them outwards to crush the surrounding particles. This was an attempt to simulate the action within a tumbling ball mill.

Over the range of specific energy explored, increase of initial particle size led to increase of energy utilization, which would be expected from the fact that larger are weaker than smaller particles and therefore produce a greater surface for a given expenditure of energy.

(iii) Although some tests, such as those summarized in Figure 6.17 for quartz and Figure 6.18 for cement clinker (both by impact crushing), do show an optimum specific energy for maximum energy utilization, most others result in relatively flat lines. Admittedly, for most of the minerals treated the range of specific energy was limited, but for the slow compression of quartz the curve remained flat over a 10 : 1 energy range. Similarly Gross and Zimmerley (1930) found a relatively flat curve from impact tests on quartz over a 100 : 1 energy range. It can be deduced that the optimum specific energy for maximum energy utilization is very much less distinct in bed breakage compared with

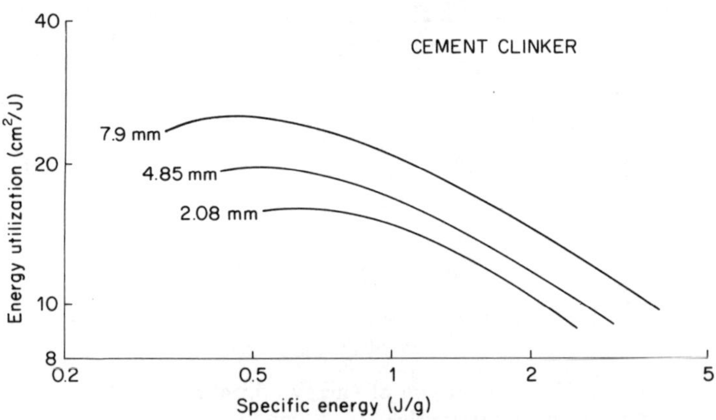

Figure 6.18. Energy utilization versus specific energy, with initial particle size as a parameter, for a bed of cement clinker crushed by a drop-weight method (Based on Rose, 1974). *Reproduced by permission of the Institution of Mining and Metallurgy*

single-particle breakage, which would be expected from particle-to-particle interaction effects within the bed.

No data were found for the effects in bed breakage of energy investment or impact velocity on applied energy utilization, and no way is known for the assessment of comminution energy in these circumstances.

6.2.3.3 Fully and partly confined bed tests

Buss *et al.* (1982) have compared results for fully and partly confined bed arrangements indicated in Figure 6.19(a). The results are shown graphically in Figure 6.19(b). For each case the specific surface and the energy utilization are plotted against the applied energy measured in joules.

For the fully confined case, the results are in line with those of other workers, exemplified in the broken line for quartz in Figure 6.17; that is, the energy utilization is independent of the applied energy over a wide range.

For the partially confined case, both the specific surface and the energy utilization fall off from the respective levels of the fully confined case, as applied energy is increased.

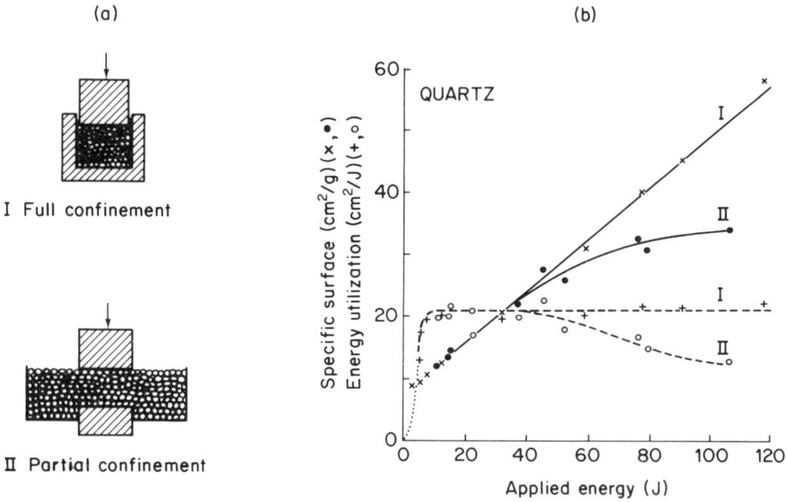

Figure 6.19. Tests with full and partial particle confinement: (a) arrangement, (b) results for quartz 2.5 mm to 3.15 mm initial size (Buss *et al.*, 1982). *Reproduced by permission of Bergakademie, Freiberg*

6.2.4 Survey and application of specialized energy test data

6.2.4.1 Tabulation of energy utilization trends

Applied energy utilization has proved to be a valuable concept in the analysis of basic test data, and the effects thereon caused by variation of other

Table 6.6. Effect on applied energy utilization of a range of variables in specialized energy tests

Variable	Effect on applied energy utilization: Single-particle tests: Slow compression	Single-particle tests: Impact	Confined particle bed tests
Increase of initial particle size at constant specific energy	Decrease (Figures 6.3, 6.20)	Increase at low specific energy (Figures 6.11, 6.13(b)) Decrease at high specific energy (Figures 6.13(a) and (b)) See also Figure 6.20	Increase (Figure 6.18)
Increase of initial particle size at constant energy investment ($E_V x$)	Constant for hard materials. Increase for soft materials because of aggregation (Figures 6.4, 6.5 6.6 and 6.7)	No clear-cut trend (Figures 6.4, 6.5 6.6 and 6.7)	No data
Increase of energy investment ($E_V x$)	Decrease (Figures 6.4, 6.5 6.6 and 6.7)	Decrease (Figures 6.4, 6.5 6.6 and 6.7)	No data
Specific energy	Reduction of specific energy increases energy utilization (Figure 6.3)	Optimum specific energy for maximum utilization (Figure 6.12)	Optimum specific energy for maximum utilization, less marked than in single-particle tests (Figures 6.17, 6.18)
Velocity of impact	Not applicable	Optimum velocity for maximum energy utilisation (Figure 6.13)	No data

parameters is summarized in Table 6.6 from §6.2.2.1(e) for single-particle slow compression, §6.2.2.2(b) for single-particle impact, and from §6.2.3.2 for particle bed tests.

6.2.4.2 Survey and application of single-particle tests

Table 6.6 clearly displays conflicting trends in the effect of variables on the energy utilization, due partly to the difficulty in size reduction exercises to isolate the influence of a single parameter, particularly under impact con-

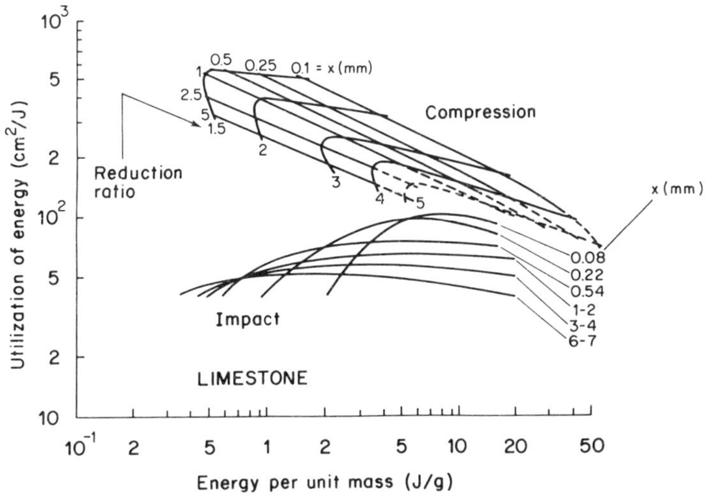

Figure 6.20. Energy utilization versus specific energy for limestone crushed by slow compression and by impact (Schönert, 1972). *Reproduced by permission of the American Institute of Mining, Metallurgical and Petroleum Engineers*

ditions. The different pattern of behaviour under slow compression and impact is well-illustrated for limestone in Figure 6.20 (Schönert, 1972), though at high values of specific energy the curves tend to merge together. For compression, at increased reduction ratio, particle size is reducing from the initial size shown and particle-to-particle friction is increasing, and hence energy utilization decreases. For impact, starting with low energy, increase in energy produces more surface, as would be expected; but as energy is increased, within the short time of impact, the same effects take place as indicated above for compression, and so an optimum is reached beyond which utilization falls.

Smith (1979) has attempted to assess the optimum speed of impact for maximum energy utilization to improve the performance of hammer mills. The very distinct optimum energy density is a prominent feature of single-particle breakage, in contrast to bed studies in which the optimum is less well-defined. This is an area where the direct application of single-particle tests to commercial equipment is valid, since no bed of particles exists in a hammer mill.

6.2.4.3 Survey and application of confined bed tests

Because of the limited extent of energy studies in beds of particles, the application to industrial machines is equally limited. The only known attempt is that of Rose (1974), who used the specialized drop weight apparatus mentioned in §6.2.3.2(ii) to obtain data which were applied to ball mill operation, but not too successfully in all aspects. For example, the prediction relative to optimum ball size distribution proved to be contrary to findings in full-scale mills.

Despite this, there is definite scope for advantageous research into the energy and particle size aspects of compression in confined beds.

6.3 LABORATORY MILL TESTS MEASURING OUTPUT SURFACE

6.3.1 Test data

Data are available for the plotting of energy utilization against specific energy, as indicated in Table 6.7.

Figure 6.21 summarizes the results of Svensson and Murkes (1957) for limestone, both wet and dry operation, and for ferrosilicon and quartz. The

Table 6.7. Test reports on laboratory mills grinding minerals for which output surface is measured

1 Reference	2 Material	3 Condition	4 Output fineness range %<75 micron	5 Measure of size reduction
Mortsell and Svensson (1951)	Quartz	Dry	17–98	Surface PE*
Svensson and Murkes (1957)	Quartz,	Dry, wet	4–86.7	Calculated for range>589 micron, Surface (PE) <589 micron
	limestone,	Dry, wet	–	
	ferro-silicon	Dry	–	
Hukki and Reddy (1967)	Quartz	Dry, wet	–	Surface PE

*PE = permeability method

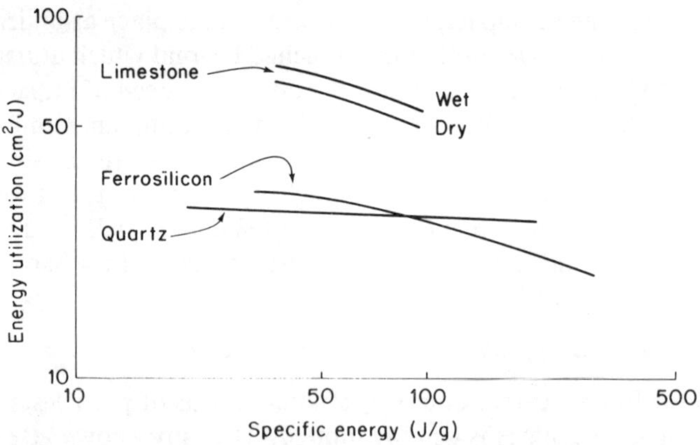

Figure 6.21. Energy utilization versus specific energy for limestone, ferrosilicon and quartz in a laboratory ball mill (Svensson and Murkes, 1957). *Reproduced by permission of Almqvist & Wiksell Förlag AB*

more easily ground limestone produces a greater output surface for a given expenditure of applied energy than the harder minerals. These results for quartz are consistent with those of Mortsell and Svensson (1951) and Hukki and Reddy (1967), as seen in Figure 6.23. A wide range of output fineness was covered as indicated in column 4 of Table 6.7 and energy utilization decreases as output fineness is increased. (see §6.3.2.1.)

6.3.2 Comparison of laboratory mill and specialized energy tests

6.3.2.1 Specific new surface versus specific applied energy

Two methods of presentation of surface/energy data have been adopted to compare specialized energy test data with that from laboratory mills.

Behrens (1965) and Schönert (1966) plotted specific surface per unit mass against specific energy (mass-related) for quartz, as given in Figure 6.22. Three features of this diagram should be noted:

(1) Apart from the single-particle slow compression data of Axelson and Piret (1950), who did not perform enough tests at each particle size to obtain adequate average specific energies, the approximation holds that increase in surface is directly proportional to the applied specific energy over a

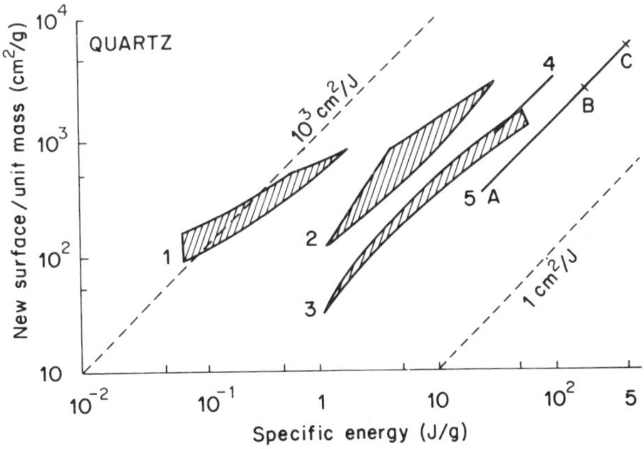

Figure 6.22. New surface per unit mass versus specific energy for quartz (Schönert, 1966). *Reproduced by permission of the author*

1	Single-particle, slow compression	Axelson and Piret (1950)	GA
2	Single-particle, compression of irregular particles	Schönert (1966)	PS + Calc
3	Single-particle, impact	Behrens (1965)	PS + Calc
4	Grindability test	Schönert (1960)	PS + Calc
5	Laboratory mill test (gross energy)	Mortsell and Svensson (1951)	PE

GA = gas adsorption; PE = permeability; PS = photosedimentation range of types of experiment. It is also true for other materials such as coal (Peters Rit and van Brakel, 1982).

(2) By comparing item 2, the single-particle compression of irregularly shaped particles, with item 5, the gross energy required in a laboratory ball mill, at the same level of new specific surface, it is seen that ten times more energy is needed in the laboratory mill than for single-particle breakage.

(3) For item 5, the points marked represent output fineness as follows:

Point	*% minus 75 micron*
A	17
B	70
C	90

Over this range of fineness there is more than 10:1 increase in specific energy.

6.3.2.2 Energy utilization versus specific surface

Based on Hukki and Reddy (1967), Figure 6.23 gives, for quartz, energy utilization against specific energy in the units used in the commercial world,

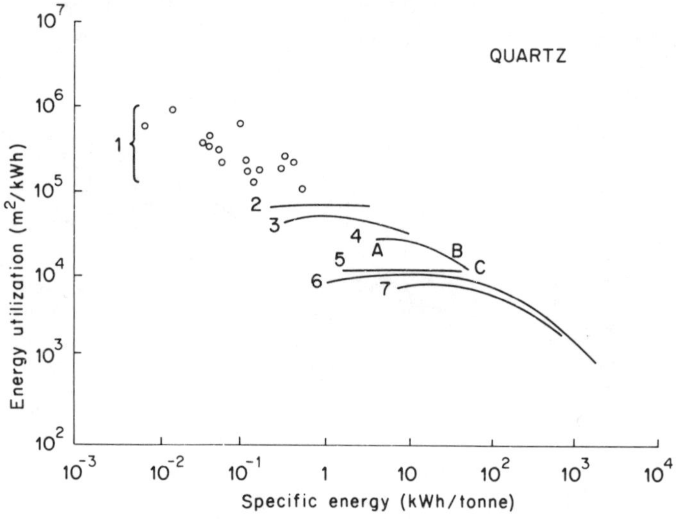

Figure 6.23. Energy utilization versus specific energy for quartz (based on Hukki and Reddy, 1967). *Reproduced by permission of DECHEMA, Frankfurt*

1	Single-particle, slow compression	Axelson and Piret (1950)	GA
2	Confined-bed, impact	Gross and Zimmerley (1930)	HF
3	Multiple-particle, slow compression	Axelson and Piret (1950)	GA
4	Laboratory ball mill—dry	Mortsell and Svensson (1951)	PE
5	Laboratory ball mill—dry	Svensson and Murkes (1957)	PE + Calc
6	Laboratory ball mill—dry	Hukki and Reddy (1967)	PE
7	Laboratory ball mill—wet	Hukki and Reddy (1967)	PE

GA = gas adsorption; HF = dissolution in HF acid; PE = permeability namely kWh/tonne. The total range of specific energies from Table 1.1 does, in fact, coincide with the test-found range, which indicates that the tests are realistic in commercial terms.

Since the methods of surface measurement in Figure 6.23 are not all consistent, it is dangerous to make detailed deductions in respect of the relative positions of the curves; but the general pattern of the graph gives valid trends as follows:

(1) The single particle tests of Axelson and Piret (1950) were not done in sufficient numbers at each particle size to give averaged data for that size; hence the scatter of points. However, it is still true that the higher energy utilization is associated with single-particle breakage and with low values of specific energy.
(2) In multiple-particle and confined bed tests, because of particle-to-particle friction and other factors (Figure 6.26), specific energy increases and energy utilization decreases compared with single particle tests.
(3) In laboratory ball mills, the same is true, but to a greater degree.
(4) It is clear that many parameters affect the energy utilization and that to attempt to single out any one would be unrealistic, though such attempts have been made. (see §6.5.)

6.4 EFFICIENCY TESTS

6.4.1 Preview

Referring to §6.1.3, two valid definitions of efficiency are there expressed, which may be summarized as

Ideal: comminution energy/applied energy
Practical: (applied energy – heat loss)/applied energy

6.4.2 Ideal efficiency

Data from Figure 6.16(b) allows the calculation of this efficiency by taking the ratio of the inverse of the ordinates for energy utilization for each of comminution and applied energies at several values of specific energy. When this is done, Figure 6.24 is the result, applying to single-particle double-impact breakage of quartz. A definite peak at about 0.3 kWh/tonne is obtained, the optimum specific energy for a maximum efficiency of 30% under these conditions.

Clearly, efficiency of breakage is very dependent on the level of specific energy employed.

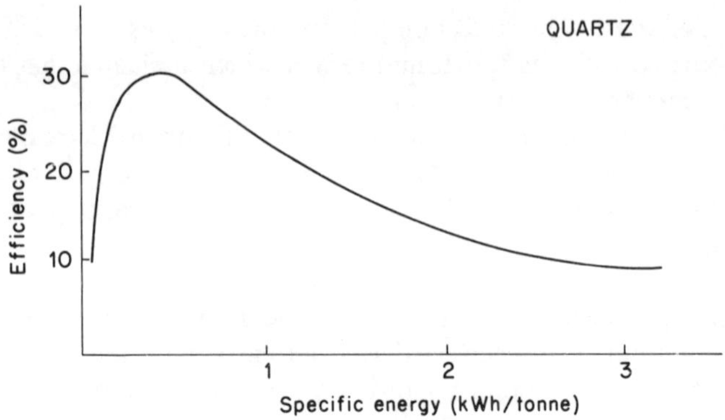

Figure 6.24. Ideal efficiency, defined as the ratio of comminution to applied energy, derived from Figure 6.16(b), for the double-impact crushing of single particles of quartz (5 mm)

6.4.3 Practical efficiency

6.4.3.1 Significant test reports

Table 6.8 lists in date order calorimeter tests designed to assess the heat loss during continuous size reduction.

Table 6.8. Significant test reports of calorimeter efficiency tests on mills

Reference	Material	Equipment	Condition	Range of efficiencies (%)
Cook (1915)	Gold-bearing ore	Commercial stamp machine	Wet	18.8–21.4
Fahrenwald *et al.* (1931)	Silica	Laboratory ball mills	Wet	7.4–19.7
Prentice (1949)	Gold-bearing ore	Commercial m/c:		
		gryatory crusher	Dry	19.9–21.3
		stamp m/c	Wet	18.6–21.0
		ball mill	Dry	17.5–22.1
Schellinger (1951)	Quartz	Laboratory ball mill	Dry, wet	8–19 (see Figure 6.25)
Schellinger and Lalkaka (1951)	Limestone	Commerical ball mills:		
		$9\frac{1}{2}$ ft × 10 ft	Wet	10.2–19.7
		9 ft × $11\frac{1}{2}$ ft	Wet	19.9
Rose (1962) (Discussion)	Quartz	Laboratory vibration mill	Dry	20–25 (early assessment)
Rose (1967)	Silica	Laboratory vibration mill	Dry	Less than 1%

6.4.3.2 *Discussion of test reports*

The Schillinger (1951) data allow thermal efficiency to be plotted against mill load in Figure 6.25 for both wet and dry conditions. An optimum load exists for a maximum efficiency of 19%.

In view of the fact that very small temperature differences are required to be measured, there is a remarkable consistency in the maximum level of efficiency (about 20%) obtained from laboratory and commercial equipment up to 1951, indicated in Table 6.8. It is to be noted, however, that the products of comminution in all these cases were relatively coarse, in the range 20–35% minus 75 microns, associated with primary crushing. (Fine grinding is liable to be less efficient than coarse owing to the greater tendency to overgrind in the former.) This maximum efficiency of 20% might be thought reasonable in the light of 30% maximum obtained in the rather ideal conditions of single-particle breakage (Figure 6.24). However, the real efficiency is likely to be well below 20% because it is difficult to be sure that all losses have been accounted for. This may be illustrated in the two results from the work of Rose shown in Table 6.8. The value of 20–25% was an early assessment only, and did not include the full losses reported in 1967. When the energy associated with phase change in the material was included, then a very low value of efficiency was the result for the vibration mill. The vibration mill, however, would be expected to have a lower efficiency than other mills, as Bond pointed out in discussion on Fuerstenau (1959), because a large number of metal-to-metal contacts lead to wasted energy owing to the small percentage of voids between

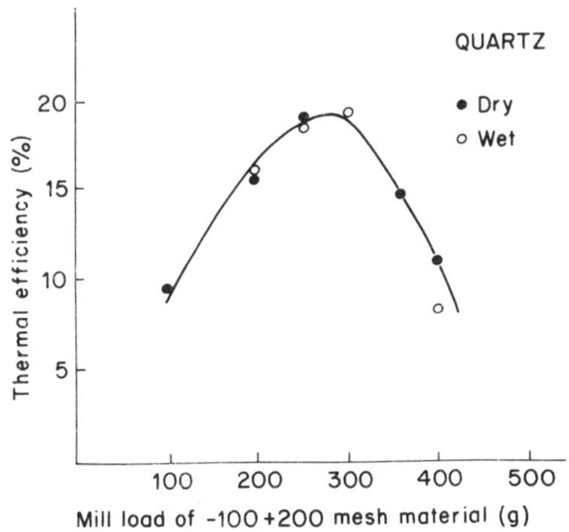

Figure 6.25. Thermal efficiency of the ball milling of quartz versus powder filling load by a calorimeter method, for both dry and wet conditions (Schellinger, 1951). *Reproduced by permission of the American Institute of Mining, Metallurgical and Petroleum Engineers*

grinding media which is occupied by powder. For tumbling mills, a value somewhat greater than 1% for thermal efficiency is likely to be found in practice, and Austin and Klimpel (1964) have proposed 3%. Planiol (1963) concluded the finer the grinding, the lower the efficiency.

Today little emphasis is placed on thermal efficiency of grinding because it is inherently low anyway; because it is applied-energy-density dependent; because its measurement is very difficult; and because efforts are better directed into more fruitful directions, such as energy utilization.

6.4.3.3 *Analysis of energy losses*

The complexity of the energy exchange during crushing and grinding has been highlighted by Rumpf (1973), as illustrated in Figure 6.26. Most of the listed effects have been discussed in Chapter 3, but their interrelations are clearly indicated in Figure 6.26. It is little wonder that attempts to measure efficiency by the losses method has produced such a wide variety of answers!

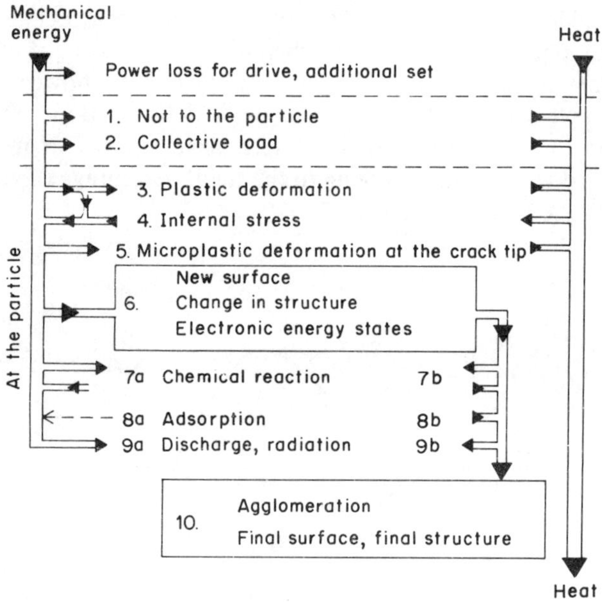

Figure 6.26. Diagram of energy exchange during comminution (Rumpf, 1973). *Reproduced by permission of Elsevier Sequoia*

6.5 ENERGY LAWS

The proposed laws relating energy for breakage to simple parameters are set out in historical sequence.

6.5.1 Rittinger's law

Rittinger (1867) postulated that the energy expended in crushing is proportional to the new surface formed, and hence on the basis of equation (4.38), assuming a constant specific surface coefficient:

$$\bar{E}_A \propto \left(\frac{1}{x_2} - \frac{1}{x_1}\right) = K_1 \Delta s \tag{6.9}$$

where x_1 and x_2 are characteristic dimensions of feed and product respectively.

Hence, in terms of energy utilization:

$$\frac{\Delta s}{\bar{E}_A} = K_1 \tag{6.10}$$

6.5.2 Kick's law

Kick (1885) stated that for any unit mass of material, the energy required to produce a reduction ratio is constant, no matter what the original size of the particle may be. Following Bond and Wang (1950), let K_2 be the specific applied energy to reduce from size x_1 to $x_1/2$. This represents one reduction step. If r steps are required for a reduction ratio of x/n, then $n = 2^r$ or $r = \log n/\log 2$. Then the energy needed is $K_2(\log n/\log 2)$, or generally for a reduction ratio of x_1/x_2:

$$\bar{E}_A = K_3 \log(x_1/x_2) \tag{6.11}$$

To express Kick's law in terms of energy utilization, from equation (6.11):

$$x_1/x_2 = e^{(\bar{E}_A/K_3)}$$

But:

$$\Delta s = K_4\left(\frac{1}{x_2} - \frac{1}{x_1}\right) = \frac{K_4}{x_1}\left(\frac{x_1}{x_2} - 1\right)$$

Hence:

$$\frac{\Delta s}{\bar{E}_A} = \frac{K_4}{\bar{E}_A x_1}[e^{(\bar{E}_A/K_3)} - 1] \tag{6.12}$$

6.5.3 Bond's law

Bond (1952) concluded that the work input to break a cube of side x_1 is proportional to the volume x_1^3 of the cube; but on formation of the first crack, the energy flows to the resultant new surfaces, such energy being proportional to x_1^2. When an irregularly shaped particle is broken, the strain energy is irregularly distributed, he argued, and hence the energy required for breakage lies between x_1^3 and x_1^2, the geometric mean being $x_1^{2.5}$, a compromise between Kick and Rittinger. Hence the number of particles of assumed similar shape per unit volume will be proportional to $1/x_1{}^3$, so that the work input to break a unit volume was taken to be $x_1^{2.5}/x_1^3 = 1/\sqrt{x_1}$. In reducing a fixed size x_1 to a product size x_2, the work done was expressed as:

$$\bar{E}_A = K_5\left(\frac{1}{\sqrt{x_2}} - \frac{1}{\sqrt{x_1}}\right) \tag{6.13}$$

Bond then introduced the term 'work index' W_i, the total energy required to break from infinite size to 80% passing 100 microns, or:

$$W_i = K_5 \frac{1}{\sqrt{100}} \tag{6.14}$$

This resulted in the well-known Bond formula:

$$\bar{E}_A = W_i\left[\frac{10}{\sqrt{x_2}} - \frac{10}{\sqrt{x_1}}\right] \tag{6.15}$$

where x_1 and x_2 are expressed in microns as the size associated with 80% passing.

As pointed out by Austin and Brame (1983), it must be noted that Bond's claim (Bond, 1961) that the energy for breakage, inversely proportional to the square-root of the crack length, is based on Griffith theory, has no foundation in fact: in other words, the Bond relation is not a fundamental law.

The Bond equation (6.13) may be expressed in terms of energy utilization (Rose, 1972) as:

$$\frac{\Delta s}{\bar{E}_A} = \frac{1}{K_6}(\bar{E}_A) + \frac{1}{K_7}\left(\frac{1}{\sqrt{x_1}}\right) \tag{6.16}$$

When equations (6.10), (6.12) and (6.16) are plotted, a graph results as in Figure 6.27. Comparing Figure 6.27 with Figure 6.23 reveals that Rittinger's law applies over the horizontal sections of the curves, Kick's law hardly seems

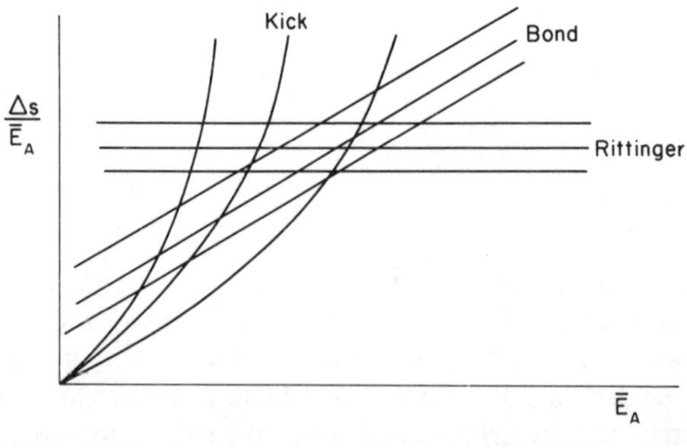

Figure 6.27. Typical plots of energy utilization versus specific energy when the laws of Rittinger, Kick and Bond are applicable (Rose, 1972). *Reproduced by permission of DECHEMA, Frankfurt*

to be appropriate to the data presented, and Bond's law pertains to the low specific energy ends of some of the curves. None of the laws applies to the high specific energy ends of the curves. This illustrates the inadequacy of these laws to express the test-found surface-energy relationships over their full range. See also §9.2.2.4.

6.5.4 Holmes' law

It was understandable in the light of these inadequacies that Holmes (1957) should introduce a variable exponent r in place of the power of 0.5 used by Bond, producing:

$$\bar{E}_{\mathrm{A}} = W_{\mathrm{i}}\left[\left(\frac{100}{x_2}\right)^r - \left(\frac{100}{x_1}\right)^r\right] \tag{6.17}$$

thus allowing lines of any desired slope on the energy utilization plot. The problem still remains, however, that in changing output fineness, the slope also changes as exemplified by A, B and C on curve 4 in Figure 6.23, along which fineness moves from 17% through 70% to 98% minus 75 micron. Since the slope of the curve changes in moving from A to C, the exponent r will also change with increase of output fineness, and thus equation (6.17) is not easily applied in practice.

6.5.5 Charles' law

Charles (1957) also incorporated a variable exponent into the energy equation but in the context of the size modulus of the Schuhmann plot. Based on Charles' concept, Schuhmann (1960) expressed the energy for breakage as:

$$\bar{E}_{\mathrm{A}} = A(x')^{-n} \tag{6.18}$$

where n is the slope of the straight-line portion of the Schuhmann plot.

Kinasevich and Fuerstenau (1964) demonstrated that equation (6.18) can apply to relatively simple comminution systems, but it is easy to find conditions for which n changes with time of grind—that is, with product fineness (Berlioz and Fuerstenau, 1967). For heterogeneous materials, the Schuhmann plots are not straight lines (Fuerstenau, 1961). Hence the argument of §6.5.4 in respect of the exponent r in equation (6.17) also applies to the exponent n in equation (6.18). Further, no provision is made in equation (6.18) for varying feed conditions.

Hence Charles' law displays the shortcomings of the other simple relationships between applied energy and some chosen size-reduction parameters. The comminution process is far too complex for such elementary concepts to be applicable (Austin and Luckie, 1972; Austin, 1973a,b)

Simple relationships other than the five cited here have been advocated, but the same shortcomings equally apply.

6.6 GRINDABILITY

6.6.1 Preview

Five simple energy relationships, which formerly have been advocated, came under review in §6.5 and were found to be inadequate in practical circumstances.

One alternative has been the pragmatic assessment of the ease or difficulty with which a material is reduced in size, i.e. its grindability, which in turn is related to mill performance. Many expressions of grindability have been proposed over the years (Chandler, 1965), but of these two have come into prominence because they have become the recognized basis for the design of certain types of mill. They are:

(1) the Hardgrove index, associated mainly with vertical spindle mills;
(2) the Bond work index, associated with tumbling mills.

6.6.2 Hardgrove index

The essence of the Hardgrove index is a measure of the mass of minus 75 micron material produced in a standardized laboratory mill operated for a

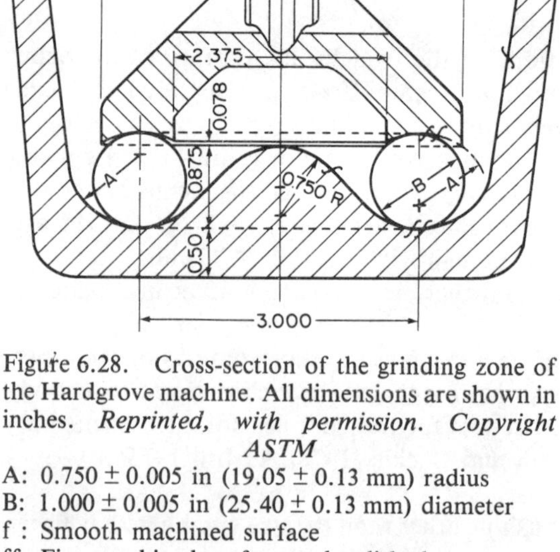

Figure 6.28. Cross-section of the grinding zone of the Hardgrove machine. All dimensions are shown in inches. *Reprinted, with permission. Copyright ASTM*

A: 0.750 ± 0.005 in (19.05 ± 0.13 mm) radius
B: 1.000 ± 0.005 in (25.40 ± 0.13 mm) diameter
f : Smooth machined surface
ff: Fine machined surface and polished

set number of revolutions under standardized batch conditions with a feed prescribed in quantity and size.

The laboratory mill was first introduced by Hardgrove (1932), and present-day details are contained in British Standard 1016 (Part 20) (1981) and ASTM D 409–71 (1978). Briefly, the grinding zone, shown in cross-section in Figure 6.28, consists of a stationary steel bowl in which rotate eight steel balls each of 25.4 mm (1 inch) diameter. The upper grinding ring is loaded with a total weight of 64 lb and is rotated at 20 rev/minute for 60 revolutions. The test sample consists of 50 g of air-dried material passing a 1.18 mm sieve but retained on a 600 micron sieve. The product after test that passes through a 75 micron sieve is weighed and the Hardgrove index is read off from a calibrated chart appropriate to each machine. Calibration to obtain the chart, given typically in Figure 6.29, is by means of four standard powders having respective indices of 40, 58, 83 and 107. The larger the index the easier the grinding, i.e. the greater the mass of material passing the 75 micron sieve.

A list of indices for many materials has been produced by Hardgrove (1938).

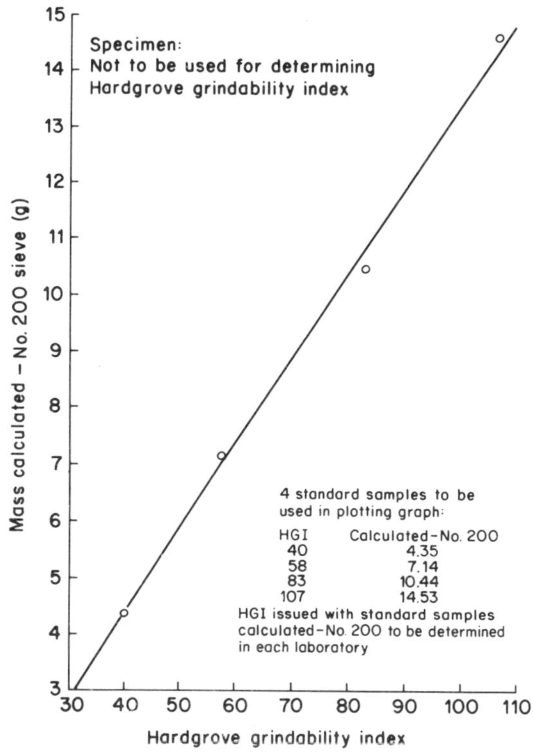

Figure 6.29. Typical calibration chart for the Hardgrove machine. *Reprinted, with permission. Copyright ASTM*

6.6.3 Bond work index

The Bond (1961) ball mill grindability test is a locked-cycle dry grinding and screening process which is continued until an equilibrium state is reached, as described below, aimed at simulating in a batch test the recirculation conditions of a full-scale mill/classifier system.

The Bond mill, having inside measurements of 305 mm (12 inches) and 305 mm diameter, without lifters, is operated at 70 rev/minute and is equipped with a revolution counter. The charge consists of 20.125 kg of iron balls in a prescribed size distribution. The feed is stage crushed, all to pass a 335 micron sieve and packed to 700 cm^3 volume using a vibrating table. The mass of this volume is the initial charge to the mill, maintained throughout the test. The first run is continued until 300–400 g of finished product is obtained which passes the x_t micron test sieve. The oversize fraction is returned to the mill for the second run together with fresh feed to make up the original mass corresponding to 700 cm^3. The mass of product per unit of mill revolution, called the ore grindability of the cycle, is then calculated and is used to estimate the number of revolutions required for the second run to be equivalent to a recirculating load of 250%. In a batch test, this is equivalent to 1/3.5 of the original ore charge. The process is continued until a constant value of the ore grindability is achieved, which is the equilibrium condition. The average of the last three ore grindabilities is taken to obtain g_b in equation (6.19) which equation expresses the Bond work index in kWh/short ton:

$$W_i = \frac{44.5}{(x_t)^{0.23}(g_b)^{0.82}}\left[\frac{10}{\sqrt{x_2}} - \frac{10}{\sqrt{x_1}}\right]^{-1} \tag{6.19}$$

where x_1 and x_2 are the sizes in microns passing 80% of the feed and product respectively.

It is to be noted that the index from equation (6.19) predicts the specific energy requirement for a 2.44 m (8 ft) inside diameter ball mill under wet grinding conditions, with energy measured at the mill pinion shaft.

Because the above process is very time consuming, various simplifications have been introduced, as summarized by Yap *et al.* (1982), who also show the degree of correlation between the indices from simpler tests and from the standard Bond test, which may be good. (see also §9.5.3.2.)

Bond (1952) has listed ball mill work indices for many ores.

There is also a rod mill test (Bond, 1961) which gives a rod mill grindability index different from the ball mill index.

6.6.4 Relation between Hardgrove and Bond grindabilities

Since the essence of both Hardgrove and Bond indices is a mass g_b of material of a given size produced per revolution of the mill, a loose relationship between them would not be unexpected. Further, since specified particle sizes

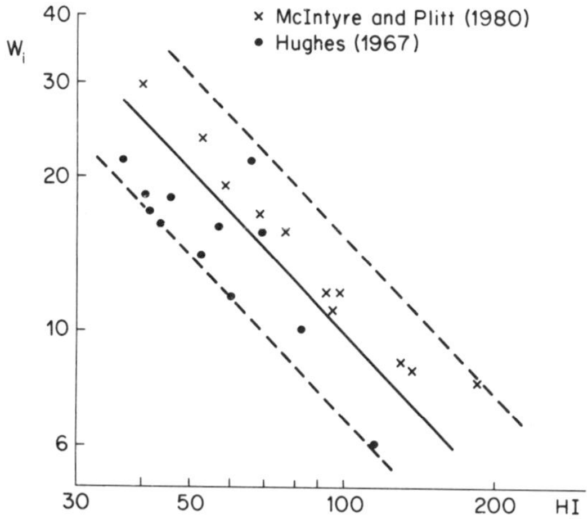

Figure 6.30. Correlation between Hardgrove and Bond grindabilities covering a wide range of minerals (McIntyre and Plitt, 1980; Hughes, 1967; together with data supplied by Warren Springs Laboratory, Stevenage, UK)

are employed for each, the following approximations may be given:

$$HI = K_1 g_b \quad \text{and} \quad W_i = K_2 / g_b$$

Hence:

$$HI \times W_i = \text{constant}$$

Otherwise, if HI is plotted against W_i on log–log paper, a straight line of slope − 1 would be expected.

Figure 6.30 shows this is true within limits of experimental scatter for available data from McIntyre and Plitt (1980) and Hughes (1967), covering a wide range of ores.

More precise relationships which have been attempted by Bond (1954), Bond (1961) and McIntyre and Plitt (1980) are not justified in the light of the differences between the test procedures.

6.7 USE OF GRINDABILITY IN MILL DESIGN

6.7.1 Hardgrove index—vertical spindle mills

Because of the differences between the batch test whereby the Hardgrove index is derived and the full-scale continuous mill with internal recirculation, the index cannot be directly used in mill design. Hence a grindability factor has to be found for a given mill type within the vertical spindle family, (see Figure 12.3) which relates mill performance to the Hardgrove index. Such a factor is,

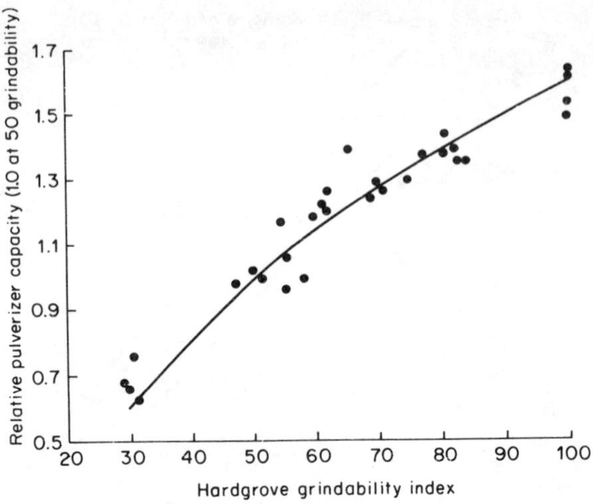

Figure 6.31. Typical relation between relative capacity (at a given output fineness) and the Hardgrove index for coal ground in a ring-ball mill (Hardgrove, 1933)

in fact, a constant-power (for a given mill motor size), constant-output-fineness capacity factor. In other words, it is a quantitative measure of the change of capacity according to the ease or difficulty with which a material is ground.

Typical grindability factors are given in Figure 6.31 for coal ground in a ring-ball mill (Hardgrove, 1933). It is noted that the factor has been arbitrarily taken as unity for a Hardgrove index of 50. Considerable scatter of experimental points exists about the mean line shown.

The specific energy required at a standard capacity, product fineness and material grindability (50) is found empirically, and hence the specific energy at other grindabilities can be assessed using a curve like Figure 6.31 assuming constant power.

6.7.2 Bond work index—tumbling ball or rod mills

Since the Bond work index as given by equation (6.19) is geared to wet closed-circuit grinding with a coarse feed in a 2.44 m (8 ft) inside diameter liners mill, factors must be applied for other conditions as follows (Rowland and Kjos, 1978):

(1) Dry grinding factor $f_1 = 1.3$.
(2) Open-circuit factors f_2 are given in Table 6.9.
(3) Diameter factor f_3 is given by:

$$f_3 = (2.44/D)^{0.2} \text{ with } D \text{ in metres}$$

$$\text{or } f_3 = (8/D)^{0.2} \text{ with } D \text{ in feet}$$

Table 6.9. Ball mill open-circuit factor, according to product fineness

Product size (% minus 75 microns)	f_2
50	1.035
60	1.05
70	1.10
80	1.20
90	1.40
92	1.46
95	1.57
98	1.70

Reproduced by permission of the American Institute of Mining, Metallurgical and Petroleum Engineers.

(4) Oversize feed factor

$$f_4 = 1 + [(W_i - 7)(x_1 - x_o)/x_o R] \tag{6.20}$$

where x_o is the optimum feed size (80% passing), x_1 is the actual feed size, coarser than optimum (80% passing), and R is the reduction ratio x_1/x_2. Clearly equation (6.20) is not applicable for $W_i < 7$, and further:

$$x_o = 4000\sqrt{(13/W_i)} \text{ for ball milling}$$

$$\text{and } x_o = 16{,}000\sqrt{(13/W_i)} \text{ for rod milling}$$

(5) Fineness of grind for factor f_5 is given by:

$$f_5 = \frac{x_2 + 10.3}{1.145 x_2} \tag{6.21}$$

and applies when the 80% passing size of the product exceeds 75 micron.

(6) Rod mill factor, for high or low reduction ratio, f_6 is given by:

$$f_6 = 1 + (R - R_o)^2/150 \tag{6.22}$$

where optimum reduction ratio R_o is given by:

$$R_o = 8 + 5L_R/D \tag{6.23}$$

where L_R is the rod length (using the same length units as for D).

(7) Ball mill factor for low reduction ratio f_7 is given by:

$$f_7 = \frac{2(R - 1.35) + 0.26}{2(R - 1.35)} \tag{6.24}$$

and it only applies if $R < 6$.

The factors having been determined, the required specific energy measured at the drive pinion in kWh/short ton, from equation (6.15), is given by:

$$\bar{E}_A = W_i\left[\frac{10}{\sqrt{x_2}} - \frac{10}{\sqrt{x_1}}\right] \times \left[\begin{matrix}\text{Product of appropriate} \\ \text{factors } f_1 \text{ to } f_7\end{matrix}\right] \tag{6.25}$$

For details of the application of the factors, reference must be made to Rowland and Kjos (1978).

The highly empirical nature of the determination of the specific energy for tumbling mills is apparent from equation (6.25).

6.8 GRINDABILITY UNCERTAINTIES

6.8.1 Survey

Because grindability is so widely used in mill design the assumption is often made that the method is valid without a sufficiently critical examination of the associated uncertainties, which can lead to large contingency factors in design. Too large an energy contingency factor can lead to a drive motor having to be operated at such a fraction of maximum load that a considerable efficiency penalty is invoked. Hence the problems of grindability inaccuracies are important, and they are discussed below relative to Hardgrove and Bond.

6.8.2 Hardgrove index deficiencies

Guner *et al.* (1964), among others, discuss factors affecting the Hardgrove index, and the following are highlighted.

6.8.2.1 Sample density

In respect of the test machine, the volume of 50 g of sample will vary according to sample density. Hence the initial level of powder in the bowl could be much lower for a very dense material than for a light material. Now the ring-ball mill is essentially a constant-volume device, in that the maximum size of particle gripped by a rolling ball is a function of the angle of friction between steel and the powder, which does not vary greatly. Hence the effective volume ground is constant provided a sufficient depth of bed is maintained. In a deeper bed, material above the critical level is just pushed to one side without being gripped. In a bed shallower than the critical, however, the bed could be not running full, with an unknown effect on the breakage pattern. To overcome this problem Agus and Waters (1971, 1972) proposed that 75 cm^3 of material, packed by a standard process, should be used instead of 50 g, and a corresponding volumetric Hardgrove index derived. With this they obtained more consistent results than with the constant-mass index, particularly for heterogeneous materials such as coal having a hard ash content.

6.8.2.2 Heterogeneous sample

Heywood (1947) highlighted the problem of the preferential grinding of the softer coal consituent of high-ash coals containing hard shale. In an industrial pulverizer, grinding is continuous and equilibrium must be established eventually between the ash contents of input and output. However, the recirculated material will have a higher concentration of hard ash than the feed, and hence the material presented to the grinding zone (a combination of a small proportion of new feed and a large proportion of recirculated material) will have a much lower Hardgrove index than that of the feed tested. If the mill was designed on the feed Hardgrove value, it could fail to meet its required capacity because of this feature.

6.8.2.3 Sample moisture content

Vecci and Moore (1978) reported the results of tests on sub-bituminous and lignitic coals at several levels of moisture, as given in Figure 6.32. These coals have high inherent moisture contents although they feel dry to the touch. There appears to be an optimum moisture content for minimum Hardgrove index. Hence it is clear that, to obtain a realistic grindability for design purposes, the degree of drying in an operational mill must be forecast so that the appropriate value of the Hardgrove index may be selected.

In view of curve B in figure 6.32, it would seem prudent also to check the grindability of bituminous coals at several moisture levels.

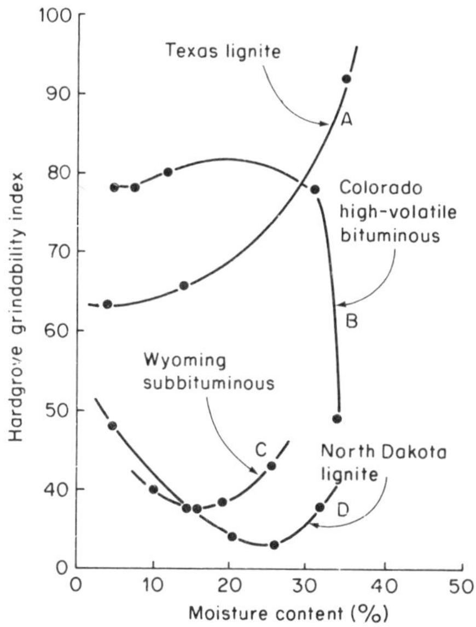

Figure 6.32. Variation of Hardgrove index with moisture content for coals (Vecci and Moore, 1978)

6.8.2.4 *Relative humidity of the air*

Meintjes (1965) studied the effect of relative humidity of the air at the time of test on both the moisture content of the sample (Figure 6.33) and the Hardgrove index (Figure 6.34). Some coals are clearly more influenced than others,

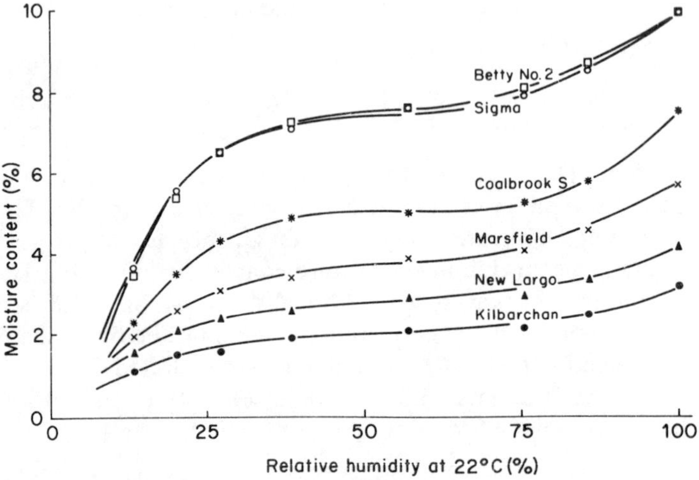

Figure 6.33. Moisture content (determined at 110°C) of six coal samples as a function of relative humidity of the air (Meintjes, 1965). *Reproduced by permission of the South African Institute of Mining and Metallurgy*

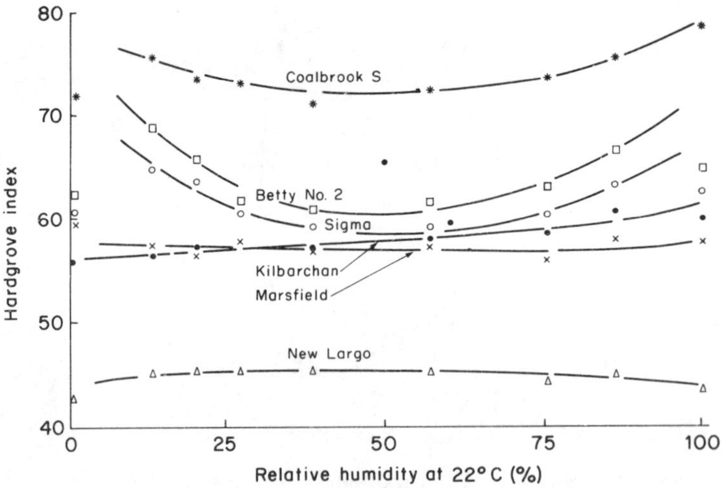

Figure 6.34. Influence of relative humidity of the air on the Hardgrove index of six coal samples (Meintjes, 1965). *Reproduced by permission of the South African Institute of Mining and Metallurgy*

but the variations are such that it would appear the 'air-dried' condition of the test specification needs to be reviewed. This is doubly so, in view of the effect of sample moisture discussed in §6.8.2.3.

6.8.2.5 Applied load

The Hardgrove index test is performed at constant pressure between grinding elements. For a series of industrial mills, where constant pressure is maintained, even if this is different from that of the test machine, the relative values of *HI* will be valid. If, however, the pressure changes for any reason, in the industrial mill series, then a test-found grindability could give rise to a discrepancy.

Items in §§6.8.2.3 and 6.8.2.4, associated with the relative humidity of the air at the time of the test, could have a direct bearing on the reproducibility of the Hardgrove index. It is therefore not surprising that Edwards *et al.* (1980) of Shell's Thornton Research Centre should contest the three-point variation in Hardgrove index, which is claimed by ASTM D 409 (1978) and British Standard 1016 (Part 20) (1981), and make a strong case for a five-point variation as being more realistic.

6.8.3 Bond work index shortcomings

Austin *et al.* (1982) and Austin and Brame (1983) have summarized the problems with the Bond grindability as follows.

(i) The method does not explicitly include the following important variables in a tumbling mill closed circuit:

(1) recycle ratio;
(2) classifier effectiveness;
(3) the optimum mixture of ball sizes in the charge, which is proving to be a most important factor;
(4) variation of the residence time distribution of particles within the mill with both particle size and mill geometry;
(5) variation of the residence time distribution of particles within the mill with slurry density (wet milling);
(6) lifter design;
(7) slurry rheology (wet milling).

(ii) It is assumed that specific energy is not a function of ball load, contrary to known fact.

(iii) Only the 80% passing size is considered, whereas it is known that the shape of the feed size distribution does have an effect on the product size distribution.

(iv) No account is taken of over- or under-filling of the mill.

(v) Causes of inefficiency in grinding are not explored.

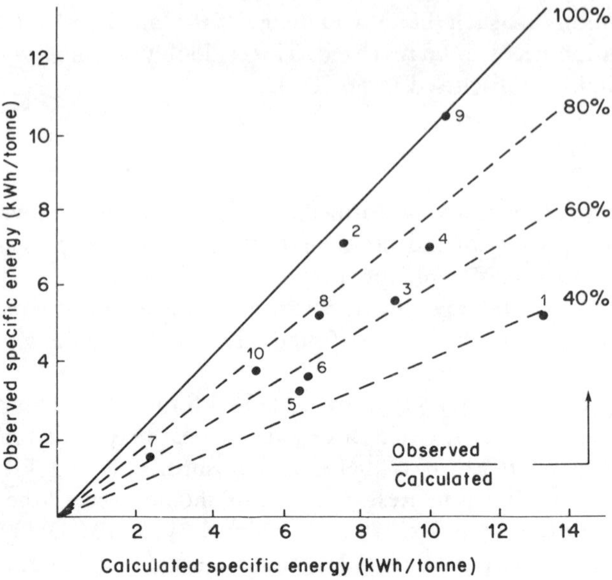

Figure 6.35. Correlation between observed and calculated specific energy for ten Australian sulphide ore plants using Bond equations (6.19) and (6.25) (Blaskett, 1970). *Reproduced by permission of the Institution of Mining and Metallurgy*

Hence it would be expected that there could be strong deviations between the Bond predicted power requirements and those actually drawn in the operating plant. Blaskett (1970) has demonstrated this for ten Australian plants operating on sulphide ores.

Using equation (6.25) with W_i as defined in equation (6.19) and three appropriate factors f_3, f_4 and f_5 together with a transmission loss factor where the mills were not directly driven, Blaskett (1970) obtained the correlation shown in Figure 6.35. This shows that points lie in the range 40–100% for the ratio of observed to calculated specific energy.

Blaskett also used a work index which was produced earlier (Bond, 1954) and recommended for homogeneous materials:

$$W_i = \frac{16}{(g_b)^{0.82}}\left(\frac{x_t}{100}\right)^{0.5} \tag{6.26}$$

This was incorporated into equation (6.15) without any factors.

Based on equations (6.15) and (6.26), Blaskett (1970) obtained the correlation given in Figure 6.36. The points lie in the range 50–120% for the ratio of observed to calculated specific energy.

In either case, the range of variation of the ratio observed to calculated specific energy gives no cause for complacency in respect of the Bond work index method of energy assessment in sulphide ore plants.

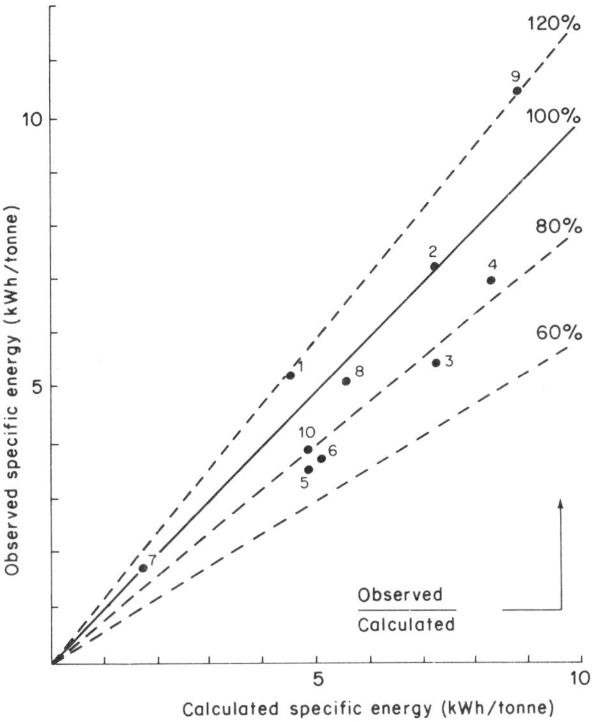

Figure 6.36. Correlation between observed and calculated specific energy for ten Australian sulphide ore plants using Bond equations (6.15) and (6.26) (Blaskett, 1970). *Reproduced by permission of the Institution of Mining and Metallurgy*

In similar vein, based on a powder survey by Smith (1959) in a cement context, Herbst and Fuerstenau (1980) have shown that actual plants operate within the range ±20% of design, based on the Bond method.

Thus from two very different industries comes the same disturbing message.

6.9 OTHER METHODS OF FORECASTING ENERGY REQUIREMENTS

Apart from grindability, discussed in §§6.6 and 6.7 which certainly has its drawbacks as outlined in §6.8, two other methods of forecasting energy requirements are extant:

(1) The use of a scaled-down version of a mill or mill/classifier system, for which overall scaling laws in terms of capacity and energy are found pragmatically. This is the subject of Chapter 7.

(2) The use of specific rate of breakage and breakage distribution functions to describe the size reduction mechanisms within a mill and their variation with particle size. To this is added residence time distribution information and classification functions as appropriate. Then the relation between these and energy requirement is established. This is at once the most satisfying and the most demanding method in terms of knowledge of the above functions, and is further studied in Chapters 8, 9 and 10.

6.10 REFERENCES

Agus, F., and Waters, P. L. (1971). 'Determination of the grindability of coal, shale and other minerals by a modified Hardgrove machine method', *Fuel* (London), **50**, 405–431.

Augs, F., and Waters, P. L. (1972). 'Predicting the grindability of coal-shale mixtures', *Fuel* (London), **51**, 38–43.

American Society Testing and Materials (ASTM) (1978). 'Standard test method for grindability of coal by the Hardgrove machine method', D 409–71.

Austin, L. G. (1973a). 'A commentary on Kick, Bond and Rittinger laws of grinding', *Powder Technol.*, **7**, 315–317.

Austin, L. G. (1973b). 'Note on Rittinger's law of grinding', *Trans. SME/AIME*, **254**, 310–311.

Austin, L. G., and Brame, K. (1983). 'A comparison of the Bond method for sizing wet tumbling ball mills with an SMB simulation model', *Powder Technol.*, **34**, 261–274.

Austin, L. G., and Klimpel, R. R. (1964). 'The theory of grinding operations', *Ind. Eng. Chem.*, **56**, (11), 18–29.

Austin, L. G., Klimpel, R. R., Luckie, P. T., and Rogers, R. S. C. (1982). 'Simulation of grinding circuits for design'. In: *Design and Installation of Comminution Circuits* (Mular, A. L., and Jergensen, G. V., Eds.), AIME, New York, 301–324.

Austin, L. G., and Luckie, P. T. (1972). 'Grinding equations and the Bond work index', *Trans. SME/AIME*, **252**, 259–266.

Axelson, J. W., Adams, J. T., Johnson, J. F., Kwong, J. N. S., and Piret, E. L. (1951). 'Basic laboratory studies in the unit operation of crushing', *Min. Eng. Trans. AIME*, **190**, (12), 1061–1069.

Axelson, J. W., and Piret, E. L. (1950). 'Crushing of single particles of crystalline quartz', *Ind. Eng. Chem.*, **42**, (4), 665–670.

Barnard, W. J. (1983). 'Primary breakage of brittle particles using the drop-shatter test apparatus', Ph.D. Thesis, University of Melbourne, Australia.

Batra, N. K., and Biswas, A. K. (1971). 'Energy comminution studies on a drop weight apparatus', *Indian J. Technol.*, **9**, (12), 448–453.

Baumgardt, S., Buss, B., May, P., and Schubert, H. (1973). 'Zum vergleich von zerkleinerungsergebnissen der einzelkornzerkleinerung bei verschiedenen beanspruchungsarten', *Powder Technol.*, **8**, 107–115.

Baumgardt, S., Buss, B., May, P., and Schubert, H. (1976). 'On the comparison of results in single grain crushing under different kinds of loads', 11th Int. Miner. Process,. Congr., Cagliari, Aug. 1975, Inst. di Arte Mineraria, Cagliari.

Beaven, C. H. J. (1980). 'How many measurements?', *Proc. Powder Europa*, Weisbaden, 22–24 June, Tech. Sessions, Vol. 1, 1–25.

Behrens, D. (1965). 'Prallzerkleinerung von glas-kulgen und unregelmassig gerformten teilchen aus schwerspat, kalkstein und quarzsand', *Chemie Ing. Techn.*, **37**, (5), 473–483.

Bergstrom, B. H. (1965). 'Crushing of single brittle particles', *Proc. 1st Tewksbury Symp.*, Melbourne, Australia, Aug. 1963, Publ. Univ. Melbourne, 329–343.

Bergstrom, B. H., Sollenberger, C. L., and Mitchell, W. (1961). 'Energy aspects of single particle crushing', *Trans. AIME*, **220**, 367–372.

Berlioz, L. M., and Fuerstenau, D. W. (1967). 'A test of Charles energy-size reduction relationship', *Trans. SME/AIME*, **238**, 282–284.

Blaskett, K. S. (1970). 'Estimation of the power consumption in grinding mills', *Proc. 9th Commonwealth Miner. Metal. Congr.*, 1969, Publ. IMM (London), **79**, 631–649.

Bond, F. C. (1952). 'The third theory of comminution', *Min. Eng. Trans. AIME*, **193**, 484–494.

Bond, F. C. (1954). 'Crushing and grinding calculations', *Canadian Min. Metal Trans.*, **62**, 286–292.

Bond, F. C. (1961). 'Crushing and grinding calculations', *Brit. Chem. Eng.*, **6**, 378–385, 543–548.

Bond, F. C., and Wang, J. T. (1950). 'A new theory of comminution', *Min. Eng. Trans. AIME*, **187**, 871–878.

Bown, R. W. (1966). 'Measurements of the energy required to break coal particles under free crushing conditions', *Trans. IMM (London)*, **76**, C163–C172.

British Standard 1016 (Part 20) (1981). 'Determination of Hardgrove grindability index of hard coal'.

Broadbent, S. R., and Callcott, T. G. (1956). 'Coal breakage processes', *J. Inst. Fuel*, **29**, 524–539.

Bull, F. A. (1955). 'The designation of the fineness of pulverised brown coal', *J. Inst. Fuel.*, **25**, 163–170.

Buss, B., Hanische, J., and Schubert, H. (1982). 'Über das zerkleinerungsverhalten seitlich begrenzter und nicht-begrenzter kornschichten bei druckbeanspruchung', *Neue Bergbautechnik*, **12**, (5), 277–283.

Buss, B., and Schubert, H. (1970–71). 'Beziehungen zwischen der korngrössenverteilung und der beanspruchungsenergie bei der einzelkornprallzerkleinerung', *Powder Technol.*, **4**, 139–145.

Carey, W. F., and Stairmand, C. J. (1953). 'A method of assessing the grinding efficiency of industrial equipment', *Trans. IMM (London)*, **62**, 117–136.

Chandler, R. L. (1965). 'Grindability tests for coal', *British Coal Utilisation Research Association Monthly Bull.*, **29**, Nos. 248, 249.

Charles, R. J. (1952). 'Energy-size reduction relationships in comminution', *Trans. AIME*, **208**, 80–88.

Cook, J. (1915). 'Investigation of ore milling to ascertain the heat developed in crushing', *Trans. IMM (London)*, **24**, 243–251.

Edser, E. (1922). British Association for the Advancement of Science, 4th Report on Colloidal Chemistry, 281.

Edwards, G. R., Evans, T. M., Robertson, S. D., and Summers, C. W. (1980). 'Assessment of the standard method of test for the grindability of coal by the Hardgrove machine', *Fuel*, **59**, 826–830.

Fahrenwald, A. W., Hammar, G. W., Lee, H. E., and Staley, W. W. (1931). 'Ball mill studies: II, Thermal determination of ball mill efficiency', *AIME Tech. Publ.* 416, 3–13.

Fuerstenau, D. W. (1959). 'Retention time in continuous vibratory ball milling', *Min. Eng.*, **214**, 1238–1243.

Fuerstenau, D. W. (1961). 'Size distributions resulting from the comminution of heterogeneous materials', *Trans. AIME*, **220**, 395–396.

Gross, J., and Zimmerley, S. R. (1930). 'Crushing and grinding: III, Relation of work input to surface produced in crushing quartz', *Trans. AIME*, **87**, (2), 35–50.

Guner, M., Lovell, H. L., and Spicer, T. S. (1964). 'Factors affecting the results of the

Hardgrove grindability test and a proposed method of correction', ASME Paper 64-WA-FU3.

Hardgrove, R. M. (1932). 'Grindability of coal', *Trans. ASME*, **54**, 37–46.

Hardgrove, R. M. (1933). 'The relation between pulverising capacity, power and grindability', ASME, Chicago, 27 June meeting.

Hardgrove, R. M. (1938). 'The effect of grindability on particle size distribution', *Trans. Am. Inst. Chem. Eng.*, **34**, 131–152.

Herbst, J. A., and Fuerstenau, D. W. (1980). 'Scale-up procedure for continuous grinding mill design using population balance models', *Int. J. Miner. Process.*, **7**, 1–31.

Hess, W., and Schönert, K. (1976). 'Die zerkleinerung von kalkstein kugeln bei kombimerter druck-und schubbeanspruchung', Dechema Monograph 79, 151–165.

Heywood, H. (1947). 'Determination of coal grindability', *Proc. 1st Pulverised Fuel Conf.*, Harrogate, Inst. Fuel, 594–611.

Hildinger, P. (1969). 'Einzelkornbeansprungung von glaskugeln sowie quarzzementklinker- und kalksteinteilchen durch fallkörper', *Chemie Ing. Techn.*, **41**, 278–285.

Holmes, J. A. (1957). 'A contribution to the study of comminution: a modified form of Kick's law', *Trans. Inst. Chem. Eng.*, **35**, 125-156.

Hughes, T. H. (1967). 'Surface-energy relationships in a laboratory grinding mill', Dechema Monograph 57, 667–693.

Hukki, R. T. (1944). 'An experimental study of the principles of comminution', D.Sc. thesis, Massachusetts Institute of Technology.

Hukki, R. T., and Reddy, I. G. (1967). 'The relationship between energy input and fineness in comminution', Dechema Monograph 57, 313–339.

Johnson, J. F., Axelson, J. W., and Piret, E. L. (1949). 'Energy-new surface relationships in the crushing of solids: the application of gas adsorption measurements to an investigation of the crushing of quartz', *Chem. Eng. Prog.*, **45**, (12), 708–715.

Kerlin, H. P. (1979). 'Zerkleinerung von kohle als einzelkorn und im kollektiv', Dissertation Technischen Universitat Braunschweig.

Kick, F. (1885). 'Das gesetz der proportionalen widerstände und seine anwendungen', Arthur Felix, Leipzig.

Kinasevich, R. S., and Fuerstenau, D. W. (1964). 'Research on the mechanisms of comminution in tumbling mills', *Canadian Metal. Q.*, **3** (1), 1–20.

Lenkewitz, K. H. (1971). 'Die prallzerkleinerung körniger massenguter durch einmalige beanspruchung in einen schlagrad bei unter-schiedlichen prallgeschwindigkeiten', *Fortschritt-Berichte VDI-Z.*, **3**, 1–14.

McIntyre, A., and Plitt, L. R. (1980). 'The inter-relationship between Bond and Hardgrove grindabilities', *CIM Bull.*, June, 149–155.

Meintjes, A. A. (1965). 'The grindability and abrasive properties of South African coals', *J. South African Inst. Min. Metal.*, July, 629–641.

Mortsell, S., and Svensson, J. (1951). In Discussion on Bloecher, F. W., 'A new surface measurement tool for mineral engineers', *Min. Eng. Trans. AIME*, **190**, 981–983.

Okuda, S., and Choi, W. S. (1980). 'Grinding characteristics of polymeric materials by target type fluid-energy mill', *Europ. Symp. Particle Technol.*, Amsterdam, 171–183.

Peters Rit, A. W. P. G., Hagg, H., and van Brakel, J. (1983). 'Characterisation of mechanical strength of coal particles', *I. Ch. EME Symp. Series* 69, POWTECH '83, 9th Int. Powder Bulk Solids Tech. Exh. & Conf., Birmingham, UK, 1–19.

Peters Rit, A. W. P. G., and van Brakel, J. (1982). 'Particle size reduction under controlled conditions'. In: *Deeltjesverkleining van Kooldeeltjes*, Technische Hogeschool, Delft, C3/1–25.

Planiol, R. (1965). 'Distribution of energy in centrifugal crushers turning on no load',

Proc. 6th Int. Congr., Cannes, 1963. In: *Mineral Processing* (Roberts, A., Ed.), Pergamon Press, 35–44.

Prentice, T. K. (1946). 'Crushing and grinding efficiency', IMM (London), Bull. 447, 7–36.

Priemer, J. (1965). 'Untersuchungen zur prallzerkleinerung einzelteilchen', *Fortschritt-Berichte VDI-Z.*, **3**, 8.

Rittinger, von P. R. (1867). 'Lehrbuch der aufbereitungskunde', Berlin.

Rose, H. E. (1962). 'Some observations on vibration mills and vibration milling', *Proc. 1st Europ. Symp. Zerkleinern*, Frankfurt, Publ. VDI-Verlag, Düsseldorf, 427–456.

Rose, H. E. (1967). 'A comprehensive theory of the comminution process', Dechema Monograph 57, 27–62.

Rose H. E. (1972). 'A new treatment of comminution and its application to the calculation of mill capacity', Dechema Monograph 69, 87–120.

Rose, H. E. (1974). 'Drop weight tests as the basis for calculations of the performance of ball mills', *Proc. 10th Int. Miner. Process. Congr.*, April 1973, London, Publ. IMM (London), 143–147.

Rowland, C. A., and Kjos, D. M. (1978). 'Rod and ball mills'. In: *Mineral Processing Plant Design* (Mular, A. L., and Bhappu, R. B., Eds.), SME/AIME, New York, 239–278.

Rumpf, H. (1959). 'Beanspruchungs theorie der prallzerkleinerung', *Chemie Ing. Techn.*, **31**, (5), 323–337.

Rumpf, H. (1965). 'Die einzelkornzerkleinerung als grundlage einer technischen zerkleinerungswissenschaft', *Chemie Ing. Techn.*, **37**, 187–202 [SLA Translation No. 66-13859].

Rumpf, H. (1973). 'Physical aspects of comminution and new formulation of a law of comminution', *Powder Technol.*, **7**, 145–159.

Rumpf, H. (1975). *Mechanische Verfahrenstechnik*, Carl Hanser Verlag, München Wein.

Schellinger, A. K. (1951). 'A calorimetric method for studying grinding in a tumbling medium', *Min. Eng. Trans. AIME*, **3**, 518–522.

Schellinger, A. K., and Lalkaka, R. D. (1951). 'Approximation to the energy efficiencies of commercial ball mills by the energy balance method', *Min. Eng. Trans. AIME*, **3**, 523–524.

Schönert, K. (1966). 'Einzelkorn-druck zerkleinerung und zerkleinerungskinetik', Dissertation Doktor-Ingenieurs, Karlsruhe, Germany,

Schönert, K. (1972). 'Role of fracture physics in understanding comminution phenomena', *Trans. SME/AIME*, **252**, 21–26.

Schuhmann, R. (1960). 'Energy input and size distribution in comminution', *Trans. AIME*, **217**, 22–25.

Schwendig, G. (1966). 'Versuche und betrachtungen zur überwalzzerkleinerung eines mahlbettes', *Aufbereit Techn.*, **7**, 489–492.

Sligar, N. J. (1979). 'Size reduction functions for a roll passing over a bed of coal', *Proc. 4th Tewksbury Symp.*, Melbourne, Australia, 21: 1–19.

Smith, G. B. (1979). 'Impact breakage in rotary crushing mills', *Proc. 4th Tewksbury Symp.*, Melbourne, Australia, 22: 1–15.

Smith, R. W. (1959). *1956 Power Survey*, US Portland Cement Assoc. Report MP73.

Svensson, J., and Murkes, J. (1957). 'An empirical relationship between work input and particle size distribution before and after grinding', *J. Trans. Int. Miner. Dressing Congr.*, Stockholm, 37–58.

Tschorbadjski, J. (1969). 'Beanspruchungsmechanismus beim schiefen stoss und einwirkung des aufprallwinkels auf die einzelkornprallwinkels auf die einzelkornprallzerkleinerung', *Aufbereit Techn.*, **10**, (1), 277–283.

Vecci, S. J., and Moore, G. F. (1978). 'Determine coal grindability', *Power*, March, 74.

Yap, R. F., Sepulveda, J. L., and Jauregui, R. (1982). 'Determination of the Bond work index using an ordinary laboratory batch ball mill'. In: *Design and Installation of Comminution Circuits* (Mular, A. L., and Jergensen, G. V., Eds.), SME/AIME, New York, 176–203.

CHAPTER 7

Scaling of capacity and energy

NOTATION FOR CHAPTER 7

A	Constant in equation (7.32)
d	Diameter of grinding element, ball or roller
d_N	Diameter of nozzle jet
d_p	Diameter of cut size particle
D	Mean track diameter of vertical spindle mill or internal drum diameter
D_o	Outlet diameter
E_c	Energy to produce unit crack length in a particle
E^*	Young's modulus
F	Force
g	Acceleration due to gravity
G	Gape of jaw crusher
h	Bed depth of vertical spindle mill
H	Depth of cylinder of jet mill
J	Fraction of volume, including voids, of ball charge

K	Constant
L	Length Subscripts: 1 model 2 prototype
M	Mass
M_B	Mass of ball charge
n	Number; exponent in equation (7.33)
N	Rotational speed
N_C	Theoretical critical speed of tumbling mill
N^*	Fraction of theoretical critical speed = N/N_c
N_{CM}	Critical speed for maximum capacity of crusher
p	Power
p_b	Power dissipated in bearings
p_c	Power dissipated in charge
p_s	Specific energy
P	Pressure
P_o	Optimum pressure between grinding elements
q	Rate of production of specific surface per unit mass of ball charge
Q	Maximum capacity of mill (mass per unit time) for given conditions of material and product fineness
r	Number of rows in swing hammer mill
R	Reduction ratio of jaw crusher = gape/set
s_o	Averaged function of the maximum and minimum openings of the discharge annulus of a gyratory crusher
s_t	Set of the discharge opening of a jaw crusher
t	Time
T	Throw of jaw crusher
v	Velocity
v_i	Mean velocity of reverse flow in fluid energy mill
v_t	Tangential component of velocity
V	Volume
V_n	Volumetric flow rate
V_p	Volume of powder in a mill
W	Width
W_i	Work index (Bond)
W_o	Half the mean circumferential discharge annulus
x	Particle size
x_{Me}	Median particle size
$\bar{x}$	Mean particle size
y	Exponent
α	Amplitude of vibration
β	Damping coefficient of vibration
γ	Exponent
ε	Parameter defining surface characteristics of a material being crushed

θ	Angle of jet with radius in jet mill
λ	Scale of length $= L_2/L_1$
ξ	Parameter defining the flow of material through a crusher at rest
ρ_b	Density of grinding element
ρ_{bm}	Buk density of balls and powder in a tumbling mill charge
ρ_f	Density of feed material
ρ_m	Bulk density of powder
ρ_p	Particle density
σ	Stress
ϕ	Angle of nip
ψ	Parameter defining the size distribution of feed to a jaw crusher
ω	Frequency of oscillation of vibration mill

7.1 INTRODUCTION

This chapter is concerned with the end-results of comminution and the effects of machine variables thereon. It does not take account of the detailed kinetics of the process, such as breakage rates or distribution functions, residence time distribution functions or classifier efficiency, each of which has its own scale-up criteria, as discussed in Chapter 10. However, the end-result scale-up laws relative to capacity, power and hence specific energy are of importance to some current methods of mill design, but detailed kinetic systems are taking over from them as the required information becomes available.

Now when many variables are involved in a complex process such as size reduction, it is usual to apply the principles of dimensional analysis to obtain dimensionless groups of variables and to perform experiments to find the relation between these groups. By this, or other means, scale relations can be developed between model and prototype machine, for which the suffix 1 applies to the model and 2 to the prototype.

In practical design, scaling must be expressed in terms of relative dimensions, masses, times, forces, velocities, rotational speeds, frequencies of vibrations etc, leading to scale ratios such as $\lambda = L_2/L_1$.

Capacity, as mass per unit time, may be expressed as $\rho_p L^3/t$, and power, which is force multiplied by velocity, may be expressed as $\sigma L^2(L/t)$. Hence specific energy may be written as σ/ρ_p, which, being independent of L, may be expected, in appropriate circumstances, to be constant over a range of geometrically similar mills operating on feed of exactly similar specification, to produce the same output grade, which similarity conditions are assumed in this chapter unless stated otherwise.

This result is clearly a simplification, but it has been found to hold in terms of net power, which is the motor power less the idle power, both measured at the motor output shaft.

Capacity, power and hence specific energy scale relations are discussed in the sections that follow.

7.2 SCALING LAWS FOR MILLS

7.2.1 Tumbling mills

7.2.1.1 Capacity scaling

Hukki (1954) has reasonably assumed that the capacity of a tumbling mill, when operated below critical speed, is proportional to the mass of the ball plus powder charge and is also some function of the peripheral speed, according to the type of drum liner/lifter employed. Then:

$$Q/L \propto \left(\frac{\pi}{4} D^2 J \rho_{\mathrm{bm}}\right) (\pi D N)^{\gamma} \tag{7.1}$$

Since a constant pattern of charge movement within the mill demands a constant fraction of theoretical critical speed (Rose and Sullivan, 1958), then:

$$N \propto \frac{N^*}{\pi}\left(\frac{g}{2D}\right)^{0.5}$$

Hence from (7.1):

$$Q/L \propto J N^* D^{2+0.5\gamma}$$

or, for constant N and J :

$$Q/L \propto K_1 D^{2+0.5\gamma} \tag{7.2}$$

If $\gamma = 1$, the exponent of D becomes 2.5, but Taggart (1945) gave a value of the order of 2.6 depending on the type of lining employed. Figure 7.1, taken from manufacturers' capacity data, supports the value of 2.5 (Marshall, 1974).

Then, for constant L/D, capacity scales as $\lambda^{3.5}$.

With respect to autogenous mills, Dettmer and Sobering (1964) report an exponent of 2.85 for D when dry grinding Labradore iron ore. However, as Turner (1982) points out, with an exponent of 2.83 for D, L is likely to vary as $D^{0.67}$ to facilitate air sweeping and hence overall, capacity scales as $\lambda^{3.5}$ under practical optimum conditions for dry grinding in autogenous mills.

7.2.1.2 Power scaling

Rose and Sullivan (1958) produced equation (7.3) for a mill with a given outlet type, ball size, no lifters and when operating with a given powder feed size:

$$\frac{p}{D^5 N^3 \rho_{\mathrm{b}}} = \left(\frac{1 + 0.4\rho_{\mathrm{m}}}{\rho_{\mathrm{b}}}\right)\left(\frac{L}{D}\right) f_1\left(\frac{N_{\mathrm{c}}}{N}\right) f_2(J) \tag{7.3}$$

Now, putting $N/N_{\mathrm{c}} = N^*$, $N_{\mathrm{c}} = (1/\pi)(g/2D)^{0.5}$ and with $\rho_{\mathrm{m}}, \rho_{\mathrm{b}} =$ constant, we have:

$$p = K_2 L D^{2.5} f_3(N^*) f_2(J)$$

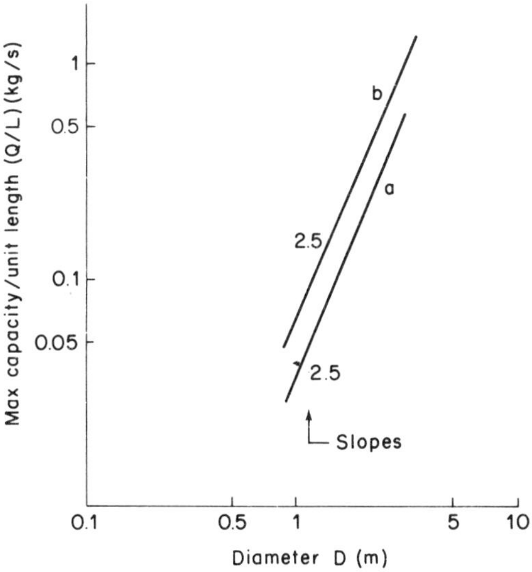

Figure 7.1. Maximum capacity per unit length of ball mill versus mill diameter from manufacturers' data. Grinding conditions: (a) Coal of 50 Hardgrove index: 70% through 75 micron sieve; (b) 'Hard material': 65% through 75 micron sieve

Hence for constant N^* and J:

$$p = K_3 L D^{2.5} \tag{7.3a}$$

which has the same form as the expression for capacity under similar conditions.

In practice, the exponent of D varies according to the type of liner used and whether some lifting of the ball charge occurs. Typical values are listed in Table 7.1.

However, taking 2.5 as the exponent for D, then for constant L/D ratio, power scales as $\lambda^{3.5}$ as does capacity.

Hence specific energy is independent of mill dimensions under the conditions stated. This is borne out in practice. One manufacturer, for example, for

Table 7.1. Exponent of D in scale equations for mill power for laboratory and commercial mills

Mill	Exponent of D	References
Laboratory	2.55	Heywood (1947)
Laboratory	2.5	Gow *et al.* (1934)
Laboratory	2.65	Fahrenwald (1934)
Commercial rod	2.6	Crocker (1952)
Commerical ball	2.5	Hancock (discussion in Gow *et al.* (1934)

$L/D = 1.5$, $J = 0.4$, $N^* = 0.8$ using steel balls and a waveform liner gives a specific energy of 18 kWh/tonne throughout a mill series when grinding a material of Hardgrove index 50 ground to an output fineness of 70% minus 75 microns.

Conditions have been reported by Turner (1982) for autogenous mills in which capacity scales as $D^{3.53}$ and power as $D^{3.50}$. Hence specific energy scales as $D^{-0.03}$, which implies reduced specific energy with increased mill size for otherwise similar operating conditions.

However, scale-up of ball mills to larger sizes has not been without its problems, as reported by Harris and Arbiter (1982) for the Bougainville copper plant. There, 18 ft × 21 ft (5.5 m × 6.4 m) mills use one-third more specific energy than would be expected from a scale-up from a 12 ft (3.7 m) mill, and the manufacturer recommends no mills larger than 16.5 ft (5 m) should be installed until the reasons are understood. Arbiter and Harris suggest that one reason is the decreased effectiveness of large-diameter mills with coarse particles (+6 mesh) for some ores. This, in turn, is related to decreased (rotational flow)/(feed flow) ratio with increasing mill diameter, reduced mixing effectiveness and reduced residence time of particles in the mill. Hence increase in N^* is proposed together with reduced ball load.

This problem highlights the advantages of SMB scale-up methods, in which detailed mill internal happenings are determined via breakage functions and residence time distributions, as discussed in Chapter 10.

7.2.2 Vibration mills

7.2.2.1 Capacity scaling

Rose and Sullivan (1961) have studied vibration mills by the method of dimensional analysis and have produced a capacity equation in the form:

$$\frac{q\rho_{\mathrm{m}}\alpha}{\omega} = K_4\left[\frac{\omega^2\rho_{\mathrm{m}}\alpha^4}{E_{\mathrm{c}}}\right]\left[\frac{d}{\bar{x}}\right]^{0.5} f_4\left[\frac{\omega^2\alpha}{g}\right] f_5[J]\, f_6[V_{\mathrm{p}}] \tag{7.4}$$

This holds for mill bodies having a circular trajectory and for $d/\bar{x} > 10$, which is the case in practice.

The optimum J for maximum $f_5[J]$ was found to be near 80%, and most mills operate at this ball filling, as if the drum is too full, vibration is impeded. Now $f_6[V_{\mathrm{p}}]$ has been demonstrated to change little between $V_{\mathrm{p}} = 0.5$ and 10, and hence is easily kept constant in practice.

$f_4[\omega^2\alpha/g]$ is seen in Figure 7.2 plotted against $\omega^2\alpha/g$ and clearly tends to a constant value in the commonly used range of 1500–3000 rev/minute.

Thus the last four factors in equation (7.4) are constant and hence, for a given powder and ball charge:

$$q = K_5\omega^3\alpha^3$$

Assuming that vertical and horizontal components of amplitude of balls are equal, Rose and Sullivan (1961) showed that the equation for maximum

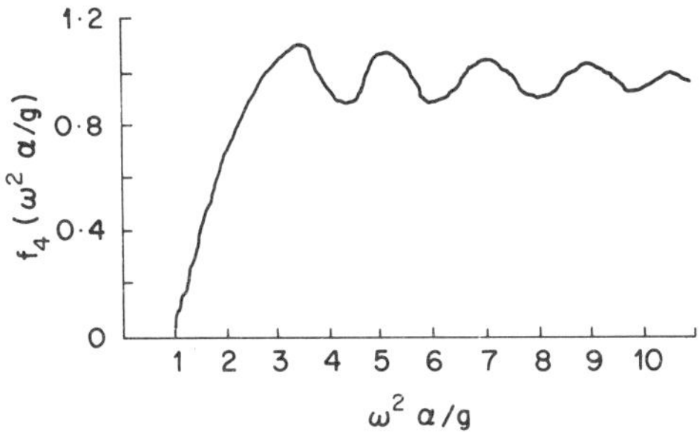

Figure 7.2. Vibration mill: $f_4[\omega^2\alpha/g]$ against $[\omega^2\alpha/g]$ (Rose, 1967)

capacity is then given by:

$$Q = K_6 M_B \omega^3 \alpha^3$$

But for constant J and ρ_p, $M_B = K_7 D^3$; i.e.:

$$Q = K_8 D^3 \omega^3 \alpha^3 \tag{7.5}$$

Thus capacity scales as λ^3 through a mill series for constant ω and α.

7.2.2.2 Power scaling

Rose and Sullivan (1961) have similarly treated the power requirements of a vibration mill, and produced the equation:

$$\frac{p_c M_B^2}{\alpha^2 \beta^3} = K_9 \left[\omega M_B \frac{N_c}{N} \right]^3 \tag{7.6}$$

For constant β and N/N_c, the power p_c dissipated in the ball charge is given by:

$$p_c = K_{10} M_B \omega^3 \alpha^2$$

or

$$p_c = K_{11} D^3 \omega^3 \alpha^2 \tag{7.7}$$

Thus power scales as λ^3 for constant ω and α.

Hence specific energy from equations (7.5) and (7.7) is given by:

$$p_s = K_{12}/\alpha \tag{7.8}$$

if constant ratios of mass charges, amplitudes and rotational speeds are maintained in a mill series, which may be difficult to achieve in practice.

7.2.3 Vertical spindle mills

7.2.3.1 Survey

Two typical cases depicted in Figure 7.3 are discussed which have respectively a sphere and a flat roller mating with a horizontal circular grinding track.

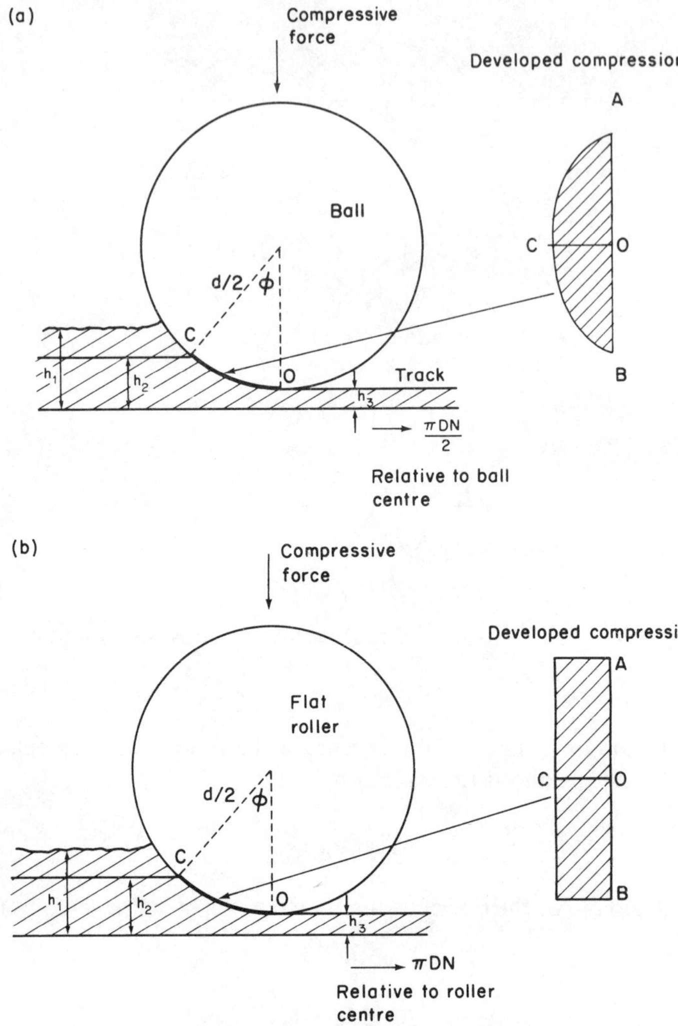

Figure 7.3. Vertical spindle mill geometry: (a) ring ball mill, (b) ring roll mill. For (a), AB = length of arc of contact between ball and track in a radial plane of the track = $\pi d/4$. For (b), AB = roller width = W (Marshall, 1974). *Reproduced by permission of the Institution of Chemical Engineers*

Other variants of these geometries follow a similar pattern of scale criteria for both capacity and power.

7.2.3.2 Capacity scaling

The quantity of material instantaneously coming under the stressing influence of a grinding element is taken to be a function of the effective contact area between element and bed, in which a particle is gripped, which in turn is a function of the angle of friction between the material and the metal of the element. According to Johanson (1965) the limiting value of ϕ is constant for a given material, and Anson (1969) shows that $h_2 \propto d$, if $h_1 > h_2$, which is the case in a practical mill. Note that the material of greater depth than h_2 is not gripped but is just pushed to one side. Since $h_2 \propto d$, it is also noted that the larger the diameter of the element the larger the feed size that can be accepted.

Now for a sphere, the contact area is proportional to d^2, if a 90° contact angle is subtended by the track at the sphere centre, which is usual in practice. For the roller, contact area is proportional to Wd.

Further, capacity is a function of the relative peripheral speed between element and track at mean track diameter, which is $\pi DN/2$ for the sphere (since the ball moves at half track speed) and πDN for the roller, assuming no slippage in each case.

Then for n elements:
for the sphere

$$Q = \frac{\pi DN}{2} n \frac{\pi}{4} d^2 = K_{13} d^2 DNn \tag{7.9}$$

for the roller

$$Q = \pi DNnWd = K_{14} dWDNn \tag{7.10}$$

In each case, maximum capacity is associated with an optimum value of N. Increasing speed from zero clearly increases capacity, but beyond the optimum speed material is thrown out of the track before it gets gripped by a passing element. This optimum speed is scaled through a series of geometrically similar mills by taking constant centrifugal acceleration at mean track diameter, that is $N \propto D^{-0.5}$. Further, d and W are proportional to D, it is assumed. Then for both ring ball and ring roll mills, maximum capacity at optimum speed, without pressure considerations, may be expressed by:

$$Q = (\text{Geometrical factor}) \times n \times D^{0.5} \tag{7.11}$$

For zero pressure between element and track, clearly no size reduction will take place. As pressure increases, grinding increases according to the pattern given in Figure 7.4, which indicates that a law of diminishing returns applies. Optimum pressure can be found by the minimum on the specific power versus pressure curve shown in Figure 7.5, but this may be modified in practice by considerations of ring element wear. Hence capacity will be a function of this optimum pressure $f_1(P_o)$

Thus the final expression for maximum capacity takes the form:

$$Q = K_{15}(\text{Geometrical factor}) nD^{2.5} f_1(P_o) \tag{7.12}$$

Hence for a given material size reduction ratio, mill geometry, number of elements and optimum pressure, capacity should scale as $\lambda^{2.5}$ and Figure 7.6 shows that this is borne out from manufacturers' data. Curve a with a slope of 2.55 appears not quite to conform to the conditions assumed above.

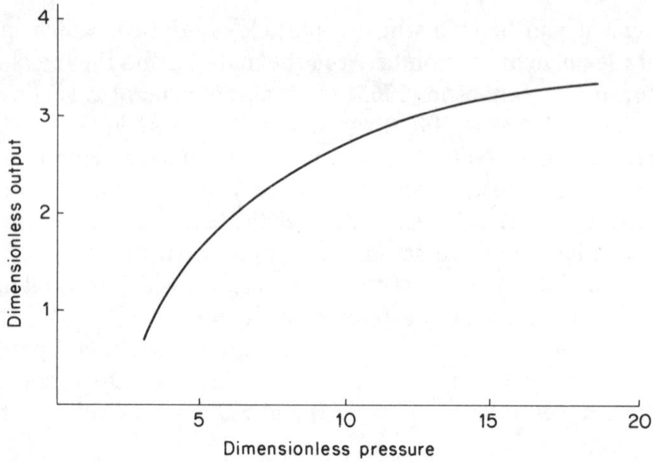

Figure 7.4. Dimensionless output versus dimensionless pressure between grinding elements for a vertical spindle mill (Marshall 1974). *Reproduced by permission of the Institution of Chemical Engineers*

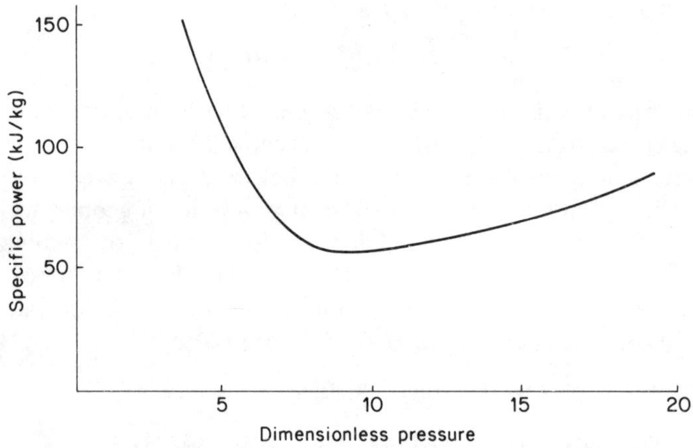

Figure 7.5. Specific power versus dimensionless pressure between grinding elements for a vertical spindle mill (Anson, 1969). *Reproduced by permission of the Central Electricity Generating Board*

7.2.3.3 *Power scaling*

The power to operate a ring-ball or ring-roll mill is given by the product of the tangential force at mean track diameter and the peripheral velocity at the same diameter. At optimum pressure P_o between grinding elements, the force per element will be the product of the contact area and $f_2(P_o)$. Hence power is given by:

For the sphere

$$p = K_{16} d^2 n f_2(P_o) \frac{\pi DN}{2}$$

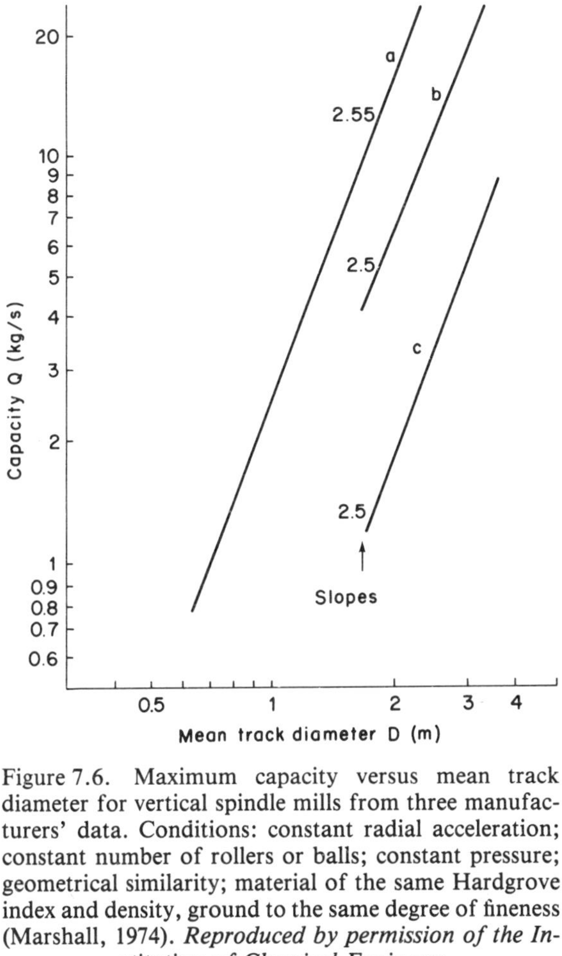

Figure 7.6. Maximum capacity versus mean track diameter for vertical spindle mills from three manufacturers' data. Conditions: constant radial acceleration; constant number of rollers or balls; constant pressure; geometrical similarity; material of the same Hardgrove index and density, ground to the same degree of fineness (Marshall, 1974). *Reproduced by permission of the Institution of Chemical Engineers*

For the roller

$$p = K_{17} W d n f_2(\mathrm{P_o}) \pi D N$$

As before, putting $N \propto /\sqrt{D}$, $d \propto D$ and $W \propto D$ for both sphere and roller, we have:

$$p = K_{18}(\text{Geometrical factor}) n D^{2.5} f_2(P_o) \tag{7.13}$$

Hence specific energy, from equations (7.12) and (7.13), is given by:

$$p_s = K_{19} f_3(P_o) \tag{7.14}$$

which should be constant for a given mill series, and this is usually borne out in practice for the conditions stated. (see §10.4.7.)

Apart from power to grind, of course, fan power is additionally required to transport material through the mill/classifier system, and this may be of the same order of magnitude as the power to grind.

7.2.4 Swing hammer mills

Following Andreasen (1957) the stresses in particles undergoing size reduction in a swing hammer mill arise from the conversion of kinetic energy of the hammer system into potential (assumed elastic) energy in the particles, which are subject to deformation δL in the direction of motion of the hammer. Equating kinetic and potential energies, we obtain:

$$Mv^2 \propto \sigma L^2(\delta L)$$

Since $M = \rho_b L^3$ and $\delta L = \sigma L/E^*$, this leads to:

$$v^2 \propto \sigma^2/\rho_b E^*$$

or

$$v \propto \sigma/(\rho_b E^*)^{0.5} \tag{7.15}$$

The denominator is constant for given materials of hammer and feed ground.

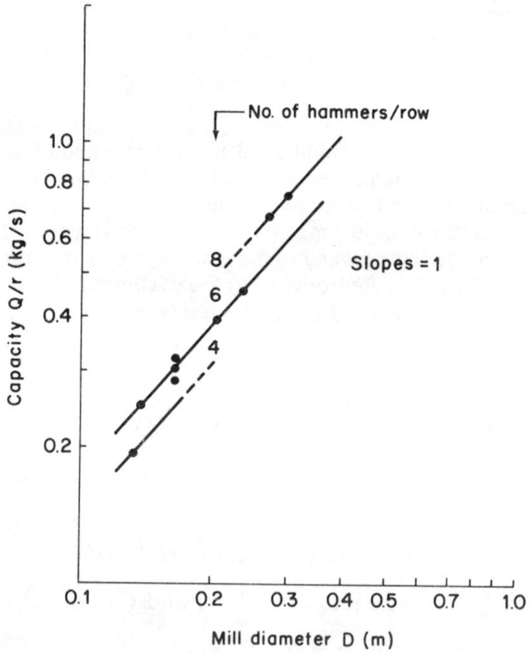

Figure 7.7. Capacity per row of hammers versus mill diameter for swing hammer mills (Marshall, 1974). *Reproduced by permission of the Institution of Chemical Engineers*

Hence on the above simplified assumptions, the scaling criterion for equivalent performance in model and prototype to obtain equal stress σ in the particles, is equal hammer peripheral velocity v.

Now since the dimensions of capacity are $\rho_p L^3/t$ and if $v = L/t = \text{constant}$, then capacity should scale as λ^2.

Figure 7.7, based on one manufacturer's data, shows a straight line of unit slope when Q/r is plotted against mill diameter D Hence:

$$Q \propto rD$$

But r, the number of rows of hammers, itself is proportional to D. Hence:

$$Q \propto D^2$$

i.e. capacity does scale as λ^2.

Power having dimensions $\sigma L^2(L/t)$, also scales as λ^2 for constant L/t. Hence specific energy should be constant throughout a similar mill series, which is usually attested in practice.

7.2.5 Crushers

7.2.5.1 Capacity scaling

For jaw crushers, the following capacity equation is used in industry:

$$Q = K_{20} v W s_t$$

where K_{20} is a constant including the bulk density of the discharge product, v is the velocity of discharge product, W is the width of the discharge opening, and s_t is the set of the discharge opening.

Similarly for gyratory crushers:

$$Q = K_{21} v W_o s_o$$

where K_{21} is a constant including the bulk density of the discharge product, W_o is half the mean circumferential discharge annulus, and s_o is an averaged function of the maximum and minimum openings of the discharge annulus.

These parameters, applying as they do to the discharge geometry and discharge product velocity, are very interrelated and are difficult to assess in terms of basic dimensional quantities.

However, Rose and English (1967) have made an analytical study of jaw crushers, and have produced an expression for capacity in terms of jaw geometry (Figure 7.8) and functions of feed and operational parameters. Their finding in relation to rotational speed is expressed in Figure 7.9, which indicates a critical speed for maximum capacity, which was deduced theoretically, then confirmed experimentally, and may be stated as:

$$N_{CM} = K_{22}\left(\frac{R-1}{RT}\right)^{0.5} \tag{7.16}$$

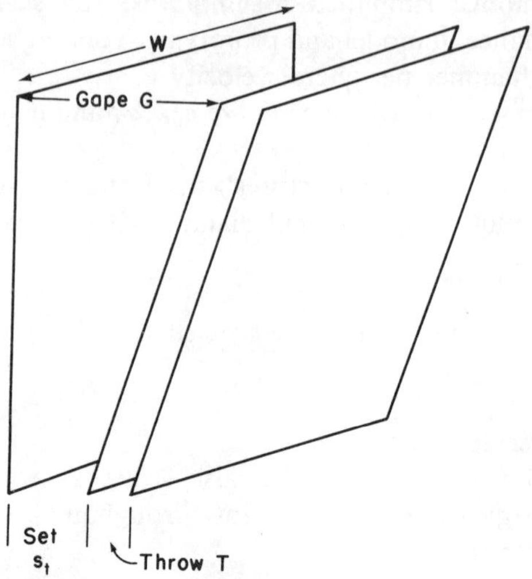

Figure 7.8. Jaw geometry for jaw crusher (Rose and English, 1967). *Reproduced by permission of the Institution of Mining and Metallurgy*

When equation (7.16) is incorporated, maximum capacity may be expressed as:

$$Q = K_{23}\varepsilon W T^{0.5}\left(s_t + \frac{T}{2}\right)\left(\frac{R}{R-1}\right)^{0.5} \rho_f f_1(\psi) f_2(\omega) \tag{7.17}$$

Hence for a given material, of constant size distribution:

$$Q = K_{24} W T^{0.5}\left(s_t + \frac{T}{2}\right)\left(\frac{R}{R-1}\right)^{0.5} \tag{7.18}$$

Since $R/(R-1)$ is dimensionless, and $(s_t + T/2)$ is a function of L, maximum capacity scales as $\lambda^{2.5}$ for a given feed size distribution. In practice larger sized machines would be associated with larger feed size distributions.

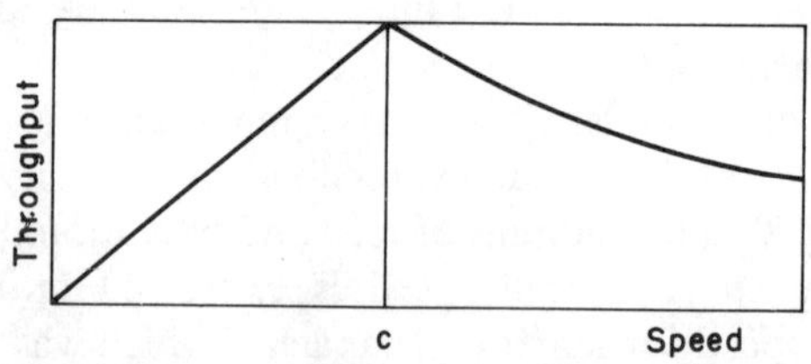

Figure 7.9. Critical speed for jaw crusher (Rose and English, 1967). *Reproduced by permission of the Institution of Mining and Metallurgy*

7.2.5.2 Power scaling

Rose and English (1967) similarly produced an expression for maximum power required at critical speed N_{CM} in terms of the Bond index W_i for the material crushed:

$$p = K_{25}\varepsilon WT^{0.5}\left(s_t + \frac{T}{2}\right)\left(\frac{R}{R-1}\right)^{0.5} \rho_f f_3(\psi) f_4(\omega) \times \left[\frac{G^{0.5} - 1.07(s_t + T)^{0.5}}{G^{0.5}(s_t + T)^{0.5}}\right] W_i \tag{7.19}$$

Hence for a given material and feed size distribution:

$$p = K_{26} WT^{0.5}\left(s_t + \frac{T}{2}\right)\left(\frac{R}{R-1}\right)^{0.5}\left[\frac{G^{0.5} - 1.07(s_t + T)^{0.5}}{G^{0.5}(s_t + T)^{0.5}}\right] \tag{7.20}$$

If the expression in the square bracket can be taken as approximately proportional to $L^{0.5}$, then power scales as λ^3 for a given feed size distribution.

From equations (7.18) and (7.20), specific energy may be expressed as:

$$p_s = K_{27}\left[\frac{G^{0.5} - 1.07(s_t + T)^{0.5}}{G^{0.5}(s_t + T)^{0.5}}\right] \tag{7.21}$$

for a given material and given feed size distribution. Specific energy, therefore, scales approximately as $L^{0.5}$. However, a larger machine would be associated with a larger feed size, and hence $f(\psi)$ must be taken into account in scaling of capacity, power or specific energy.

Thus there are no simple scale relations for jaw crushers.

7.2.6 Fluid energy mills

7.2.6.1 Theory for micronizer type of fluid energy mill

Tanaka (1972) has studied the micronizer type of fluid energy mill shown diagrammatically in Figure 7.10.

Rumpf (1960) has assumed constant tangential air velocity v_t, and that the flow path is an Archimedes spiral, for which condition centrifugal force on a particle can be equated to the drag force to give an expression for tangential velocity in terms of system geometry:

$$v_t = K_{28}(V_n/D^2)(D/d_n)/[(1 - D_o/D) - 2K_{29}(d_N/D)\cos\theta] \tag{7.22}$$

Hence to attain constant v_t, not only must the geometric relations D/d_N, D_o/D and θ be kept constant, but also V_n/D^2.

The flow pattern within the mill has been observed to have a reverse flow in the spaces between a jet and the top and bottom walls (Kürten and Rumpf, 1966) induced by the jet action on the surrounding air. This reverse flow was seen to convey oversize particles in the classifying zone back into the crushing zone. The mean velocity v_i of this reverse flow has been expressed by Tanaka

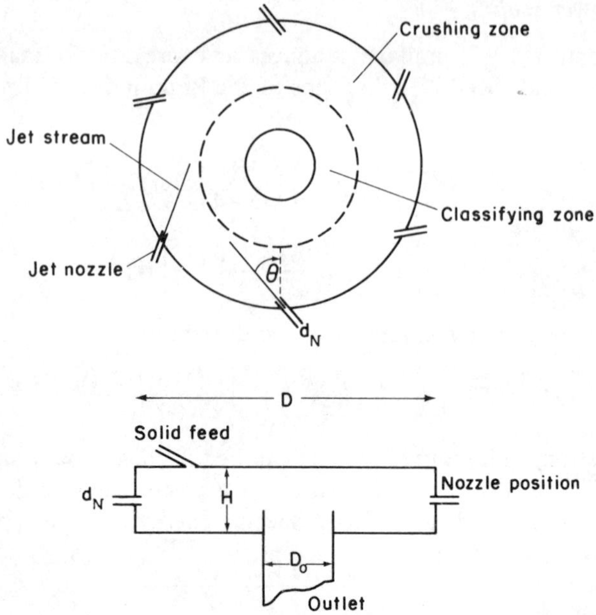

Figure 7.10. Diagram of micronizer type of impinging jet fluid energy mill (Tanaka, 1972). *Reprinted with permission from* Ind. Eng. Chem. Process. Des. Devel., *11, 238–241. Copyright 1972 American Chemical Society*

(1972) as:

$$v_i = (V_n/D^2)(D/d_N)/[(H/D)(D/d_N) - K_{30}] \quad (7.23)$$

Hence to attain constant v_i, additional to constant V_n/D^2 and D/d_N required for constant v_t, there is also the requirement for constant H/D.

Now a true scale-up should be based on constant cut size particle d_p, not on constant velocity, and Tanaka (1972) has derived an equation for d_p as follows:

$$d_p = [(D^3/V_n)(d_N/D)]^{0.5}[(1 - D_o/D) - 2K_{31}(d_N/D)\cos\theta] \quad (7.24)$$

Comparing the conditions for constant v_t, and d_p, (see Table 7.2) it is clear that the only different factor is V_n/D^3 for d_p, but V_n/D^2 for v_t.

For constant geometrical relations, equation (7.24) may be expressed as:

$$d_p \propto [D^3/V_n]^{0.5} \propto [D(D^2/V_n)]^{0.5} \quad (7.25)$$

Hence if the condition for constant radial acceleration v_t is applied d_p will increase as D increases, i.e. size scale-up will coarsen the product.

To expect exactly similar behaviour of the recycled oversize particles to the crushing zone, the conditions for constant v_i must be considered. Then for

Table 7.2. Deductions from equations (7.22)–(7.24)

Condition for constant	Constant				
v_t	V_n/D^2	D/d_N	D_o/D	—	θ
v_i	V_n/D^2	D/d_N	—	H/D	—
d_p	V_n/D^3	D/d_N	D_o/D	—	θ

constant d_N/D and v_i in equation (7.23):

$$H/D \propto V_n/D^2 + \text{constant} \propto D(V_n/D^3) \tag{7.26}$$

Hence, if constant V_n/D^2 appropriate to constant v_t is used then H/D is constant, or $H \propto D$. But if constant V_n/D^3 appropriate to constant d_p is used, then $H/D = K_{32}D + K_{33}$ gives the relation between H and D that must be maintained.

7.2.6.2 *Capacity scaling*

On the assumption that the grinding rate is a function of x^y, Tanaka (1972) derived an expression for the feed rate Q based on equation (7.22), i.e. that V_n/D^2 is constant, as follows:

$$Q \propto V_n{}^2[(D^2/V_n)D]^{0.5y}$$

$$\propto V_n{}^2[(D^2/V_n)(V_n)^{0.5}]^{0.5y}$$

Hence for constant $V_n/D^{2:}$

$$Q \propto V_n^{2+0.25y} \tag{7.27}$$

Now Mori and Jimbo (1961), without maintaining constant d_p, have reported empirically found exponents of V_n in the range of 2.2 to 3.3, corresponding to values of y from 1 to 5. Taking $y = 1$ then:

$$Q \propto Vn^{2.25} \tag{7.28}$$

would be expected to be a safe scale relation for constant v_t.

On the other hand, if equation (7.24) is accepted as the basis for capacity determination, Tanaka (1972) obtained:

$$Q \propto V_n^2[(D^3/V_n)(d_N/D)]^{0.5y}$$

Then with (d_N/D) and V_n/D^3 constant:

$$Q \propto V_n^2 \tag{7.29}$$

would be the expected scale relation for constant d_p. No experimental data have been found for constant d_p.

Tanaka *et al* (1972) took account of classification which proceeds concurrently with grinding in a fluid energy mill (see §10.6.1) which Mori and Jimbo (1961) failed to do. By a very different route from that of Mori and Jimbo involving bold assumptions, Tanaka *et al* concluded that feed make-up rate is proportional V_n^3, which falls within the range proposed by Mori and Jimbo, indicated above, and provides interesting confirmation of that range.

To indicate the complexity of the system, Mohanty and Narasimhan (1982) have found an optimum feed make-up rate, different for each material ground, at which size reduction is maximized, and they were able to produce an empirical expression relating grinding rate with new surface produced.

Skelton *et al*. (1980) have provided design correlations in terms of the dimensionless groups: the nozzle Reynolds number, the fluid-to-solid mixing ratio, and the particle-to-chamber diameter ratio.

7.2.6.3 *Power scaling*

If, for the same nozzle, $p \propto V_n$, then for constant v_t, specific energy is given by:

$$p_s \propto V_n/V_n^{2.25} \propto V_n^{-1.25} \tag{7.30}$$

and for constant d_p:

$$p_s \propto V_n/V_n{}^2 \propto V_n{}^{-1} \tag{7.31}$$

In either case, the remarkable result is noted that increase in airflow V_n reduces specific energy for the theoretically perfed flow and similarity assumptions made in the study. However, the study is incomplete, and more work requires to be directed towards such problems as the variation of flow pattern within a mile on scale-up and the effect of enlarged nozzles with scale-up (nozzles assumed constant diameter in the study).

7.2.7 Attrition mills

7.2.7.1 *Mill features*

One type of attrition mill, the sand (Perl) mill, also known as the stirred ball mill, is given in Figure 7.11. The ball charge is agitated continuously by means of a rotating paddle and a fluid medium is continuously passed through the charge to remove fines as they are formed. Another type has a rotor mounted within a stator (Sadler *et al*., 1975), but is otherwise similar in operation. (see §11.2.3.)

The advantage of the attrition mill is its capability for producing an ultrafine product with less specific energy than either a vibration or a conventional ball mill, as seen graphically in Figure 7.12. This advantage is attributed to intense shear stresses not attained by the other two mills.

7.2.7.2 *Power scaling*

Sadler *et al* (1975) performed more than twenty experiments to obtain the following relation for the power required by an attrition mill:

$$p = 2.55 \times 10^{-5} V^{1.75} N^{1.37} d^{0.48} \rho_b{}^{1.09} \tag{7.32}$$

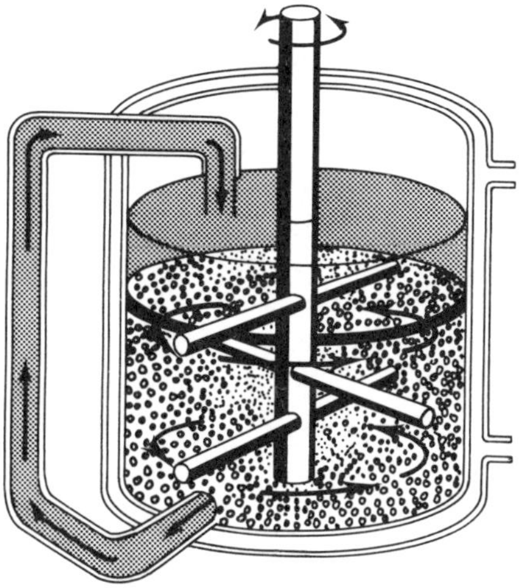

Figure 7.11. Typical sand (Perl) mill arrangement (Herbst and Sepulveda, 1978)

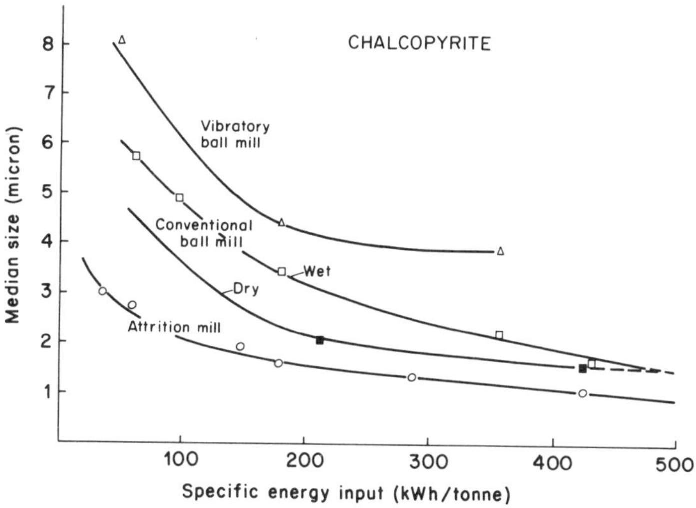

Figure 7.12. Specific energy versus product median size for three types of mill (Herbst and Sepulveda, 1978)

where p is the power measured at the shaft (kW), V is the attritor volume (gallons), N is the shaft rotational speed (rev/minute), d is the diameter of grinding balls (inches), and ρ_b is the density of grinding balls (g/cm^3).

7.2.7.3 *Scaling of specific energy*

Figure 7.13 gives the relation between specific energy and the median particle size of the product on a log–log plot, and shows a straight line over the range

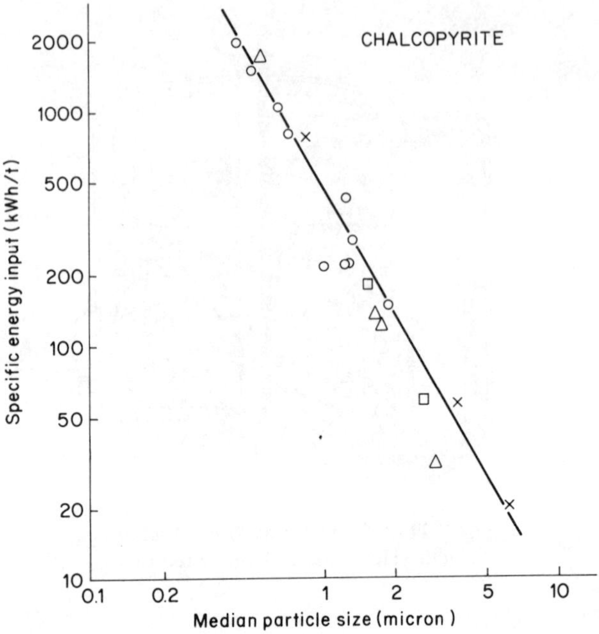

Figure 7.13. Energy–size reduction relationship for the attrition milling of chalcopyrite in water for four mills. Attritors: ○ Pima S-1, □ Pima C-3, △ Pima C-10, × Anamax S-1 (Herbst and Sepulveda, 1978)

of particle size considered for three mill sizes. Hence Charles' law applies and equation (6.18) may be expressed as:

$$p_s = A(x_{me})^{-n} \tag{7.33}$$

where p_s is specific energy, A is a constant for a given material, x_{me} is the median particle size of product (micron), and n is the slope of the line in Figure 7.13.

Hence when these parameters are known for a given material, since specific energy is independent of mill size, scaling is made easy.

7.2.7.4 Application of scaling laws

Following Herbst and Sepulveda (1978), chalcopyrite was fed at a rate of 6 lb/hour to an attrition mill with a volume of 10 gallons, operated at a speed of 175/rev/minute with 3/32 inch steel balls and gave a predicted power from equation (7.32) of 5.31 kW. This compared well with an observed value of 5.25 kW.

Assuming 2000 lb = 1 ton, the application of equation (7.33) gave:

$$p_s = \frac{5.31}{(6/2000)} = 1770 \text{ kWh/t} = 460\ x_{me}{}^{-1.79}$$

or $x_{me} = 0.47$ micron. This agreed with the observed value of 0.46 micron.

7.2.8 Summary of scaling criteria

Table 7.3 is a summary of the scaling criteria for the mills studied.

Table 7.3. Summary of scaling criteria for mills

Mill	Basic equations	Conditions	Scale law	Comments
Tumbling	$Q \propto LD^{2.5}$ $p \propto LD^{2.5}$ p_s independent of D	Constant N/N_c Constant J (with optimum ball sizes)	$\lambda^{3.5}$ for $L \propto D$	Variations possible for autogenous mills—see text
Vibration	$Q \propto D^3$ $p \propto D^3$ p_s independent of D p_s independent of D	Constant ω (>1500 r.p.m.) Constant J (0.8) Constant α $V_p > 0.5$	λ^3	Maximum capacity when vertical and horizontal components of amplitude are equal
Vertical spindle	$Q \propto D^{2.5}$ $p \propto D^{2.5}$ p_s independent of D	Constant geometry Constant n Constant N/N_c Constant P_o	$\lambda^{2.5}$	Since $h_2 \propto d$, larger machines can accept larger feed size
Swing hammer	$Q \propto D^2$ $p \propto D^2$ p_s independent of D	Constant peripheral speed of hammers	λ^2	—
Jaw crushers	Complex relations for Q, p and p_s (see text)		—	Larger machines associated with larger feed sizes
Fluid Energy	$Q \propto V_n{}^{2.25}$ $p \propto V_n$ $p_s \propto V_n{}^{-1.25}$	Constant v_t	—	Equations apply only to theoretically perfect assumptions
	$Q \propto V_n{}^2$ $p \propto V_n$ $p_s \propto V_n{}^{-1}$	Constant d_p		
Attrition	$p \propto V^{1.75}$ p_s independent of D	Constant N/N_c, d, ρ_b	—	—

7.3 USE OF SCALING LAWS

7.3.1 Preview

In §6.7, mill design by means of grindability was discussed, exemplified in the Hardgrove index applied to vertical spindle mills and the Bond work index applied to tumbling ball mills. In each case the index could not be used directly but had to be supported by empirically found factors according to the mill concerned.

Scaling laws provide one alternative to the grindability method, as indicated in §6.9. Instead a scaled-down version of the mill is operated, which has the distinct advantage of requiring a much smaller feed rate than the prototype. On the other hand, because of the problems inherent in scaling, outlined in §7.4, the largest model that can be accommodated should be used to reduce the extent of the scale-up.

In the following sections examples are given of the application of scaling laws to well-known mill types, models of which may be operated in a batch or a continuous mode and for the latter in either open- or closed-cycle conditions.

Batch tests are experimentally simpler, and they may also obviate the need to consider scaling of airflow, which involves different laws from the grinding process itself.

7.3.2 Application to tumbling mills

7.3.2.1 'Rate of production' sizing

Austin (1973) has described this method, based on a batch test, in which the mill is considered as a reactor which transforms feed into product. In the reactor it is assumed that the rate of production of acceptable product does not vary substantially with the method of circuit operation: for example, that the internal flow pattern is similar and that the effects are negligible. The only pertinent variable is taken to be the internal drum diameter D. It has been observed empirically that approximately, for otherwise similar conditions, the rate of production of minus x_{80} material per unit volume of mill is proportional to $D^{0.6}$, and hence crude scaling is possible on this basis.

7.3.2.2 Constant specific energy sizing

For the stated conditions, constant specific energy throughout a mill series has been noted in §7.2.1.2 for ball and rod mills (not autogenous mills). This feature has often been made the basis of mill design; but Austin (1973) has shown that it involves the same uncertainties as the rate of production method when used to scale-up from laboratory to commercial size equipment.

7.3.3 Application to vertical spindle mills

A continuously operated mini-mill of the vertical spindle type has been used to forecast results for larger mills of similar geometrical form. Capacity was scaled from equation (7.12) which implied increased speed according to the relation $N \propto D^{-0.5}$, the same number of balls n and the same optimum pressure P_o as in the prototype. A problem arose over the feed size since the angle of nip ϕ demanded a very much smaller maximum particle size than in the prototype for which the associated surface was significantly greater than for the feed in the prototype. Thus a correction factor had to be introduced to account for the reduced feed size in the model. Under these conditions scaling of capacity when grinding coal was achieved over a wide range of D.

When idle power was subtracted from operating power, each at motor output, the resulting net specific energy for grinding coal remained constant over a wide range of D, according to equation (7.14).

7.4 SHORTCOMINGS OF END-RESULT SCALING LAWS

The scaling laws discussed in this chapter relate variables to the end result only of a complex process, thereby masking scaling effects due to specific areas of breakage or classification within the mill. Hence there is a requirement to identify distinct areas of activity, examples of which follow.

A quarter of a century ago, Callcott (1960) divided the operation of the swing hammer mill as follows (see figure 7.14):

(1) primary impact zone;
(2) zone adjacent to the breaker plate;
(3) zone adjacent to the breaker bars;
(4) zone adjacent to the discharge grid;
(5) zone of recycled material.

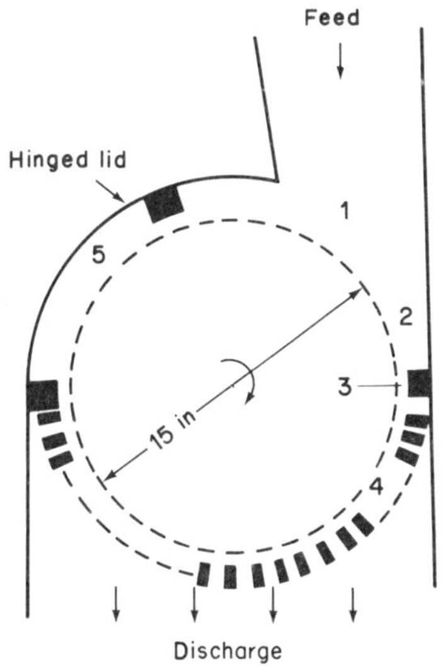

Figure 7.14. Size reduction and size separation zones within a swing-hammer mill. Hammer tip speed = 11,000 ft/minute. Mill width = 8 inches. Rotor carries four rows of rectangular hammers: two rows of three hammers and two rows of two hammers. Zones: 1 primary impact, 2 breaker plate, 3 breaker bar, 4 grid, 5 return (Callcott, 1960). *Reproduced by permission of the Institute of Energy*

The aid of cinephotography was invoked to clarify the mechanisms in the above areas and of matrix analysis to quantify the observed breakage and classifying processes.

More recently, Austin and Luckie (1982) have similarly treated a vertical spindle mill, because their analysis of internal mechanisms implies the following zones (see Figure 5.22) :

(1) grinding zone;
(2) elutriation zone (treated as internal classification);
(3) zone of transport to main classifier;
(4) main classifying zone.

The mass balances for these individual zones have not always been obtained. The main classifier flows and grades have been assessed by Babcock Power Research by the arrangement shown in Figure 7.15. The classifier within the mill was replaced by a dummy cone of the same external dimensions so that the airflow pattern within the mill remained unchanged.

However, the mass balance for the elutriation zone is very difficult to obtain because of sampling problems, and it remains a challenge to enterprising investigators. Austin and Luckie (1982) bypassed direct measurement by obtaining an expression for the product of recycled quantity and the fraction of a given size grade in terms of measureable variables, which they used in their analysis. (see §10.4.1.)

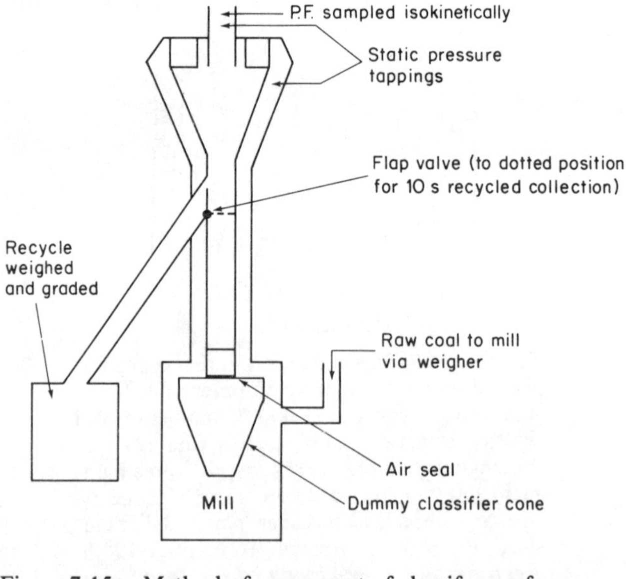

Figure 7.15. Method of assessment of classifier performance in a vertical spindle mill (Babcock Power Research)

Tumbling mills, because of their very widespread use, have been the object of more detailed study than any other type, and results are available in the following areas:

(1) mill speed and related ball motion changes;
(2) mill shell geometry;
(3) grinding media and powder charge;
(4) operational effects, such as axial and radial mixing, residence time distribution, and slurry rheology (wet milling).

All such factors have to be taken into account if accurate mill design is to be achieved. They are listed as above to illustrate how much is missed by end-result scaling, and they are studied in more detail in Chapters 8, 9 and 10.

7.5 REFERENCES

Andreasen, A. H. M. (1957). 'The principle of similarity applied to mills and to the grinding process', *Trans. Inst. Chem. Eng.*, **35**, 104–110.

Anson, B. (1969). 'Kinetics of grinding in vertical spindle mills', Central Electricity Generating Board Report SSD/NE/R169.

Arbiter, N., and Harris, C. C. (1982). 'Scale-up and dynamics of large grinding mills—a case study'. In *Design and Installation of Comminution Circuits* (Mular, A. L., and Jergensen, G. V., Eds.), SME/AIME, New York, 491–508.

Austin, L. G. (1973). 'Understanding ball mill sizing', *Ind. Eng. Chem. Process. Des. Devel.*, **12**, (2), 121–129.

Austin, L. G., and Luckie, P. T. (1982). 'An analysis of ball and race milling: Part II, The Babcock 1.7E mill', *Powder Technol.*, **33**, 113–125.

Callcott, T. G. (1960). 'A study of the size reduction mechanisms of swing hammer mills', *J. Inst. Fuel*, **33**, (238), 529–539.

Crocker, B. S. (1952). 'Screened ore used for fine grinding at Lake Shore mines', *Min. Eng. Trans. AIME*, 1069.

Dettmer, P. B., and Sobering, A. (1964). 'Pilot and commercial dry autogenous grinding of Labradore iron—Carol Lake'. *Proc. 7th Int. Miner. Process. Congr.*, New York, 573–593.

Fahrenwald, A. W. (1934). 'Some fine grinding fundamentals', *Trans. AIME*, **112**, 88–115.

Gow, A. M., Guggenheim, M., Campbell, A. B., and Coghill, W. H. (1934). 'Ball milling", *Trans. AIME*, **112**, 24–78.

Harris, C. C., and Arbiter, N. (1982). 'Grinding mill scale-up problems', *Min, Eng.*, Jan., 43–46.

Herbst, J. A., and Sepulveda, J. L. (1978). 'Fundamentals of fine and ultrafine grinding in a stirred ball mill', *Proc. Int. Conf. Powder Bulk Handling Process.*, Rosemont, Ill, 1978, Ind. Sci. Conf. Management, Chicago, 452–470.

Heywood, H. (1947). 'Determination of coal grindability', *Proc. 1st Conf. Pulverised Fuel*, Harrogate, Inst. Fuel, 594–611.

Hukki, R. T. (1954). 'Tumbling mill capacity and power consumption as related to mill speed', *Min. Eng.*, **196**, 728–730.

Johanson, J. R. (1965). 'A rolling theory for granular solids', *J. App. Mech.* (*ASME*), **32**, 842–848.

Kürten, H. and Rumpf, H. (1966). 'Strömungsverlauf und zerkleinerungsbedingungen in der spiralstrahlmühle', *Chemie Ing. Tech.*, **38**, 1187–1192.

Marshall, V. C. (Ed.) (1974). 'Comminution: a report to the Inst. Chem. Eng. Working Party'.

Mohanty, B., and Narasimhan, K. S. (1982). 'Fluid energy grinding', *Powder Technol.*, **33**, 135–141.

Mori, Y., and Jimbo, G. (1961). 'Studies on the circulating type jet pulveriser', *Kagaku Kogaku*, **25**, 477–486.

Rose, H. E. (1967). 'Some observations on the application of vibration mills', Paper to the Chemical Engineering Group of Society of Chemical Industry, London, 15 March.

Rose, H. E., and English, J. E. (1967). 'Theoretical analysis of the performance of jaw crushers', *Trans. IMM (London)*, **76**, C32–C43.

Rose, H. E., and Sullivan, R. M. (1958). *Treatise on the Internal Mechanics of Ball, Tube, and Rod Mills*, Chemical Publishing Co, New York.

Rose, H. E., and Sullivan, R. M. (1961). *Vibration Mills and Vibration Milling*, Constable, London.

Rumpf, H. (1960). 'Versuche zur bestimmung der teilchenbewegung in gasstrahlen und beanspruchungsmechanismus strahlmühlen', *Chemie Ing. Tech.*, **32**, (29), 335–342.

Sadler, L. Y., Stanley, D. A., and Brooks, D. R. (1975). 'Attrition mill operating characteristics', *Powder Technol.*, **12**, 19–28.

Skelton, R., Khayyat, A. N., and Temple, R. G. (1980). 'An investigation of microniser performance—design and operational parameters', *Proc. Europ. Symp. Particle Technol.*, Amsterdam, 410–423

Taggart, A. F. (1945). *Handbook of Mineral Dressing*, J. Wiley, New York.

Tanaka, T. (1972). 'Scale-up theory of jet mills on basis of comminution kinetics', *Ind. Eng. Chem. Process. Des. Devel.*, **11**, (2), 238–241.

Tanaka, T., Nakajima, Y., and Furuya, M. (1972). 'Closed circuit grinding theory based on comminution kinetics and its application to jet milling mechanism', Dechema Monograph 69, 515–526.

Turner, R. R. (1982). 'Selection of primary autogenous and semi-autogenous grinding mills. Part A: Design and operation'. In: *Design and Installation of Comminution Circuits*, (Mular, A. L., and Jergensen, V. J., Eds.), SME/AIME, New York, 474–490.

CHAPTER 8

Basics of mathematical simulation

NOTATION FOR CHAPTER 8

$a_{i,j}$	Parameter in equation (8.59)
a'	Proportion of classified stream reporting to coarse stream independently of size
a^*	Parameter in equation (8.2)
A	Parameter given by $\Delta M/M_1(0)$ in equation (8.25)
$b_{i,j}$	Mass fraction of material broken from interval j which appears in interval i

$b(x,y)$	Mass fraction of material broken from size y which appears in size x
$\mathbf{b}$	Column vector form of b_{i-j}
$\mathbf{b}$	Column–row matrix form of $b_{i,j}$
$\mathbf{B}_{i,j}$	Cumulative mass fraction of material broken from interval j which appears below size x_i (breakage distribution function)
$B(x,y)$	Cumulative mass fraction of material broken from size y which appears below size x
$\boldsymbol{B}$	Column vector form of B_{i-j}
$\mathbf{B}$	Column–row matrix form of $B_{i,j}$
C	Concentration by mass
$C(t)$	Concentration at time t
C°	Concentration of tracer, if uniformly distributed throughout a system
$C^{\circ}_m(t)$	Tracer concentration at time t in the mth reactor in equation (8.52)
d	Ball diameter
D	Mill internal diameter
D_{eff}	Effective internal diameter
D^*	Mixing factor due to ball action within a tumbling mill
f_c	Fraction of mill occupied by powder charge = 0.4 JU
$\boldsymbol{f}$	Column vector for feed
F_i	$= S_jB_{i,j}$ approximately
$F_i(t)$	Cumulative mass fraction less than size x_i at time t
$F(x,t)$	Cumulative mass fraction less than size x at time t
$F_n(x)$	Cumulative mass fraction less than size x after n stages
F^*	Feed rate (mass/unit time) (F_1^* corresponds to $U=1$)
G	Intercept on ordinate at zero time in Figure 9.2
h	Height
$h_{i,j}$	Parameter in equation (8.17)
$h'(t-\eta_c\tau)$	Unit (positive) step function associated with time delay in the plug flow reactor of a finite-stage model
H	Slope of line in Figure 9.2
$\mathbf{I}$	Unit matrix
J	Fraction of drum volume occupied by ball charge, including voids
J_T	Value of J near gate discharge
k	Fraction of flow in active reactor which passes through the stagnant reactor in a finite-stage model
k_o	Overfilling factor (Figure 8.13)
K	Constant
l	Distance along ball mill axis from the feed end
L	Mill length
m	Number of reactors in series

$M_i(t)$	Mass fraction in size interval i at time t Subscripts: obs observed calc calculated
$\mathbf{M}(t)$	Column vector form of $M_i(t)$
$M_{i,n}$	Mass fraction in size interval i after n stages
$M_n(x)$	Mass fraction of size x after n stages
M_k	Mass fraction of balls in the size interval k
$M(x,t)$	Mass fraction of size x at the time t
ΔM	Mass fraction taken from interval 1
n	Number
N	Rotational speed
N_c	Theoretical critical speed of rotation
N^*	Fraction of theoretical critical speed of rotation $= N/N_c$
p_i	Total fraction of mill product in size interval i
$\mathbf{p}$	Column vector for mill product
p_{SS}	Parameter describing steady-state product
$p(t)_{BATCH}$	Parameter describing batch process product at time t
Pe	Peclet number $= \bar{v}L/D^*$
q	Steady-state volumetric flow rate at entry and exit
r	Parameter in equation (8.26)
r'	Radius
R	Geometric ratio of sieve sizes
S_i	Specific rate of breakage of particles in interval i
$S_{i,k}$	Specific rate of breakage of particles in interval i by balls in the size interval k
$\bar{S}_i$	Weighted mean value of $S_{i,k}$ in equation (8.6)
S_i'	$\Delta t S_i$
$S(x)$	Specific rate of breakage of particles of size x
$S(y,n)$	Specific rate of breakage of size y particles in the nth stage
$\mathbf{S}$	Column–row matrix form of S_i
t	Time
t^*	False time defined by $t^* = \int_0^t \varkappa(t)\, dt$ (slowing-down effect)
T	Thickness
U	Fraction of powder filling of the interstices between the balls in the ball charge Subscripts: 1 Corresponds to $U = 1$ C Corrected for higher level of charge at grate outlet than over rest of the mill
$\bar{v}$	Average velocity
V	Total system volume
w	Width
W	Mass of material ground in a batch process
W^*	Mass of hold-up in a continuous process (W_1^* corresponds to $U = 1$)

x	Particle size Subscripts: M Corresponding to value for which S_i is a maximum 0 Reference size
x_i	Top size of interval i
x_{i+1}	Bottom size of interval i
$\bar{x}_i$	Mean size of interval i
x^*	Value of x for which the denominator of equation (8.3) has a value of 0.5
y	Particle size greater than x
z	Number of time intervals
α	Exponent in equation (8.2)
β	Exponent in equation (8.13)
γ	Exponent in equation (8.13)
δ	Sieve correction
Δ_2	Function given by equation (8.25)
ε	Exponent in equation (8.24)
ζ	Parameter in equation (8.29)
η	Fraction of total system volume in finite-stage model Subscripts: a active fully mixed reactor b stagnant fully mixed reactor c plug flow reactor
θ	Dimensionless time $= t/\tau$
$\varkappa$	Slowing-down effect factor
λ_j	Algebraic expression involving six parameters in the finite-stage model, equation (8.53)
Λ	Exponent in equation (8.3)
μ	Median in log-normal distribution
ξ_1	Exponent in equation (8.4a)
ξ_2	Exponent in equation (8.5)
ρ_s	Solids density in a slurry
σ	Standard deviation in log–normal distribution
$(\sigma_t)^2$	Variance given by equation (8.32)
$(\sigma_\theta)^2$	Dimensionless variance given by equation (8.40)
τ	Mean residence time Subscripts: B bulk mean residence time $= W^*/F$ L large fully mixed reactor P plug flow reactor S small fully mixed reactor
τ_1	$= W_1^*/F_1^*$
$\bar{\tau}$	$= \tau/m$
$\phi(t)$	Residence time distribution function
$\phi_m(t)$	RTD function for mill
$\phi_s(t)$	RTD function for sump
$\phi(\theta)$	Dimensionless RTD function

Φ	Parameter in equation (8.13)
ψ	Exponent in equation (8.14)
Ω	$\tau_L/(\tau_L - \tau_S)$ in equation (8.48)

8.1 INTRODUCTION

The very real shortcomings of past and some present methods of mill design have been underlined in previous chapters. The energy 'laws' are now recognized as applicable to very limited ranges of energy density (see §6.5). Similarly, grindability uncertainties (see §6.8) can lead to significant discrepancies between calculated and operationally found power requirements. Equally, end-result scaling criteria for capacity and energy, though very useful within their limitations, have been demonstrated in Chapter 7 to provide too crude a basis for accurate design and have shown the need for the study of the finer details of mill operation to ensure confident scaling over a sufficiently wide range of mill sizes. Hence close investigation must be made of the size reduction process, and that for each particle size range, since behaviour is different, as size becomes smaller.

Already in §4.3.8.5 the concepts of specific rate of breakage (selection function) and breakage distribution (breakage function) have been introduced. It remains to define them more specifically, show how they may be assessed and how they are related to the size–mass balance (SMB) equation for the grinding zone of a mill.

In a tumbling mill, beside size reduction, there is also mass transport from entrance to exit. Thus the residence time distribution function has come to be important to the context of continuous, as opposed to batch, operation.

This chapter lays the foundation for mathematical simulation by discussing these basic concepts, chiefly in relation to the tumbling mill operated in open-circuit, but also for other mill types where information is available.

8.2 BREAKAGE FUNCTIONS

8.2.1 Survey

Coarse material of a whole range of particle sizes is fed into a mill and there appears as product from the grinding zone a finer material, again covering a whole range of particle sizes. Hence, in exploring the process whereby feed becomes product, it is not only necessary to know the rate at which particles break but also to describe their overall sizes. These two requirements are reflected in two breakage functions, originally conceived by Epstein (1948), namely the specific rate of breakage and the breakage distribution functions.

8.2.2 Specific rate of breakage, S

8.2.2.1 Definition

The rate at which a particle breaks depends on the size of the particle, because a smaller one is stronger than a larger one. If then S_j is the specific rate of breakage of material of size j and $M_j(t)$ is the mass fraction at time t of size j of the total mass W of particles in a batch test, then the rate of breakage of material of size j is $S_jM_j(t)W$. This indicates how the specific rate of breakage S is used. It was formerly called the selection function, following the concept that certain particles will, in a single pass through a mill, move through without size reduction, and hence only certain ones will be selected for breakage. However, specific rate of breakage is a better term.

It is usual to divide the total particle size range into geometric screen intervals, numbered 1 for the top size, 2 for the next and so on down to j . . ., down to i . . ., and finally down to n, which may be 38 microns, the smallest practical sieve size. See Figure 8.1 for the nomenclature diagram.

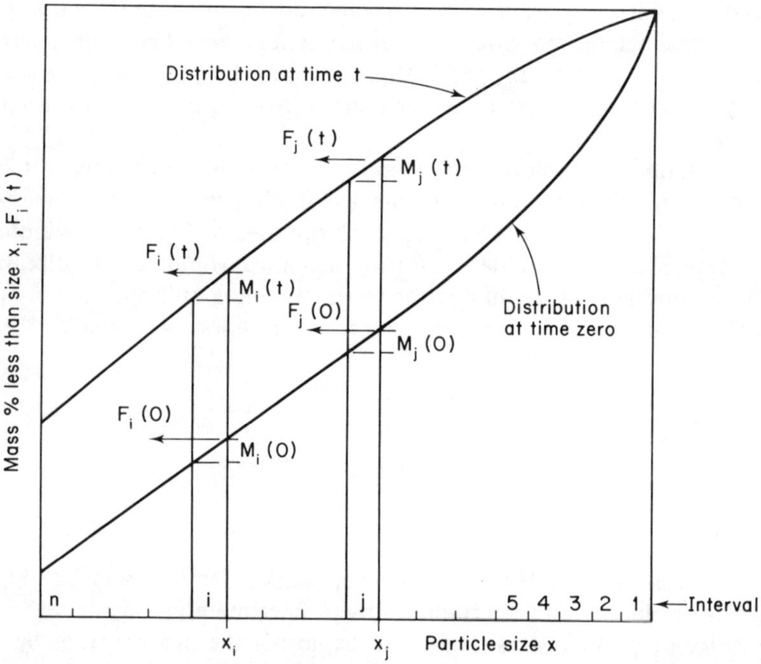

Figure 8.1. Nomenclature diagram for the discrete size, time continuous concept (Based on Klimpel and Austin, 1970). *Reprinted with permission form* Ind. Eng. Chem. Fundamentals, *9, 230–237. Copyright 1970 American Chemical Society*

8.2.2.2 *Linearity of breakage rate*

It has been postulated as reasonable, and demonstrated by Sedlatschek and Bass (1953) as being in accord with probability theory, that the rate of disappearance of size j material due to breakage is proportional to the amount of size j material present to be broken. Thus:

$$-\frac{d}{dt}[M_j(t)W] = S_jM_j(t)W$$

If S_j is constant with time, this can be expressed as:

$$\log\,[M_j(t)/M_j(0)] = -\frac{1}{2.3}(S_jt) \tag{8.1}$$

where $M_j(0)$ is the mass fraction of material in interval j at zero time (feed size), and $M_j(t)$ is the mass fraction of material in interval j at time t. Hence if $\log M_j(t)$ is plotted against t, a straight line of slope $S_j/2.3$ should result, typically as in Figure 8.2.

Tests have demonstrated this linearity in many cases, as indicated below. Austin *et al.* (1976a) have proved it under certain conditions for tumbling ball mills, vibration and hammer mills and shredder cutters. Shoji and Austin (1974) have shown it to be not quite true for tumbling rod mills, and Austin *et al.* (1981) found it only true for coal in a Hardgrove ring-ball mill when overgrinding was avoided. The linearity was demonstrated for a small vertical spindle mill used in a batch manner without airflow, by Austin *et al.* (1982b), (see Figure 8.3) but more scatter existed than with the corresponding tests on the Hardgrove machine.

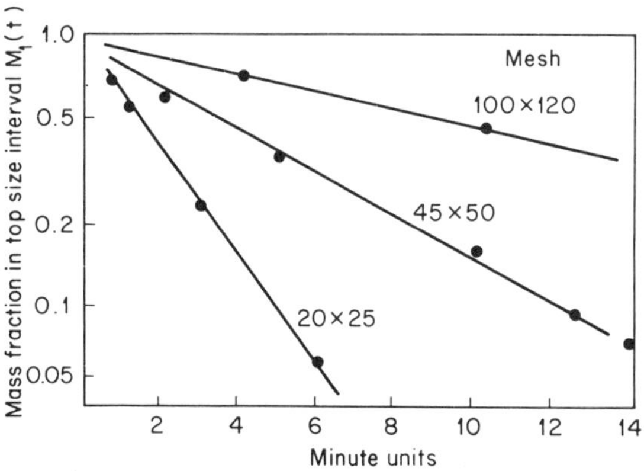

Figure 8.2. Linear (first-order) breakage plots for dry batch grinding of silicon carbide in a laboratory ball mill (Austin *et al.*, 1976a). *Reprinted with permission from* Ind. Eng. Chem. Process. Des. Devel., *15, 187–196. Copyright 1976 American Chemical Society.*

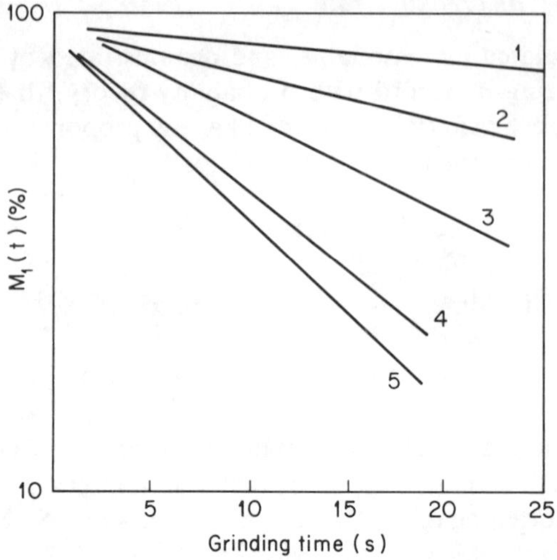

Figure 8.3. Linear (first-order) breakage plot for batch grinding of anthracite in a Babcock 1.7E mill, without airflow. Mesh: 1 50 × 70, 2 16 × 20, 3 0.75 × 0.53 inch, 4 and 5 0.265 inch × 4. Mean lines through experimental points (Austin *et al.*, 1982b). *Reproduced by permission of Elsevier Sequoia*

In the literature, linearity of breakage rate, as discussed above, is also described as 'linear kinetics', 'first-order grinding' or 'the first-order hypothesis'. Where the linearity holds, simplification of the mathematics of batch grinding is possible, such as 'normalization', shown in §8.2.3.3.

When linearity does not hold, the deviation therefrom may be to one side or the other, indicated by *ac* and *ad* in Figure 8.4. The curve *ac* implies an increased rate of breakage compared with *ab*, which may be due to (i) removal by breakage of larger particles which were initially shielding the smaller, (ii) the material becoming progressively weaker with time because of impacts which reduce the strength of a particle without actually breaking it, or (iii) the mechanics of the mill may become more effective (usually higher power) as fines increase.

Similarly, the curve *ad* indicates a slower rate of breakage than *ab*, which may be due to (i) the accumulation of stronger fractions of a heterogeneous material in the unbroken fragments as grinding proceeds, (ii) a cushioning action of fines around coarse material, or (iii) the mill mechanics may become less efficient with time (Austin, 1971/72). A further example of *ad* is found in coal–water slurry grinding (Klimpel, 1982), but return to *ab* is possible by the addition of the chemical additive polycarboxylate which improves the fluidity of the slurry.

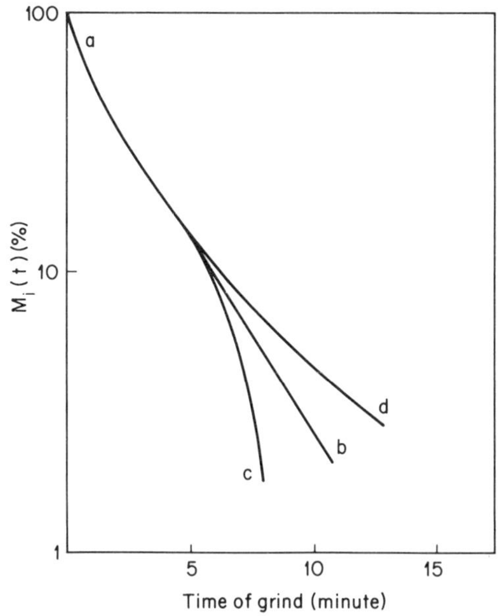

Figure 8.4. Derivations from linearity of breakage rate (Austin, 1971/72). *Reproduced by permission of Elsevier Sequoia*

8.2.2.3 Variation of S with particle size

Figure 8.5, based on results from grinding in a laboratory ball mill, is a log–log plot of S against particle size for several minerals which show a linear relationship (Austin *et al.*, 1976a). For digital computer analysis it is necessary to obtain a mathematical expression, which clearly takes the form:

$$S_i = a^*(x_i/x_0)\alpha \tag{8.2}$$

where a^* and α are constants for a given material ground in a particular mill under stated conditions of operation, and x_0 is a references size, usually 1 mm.

More recent work reported by Austin *et al.* (1982a) has demonstrated that, for larger particle sizes, the straight-line relationship no longer holds, and the pattern of curve 6 in Figure 8.5 is then typical. The falling off in the specific rate of breakage is due to two factors: the larger particles are not readily nipped between balls, and they tend to be cushioned by a layer of fines. To allow for the slower mean rates in larger particle sizes, it has become customary to introduce a multiplying factor in the right-hand side of equation (8.2) as follows:

$$S_i = \frac{a^*(x_i/x_0)^\alpha}{[1 + x_i/x^*]^\Lambda} \quad \Lambda > 0 \tag{8.3}$$

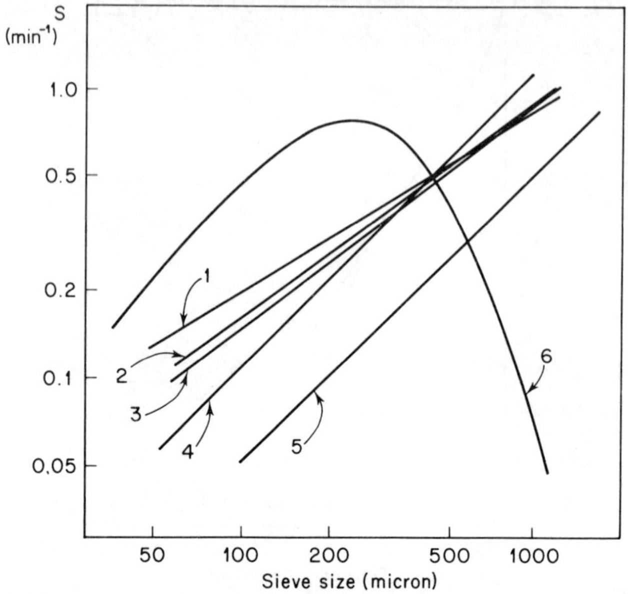

Figure 8.5. Variation of specific rate of breakage S for several materials in a dry laboratory ball mill: 1 limestone, 2 cement clinker II, 3 quartz, 4 cement clinker I, 5 silicon carbide 6 see text (Austin *et al.*, 1976a). *Reprinted with permission from* Ind. Eng. Chem. Process. Des. Devel., *15, 187–196. Copyright 1976 American Chemical Society*

where x^* is the particle size at which the denominator is 0.5 and Λ is an index of how rapidly the rate of breakage falls away.

The parameter x^* may be obtained in terms of x_M, which corresponds to the peak value of S, by differentiating equation (8.3) with respect to x and setting $dS/dx = 0$ at $x = x_M$. This gives:

$$x^* = x_M \left[\frac{\Lambda - \alpha}{\alpha}\right]^{1/\Lambda} \quad \Lambda > \alpha \tag{8.4}$$

For typical values of Λ and α, $x^* = 1.4\ x_M$.

Austin *et al.* (1982a) have shown that x^* needs to be scaled according to the square of ball diameter, and with respect to drum diameter according to:

$$x^* \propto D^{\xi_1} \tag{8.4a}$$

where $0.1 < \xi_1 < 0.2$.

Equation (8.3) has been shown to be more amenable to analysis than the formerly proposed log–normal distribution, also involving two parameters: the median μ and the standard deviation σ (Austin *et al.*, 1976a).

Thus the parameters associated with S in the mathematical representation are four: either a^*, α, μ and σ if the log-normal distribution is employed, or a^*, α, x^* and Λ if equation (8.3) is used.

8.2.2.4 *Variation of S with tumbling mill speed*

Figure 8.6 shows three distinct types of movement of the ball charge in a tumbling mill:

(a) The rising balls tend to be locked together, depending on the amount of slippage allowed by the liner geometry, and hence little size reduction takes place.
(b) The balls emerge at the inclined surface of the ball bed and in tumbling make many impacts with the powder charge. This is called cascading, effective for the fine grinding of smaller particles, depending on ball size.
(c) At higher fractions of critical speed than for b), with a given liner and ball and powder charge, the balls are thrown clear of the rising bed surface and fall into the toe of the mill. This is called cataracting, effective in breaking larger lumps of material, exemplified in autogenous mills. (see Figure 12.2.)

The fractions of critical speed at which cascading changes to cataracting varies with the ball filling fraction J and the powder filling fraction U. Hence in reporting test results, it is most important to specify the applicable ranges of N^*, J and U, and the liner configuration.

This change from cascading to cataracting pattern is the cause of the peak of the feed size specific rate of breakage S_1 in Figure 8.7. In this case, $J = 0.5$ and $U = 1$ in the dry batch grinding of dolomite. (Herbst and Fuerstenau, 1972). For these conditions, and with the eight 0.25 inch lifter bars on the 10 inch diameter shell (Yang *et al.*, 1967), the change occurred at $N^* = 0.75$.

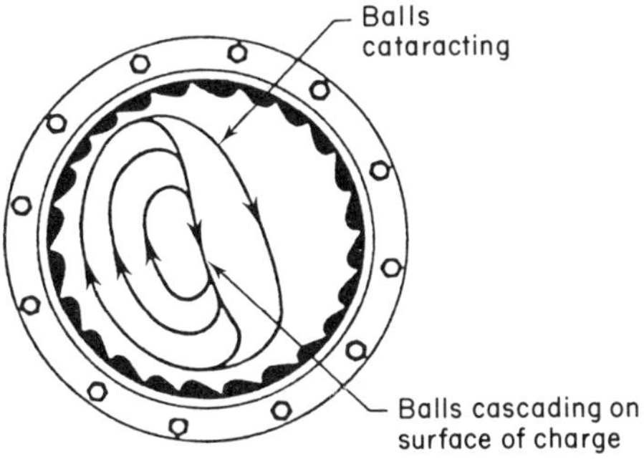

Figure 8.6. Patterns of ball charge movement within a tumbling mill (Rogers *et al.*, 1982b). *Reproduced by permission of the Canadian Institute of Mining & Metallurgy*

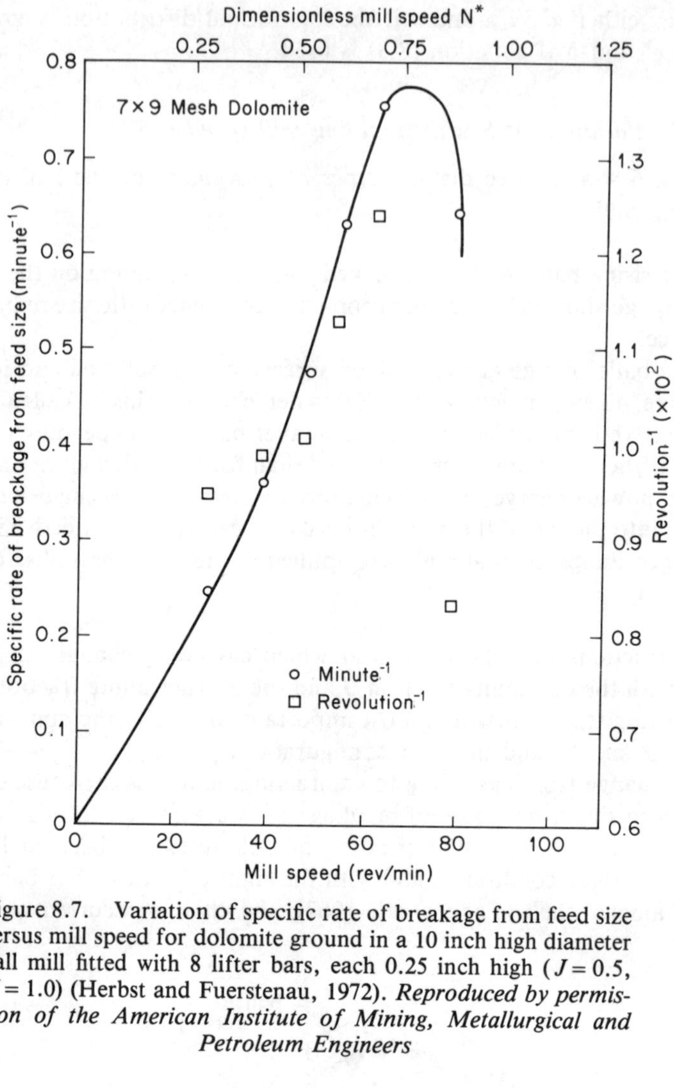

Figure 8.7. Variation of specific rate of breakage from feed size versus mill speed for dolomite ground in a 10 inch high diameter ball mill fitted with 8 lifter bars, each 0.25 inch high ($J = 0.5$, $U = 1.0$) (Herbst and Fuerstenau, 1972). *Reproduced by permission of the American Institute of Mining, Metallurgical and Petroleum Engineers*

Further, the change to cataracting is accompanied by a reduction in S_1 and hence cascading is the major action within the mill for the conditions stated.

Austin and Brame (1983) have attempted to quantify the correction to a^* in equation (8.2), and hence to S, for mill speed by:

$$a^* \propto (N^* - 0.1)/\{1 + \exp[15.7(N^* - 0.94)]\} \tag{8.4b}$$

but admit that this expression will vary with type of lifter.

8.2.2.5 *Variation of S with tumbling mill geometry*

(a) Mill diameter Malghan and Fuerstenau (1976) and Austin *et al.* (1982a)

agree that the specific rate of breakage varies approximately as $D^{0.5}$ for otherwise similar conditions with $D < 12.5$ ft (3.81 m), but approximately as $D^{0.3}$ with $D > 12.5$ ft (Austin and Brame, 1983).

(b) Mill length Based on the assumption, for any size fraction, of constant absolute rate of breakage per unit mill length, $S_i W^*/L$, in a grate discharge mill, Karra *et al.* (1980) obtained only a fair degree of correlation between calculated and experimental output size distribution, using a simplified SMB method. Hence it would appear that small changes to S_i occur along the length of mill during continuous operation.

(c) Drum liner Three very different types of drum liner, shown in Figure 8.8, were fitted in turn to a laboratory ball mill used for the wet grinding of quartz. The cross-section of liner *a* applies to only one of multiple slices lying in planes perpendicular to the mill axis. Each slice is angularly displaced relative to its adjacent slice, causing the rotating ball charge to spiral in such a way that larger balls tend to move toward the feed end. Operating conditions were $J = 0.2$, $U = 1.0$ and $N^* = 0.7$.

The effect on the specific rate of breakage is seen in Figure 8.9. As would be expected, liner *c* caused the highest rate of breakage of small sized particles and the lowest rate for larger particles, but it was surprising that liners *a* and *b* produced almost identical results. It was deduced by Rogers *et al.* (1982b) that the internal ball movement pattern must have been similar for *a* and *b*.

The importance of obtaining the same pattern of ball and powder charge

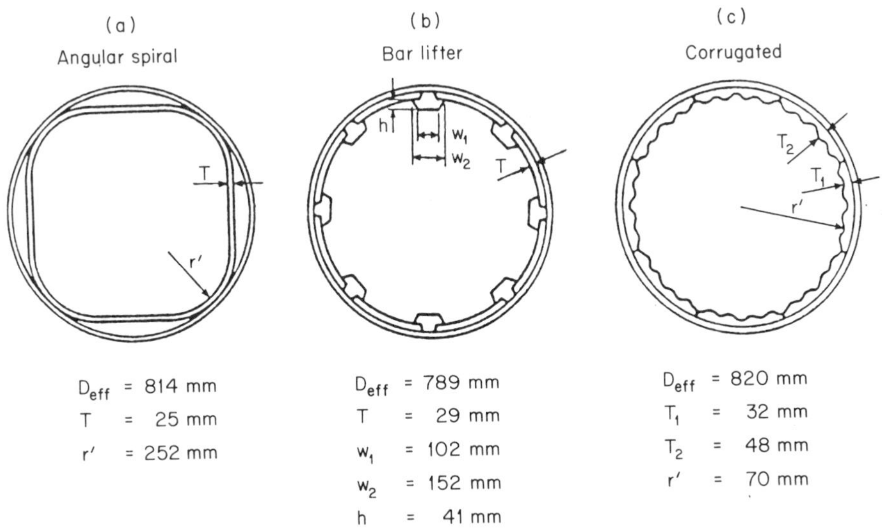

Figure 8.8. Cross-sections of three liners used in a 0.91 m diameter × 1.52 m ball mill (Rogers *et al.*, 1982b). *Reproduced by permission of the Canadian Institute of Mining and Metallurgy*

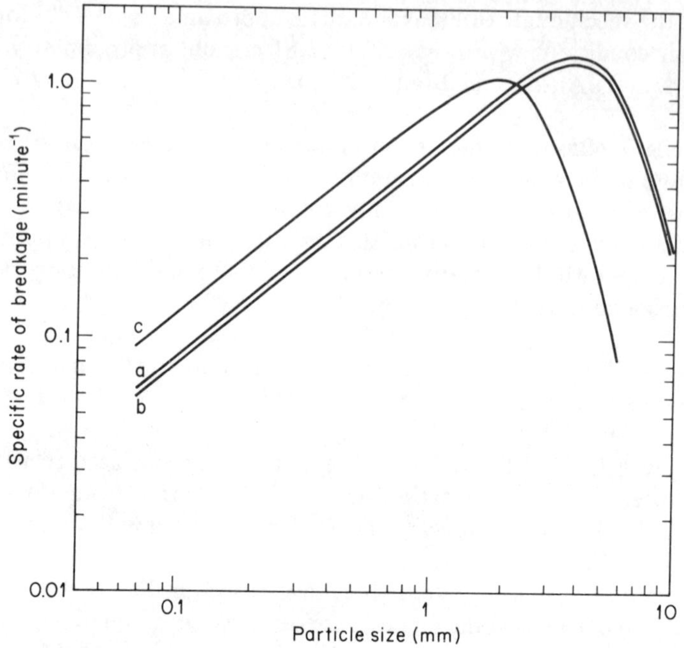

Figure 8.9. Specific rate of breakage versus particle size for the three liners shown in Figure 8.8 ($N^* = 0.7$, $J = 0.2$, $U = 1.0$) (Rogers *et al.*, 1982b). *Reproduced by permission of the Canadian Insititute of Mining Metallurgy*

movement within a laboratory mill and the commercial mill from which it is scaled, has not always been fully realized. For example, Shoji *et al.* (1982) concluded that their former laboratory work (Shoji *et al.*, 1980) was unacceptable because of slippage between liner and charge, which was not occurring in the large-scale mill. When the experiment was repeated with adequate lifters to simulate the full-size mill ball charge movement, significantly different results were obtained, as given in §8.2.2.7.

Equation (8.3) was found to fit very closely the curves of Figure 8.9.

8.2.2.6 *Variation of S with tumbling mill grinding media*

(a) Ball size Following Austin *et al.* (1982a) the relation between ball diameter d and a^* has been shown to be:

$$a^* \propto 1/d^{\xi_2} \tag{8.5}$$

Austin *et al.* (1976b) indicated that ξ_2 approximated to unity; but for the wet grinding of quartz, Kelsall *et al.* (1967/68) gave a value of $\xi_2 = 1.5$.

This reduction of a^* with increase in d is seen in Figure 8.10 at small values of particle size, but increase in d also displaces x_m (corresponding to the peak

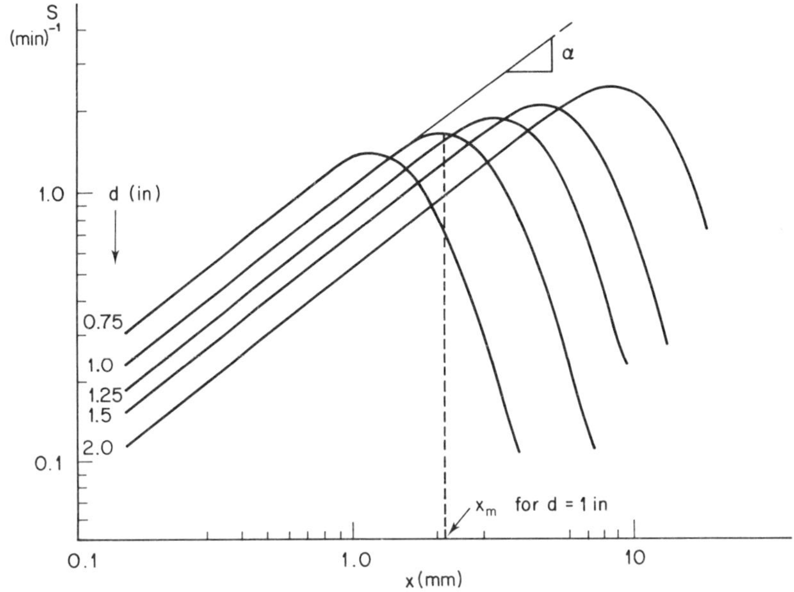

Figure 8.10. Typical variation of specific rate of breakage versus particle size for various ball diamters in a tumbling mill (Austin *et al.*, 1982a). *Reproduced by permission of the American Institute of Mining Metallurgical and Petroleum Engineers*

of the curves) to higher values of particle size. Gupta and Kapur (1974) made similar findings.

Thus two opposing trends are noticed in the effect of ball size on the specific rate of breakage, and it is not surprising that several investigators have found an optimum size for maximum capacity at a given output fineness. An example is Kelsall *et al.* (1967/68).

It is more usual, however, to adopt a size distribution of balls according to the material ground and the required product distribution. For this situation, Austin *et al.* (1982a) propose an overall weighted mean value $\bar{S}_i$ given by:

$$\bar{S}_i = \sum_k S_{i,k} M_k \tag{8.6}$$

where M_k is the mass fraction of balls in the size interval denoted by k, and $S_{i,k}$ is the specific rate of breakage of particle size i by balls of size k.

There would appear to be considerable scope for the use of SMB methods in the prediction of the optimum ball size distribution to attain a required product sizing.

(g) Ball density Seebach (1969) found, by using hollow steel balls of varying thickness, that for dry grinding of cement clinker, the specific rate of breakage

was directly proportional to the ball density. However, Austin *et al.* (1982a), using alloy steel then ceramic balls (density ratio = 2), obtained a ratio of only 1.75 for the resulting S_i values, which is less than the density ratio. Hence it would seem necessary to determine S_i separately for each ball material.

Ball hardness Rose and Sullivan (1958) showed that, above a minimum hardness desirable from a wear point of view, increase in hardness does not affect mill capacity.

8.2.2.7 *Variation of S with tumbling mill ball and powder charge*

Many facets of this subject have been studied over the years, by for example Austin and Klimpel (1964), Austin *et al.* (1967), Mika *et al.* (1967), Kelsall *et al.* (1968/69) and Kapur and Agrawal (1970). However, the interrelations of the effects of ball and powder charge on the specific rate of breakage are perhaps best summarized in Figures 8.11 and 8.12 due to Shoji *et al.* (1982). These results apply to laboratory mills of mean internal diameter 0.195 m and 0.6 m with adequate lifters to prevent slippage between liner and charge. Throughout $N^* = 0.7$.

The variables used in these graphs are related by:

$$f_c = 0.4\ JU \tag{8.7}$$

where f_c is the fraction of mill volume occupied by powder charge, J is the fraction of mill volume occupied by ball charge, and U is the fraction of interstitial voids in the ball charge occupied by the powder charge. Ball charge porosity is taken to be 0.4. When $U = 1.0$ the ball charge is said to be saturated.

Figure 8.11 applies to batch dry grinding of quartz, and the solid lines are computed from:

$$S = K_1[1/(1 + 6.6\ J^{2.3})]\exp(-1.2U) \tag{8.8}$$

with $0.2 \leqslant J \leqslant 0.5$ and $0.5 \leqslant U \leqslant 1.5$.

For wet grinding a very similar expression to equation (8.8) holds, except that exp $(-1.2U)$is replaced by exp $(-1.32U)$.

The product of S and f_c is called the 'relative absolute rate of breakage', important because it is proportional to mill capacity. In Figure 8.12, Sf_c/K_1 is plotted against f_c over the ranges $0.1 < J < 0.5$ and $0.5 < U < 1.5$ for dry grinding of quartz. Throughout, the optimum range of U is $0.6 < U < 1.1$. In fact, the absolute breakage remains virtually constant over the range $0.5 < U > 1.0$, but for $U > 1.0$ the breakage rate falls quite dramatically, as seen in Figure 8.13, and corresponds to mill overfilling. An overfilling factor k_o is introduced in Figure 8.13. Since power is found not to be reduced by overfilling, this condition implies lost efficiency.

A similar set of curves to that of Figure 8.12 was produced for wet grinding of quartz, and the optimum value of U for maximum capacity was found to

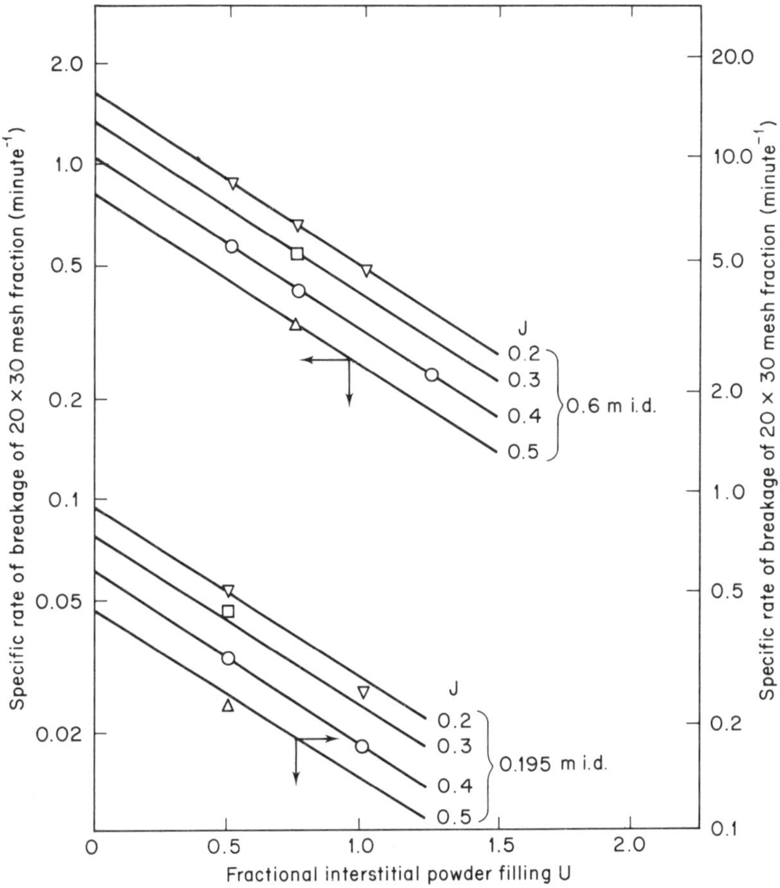

Figure 8.11. Typical variation of specific rate of breakage with ball and powder filling (Shoji *et al.*, 1982). *Reproduced by permission of Elsevier Sequoia*

be 0.83 for dry and 0.76 for wet grinding, each with $J = 0.4$ for these laboratory size mills. In respect of wet grinding, Austin *et al.* (1982a) acknowledge the need for more research to quantify the effects of change of slurry density and rheology on the specific rate of breakage.

8.2.2.8 Slowing-down effect at long times of grinding

At long times of grinding, simulation predicts a finer product than that obtained experimentally. This results from a slowing down of the breakage rate as the charge becomes finer owing to the development of high effective viscosity of the slurry in the case of wet grinding (Klimpel and Austin, 1982). Such slowing down corresponds to a line *ad* in Figure 8.4.

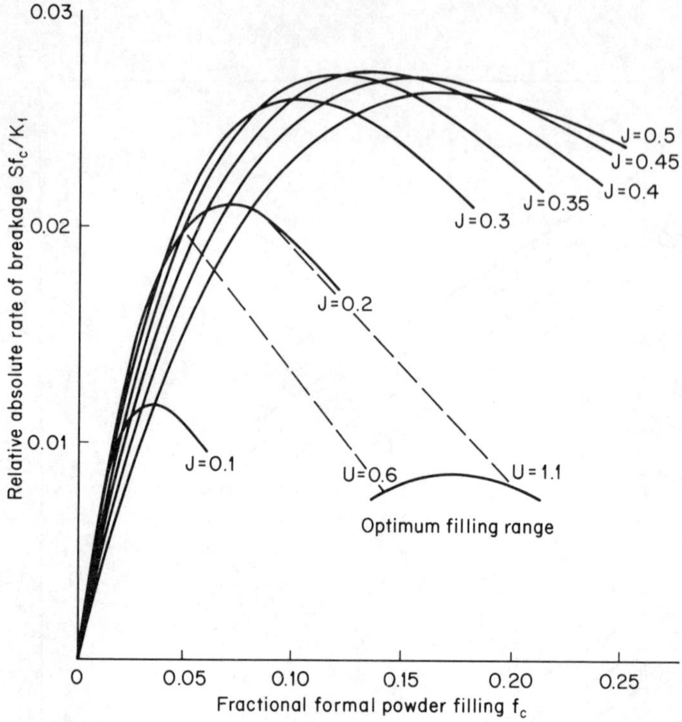

Figure 8.12. Typical variation of the function (Sf_c/K_1) with f_c for a tumbling ball mill (Shoji *et al.*, 1982). *Reproduced by permission of Elsevier Sequoia*

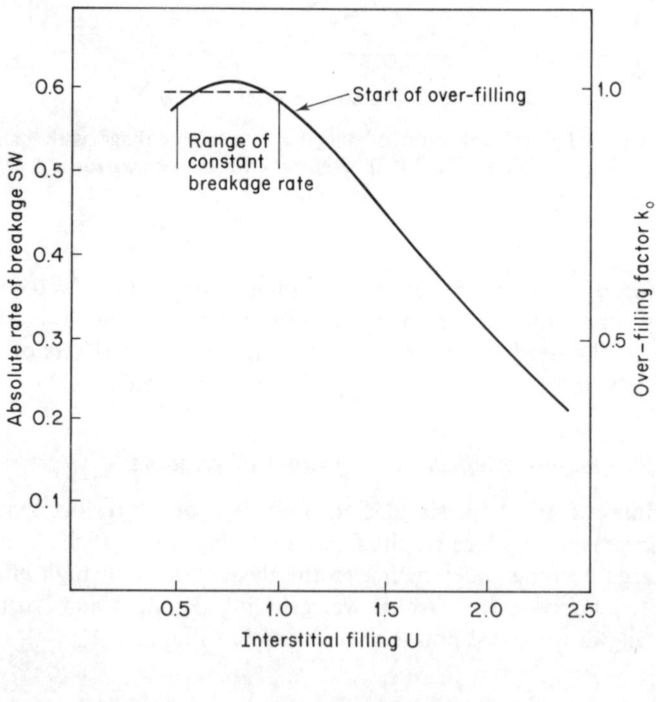

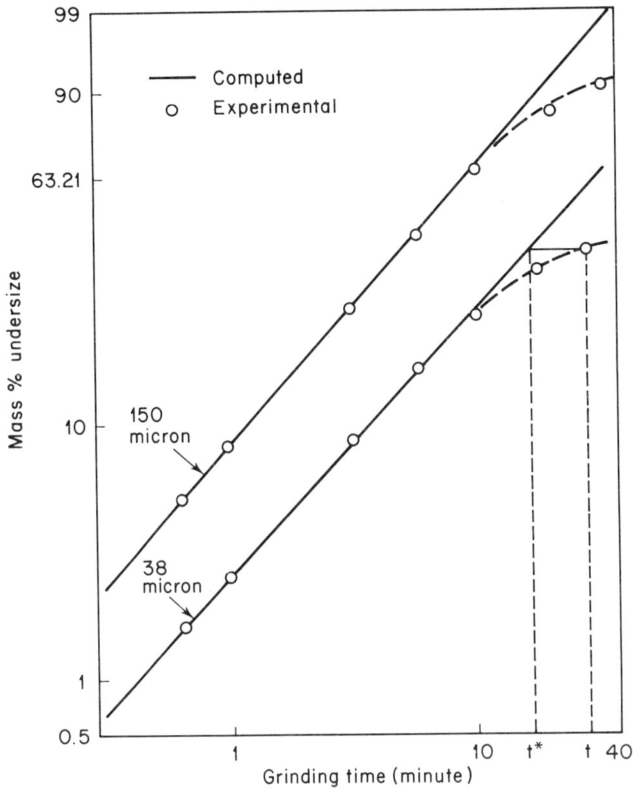

Figure 8.14. Weibull plot for 16 × 20 mesh cement clinker ground for different times in a ball mill (Austin and Bagga, 1981). *Reproduced by permission of Elsevier Sequoia*

Austin and Bagga (1981), in the context of dry grinding, have quantified the effect by using a slowing down factor $\varkappa$ defined by:

$$S_i(t) = \varkappa(t)S_1(0) \qquad 0 \geqslant \varkappa(t) \geqslant 1 \tag{8.9}$$

Further, to restore the linear relation of the line *ab* in Figure 8.4, so that equations assuming linear breakage may be applied, a false time of grind t^* is defined by $t^* = \int_0^t \varkappa(t)\mathrm{d}t$ so that $\mathrm{d}t^* = \varkappa(t)\mathrm{d}t$. The ratio t^*/t is a measure of the slowing down of the breakage rate between times 0 and t.

Figure 8.14 illustrates how t^* may be estimated from a plot of mass per cent undersize against grinding time (Weibull plot).

Figure 8.13. Variation of absolute rate of breakage SW and over-filling factor k_o with interstitial filling U (Austin and Brame, 1983). *Reproduced by permission of Elsevier Sequoia*

8.2.2.9 *Variation of S in other than tumbling mills*

(i) A laboratory hammer mill was modified for the batch grinding of maize (Jindal and Austin, 1976). First-order kinetics were observed. Above a certain minimum speed, specific rate of breakage increased linearly with speed, and the pattern of variation of the absolute rate of breakage with mill load was given in Figure 8.15. This is similar to that reported for anthracite ground in a ball mill by Austin *et al.* (1967). At low load, the rate of breakage increased with load, then passed through a maximum as the mill became choked.

(ii) In a fully mixed rotary cutter used for size reduction of plastics, Klimpel and Austin (1972) found the variation of absolute rate of breakage *SW* with mill load as seen in Figure 8.16. In the operating range, *SW* was constant. Further, specific rate of breakage varied linearly with particle size.

(iii) For ring-ball mills the absolute rate of breakage versus particle size is shown in Figure 8.17, due to Austin *et al.* (1982b). The graphs applied to the Hardgrove machine and to the Babcock 1.7E mill modified to be used batch-wise without air sweeping and at low speeds, and showed a similar pattern for

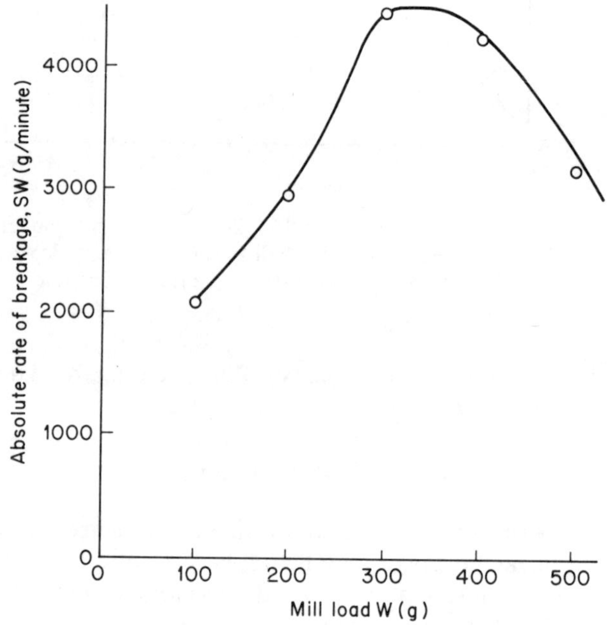

Figure 8.15. Typical variation of the absolute rate of breakage *SW* with mill load for a laboratory hammer mill grinding maize (Jindal and Austin, 1976). *Reproduced by permission of Elsevier Sequoia*

Figure 8.17. Variation of absolute rate of breakage for ring ball mills used in a batch manner (Austin *et al.*, 1982b). *Reproduced by permission of Elsevier Sequoia*

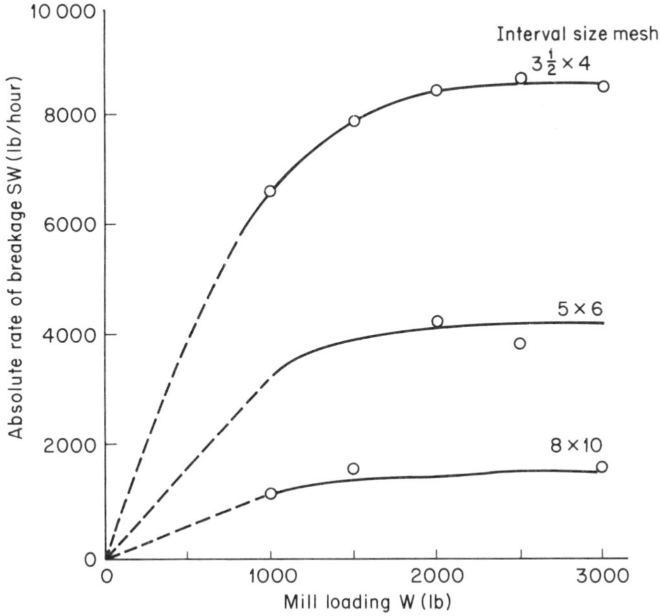

Figure 8.16. Variation of the absolute rate of breakage *SW* versus mill load for a rotary cutter operating on plastics, for three initial size intervals (Klimpel and Austin, 1972). *Reproduced by permission of DECHEMA, Frankfurt*

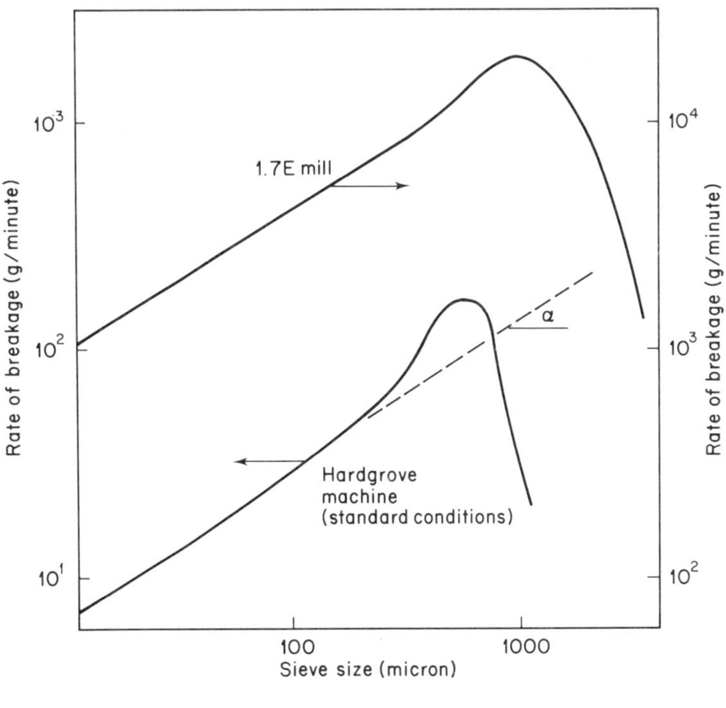

the two cases. The straight-line portion allows the application of equation (8.2), and α appears to be the same in each mill. Large-sized particles cannot be gripped by the balls, which accounts for the fall-off in breakage rate.

8.2.3 Breakage distribution function, *B*

8.2.3.1 Definition

When material of size j breaks, it produces a suite of daughter particles. This suite, before rebreakage occurs, is called the primary progeny distribution and is given the symbol $B_{i,j}$ in cumulative form, which is the cumulative weight fraction of material broken from interval j which appears below size x_i, the upper size of interval i.

The fraction of primary breakage products from interval j which goes into interval i is denoted by the symbol $b_{i,j}$ and it may be calculated from:

$$b_{i,j} = B_{i,j} - B_{i+1,j} \tag{8.10}$$

See Figure 8.19 for a diagrammatical representation of a particular case.

B was formerly loosely called the breakage function, but it is more correctly termed the distribution function, and it strictly applies to primary progeny before rebreakage because the comminution process is just a succession of such events, the summation of which describes the whole, if the size intervals chosen are sufficiently small e.g. $2^{1/4}$ or at least $2^{1/2}$ (Austin and Luckie, 1970/71).

8.2.3.2 Variation of B with particle size

Typical variation of *B* with particle size is given in Figure 8.18 covering a laboratory ball mill, a hammer mill and a vibration mill, and in Figure 8.19 for a Hardgrove mill grinding in turn cement clinker, anthracite, quartz and silicon carbide. The form of the curves is similar in all cases (Austin *et al.*, 1976a).

8.2.3.3 Normalization of B

There is experimental evidence, in many cases, that *B* is a normalizable function, which means that $B_{i,j} \propto x_i/x_j$; that is, it is proportional to the ratio of the top sizes of the intervals i and j. If intervals are in geometric ratio, as is usual:

$$x_i = 1$$

$$x_2 = Rx_1$$

$$x_3 = R^2 x_1$$

$$x_j = R^{j-1} x_1$$

$$x_i = R^{i-1} x_1$$

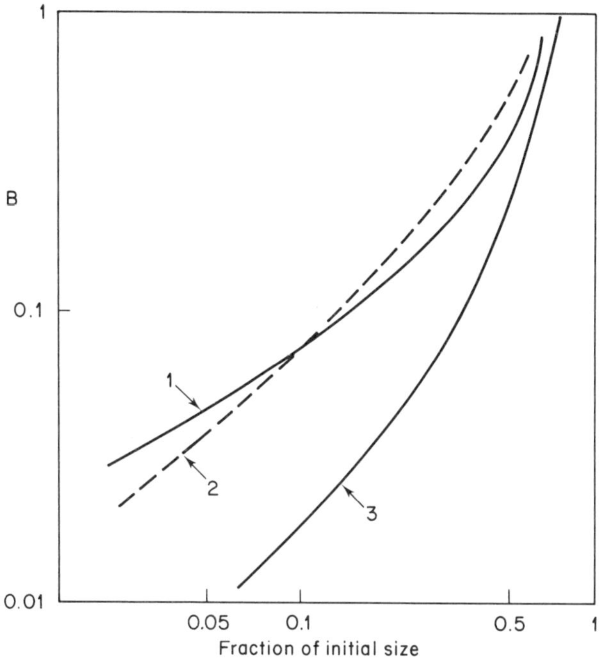

Figure 8.18. Variation of breakage distribution function B with particle size for various mills: 1 cement clinker ground in laboratory ball mill, 2 corn ground in hammer mill at 3000 rev/minute, 3 silicon carbide ground in wet vibration mill (Austin *et al.*, 1976a). *Reprinted with permission from* Ind. Eng. Chem. Process. Des. Devel., ***15,*** *187–196. Copyright 1976 American Chemical Society*

so that $x_i/x_j = R^{i-j}$. Hence:

$$B_{i,j} \propto \frac{x_i}{x_j} \propto R^{i-j} \propto B_{i-j} \tag{8.11}$$

i.e. the matrix **B** can be replaced by the column vector ***B***.

To exemplify equation (8.11), if $i-j=1$, then $B_{2,1}=B_{3,2}=B_{4,3}$ etc. Similarly $b_{i,j}=B_{i,j}-B_{i+1,j}$. Using equation (8.11):

$$b_{i,j}=B_{i-j}-B_{i-j+1}=b_{i-j} \tag{8.12}$$

i.e. the matrix **b** can be replaced by the column vector ***b***.

Hence for either cumulative or interval form of the breakage distribution function, if normalizable, it can be expressed by one vector for all values of the function, which is very useful. (Austin and Bhatia, 1971/72) (see also §9.4.2)

8.2.3.4 *Mathematical representation of $B_{i,j}$.*

For analysis by digital computer, it is necessary to express the curves shown

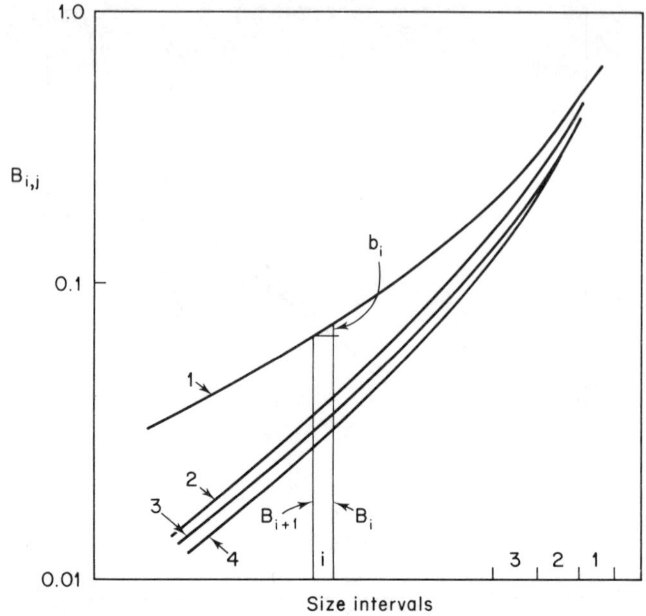

Figure 8.19. Variation of breakage distribution function B with particle size for the Hardgrove machine: 1 cement clinker, 2 silicon carbide, 3 anthracite, 4 quartz (Austin *et al.*, 1976a). *Reprinted with permission from* Ind. Eng. Chem. Process. Des. Devel., *15, 187–196. Copyright 1976 American Chemical Society*

in Figures 8.18 and 8.19 by a mathematical equation which may take the form of:

$$B_{i,j} = \Phi_j\left(\frac{x_i - 1}{x_j}\right)^{\gamma} + \left(1 - \Phi_j\right)\left(\frac{x_i - 1}{x_j}\right)^{\beta} \qquad (8.13)$$

with $0 \leqslant \Phi_j \leqslant 1$ and $n \geqslant i \geqslant j$.

Equation (8.13) is explained in Figure 8.20. The curve is replaced by the sum of two straight lines on a log–log plot, one with a slope of β and the other with a slope of γ. The intercept on the ordinate at the maximum value of x_{i-1}/x_j by the line of slope γ gives the value of Φ_j (Austin *et al.*, 1982a).

For non-normalized B values, Klimpel and Austin (1984) give:

$$B_j = B_1(x_j/x_1)^{-\psi} \qquad \psi > 0 \qquad (8.14)$$

Thus the parameters associated with B in the mathematical representations are four: Φ, β and γ together with ψ for the case of non-normalization.

Earlier, simpler expressions had been tried for the representation of B, including that of Broadbent and Callcott (1956):

$$B(x, y) = (1 - e^{-x/y})/(1 - e^{-1})$$

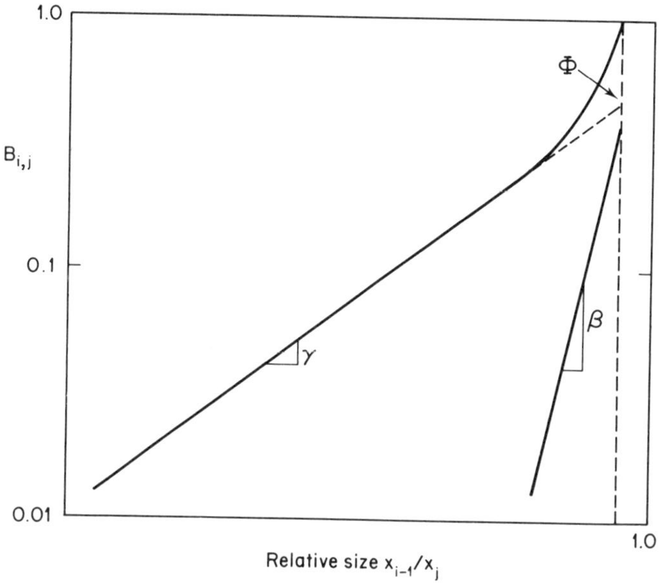

Figure 8.20. Mathematical representation of $B_{i,j}$ on a log–log plot (see equation (8.13)) (Austin *et al.*, 1982a). *Reproduced by permission of the American Institute of Mining, Metallurgical and Petroleum Engineers*

but experience has demonstrated their inadequacies in certain situations.

8.2.3.5 Variation of B with tumbling ball mill variables

The breakage distribution function has been shown to be independent of the following variables in tumbling ball mills:

mill diameter	Malghan and Fuerstenau (1976)
ball density	Kelsall *et al*. (1973a)
ball load (limited range)	Malghan and Fuerstenau (1976)
	Herbst and Fuerstenau (1972)
powder charge	Shoji *et al.* (1980)
	Herbst and Fuerstenau (1973)
hold-up	Kelsall *et al.* (1968/69),
	Mika *et al.* (1967)

With respect to ball size, Kelsall *et al.* (1967/68) reported no effect on B; but later work by Austin *et al.* (1982a) demonstrated for quartz that B altered in a systematic manner expressed in terms of γ and Φ_j in equation (8.13), with β remaining unchanged (see Table 8.1.)

Table 8.1. Variation of B with γ and Φ_j

Ball size (mm)	γ	Φ_j
19	1.02	0.81
22	1.01	0.92
25	1.00	1.00
32	0.97	1.08
38	0.93	1.10
44	0.88	1.11
51	0.81	1.11
64	0.72	1.11

Reproduced by permission of the American Institute of Mining, Metallurgical and Petroleum Engineers.

These trends are in keeping with the concept of the greater impact force of a collision involving a larger ball producing a somewhat larger proportion of fines than with a smaller ball, i.e. reduced γ and increased Φ_j (see Figure 8.20).

The mean value of $\bar{B}_{i,j}$ for a mixture of ball sizes may calculated from :

$$\bar{B}_{i,j} = \sum_k M_k S_{j,k} B_{i,jk} / \bar{S}_j \tag{8.14a}$$

where $B_{i,jk}$ is the cumulative breakage distribution for size k balls acting on size j material, and $\bar{S}_j$ is found from equation (8.6).

Increasing the speed of rotation, yet keeping below the critical speed, tends towards a coarser primary breakage product (Herbst and Fuerstenau, 1972)

B also appears to vary a little (for the same material ground) between wet and dry operation and is affected by lifter geometry (Shoji *et al.*, 1982) and by the shape of the grinding media (Kelsall *et al.*, 1973b).

However, for a given mill geometry, operating at a given fraction of critical speed, with a given liner geometry and constant ball and powder filling ratios, the scale-up of a batch process does not involve change of B for a given material ground. This is an important finding. In studies of continuously operated mills, by contrast, problems have arisen, one cause of which may be a small change of B during scale-up (Klimpel and Austin, 1984) and is still under investigation.

Overall, B is far less affected by change of mill variables than is S.

8.3 SIZE–MASS BALANCE EQUATION FOR BATCH GRINDING

Using the notation of §8.2 and assuming no aggregation of particles, which applies throughout, the rate of increase of size i may be written:

$$\frac{\mathrm{d}}{\mathrm{d}t}\,[M_i(t)W]$$

this must be equal to:

$$\left\{\begin{array}{l}\text{Summation of production of size } i\\ \text{material from all larger sizes}\end{array}\right\} \text{minus} \left\{\begin{array}{l}\text{Rate of breakage of size } i\\ \text{to smaller sizes}\end{array}\right\}$$

Table 8.2. Alternative forms of the fully mixed batch grinding equation (Austin, 1971/72)

Size	Time	Form	Equation	Equation Ref.
Discrete	Continuous	Summation	$\frac{d}{dt}[M_i(t)] = -S_i M_i(t) + \sum_{\substack{j=1 \\ i\geq 1}}^{i-1} b_{i,j} S_j M_j(t)$	(8.15)
		Matrix	$\frac{d}{dt}[M(t)] = -[\mathbf{I} - \mathbf{B}]\mathbf{S} M(t)$	(8.16)
	Discrete†	Summation	$M_{i,n} - M_{i,n-1} = -S_i' M_{i,n-1} + \sum_{\substack{j=i-1 \\ i\geq 1}}^{1} b_{i,j} S_j' M_{j,n-1}$	—
		Matrix	$p_n = f[\mathbf{BS} + \mathbf{I} - \mathbf{S}]^n$	(4.37a)
Continuous	Continuous	Cumulative	$\frac{d}{dt}[F(x,t)] = \int_{y=x}^{x_{max}} S(y)B(x,y)\frac{d}{dy}[F(y,t)]\,dy$	(9.19)
		Density	$\frac{d}{dt}[M(x,t)] = -S(x)M(x,t) + \int_{y=x}^{x_{max}} S(y)b(x,y)M(y,t)\,dy$	(9.16)
	Discrete†	Cumulative	$F_n(x) = F_{n-1}(x) + \int_{y=x}^{x_{max}} S^*(y,n)B(x,y)\,dn_{n-1}(y)$	—
		Density	$M_n(x) - M_{n-1}(x) = -S_{n-1}^* M_{n-1}(x) + \int_{y=x}^{x_{max}} S^*(y,n)b(x,y)M_{n-1}(y)\,dy$	—

†For n stages of operation or time intervals
Note $S^*(y,n)$ is the specific rate of breakage of size y in the nth stage and $S' = \Delta t.S$
Reproduced by permission of Elsevier Sequoia

or

$$\frac{d}{dt}[M_i(t)W] = \sum_{\substack{j=1 \\ i \geqslant 1}}^{i-1} b_{i,j}S_jM_j(t)W - S_iM_i(t)W$$

that is

$$\frac{d}{dt}[M_i(t)] = -S_iM_i(t) + \sum_{\substack{j=1 \\ i \geqslant 1}}^{i-1} b_{i,j}S_jM_j(t) \tag{8.15}$$

Broadbent and Callcott (1957) put equation (8.15) in matrix form:

$$\frac{d}{dt}[\boldsymbol{M}(t)] = -[\mathbf{I} - \mathbf{B}]\mathbf{S}\boldsymbol{M}(t) \tag{8.16}$$

where $\boldsymbol{M}(t)$ is a column vector and **I**, **B**, **S** are column–row matrices. However, it is more usual today to work with the form given in equation (8.15), which appears widely in the literature.

It is the discrete-size, time-continuous form of the SMB equation for fully mixed batch grinding. The discrete size arises from the fact that it is common practice to sub-divide size distributions into intervals defined by sieve sizes. In theory, however, it is possible to conceive of continuous change of size, leading to a 'size continuous, time continuous' form, given as equation (9.16).

Similarly, distinct stages of grinding, and hence distinct time intervals, may be considered as in §4.3.8.5 with which may be associated either discrete size or size continuous concepts.

Thus there are four basic ways of expressing the size–mass balance:

(1) discrete size, time continuous
(2) discrete, size, discrete time
(3) size continuous, time continuous
(4) size continuous, discrete time

Austin (1971/72) has well summarized the appropriate equations as given in Table 8.2 and has treated both their history and the method of changing from one to the other.

8.4 SOLUTION TO THE BATCH GRINDING EQUATION

8.4.1 Survey

Three types of solution to equation (8.15) are available, according to the prevailing conditions: analytical, matrix and numerical analysis, which are exemplified in turn.

8.4.2 Analytical solutions

Reid (1965) has given an analytical solution to equation (8.15) assuming S and b for a given particle size do not vary with the degree of grinding. When this is so, we have:

$$M_1(t) = M_1(0)e^{-S_1 t}$$

$$M_2(t) = M_2(0)\frac{Sb_{2,1}}{S_2 - S_1}e^{-S_1 t} + \left[M_2(0) - \frac{S_1 b_{2,1}}{S_2 - S_1}M_1(0)\right]e^{-S_2 t}$$

Hence the general term $M_i(t)$ can be expressed as

$$M_i(t) = \sum_{j-1}^{i} h_{i,j}e^{-S_j t} \tag{8.17}$$

$$\text{where } h_{i,j} = \begin{cases} 0 \quad \text{for} \quad i < j \\ M_i(0) - \sum_{k=1}^{i-1} h_{i,k} \quad \text{for} \quad i = j \\ \dfrac{1}{S_i - S_j}\sum_{k=j}^{i-1} S_k b_{i,k} h_{k,j} \quad \text{for} \quad i > j \end{cases}$$

See equations (9.87) to (9.88).

Since the solution contains the term $(S_i - S_j)$, there are practical problems if S passes through a maximum when plotted against particle size (see Figure 8.5, curve 6). However, when S continues to increase with increasing particle size, independently of change in particle environment, the Reid solution is viable.

8.4.3 Matrix solution

A matrix solution to equation (8.16) has been developed by Broadbent and Callcott (1957) on the same conditions as applied to the Reid solution. They defined $S' = S\Delta t$, where Δt is a time increment. Then their solution is:

$$\boldsymbol{M}(t) = (\mathbf{I} + \mathbf{S}'\mathbf{B} - \mathbf{S}')\boldsymbol{M}(0) \tag{8.18}$$

where

$\boldsymbol{M}(0)$ = vector of the feed distribution (time = zero)

$\boldsymbol{M}(t)$ = vector of the distribution at time t

Meloy and Bergstrom (1965) demonstrate that, by applying equation (8.18) over a succession of equal short intervals, the solution becomes:

$$\boldsymbol{M}(t) = [\mathbf{I} - \mathbf{S}'\mathbf{B} - \mathbf{S}']^n\boldsymbol{M}(0) \tag{8.19}$$

This equation is explicit in $\boldsymbol{M}(0)$ as distinct from the Reid solution, which is explicit in t. This can clearly affect the choice of solution for a particular problem (Snow, 1973).

8.4.4 Numerical analysis solution

When S and b are not invariant with change of environment, then a complex numerical analysis has to be undertaken to solve equation (8.15). For the bearing of this on the assessment of S and b by the use of equation (8.15), see Klimpel and Austin (1970).

8.5 MEASUREMENT OF BREAKAGE FUNCTIONS

8.5.1 Survey

There appear to be three broad types of experimental measurement of the breakage functions S and B (Herbst and Mika, 1970):

(1) direct measurement with a full batch test for each size fraction;
(2) indirect estimation by back calculation from very abbreviated batch tests;
(3) combined limited direct measurement and limited back calculation.

Method 1 is very time consuming. Method 2 is an attempt to reduce experimentation and so save time, but accuracy can suffer. It is the most commonly used group of methods to date. Method 3 is a compromise between 1 and 2.

8.5.2 Direct measurement

8.5.2.1 Summary of available methods

Direct measurement methods may be divided into three types:

(1a) those which use a single sized feed fraction;
(1b) those which use the full feed size distribution but with replacement of a single size fraction by:
   (i) one of different chemical composition (e.g. quartz in a calcite environment)
   (ii) one labelled with a radioactive tracer;
(1c) those which use a full feed size distribution but with, say, six size fractions simultaneously tagged by a nuclear activation analysis technique.

All four methods are effective in a linear grinding rate situation, but when non-linear conditions prevail, only 1b(ii) and 1c are valid. Because of the experimental problems, 1b(ii) and 1c methods have not been widely advocated. However, it appears (see Table 8.3), that radioactive tracer methods may be necessary to facets of continuous grinding and there may well be a return to a greater use of 1b(ii) for determination of breakage functions.

The direct measurement methods are detailed in the following sections.

8.5.2.2 *Method 1a: single sized feed fraction (top size only)*

(a) Austin and Luckie method I for assessment of B Austin and Luckie (1971/72a) have developed a method (their method I) which requires measurement of the size distribution from primary breakage of a feed restricted to size interval 1 (see Figure 8.21). It is necessary to use a sieving correction δ due to particle breakage during sieving. Then $M_1(0) = 1 - \delta$ and $M_2(0) = \delta$.

$$\text{Fractional mass broken out of size } 1 = M_1(0) - M_1(t)$$

$$= 1 - \delta - M_1(t) = \Delta M$$

$$= F_2(t) - F_2(0)$$

$$\text{Mass arriving at size } 2 = M_2(t) - \delta$$

Hence

$$b_{2,1} = \frac{\text{Mass arriving at size 2}}{\text{Mass broken from size 1}} = \frac{M_2(t) - \delta}{\Delta M}$$

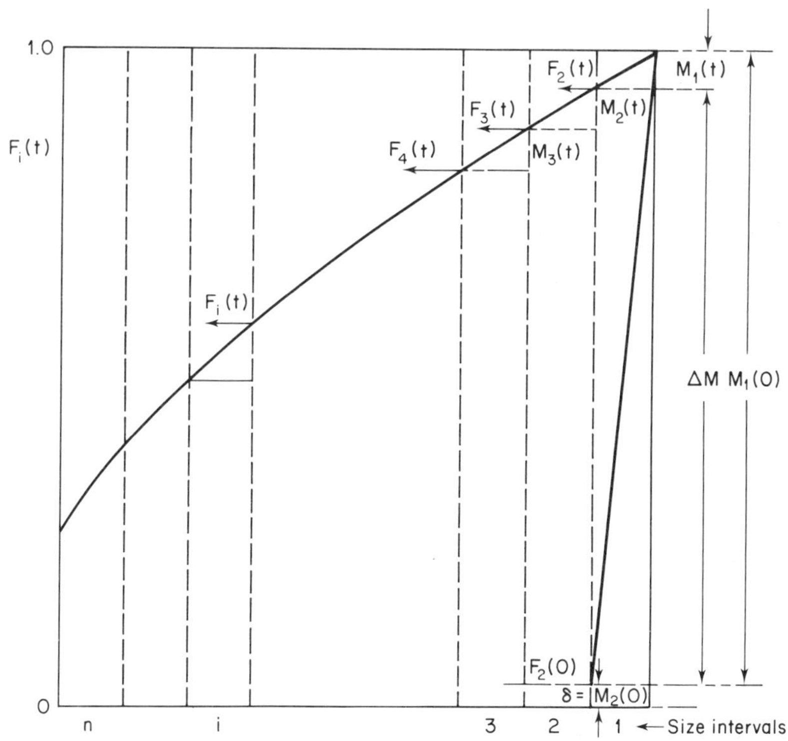

Figure 8.21. Diagram to illustrate single-sized feed fraction confined to size interval 1

Since no re-breakage occurs:

$$M_3(0) = M_4(0) \text{ etc.} = 0$$

$$b_{3,1} = M_3(t)/\Delta M \text{ etc.}$$

Hence

$$M_{i,1} = M_i(t)/\Delta M$$

In cumulative B notation, this is:

$$B_{1,1} = 1$$

$$B_{2,1} = 1 \text{ (except for } \delta \text{ correction)}$$

$$B_{3,1} = F_3(t)/\Delta M$$

or generally:

$$B_{i,1} = \frac{F_i(t)}{F_2(t) - F_2(0)} \quad \text{or} \quad \frac{F_i(t)}{1 - \delta - M_i(t)} \tag{8.20}$$

which is the working equation for $B_{i,1}$.

The problem with this method is that it is often very difficult to get accurate data with the very short grinding times involved.

(b) Austin and Luckie method II for assessment of B Austin and Luckie (1971/72a) have further developed another system (their method II), using a feed restricted to size interval 1, but allowing some rebreakage of primary products to occur, by not restricting the grinding to the very short times required by their methed I.

Assuming the product $S_jB_{i,j}$ is approximately constant for a given value of i, an analytical solution to equation (8.15) is given by:

$$1 - F_i(t) = [1 - F_i(0)]\exp(-S_jB_{i,j}t) \quad \text{for} \quad j < 1$$

Applying this for relatively short grinding times to a feed in the top size fraction, we obtain:

$$1 - F_i(t) = [1 - F_i(0)]\exp(-S_1B_{i,1}t)$$

Assuming that the initial distribution at zero time is $(1 - \delta)$ in size 1, then $F_2(0) = \delta$ and $B_{2,1} = 1$; i.e.:

$$1 - F_2(t) = (1 - \delta)\exp(-S_1t)$$

$$B_{i,1} = \log\left[\frac{1 - F_i(0)}{1 - F_i(t)}\right] \Big/ \log\left[\frac{1 - F_2(0)}{1 - F_2(t)}\right] \tag{8.21}$$

which is the working equation of $B_{i,1}$. Austin and Bhatia (1971/72) give an example of the use of equation (8.21) for cement clinker.

(c) Kapur's modified G–H method Kapur (1982) has criticized the methods described under (a) and (b) in that strictly they demand an infinitesimal time

interval so that particles undergo single-event breakage only, which is unattainable in reality. Other problems include time-lag errors due to inertia of the tumbling mill charge, and insufficient mass of ground particles for accurate sizing of the product, particularly in the fine range.

No restriction on rebreakage is a feature of an improved method proposed by Kapur (1982). This is based on a solution to the discrete size batch grinding equation in the form (Kapur, 1972) given as equation (9.151), and seen graphically depicted in Figure 9.2. There G is the intercept on the zero time ordinate and $H/2$ is the slope of a line.

For a single sized feed, $M_i(0) = 1$ for $i = 1, 2, 3, \ldots$, and hence equation (9.151) becomes:

$$G_1 = -S_1$$

and

$$G_i = -SB_{i,1} \quad \text{for} \quad i = 1, 2, 3, \ldots \tag{8.22}$$

Therefore the feed size cumulative breakage parameters become:

$$B_{i,1} = G_i / G_1 \tag{8.23}$$

Then for a single sized feed it can be shown that $H = 0$; that is, the lines in Figure 9.2 are parallel to the abscissa and the intercepts on the ordinate at zero time give the values of B. Alternatively, they can be obtained by algebraic means, and Figure 8.22 shows three sets of data so computed, compared with experimentally found values. The correlation is good.

Kapur (1982) also shows that methods (a) and (b) above are but special cases of his more general G–H method.

8.5.2.3 *Method 1b(i): quartz in calcite environment*

In the context of continuous wet grinding in an open-circuit laboratory ball mill, Kelsall (1965) developed the technique of replacing one fraction of a standard calcite feed by quartz. The whole product was checked for size distribution after grinding, then the calcite was dissolved out and the remaining quartz size distribution found. By this means linear breakage conditions were established (Kelsall *et al.*, 1966) and breakage functions assessed by analogue computer means using the Reid solution to the grinding equation together with a residence time distribution function because the grinding was continuous, not batch (see Kelsall and Reid, 1965). The agreement between calculated and experimentally found product size distributions was good for quartz feed size 590 microns and below, but diverged quite markedly for 833 microns and greater. Later work (Kelsall and Reid, 1969) showed that S plotted against x passed through a maximum at 833 microns.

It is essential to point out a fundamental distinction between the above technique and that of a batch test. In the batch test the environment in which a given size fraction is reduced in size is changing with time, whereas in the

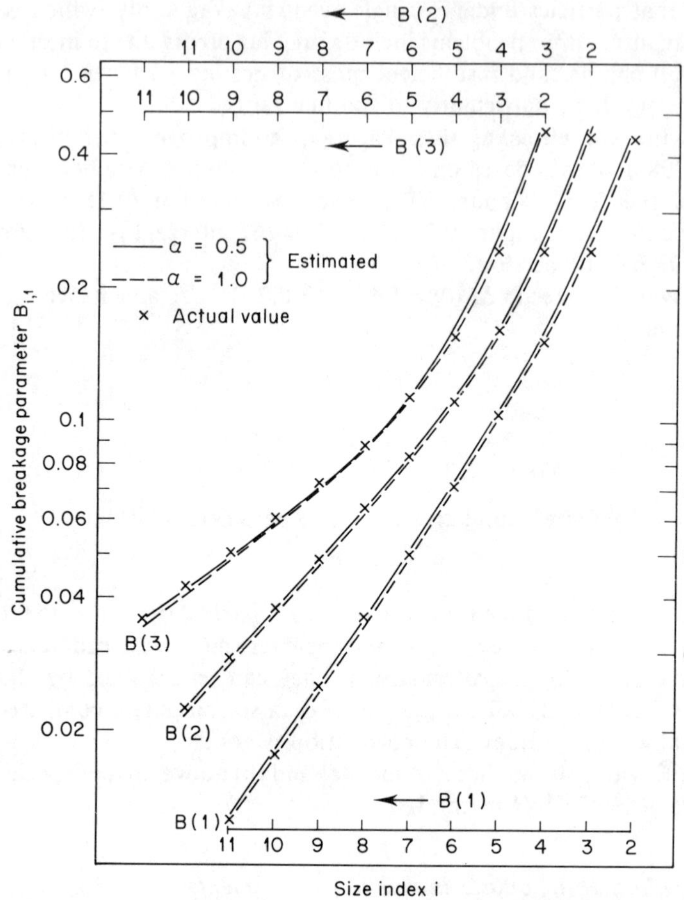

Figure 8.22. Feed-size cumulative breakage distribution functions B(1), B(2) and B(3) by the G-M method from computer-generated data. In each case two sets of data were generated with $\alpha = 0.5$ and 1.0 in equation (8.2) (Kapur, 1982). *Reproduced by permission of Elsevier Sequoia*

quartz in calcite continuous grinding test the environment is nearly constant, as quartz concentrations are demonstrated to be low enough not to change the calcite environment appreciably.

However, method 1b(i) has been discontinued because it required considerable objective judgement in its application (Klimpel and Austin, 1984).

8.5.2.4 Method 1b(ii): radioactive tracer

Quite early in the investigation of breakage functions, Gardner and Austin (1962) successfuily applied a radioactive tracer technique to the determination of breakage functions. Successively, batch tests on the Hardgrove machine were performed having different feed size fractions irradiated, and size distributions were taken after a short time of grinding. By applying equation (8.1), S_j values were calculated. $B_{i,j}$ values were found from an equivalent of equation (8.20) which involved a sieving correction. Results for three coals are given in Figures 8.23 and 8.24. The reason for the abrupt change in S at about 500 microns in Figure 8.22 was thought to be associated with smaller particles being much stronger owing to breaking out of flaws in the heterogeneous substance which is coal. More homogeneous materials show smoother curves, as given in Figure 8.19.

After 1962, there was a swing away from radioactive tracer methods (because of the large amount of experimental work involved) in favour of back calculation systems, but these have their problems. (see §8.5.3.3.)

More recently, Gardner and Sukanjnajtee (1972) have put the case for a return to the radioactive tracer because it cannot be avoided for the determination of residence times when continuous mills are simulated, and hence could well be also used in the context of breakage parameters. (see §8.5.4.)

This is partly borne out by the appearance of reports of recent studies of radioactive tracer methods, exemplified in Gardner *et al.* (1978, 1980, 1982)

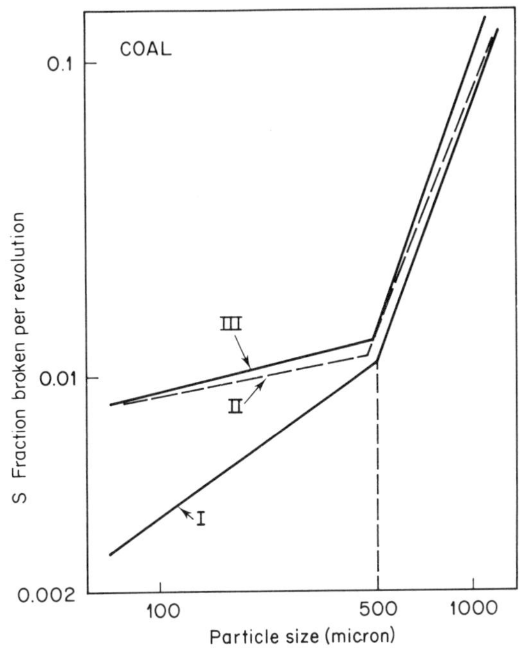

Figure 8.23. Variation of specific rate of breakage S with particle size for three coals using a radioactive tracer technique (Gardner and Austin, 1962). *Reproduced by permission of DECHEMA, Frankfurt*

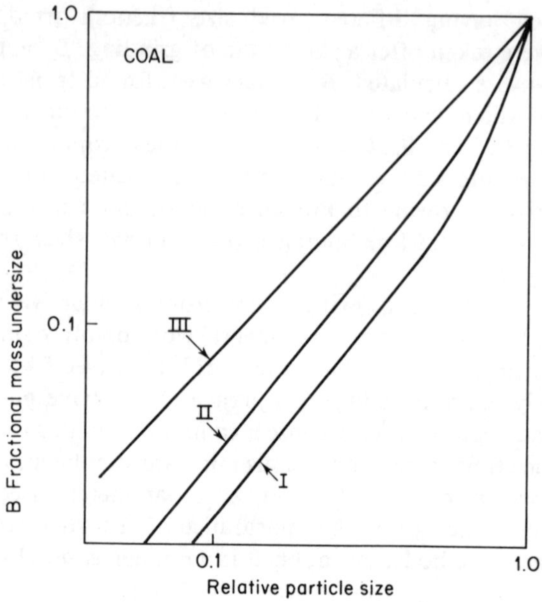

Figure 8.24. Variation of breakage distribution function B with relative particle size for the same three coals as in Figure 8.23, using radioactive tracer technique (Gardner and Austin, 1962)

Back calculation work continues, however, by those who believe that industry will be loath to become involved in radioactive methods preparatory to mill design.

8.5.2.5 *Method (1c): nuclear activation analysis technique*

Snow and Meloy (1972) discovered tags which produce peaks of emitted gamma ray energy at distinctive and different wavelengths and which have associated noise at a sufficiently low level not to obscure the distinctive peaks of the others. Six such materials are listed, allowing up to six feed size fractions in a batch grinding test to be assessed simultaneously.

Breakage functions are found in a similar manner to that described for radioactive tracers.

Because of the very extreme safety precautions required for this method it is unlikely to be used widely, and no evidence of its application other than by Snow and Meloy has been found to date, and no breakage function data from the method are known.

8.5.3 Indirect measurement: back calculation methods

8.5.3.1 Summary of available methods

Two main types of back calculation may be distinguished: those which are directly predictive, and those which involve some trial and error or optimization. In each of these broad categories, sub-division may be made with respect to the treatment of the breakage functions, which may be lumped together or used separately. Similarly, in treatment of the solution of the grinding equation, sub-division may be made into analytical matrix and fully numerical systems. However, it is not possible or desirable to exemplify every category.

8.5.3.2 Predictive systems

(a) Lumped parameters S and B Herbst and Fuerstenau (1968) developed a method using a feed confined to the top size interval only, assuming B was normalizable and invoking a treatment similar to the Reid solution of the batch equation. By these means it was demonstrated that

$$S_jB_{i,j} \simeq F_i = \frac{\mathrm{d}}{\mathrm{d}t}\left[M_i(t)\right]$$

From a linear plot of $M_i(t)$ against t for a series of particle sizes x, the slopes gave F_i and so the product $S_jB_{i,j}$. Values of S_j were obtained from:

$$S_j = S_1\left(\frac{x_jx_{j+1}}{x_1x_2}\right)^{\varepsilon/2} \tag{8.24}$$

where ε was given by the slope on a log–log plot of F_i against x, and S_1 was given by the slope of a semi-log plot of $M_i(t)$ against t for the top size feed. Then $B_{i,j}$ followed from F_i/S_j.

The derivation of the equations quoted may be obtained from Herbst and Fuerstenau (1968) and Herbst and Mika (1970).

The method makes assumptions which later studies have proved not to be always justified, as indicated in Austin and Bhatia (1971/2) and Austin *et al.* (1974), but it was a good early attempt to assess breakage functions without too much experimental work.

(b) Separate treatment of S and B Austin and Luckie (1971/72a) outlined their method III, which demands the initial estimate of the specific rate of breakage S (but it need not be a very accurate estimate) as given, for example, by the slope of linear breakage plots in Figure 8.2.

The general concept is to back calculate B, knowing S, but the back-computation is non-tractable except for the first two intervals:

For interval 1, fraction less than size $i = B_{i,1}[M_1(0) - M_1(t)]$ for $i \geqslant 1$

$= B_{i,1}\Delta M$ (see §8.5.2.2(a))

For interval 2, fraction less than size $i = B_{i,2}\int_0^t S_2 M_2(t)\,dt$

$= B_{i,1}\Delta_2$ (say) for $i > 2$

Now $M_2(t)$ is known from the Reid solution, from which it may be shown that:

$$\Delta_2 = \frac{M_2(t) - M_2(0)(1-A)^r}{A + (1-A)^r - 1}[1 - (1-A)^r - rA] + [1 - (-A)^r M_2(0)] \tag{8.25}$$

where

$$A = \frac{\Delta M}{M_1(0)}$$

$$r = S_2 t/\ln\{1/[1 - \Delta M/M_1(0)]\} \tag{8.26}$$

The invokation of the Reid solution for the third and higher intervals becomes mathematically intractable, and hence a simple linear mean rate is used. In fact the fractional amount broken to less than size i from the breakage of j material is approximated by $\frac{1}{2}B_{i,j}S_j t[M_j(0) + M_j(t)]$ for $i > j \geqslant 3$.

Now the fractional amount below the upper size of size 3 produced during grinding is the sum of the contributions from size 1 and size 2. Therefore:

$$F_3(t) - F_3(0) = B_{3,1}\Delta M + \Delta_2$$

i.e.

$$B_{3,1} = [F_3(t) - F_3(0) - \Delta_2]/\Delta M \tag{8.27}$$

Similarly:

$$F_4(t) - F_4(0) = B_{4,1}\Delta M + B_{4,2}\Delta_2 + \frac{S_3 t}{2}\left[M_3(0) + M_3(t)\right]$$

If B is normalizable, $B_{i,j} = B_{i-j,1}$ and if it is not, the error in assuming it has been shown to be small. Thus:

$$B_{4,1} \simeq \frac{F_4(t) - F_4(0)}{\Delta M} - B_{3,1}\frac{\Delta_2}{\Delta M} - \frac{S_3 t}{2\Delta M}\left[M_3(0) + M_3(t)\right]$$

where $B_{3,1}$ is known from equation (8.27).

Hence in general, with $n \geqslant i \geqslant 4$:

$$B_{i,1} \simeq \frac{F_i(t) - F_4(0)}{\Delta M} - B_{i-1,1}\frac{\Delta_2}{\Delta M} - \sum_{k=i-1}^{3}\frac{B_{i-k+1,1}S_k t[M_k(0) - M_k(t)]}{2\Delta M} \tag{8.28}$$

By the use of equations (8.25), (8.26), (8.27) and (8.28), the required values of B can thus be calculated, and Pennsylvania State University has a computer program which is available. Austin and Bhatia (1971/72) give an example of the use of this method, tabulate relevant data and give plots of B against x for cement clinker.

8.5.3.3 *Optimization systems*

(a) Lumped parameters S and B Cutting (1976) gives an example of an optimization technique using S and B lumped into a single 'transmission' function and applied to wet grinding of crushed slag in a batch rod mill. The transmission function is expressed as a single matrix, and the form of the function is chosen to be able to simulate the changes that take place in the frequency size distribution of product from a grinding zone after regularly spaced time intervals.

A minimization function, assuming errors are normally distributed, and incorporating a least-squares approach, is used to compare computed with practical results in a converging manner. So much trial and error is involved that this method has not found general acceptance. Further, the individual movements of S and B separately are found in practice to be important, particularly if mill design is the end product, as distinct from mill control. Hence the formulae used for transmission and error minimization functions are not quoted here.

(b) Separate treatment of S and B using the Reid solution Klimpel and Austin (1970) have proposed a method which involves the determination of an optimal set of S values to make experimental results follow the equation of batch grinding assuming that B is normalizable, and have successfully applied it to the prediction of product size distribution for anthracite and limestone ground in a laboratory ball mill.

B values were first determined by the method of §8.5.2.2(a) or (b). They were then plotted against x on a log–log grid to ensure consistency and, if necessary, adjusted to give a smooth curve, from which $b_{i,j}$ values were read off.

S was then calculated approximately using the Reid solution (see §8.4.2) and optimization proceeded via the least-squares formulation to match $M_i(t)_{obs}$ with $M_i(t)_{calc}$ for successive values of S and B, until the best values of breakage functions were obtained.

The practical problems with this approach were:

(1) It was found necessary to use size interval weighting factors on the individual breakage functions so as to give reasonably continuous plots of S and B against particle size, and this was very time consuming.
(2) There was difficulty with the $(S_i - S_j)$ term of the Reid solution, already alluded to in §8.4.2, when the S versus x curve was other than continuously increasing with increasing x.

(c) Separate treatment of S and B using the numerical analysis solution Because of the problems of the analytical Reid solution, Klimpel and Austin (1970) used a numerical analysis method of back calculation of S, given $b_{i,j}$, using a non-linear optimization technique. In essence, it was necessary to find

the best set of parameters ζ_1, ζ_2, ζ_3 etc. in:

$$\log S_j = \zeta_1 + \zeta_2(\log \bar{x}_i) + \zeta_3(\log \bar{x}_i)^2 + \ldots \tag{8.29}$$

to give the best fit of the experimental data, using the Goldfarb and Lapidus (1968) algorithm, which proved to be the best of several that were tried (where $\bar{x}_i$ is the mean size of an interval i).

Klimpel and Austin (1977) later extended the above technique to incorporate the four parameters associated with S, namely a^*, α, μ and σ (see §8.2.2.3) and four parameters associated with B, namely Φ, β, γ and ψ (see §8.2.3.4), making a total of eight.

Optimal values of this set having been obtained, optimal S_j and $b_{i,j}$ could be put into equation (8.15) to update $M_i(t)$ values, which were used to calculate new $M_i(t)$ values. The search for optimal values of the set of eight was then repeated, holding ΔM constant, and the whole process was repeated until the vectors of S and B values did not change appreciably.

Klimpel and Austin (1977) must be studied for details of the system (but the above is sufficient to indicate the drift of the method) and for examples of its use on gold ore and quartz grinding.

The trouble with back calculation methods is that they may give breakage parameters which are unreal and yet combine to give a good simulation of the size distribution of the product; hence checks should be made with valid experimental data where possible before a system is accepted (Klimpel and Austin, 1984).

8.5.4 Combined direct measurement and back calculation methods

Gardner and Sukanjnajtee (1972) have proposed a method which is a compromise between the direct and the indirect methods of assessment of breakage functions.

A grinding charge consisting of normal distribution of interest would be prepared with the top size tagged with a radioactive tracer. A small increment of grinding time would be employed so that about 10% of the top size is broken. The subsequent amount of tracer in each size could be used to determine the distribution function $B_{i,1}$.

A feed size distribution was then prepared identical to that used above, but having, say, the fifth size tagged with a tracer. Grinding was carried out for several time increments and the mass–size and trace–size distributions for each were determined, allowing back calculation of S values, assuming normalizable B. (Trace values were given priority over mass values, since they should contain smaller accumulated error.) The process was repeated for the ninth size fraction.

By this means the experimental effort was greatly reduced and accuracy was not sacrificed because back calculation was only over a limited range of size fractions—four in the case outlined above. Further, the tracer method served as a check on the back calculated value for the fourth and ninth intervals, etc.

Moreira *et al.* (1972) gave the application of the system to the comminution of hematite ore in a ball mill, using Reid's solution of the batch equation (see §8.4.2) and a Monte Carlo method for evaluating the compound errors which were inherent in breakage function assessment. For a description of the Monte Carlo method see Gardner and Sukanjnajtee (1972).

The hematite ore proved to be a severe test for the method, as B was found to be non-normalizable and grinding was not linear; but accuracy with a minimum of experimental effort was demonstrated, and further, the method was successfully applied to continuous grinding.

8.6 RESIDENCE TIME DISTRIBUTION (RTD) IN TUMBLING MILLS

8.6.1 RTD function in a horizontal rotating cylinder without grinding

Continuous, as distinct from batch, grinding in a tumbling mill involves the transport of powder from inlet to outlet concurrent with the grinding action.

Consider first the case of powder movement, without ball charge, in a laboratory cylindrical horizontal rotating drum, studied by Abouzeid (1973) for example, by means of injection of a trace material at the inlet, and measuring the concentration $C(t)$ as a function of time t at the outlet. He obtained a pattern of concentration change given in Figure 8.25, called the C-curve.

If the RTD function is denoted by $\phi\ (t)$, then:

$$\phi(t) = C(t) \Big/ \int_0^\infty C(t)\,\mathrm{d}t \qquad (8.30)$$

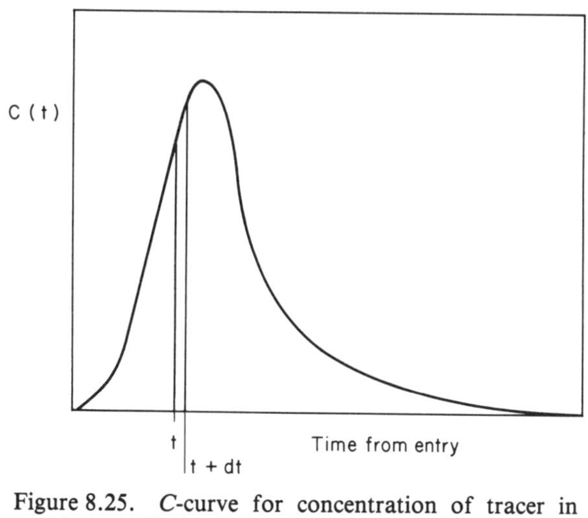

Figure 8.25. C-curve for concentration of tracer in powder at the outlet of a continuously operated horizontal rotating drum resulting from a pulse injection of tracer at the inlet (Abouzeid, 1973). *Reproduced by permission of the author*

Now $\phi(t)$ has units of 'fraction per unit time', and hence $\phi(t)\mathrm{d}t$ represents a fraction of material which has resided in the mill for time t and appears at the outlet between t and $(t + \mathrm{d}t)$.

Early workers such as Smith (1942), who seems to be the first to have applied the RTD to tumbling mills, took the time corresponding to the peak of the C-curve as indicative of the mean residence time τ of the material in the drum. However, because of the skew nature of the distribution given in Figure 8.25, this was inaccurate. Thus mean residence time is accurately given by:

$$\tau = \int_0^\infty tC(t)\,\mathrm{d}t \Big/ \int_0^\infty C(t)\,\mathrm{d}t$$

or from equation (8.30):

$$\tau = \int_0^\infty t\phi(t)\,\mathrm{d}t \tag{8.31}$$

The other important feature of the C-curve is the spread about the mean given by the variance σ_t^2, where:

$$\sigma_t^2 = \int_0^\infty (t-\tau)^2 C(t)\,\mathrm{d}t \tag{8.32}$$

The variance is a function of the mixing which occurs in the drum.

The bulk mean residence time τ_B must be distinguished from τ and is given by:

$$\tau_B = W^*/F \tag{8.33}$$

In principle, τ and τ_B should be the same, but a difference may be indicative of tracer stagnation or stratification within the powder bed.

Some workers have given the effect of operating variables on the RTD function in terms of its first and second moments τ and σ^2, rather than by the full curve of Figure 8.25.

8.6.2 RTD function in a horizontal rotating cylinder with grinding

For dry conditions, Hogg *et al.* (1975) found the addition of grinding media to a powder charge increased the mean residence time of the powder by reducing the axial displacement per cyclic movement of particles compared with the case of powder only. However, the density distribution form of the C-curve was not basically altered, as was demonstrated by Mardulier and Wightman (1971).

Similarly for wet conditions, Kelsall *et al.* (1969/70) using a laboratory mill, Rogers *et al.* (1982) using a pilot scale mill, and Marchand *et al.* (1980) using full-scale commercial mills, all confirmed the same density distribution as given in Figure 8.25. This constancy of pattern over such a wide size range is most helpful in the scale-up of the RTD function.

Kelsall *et al.* (1969) have pointed out that this pattern is much closer to that of a perfect mixer than to that of plug flow, which latter would imply no mixing or dispersion, and all particles would reside for the same time in the mill. This is made clear in Figure 8.26, which shows both density and cumulative forms of the RTD function. However, to assume a fully mixed situation is clearly an approximation.

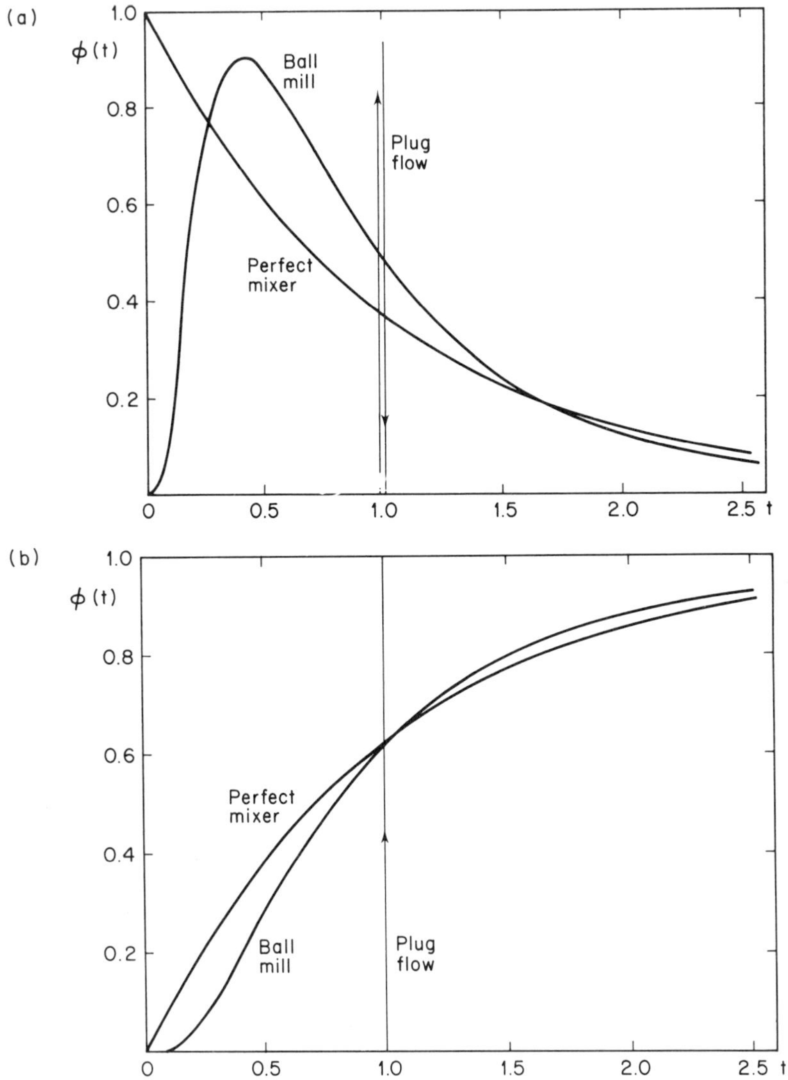

Figure 8.26. RTD function versus time compared with the distribution pattern for a perfect mixer and with plug flow for a tumbling mill: (a) density distribution, (b) cumulative distribution (Kelsall *et al.*, 1969). *Reproduced by permission of the Institution of Engineers, Australia*

Equations (8.31), (8.32) and (8.33) for mass flow without grinding in open circuit apply equally in principle to mass flow with grinding.

Further, a dimensionless form $\phi(\theta)$ of the RTD function is advantageous, expressed in terms of dimensionless time $\theta = t/\tau$. Then:

$$\phi(\theta) = \tau\phi(t) \tag{8.34}$$

$$\sigma_\theta^2 = \sigma_t^2/\tau^2 \tag{8.35}$$

where the dimensionless variance is σ_θ^2.

Austin *et al.* (1982a) acknowledge a gap in present knowledge in respect of the effect of change of mill dimensions on $\phi(\theta)$.

8.6.3 Measurement of the RTD function

Table 8.3 lists four methods of measurement of the RTD function which have been tried, together with comments on their usefulness.

For the SLRT method, a sample is taken, either mono size or of known grade, irradiated in a nuclear reactor and injected as a pulse into the steady-state mill feed. The mill product is monitored for radioactivity using a detector such as the sodium iodide scintillation type, which is sensitive to the gamma radiation of interest. Details of the necessary corrections for background conditions and for radioactive decay may be obtained from Rogers and Austin (1983).

For dry milling, the mill product is sampled at suitable times, with radiation detection being accomplished by placing each sample or portion thereof, in a container to give a constant bulk density and a constant geometry relative to the detector. In the case of a wet mill, a suitably waterproofed detector can be immersed in the mill product pumping sump (see Figure 8.33(a)) since the

Table 8.3. Comparison of methods of measurement of the RTD function

Method	Reference	Comments
Quartz in calcite	Kelsall *et al.* (1969/70)	See §8.5.2.3
Fluorescein dye (for dry grinding)	Mardulier and Wightman (1971)	Relies on micropores of mineral to have uniform dye penetration, which is not always so
Soluble salt or radioactive tracer (for wet grinding)	Marchand *et al.* (1980)	(i) Assumes slurry density within the mill is equal to that entering and leaving, (but in fact it is greater by about 15%) (ii) Assumes no absorption of tracer in the mineral solids
Short-lived radioactive tracer (SLRT) (for wet or dry milling)	Gardner *et al.* (1980)	Best method, as does not suffer from the above disadvantages

net counting rates, given as counts per unit time, will be proportioned to the amount of radioactivity in the sump at any time.

An alternative detection method which can be used for either wet or dry mills is to mount a detector collimated with lead plates adjacent to the mill discharge or at the discharge end of the mill, to monitor the tracer as it leaves the mill.

All these methods have given satisfactory data without interruption of the normal process operation and with complete safety, when the relevant precautions are taken.

The above relates to open-circuit grinding, which is the primary concern of this chapter; but in the context of measurement of RTD by SLRT it is convenient to mention the closed circuit also. The extra problems in a closed circuit are caused by the concentration of the tracer in the recycle varying with the particular size of the tracer in the feed to the classifier, and by the action of the classifier on different particle sizes, as shown by Rogers *et al.* (1982).

To accommodate this greater complexity, Gardner *et al.* (1982) propose a two-detector system which involves simultaneous measurement of SLRT at inlet and outlet, together with deconvolution, to obtain the primary RTD.

Rogers *et al.* (1982) have pioneered a single-detector method which uses an SLRT feed of sufficiently small size to be recycled in the classifier apparent bypass a' (see §5.2.6.2). This is a common form of separation inefficiency which is harnessed to good purpose by simplifying the measurement of the RTD function when a classifier is operating in a wet system.

8.6.4 Effect of operating variables on the RTD function

8.6.4.1 Variables having negligible effect

Table 8.4 groups together the variables which have been found to have little or no effect on the pattern of the RTD function in open-circuit wet grinding.

With regard to item 1 in Table 8.4, for *small* mills, different particle sizes move through continuously operated mills at different velocities. Larger particles stay in the mill for longer times than smaller, owing to preferential settling in the liquid pool. If S_i values are back calculated, assuming the same RTD for all sizes, then larger sizes will have false values of S_i. This is not so marked for *large* mills.

However, since scale-up is often done from small mill tests, this is one possible explanation why there is a lack of similitude between small and large mill S values, as found by Austin and Weller (1982).

Item 2 must be qualified by the statement of Austin *et al.* (1982a) that there is, to date, no satisfactory theory for the variation of normalized RTD with mill dimensions.

With regard to item 5, it would appear that no change from cascading to cataracting was experienced within the speed range. (see §8.6.4.2(a).)

Table 8.4. Operating variables having negligible effect on the RTD pattern for open-circuit wet ball milling

Variable	Mill diameter (m)	Approximate range	References
(1) Particle size of feed	0.3	100–1165 micron	Kelsall *et al.* (1969/70)
	0.91	4000–9250 micron	Rogers *et al.* (1982)
	1.33	Wide range	Gardner *et al.* (1980)
(2) Mill L/D ratio	0.3	0.5–1.5	Kelsall *et al.* (1969/70)
(3) Ball size	0.3	0.5–1.5 inch	Kelsall *et al.* (1969/70)
(4) Relative ball load	0.3	0.45–1.0†	Kelsall *et al.* (1969/70)
(5) Mill speed	0.3	$N^* = 0.65–0.90$	Kelsall *et al.* (1969/70)
(6) Slurry relative to carried solids	0.3	Solids size range 328–1310 micron	Kelsall *et al.* (1966)
(7) Feed rate	0.91	2 : 1 ratio	Rogers *et al.* (1982)

†1.0 relates to a ball charge of 633 × 1 inch balls in a drum 0.3 m diameter × 0.3 m.

Item 6 implies that the dimensionless RTD for the slurry coincides with that for the carried solids over the range indicated, demonstrated in Figure 8.27 for a solids size of 590 microns; but see §8.6.4.2(b).

Item 7 is further discussed in §8.6.4.2(c).

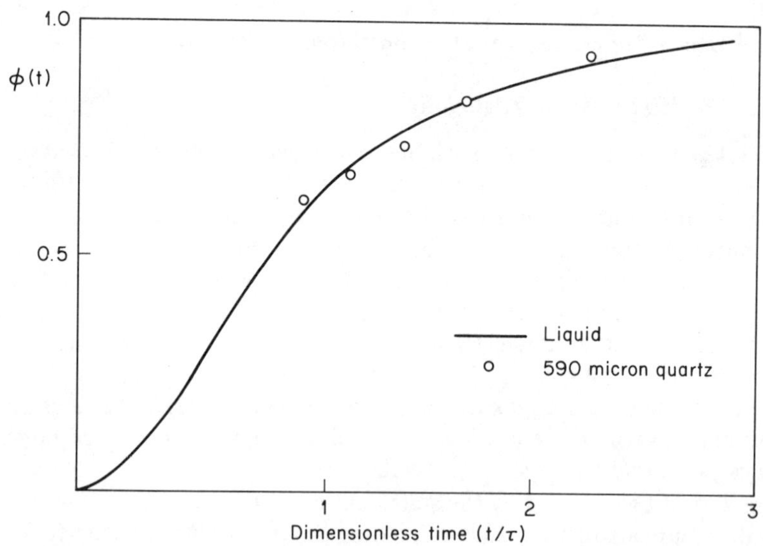

Figure 8.27. RTD function versus dimensionless time for liquid and carried solids in wet grinding (Kelsall *et al.*, 1966). *Reproduced by permission of the Australasian Institute of Mining & Metallurgy*

8.6.4.2 *Variables having a significant effect on the RTD function*

(a) Mill Speed Abouzeid *et al.* (1974) showed that considerable change to the RTD function and to the mean residence time occurred when speed changes caused a move from cascading to cataracting within a rotating drum. Similarly, the amount of cararacting can alter, with change of liner pattern, even at constant fraction of critical speed, as evidenced by Figure 8.28 which shows variation of the dimensionless RTD function with each of the liners depicted in Figure 8.8. Otherwise, for a given liner and keeping within the cascading speed range, change of speed has little effect, as shown in table 8.4.

(b) Slurry relative to carried solids By SLRT techniques, residence time for solids in open-circuit wet grinding has been shown to be longer than for the carrying fluid, even though the shape of the dimensionless RTD function remains unaltered, as indicated by item 6 in Table 8.4. Rogers and Austin (1983) point out that this finding might be thought to be in conflict with item 1 in Table 8.4, since fines would be expected to flow with the fluid. However, Hogg and Rogovin (1982) have concluded that reasonable differences in flow rate of different particle sizes would be difficult to detect by RTD measurements on traced material which is grinding finer as it passes through the mill. The longer residence time of solids over fluid means that solids hold-up estimated from the residence time of the fluid is consistently low by 10–15%.

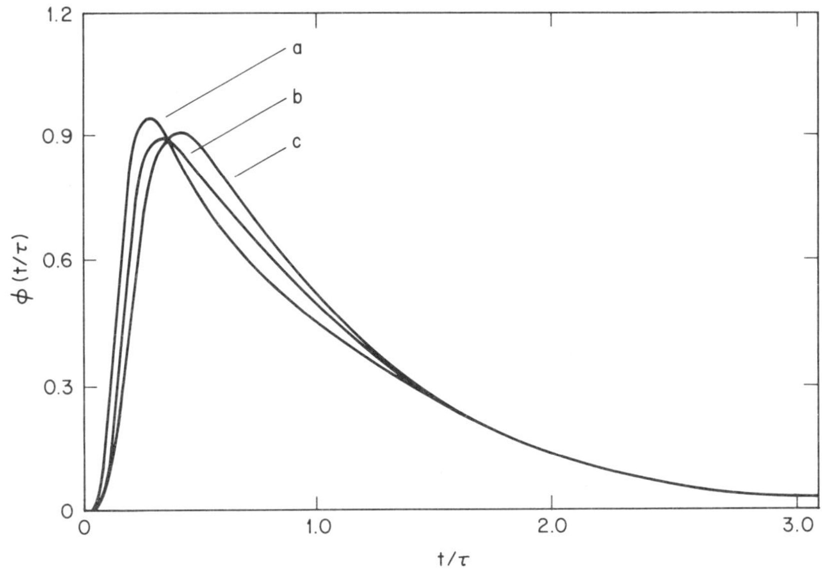

Figure 8.28. Variation of dimensionless RTD function with dimensionless time for the three liners shown in Figure 8.8: σ^2/τ^2 : a 0.60, b 0.53, c 0.48 ($N^* = 0{\cdot}7$ open circuit) (Rogers and Austin, 1984). *Reproduced by permission of Hemisphere Publishing Corporation*

(c) Hold-up and feed rate The RTD function is sensitive to changes in solids hold-up within the mill (Kelsall *et al.*, 1968/69). Now for a commercial mill, hold-up is difficult to measure directly, and hence the RTD function has been used to assess hold-up via solids residence time and equation (8.33), giving:

$$W^* = F^* \int t\phi(t)\,\mathrm{d}t \tag{8.36}$$

This use of the RTD function was first practised by Mardulier and Wightman (1971) for dry grinding of cement clinker.

Although the pattern of the dimensionless RTD function tends to be invariant over quite large changes in feed rate, as indicated in item 7 in Table 8.4, the actual residence time and hold-up change quite markedly with feed rate. Mori (1973), for example, found over the higher range of feed rates that:

$$W^* \propto (F^*)^{0.5} \tag{8.37}$$

for dry grinding in a laboratory ball mill. This relation was later confirmed in a commercial wet mill by Marchand *et al.* (1980) for constant slurry rheology and mill outlet geometry.

Combining equations (8.33) and (8.37) gives:

$$\tau_B \propto (F^*)^{-0.5} \tag{8.38}$$

which applies only for the stated conditions of slurry and mill outlet geometry. It is an outstanding exercise to quantify the effects on residence time of change of slurry properties and of mill outlet geometry.

However, for overflow mills with a ball retaining gate, investigation by Austin and Brame (1983) established that if F_1^* is the feed rate corresponding to the interstitial filling fraction $U = 1$, then for $F^* < F_1^*$, U is unaffected by flow rate, but for $F^* > F_1^*$, U varies as $(F^*)^{0.5}$. This is seen in Figure 8.29 for a particular mill with solids flow given in volumetric terms to allow comparison of materials (coal and limestone) of different density (Rogers and Austin, 1984).

In order to compare data for different mill sizes and conditions, a dimensionless flow rate has been developed based on the finding by Hogg *et al.* (1975) that, for a given fractional filling, $F^* \propto ND^4$, or since $N = N^*/\sqrt{D}$:

$$F^* \propto N^* D^{3.5} \tag{8.39}$$

which applies for a given slurry density and L/D ratio.

More generally, equation (8.39) may be expressed as:

$$F^* = K_2 \rho_s N^* D^{3.5} (L/D) \tag{8.40}$$

Hence, when operating at a given fraction of critical speed N^*, the dimensionless flow rate may be given in terms of $F^*/\rho_s D^{3.5}(L/D)$ as in Figure 8.30, which shows the same trend as Figure 8.29 but over a range of mill geometries. The ordinate in Figure 8.30 is the fractional interstitial ball filling corrected for the fact that the ball load was higher at the retaining gate of the overflow than

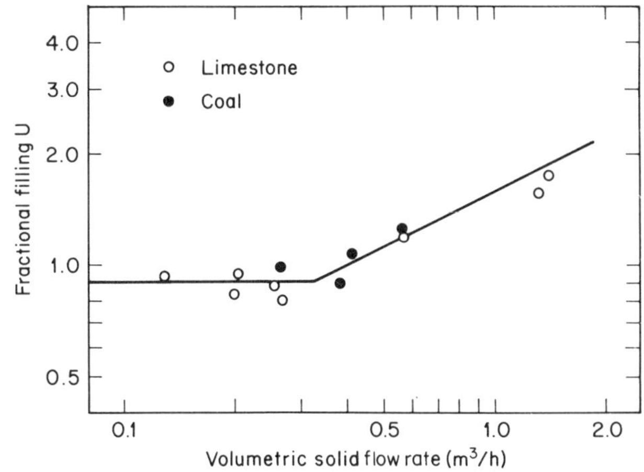

Figure 8.29. Fractional interstitial filling U as a function of solids flow rate for a pilot plant: $D = 0.82$ m, $L = 1.5$ m (Rogers and Austin, 1984). *Reproduced by permission of Hemisphere Publishing Corporation*

over the rest of the charge. Then corrected U_c is given by UJ/J_T, where J_T is the ball load corresponding to the discharge level.

From equation (8.39) it may be deduced that the flow rate per unit of mill volume is proportional to $D^{0.5}$ to give a given particular fractional filling. Now, also, from §8.2.2.5(a) it is seen that specific breakage rate is proportional to $D^{0.5}$. Hence there would appear to be justification for the conclusion that a small mill grinding a given feed will have the same fractional filling

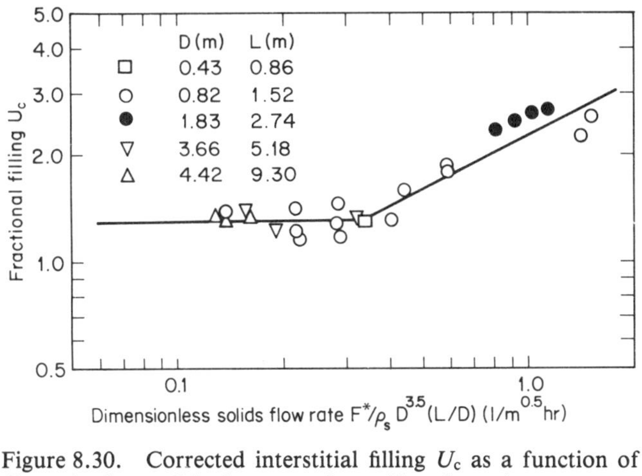

Figure 8.30. Corrected interstitial filling U_c as a function of dimensionless solids flow rate for mills of varying geometry (Rogers and Austin, 1984). *Reproduced by permission of Hemisphere Publishing Corporation*

(hold-up) as a larger if $\tau = 1/D^{0.5}$, all other conditions being identical. Assuming this can be accepted, and in the absence of any rigorous theory of mass transfer in mills, Austin *et al.* (1982a) have expressed the above concepts in the form:

$$
\begin{aligned}
U &= (F^*/F_1^*)^{0.5} \\
&= \left(\frac{F^*}{W_1^*}\Big/\frac{F_1^*}{W_1^*}\right)^{0.5} \\
&= \left(\frac{\tau_1}{W_1^*}F^*\right)^{0.5} \quad \text{for } F^* \geqslant F_1^*
\end{aligned}
\tag{8.41}
$$

where $\tau_1 = W_1^*/F_1^*$ and W_1^* the hold-up corresponding to $U = 1$.

Equation (8.41) is one way of expressing the relation between feed rate F^* and fractional hold-up U, but it may be equally expressed in terms of mill variables, following Rogers and Austin (1984). They point out that for the constant value of $N^* = 0.7$ at which the data of Figure 8.29 apply, $K_2 = F^*/\rho_s D^{3.5}(L/D)$ has a value of the order of 0.5 for $U = 1$, or:

$$F_1^* = 0.5\rho_s N^* D^{3.5}(L/D) \tag{8.42}$$

Hence

$$U = (J_T/J)\{F^*/[0.5\rho_s N^* D^{3.5}(L/D)]\}^{0.5} \quad \text{for } F^* \geqslant F_1^* \tag{8.43}$$

These expressions are very valuable in the framing of equations of mass transfer within the mill, and the RTD function has played an important part in arriving at such equations. However, it is accepted that some aspects are rather tentative and more work is called for to put the relations on a firmer footing.

8.6.5 Mathematical representation of the RTD function

8.6.5.1 Survey

Three levels of sophistication are recognized in the mathematical models which have been advanced to date for the simulation of the RTD function:

(1) relatively simple two-parameter models: (i) log-normal probability, (ii) axial dispersion, (iii) equal fully mixed reactors in series, and (iv) one large and two small fully mixed reactors in series;
(2) relatively simple models having more than two parameters: (i) one large and two small fully mixed reactors and one plug flow reactor in series, (ii) finite-stage models;
(3) more complete models of continuous flow as summarized by Mika (1971).

Item 3 is not considered further because the equations are difficult to solve, but items 1 and 2 are discussed below, for which it must be appreciated that the models give no guidance as to what is really happening within the mill: they are just means to model the RTD function as measured at the mill outlet.

8.6.5.2 *Two-parameter models*

(a) Log–normal probability It has been noted by Kelsall *et al.* (1969/70) and by Mori (1973) that the RTD function can be adequately described by the log–normal distribution, since when plotted against time on a log–normal grid, a straight line results. Hence equation (4.17) is applicable. Mori *et al.* (1964) report results using the log–normal probability model which compare favourably with those using the axial dispersion model. Again, Hodouin *et al.* (1978) have employed the log–normal distribution model in the context of continuous size reduction of minerals, but most other workers have selected other models as best suiting their needs.

(b) Axial dispersion Austin *et al.* (1971), Mori (1973) and Karra and Fuerstenau (1977) have all proposed models which make the assumption for tumbling mills that mixing in the radial plane is very rapid and only that in the horizontal axial direction need be considered. It is also assumed that the material 'diffuses' from high to low concentration C along the x-axis at a rate, according to Fick's law, of rate $= -D^* \; \partial C/\partial x$, where D^* must be taken to be an effective mixing factor due to ball action, rather than a time diffusion coefficient.

By taking a mass balance on a cross-sectional element, the basic diffusion model equation is obtained:

$$\frac{\partial C}{\partial t} = D^* \frac{\partial^2 C}{\partial l^2} - \bar{v} \frac{\partial C}{\partial l} \tag{8.44}$$

where $\bar{v}$ is the average velocity of particles in the x direction, defined by $\bar{v} = L/\tau_B$, and l is the distance from the feed end of the mill. By solving equation (8.44), Mori (1973) obtained an expression for the dimensionless RTD function:

$$\phi(\theta) = 0.5\left(\frac{Pe}{\pi\theta}\right)^{0.5} \exp\left[-\frac{Pe}{4\theta}(1-\theta)^2\right] \tag{8.45}$$

where Pe is the Peclet number $(= \bar{v}l/D^*)$ and L is the mill internal length.

Figure 8.31 indicates a reasonably good fit of equation (8.45) to test-found data for silica sand ground in an open-circuit laboratory mill.

With dry grinding, the value of D^* was found by Mori *et al.* (1964) to increase with mill diameter and with mill speed ($0.3 \leqslant N^* \leqslant 0.7$).

For a given mill operating at constant speed, wet grinding, Austin *et al.* (1983) demonstrated that D^* remains sensibly constant along the length of the mill.

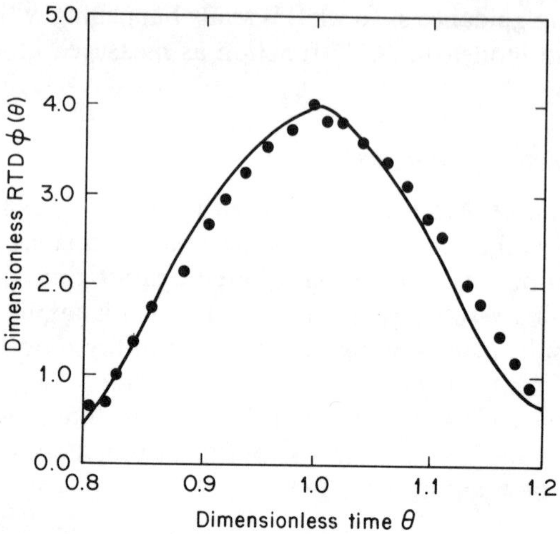

Figure 8.31. Dimensionless RTD function $\phi(\theta)$ versus dimensionless time θ using equation (8.45) compared with experimental data for silica sand ground in a 203 mm diameter × 1000 mm ball mill. Speed of rotation 37.5 rev/minute; feed rate 5.34 g/s (Mori, 1973). *Reproduced by permission of IIT Research Group*

(c) Equal fully mixed reactors in series It has been the practice in the process field for some time to simulate RTD by m equal fully mixed reactors or tanks in series (Barnes and Ellis, 1963). Then it may be shown (Levenspiel, 1962) that:

$$d(t) = (1/\tau^{-m}) \frac{t^{(m-1)}}{(m-1)!} \exp(-t/\bar{\tau}) \tag{8.46}$$

Clearly the model has little physical meaning in the context of tumbling mills, but it is a way of expressing the RTD function with only two parameters (m and τ). Practical experience reveals that good fits to mill data are not obtained with the equal-reactors-in-series model (Weller, 1980).

(d) One large and two small fully mixed reactors in series For ball mills, Weller (1980) and Marchand *et al.* (1980) have demonstrated the validity of a model having one large fully mixed reactor (mean residence time τ_L) followed by two equal smaller fully mixed reactors (mean residence time τ_S), the whole having a total mean residence time $\tau = \tau_L + 2\tau_S$. Then:

$$\phi(t) = \frac{\tau_L}{(\tau_L + \tau_S)^2}\left[\exp\left(-\frac{t}{\tau_L}\right) - \exp\left(\frac{t}{\tau_S}\right)\right] - \frac{t}{(\tau_L - \tau_S)^2}\exp\left(-\frac{t}{\tau_S}\right) \tag{8.47}$$

For normal L/D ratios, good approximations for τ_L/τ and τ_S/τ are respectively 0.7 and 0.15 (Austin *et al.*, 1982a). These may be used, if actual RTD information is not available, because it is known that the SMB equations are not too sensitive to the shape of the RTD function, particularly in closed circuit. Hence there can be occasions on which equation (8.47) is sufficiently accurate, though it is never as accurate as models with more than two parameters, as demonstrated in Figure 8.34

8.6.5.3 *Models having more than two parameters*

(a) One large and two small fully mixed reactors plus a plug flow reactor in series This three-parameter model has been developed by Marchand *et al.* (1980), in the context of large industrial wet tumbling mills. The RTD function may be expressed as:

$$\phi(t) = \frac{K_3}{\tau_S + \tau_L}\left\{\left[\frac{-(t-\tau_P)}{\tau_S} - \Omega\right]\exp\left[\frac{-(t-\tau_P)}{\tau_S}\right] + \Omega\,\exp\left[\frac{-(t-\tau_P)}{\tau_L}\right]\right\} \tag{8.48}$$

where τ_P, τ_S and τ_L are the residence times in, respectively, the plug flow reactor, in each small fully mixed reactor, and in the large fully mixed reactor; and $\Omega = \tau_L/(\tau_L - \tau_S)$.
The overall mean residence time, τ is also given by:

$$\tau = \tau_P + 2\tau_S + \tau_L \tag{8.49}$$

and the variance σ^2 by:

$$\sigma^2 = 2\tau_S^2 + \tau_L^2 \tag{8.50}$$

Marchand, *et al.* (1980) found this three-parameter model sufficiently flexible for closed-circuit application.

(b) Finite-stage models Adler and Hovorka (1961), Rogers and Gardner (1979b), Gardner *et al.* (1980), Rogers *et al.* (1982) and Gardner *et al.* (1982) have all recommended a finite-stage model of the RTD function, the basic stage of which is depicted in Figure 8.32. The stage consists of two interconnected fully mixed reactors and one plug flow reactor. The lower fully mixed reactor approximates stagnant flow regions, whereas the upper approximates active flow regions. The plug flow reactor represents a time delay. The basic stage may be repeated m times, where m is an integer.

The RTD function is given by:

$$\phi(t) = C_m^{\circ}(t) \Big/ \int_0^{\infty} C_m^{\circ}(t)\,\mathrm{d}t \tag{8.51}$$

where

$$C_m^{\circ}(t) = C_m(t - \eta_c\tau)h'(t - \eta_c\tau) \tag{8.52}$$

and $h'(t - \eta_c\tau)$ is the unit (positive) step function associated with the time delay of the plug flow reactor. Also, τ is the average residence time in the

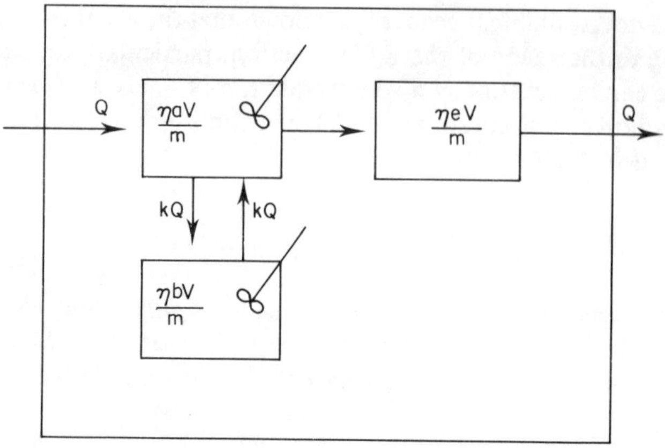

Figure 8.32. Schematic of the basic stage of the finite-stage model of the RTD function (Rogers and Gardner, 1979b)

system, and η_c is the fraction of total system volume in the plug flow reactor. $C_m(t)$ is given by:

$$C_m(t) = \left(\frac{C^\circ V}{q}\right)\left(\frac{m}{\eta_a \tau}\right)^m \sum_{j=1}^{2} \sum_{l=1}^{m} \frac{\phi_l(\lambda_j) t^{m-1} e^{\lambda_j t}}{(m-l)!(l-1)!} \tag{8.53}$$

where η_a is the fraction of total system volume in the upper fully mixed reactor, $\phi_l(\lambda_j)$ and λ_j are algebraic expressions involving the six parameters of the model, q is the steady-state entry and exit volumetric flow rates to the ith stage (volume/unit time), and C° is the concentration of the tracer if uniformly distributed throughout the system.

The form of λ_j and $\phi_l(\lambda_j)$ and a derivation of equation (8.51) are given in Appendix A of Rogers and Gardner (1979b). (see also Gardner *et al.* (1982)).

The variance of σ^2 of the distribution of equation (8.51) is given by:

$$\sigma^2 = \left(\frac{\tau^2}{m}\right)\left[(\eta_a + \eta_b)^2 + 2\eta_b^2/k\right] \tag{8.54}$$

where η_b is the fraction of total system volume in the lower fully mixed reactor, and k is the fractional cross-flow between upper and lower fully mixed reactors. Note that there are just five independent parameters involved: n, k, τ together with two of η_a, η_b and η_c, since $\eta_a + \eta_b + \eta_c = 1$.

As would be expected with a model using five independent parameters, very accurate fits to experimental RTD data are possible, evidence for which is provided in Figure 8.33. This diagram also shows the experimental arrangement, which included a sump, modelled as a single perfectly mixed reactor. The RTD measured was thus a convolution of the mill and sump RTDs given by:

$$\phi(t) = \int_0^t \phi_m(t')\phi_s(t-t')\,dt' \tag{8.55}$$

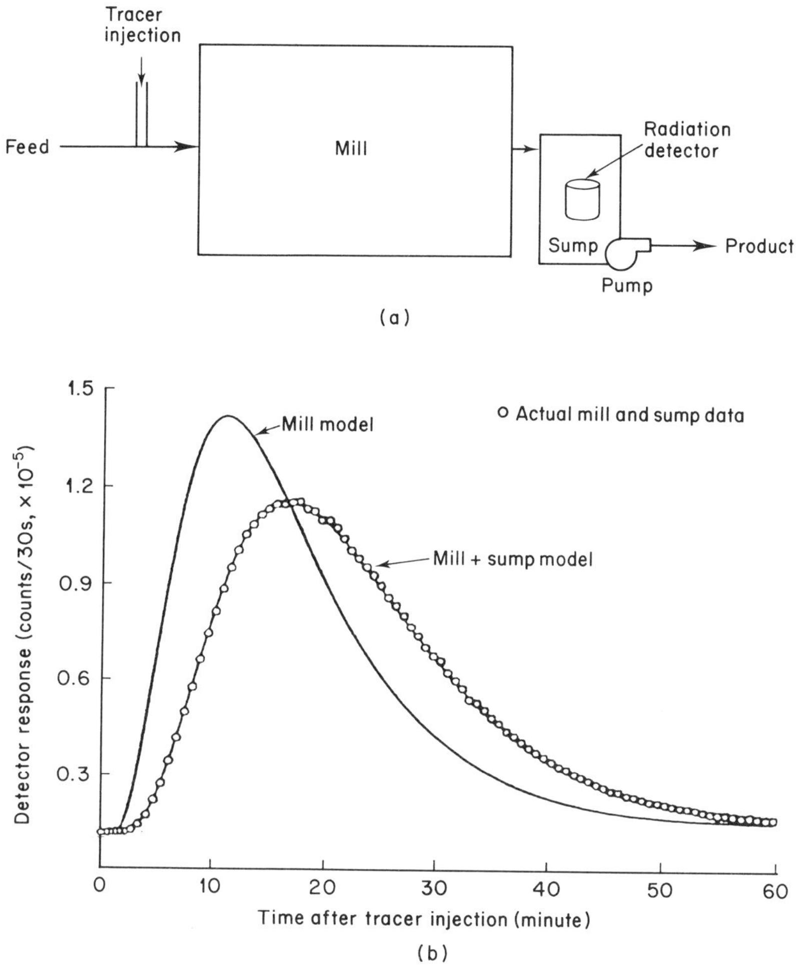

Figure 8.33. Fit of a finite-stage model to test data for wet grinding in an open-circuit ball mill provided with a sump at exit: (a) experimental arrangement, (b) fit of mill and sump data to finite-stage model (Rogers and Gardner, 1979b). *Reproduced by permission of the American Institute of Chemical Engineers*

where $\phi_m(t)$ is the RTD function for the mill and $\phi_s(t)$ is the RTD function for the sump.

The principle can be adapted to make a correction for tracer recycled in a closed circuit: it is then known as deconvolution, as exemplified in Austin *et al.* (1971) who performed the process by numerical analysis. Use of RTD functions, however, is subject to fewer errors than numerical analysis.

Figure 8.34 compares an experimentally found RTD function with those predicted in turn by finite-stage, axial-dispersion and reactor-in-series models:

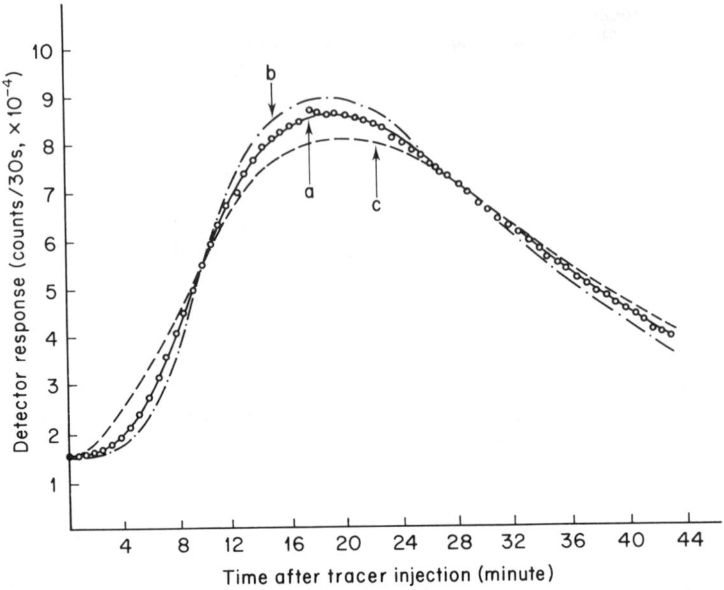

Figure 8.34. Comparison of fits of finite-stage, axial dispersion and reactors-in-series models to the same experimental data. ○ Actual mill plus sump RTD data. Sump modelled as one perfectly mixed tank. Mill modelled with: a finite-stage model, b axial dispersion model, c tanks-in-series model (Rogers and Gardner, 1979b). *Reproduced by permission of the American Institute of Chemical Engineers*

the finite-stage model follows the test data much more closely than do the two parameter models.

It is to be noted that the finite-stage model:

(1) reduces to the reactors-in-series model when the lower, fully mixed reactor and the plug flow reactor are omitted from each stage;
(2) can be used on full-scale mills with SLRT techniques without interruption of their normal operation.

8.6.6 Incorporation of RTD function into batch grinding equation

8.6.6.1 Segregated flow concept

It has been demonstrated that it is virtually impossible to measure the breakage functions S and B in full-scale tumbling mills using techniques valid for laboratory mills—see Austin and Bhatia (1971/72), Austin and Luckie (1971/72a) and Austin and Luckie (1971/72b). Hence there is a great incentive to use batch-found breakage functions in continuous mill processes and to apply the batch grinding equation in the form given in (8.15). It was Reid

(1965) who first applied the RTD function to a steady-state process. Then $\phi(t)\mathrm{d}t$ represents a fraction of a particular size of material which has resided in the mill for the time t and appears as the product at the outlet between t and $(t+\mathrm{d}t)$. This fraction will be broken as if it were batch ground for time t and its size distribution will be $M_i(t)$. Simultaneously, material which has been resident for a different time will also be leaving the mill. Thus at steady state, the total fraction p_i of product in interval i which leaves the mill will be the weighted average of all fractions of size i product. Then:

$$p_i F^* = \int_0^\infty M_i(t)\phi(t)F^*\,\mathrm{d}t$$

$$p_i = \int_0^\infty M_i(t)\phi(t)\,\mathrm{d}t \tag{8.56}$$

Equation (8.56) is, in fact, an example of the general segregated flow principle (or otherwise called the separation of phenomena concept) expressed mathematically (Himmelbau and Bischoff, 1968) as:

$$p_{ss} = \int_0^\infty [p(t)]_{\mathrm{BATCH}}\phi(t)\,\mathrm{d}t \tag{8.57}$$

where p_{ss} is the product mass size distribution in a steady-state mill, $[p(t)]_{\mathrm{BATCH}}$ is the mass size distribution in an identical mill operated in the batch mode for time t, and $\phi(t)$ is the RTD function of the continuous mill.

For equation (8.56) to be valid, as pointed out by Rogers and Austin (1983), three conditions must be satisfied:

(1) all particle sizes move through the mill with the same RTD;
(2) no classification of mill sizes occurs at the mill outlet;
(3) linear (first-order) breakage occurs within the mill.

These conditions can often be satisfied in mill practice. Although it is true that by using an RTD function measured only at the mill exit, what is happening within the mill is not known, it is still a useful device to reduce parameters to a manageable number.

8.6.6.2 Application of segregated flow concept

Equation (8.56) requires the knowledge of both $M_i(t)$ and $\phi(t)$. The former may be obtained from equation (8.17), for example, the Reid solution to the batch equation (8.15). In turn, equation (8.17) requires the knowledge of the specific rate of breakage S and the breakage distribution function b, each of which must be ascertained over the size range of interest. $\phi(t)$, assumed to be particle size invariant, is found via a selected equation from (8.45), (8.46), (8.47), (8.48) or (8.51) according to the RTD model chosen. Then, equation (8.56) may be evaluated by integration over the appropriate time range. For practical purposes, the upper limit of infinity is taken as a time sufficiently long to include all significant contributions to p_i. This time should not be less than 3τ (Austin and Brame, 1983).

8.7 NON-LINEARITY BREAKAGE PROBLEMS

Some of the general causes of deviation from linearity of breakage are noted in §8.2.2.2.

More specifically, §8.2.2.8 deals with a true slowing down of the breakage rate due to the production of fines, which is an environmental effect, and a device is described whereby linearity may be restored.

However, there have also been found material effects, as reported by Austin *et al.* (1977, 1982c) in the context of tumbling mill operation, occurring when particles are too large to be properly nipped by the balls. Then the material behaves as if some particles of a given size are relatively weak while others are stronger. As a reasonable approximation, the breakage can then be treated as a fraction $(1-\phi)$ of weak breakage function S_{Ai} and a fraction ϕ of stronger material with a breakage function S_{Bi}. These are incorporated into the appropriate SMB equations, and since the effect is only apparently non-linear, equation (8.56) may be applied. For the resulting simulation and its application to the grinding of quartz, Austin *et al.* (1982c) should be consulted. They show that two distinct assumptions may be made: (a) that fast-breaking material produces only fast-breaking material and slow-breaking material produces only slow-breaking material, or (b) both fast- and slow-breaking materials produce fragments which in turn can break fast or slowly. Depending on which assumption is made, the outcome can be significantly different, but work still remains to be done before secure deductions can be made.

For further reading in relation to non-linearity and distributed parameter models to accommodate the same, the following papers may be consulted: Herbst and Mika (1973), and Mika (1976).

One method of dealing with non-linearity, which allows for the change of S with time, is discussed by Luckie and Austin (1972) who adopt the finite-difference solution to the batch grinding equation.

Then equation (8.15) may be written, for a short interval Δt of grind time:

$$\left[M_i(\Delta t) - M_i(0)\right] \Big/ \Delta t = \sum_{\substack{j=1 \\ i>1}}^{i-1} S_j b_{i,j} M_j - S_i M_i \tag{8.58}$$

where Δt is chosen so that $M_i = [M_i(\Delta t) + M_i(0)]/2$. Experience suggests that a suitable value of Δt to satisfy this condition is $0.05/S_{\max}$, where $S_{\max}$ is the maximum value of S for the set of data considered. Solving equation (8.58) for $M_i(\Delta t)$, we have:

$$M_1(\Delta t) = \frac{2 - S_1\Delta t}{2 + S_1\Delta t} M_1(0)$$

$$M_2(\Delta t) = \frac{2 - S_2\Delta t}{2 + S_2\Delta t} M_2(0) + \frac{S_1\Delta t b_{21}}{2 + S_2\Delta t} - \frac{4}{2 + S_1\Delta t} M_1(0)$$

$$\vdots$$

$$M_i(\Delta t) = \sum_{j=1}^{i} d_{i,j}(\Delta t) M_j(0) \tag{8.59}$$

where

$$d_{i,j} = \begin{cases} 0 & \text{for} \quad i < j \\ \dfrac{2 - S_j\Delta t}{2 + S_i\Delta t} & \text{for} \quad i = j \\ a_{i,j} & \text{for} \quad i > j \end{cases}$$

$$a_{i,j} = \begin{cases} 0 & \text{for} \quad i < j \\ 4(2 + S_i t) & \text{for} \quad i = j \\ [1/(2 + S_i\Delta t)] \displaystyle\sum_{k=j}^{i-1} S_k\Delta t b_{i,k} a_{k,j} & \text{for} \quad i > j \end{cases}$$

Equation (8.59) is repeated, using $M_i(\Delta t)$ as the new feed to give $M_i(z\Delta t)$ and so on for z time intervals.

Alternatively, a back calculation may be performed and Luckie and Austin (1972) should be consulted for the appropriate equation and the full computer program which embodies it as a sub-routine, as it also similarly encompasses equation (8.59).

As indicated in §8.2.2.2, non-linearity of breakage is a feature of rod mill operation owing to preferential grinding of large sized particles. By application of equation (8.59), Shoji and Austin (1974) were able to forecast the product grading with reasonable accuracy for the dry batch grinding of limestone.

For further discussion of non-linear breakage in the context of laboratory rod mill grinding, see also Grandy and Fuerstenau (1970), Stewart and Restarick (1971), Heyes *et al.* (1973) and Austin *et al.* (1976a).

8.8 SUMMARY

The basic concepts of mathematical simulation of particle size reduction have been set down in Chapter 8, with emphasis on tumbling mills operated in open circuit. The important parameters have been found to be the specific rate of breakage S, the breakage distribution parameter B and, for continuously operated mills, the RTD function $\phi(\bar{t})$. The batch size–mass balance equation involving S and B has been stated, and means of solution have been proposed, together with a method of incorporating $\phi(t)$ for application to continuous systems when linear breakage conditions hold.

Methods of determination of the three main parameters S, and B and $\phi(t)$ have been described.

These basic concepts having been surveyed, it remains (i) to look further into the mathematics of the batch grinding equation, which is the subject of Chapter 9, and (ii) to consider closed-circuit mill operation, which is dealt with in Chapter 10.

8.9 REFERENCES

Abouzeid, A.-Z. M. A. (1973). 'Transport and mixing behaviour of particulate solids through rotary drums', Ph.D. Thesis, University of California, Berkeley.

Abouzeid, A.-Z. M. A., Mika, T. S., Sastry, K. V., and Fuerstenau, D. W. (1974). 'The influence of operating variables on the RTD for material transport in a continuous rotary drum', *Powder Technol.*, **10**, 273–288.

Adler, R. J., and Hovorka, R. B. (1961). 'A finite stage model for highly asymmetric RTDs', Digest of Technical Paper, 2nd Joint Auto. Control Conf.

Austin, L. G. (1971/72). 'A review: introduction to the mathematical description of grinding as a rate process', *Powder Technol.*, **5**, 1–17.

Austin, L. G., and Bagga, P. (1981). 'An analysis of fine dry grinding in ball mills', *Powder Technol.*, **28**, 83–90.

Austin, L. G., and Bhatia, V. K. (1971/72). 'Experimental methods for grinding studies in laboratory mills', *Powder Technol.*, **5**, 261–266.

Austin, L. G., and Brame, K. (1983). 'A comparison of the Bond method for sizing wet tumbling mills with a size-mass balance simulation model', *Powder Technol.*, **34**, 261–274.

Austin, L. G., and Klimpel, R. R. (1964). 'The theory of grinding operations', *Ind. Eng. Chem.*, **56**, 18–29.

Austin, L. G., Klimpel, R. R. and Beattie, A. N. (1967). 'Solution of equations of grinding', Dechema Monograph 57, 281–312.

Austin, L. G., Klimpel, R. R., Luckie, P. T., and Rogers, R. S. C. (1982a). 'Simulation of grinding circuits for design'. In: *Design and Installation of Comminution Circuits* (Mular, A. L., and Jergensen, G. V., Eds.), AIME, New York, 301–324.

Austin, L. G., and Luckie, P. T. (1970/71). 'Note on the influence of interval size on the first order hypothesis of grinding', *Powder Technol.*, **4**, 109–110.

Austin, L. G., and Luckie, P. T. (1971/72a). 'Methods for determination of breakage distribution parameters', *Powder Technol.*, **5**, 215–222.

Austin, L. G., and Luckie, P. T. (1971/72b). 'Estimation on non-normalised breakage distribution parameters', *Powder Technol.*, **5**, 267–277.

Austin, L. G., Luckie, P. T., and Ateya, B. E. (1971). 'Residence time distribution in mills', *Cem. Conc. Res.*, **1**, 241–256.

Austin, L. G., Luckie, P. T., and Shoji, K. (1982b). 'An analysis of ball and race milling. Part II: The Babcock 1.7E mill', *Powder Technol.*, **33**, 113–125.

Austin, L. G., Rogovin, Z., Rogers, R. S. C., and Trimarchi, R. (1983). 'The axial mixing model applied to ball mills', *Powder Technol.*, **36**, 119–126.

Austin, L. G., Shah, J., Wang, E., Gallacher, E., and Luckie, P. T. (1981). 'An analysis of ball and race milling. Part I: The Hardgrove mill', *Powder Technol.*, **29**, 263–275.

Austin, L. G., Shoji, K., Bell, D. (1982c). 'Rate equations for non-linear breakage in mills due to material effects', *Powder Technol.*, **31**, 127–133.

Austin, L. G., Shoji, K., Bhatia, V. K., and Aplan, F. F. (1974). 'Extension of the empirical Alyavdin equation for representing batch grinding data', *Int. J. Miner. Process.*, **1**, 107–123.

Austin, L. G., Shoji, K., Bhatia, V. K., Jindal, V., Savage, K., and Klimpel, R. R. (1976a). 'Some results on the description of size reduction as a rate process in various mills', *Ind. Eng. Chem. Process. Des. Devel.*, **15**, (1), 187–196.

Austin, L. G., Shoji, K., and Luckie, P. T. (1976b). 'The effect of ball size on mill performance', *Powder Technol.*, **14**, 71–79.

Austin, L. G., Trimarchi, T., and Weymont, N. P. (1977). 'An analysis of some cases of non-first order breakage rates', *Powder Technol.*, **17**, 109–113.

Barnes, R. K., and Ellis, S. C. (1963). 'Radioactive tracer experiments, Park Mine project', Final Report 2.6, Warren Springs Lab., DSIR (GB).

Broadbent, R. A., and Callcott, T. G. (1956). 'Coal breakage processes', *J. Inst. Fuel*, **29**, 524–539.

Broadbent, R. A., and Callcott, T. G. (1957). 'Coal breakage processes', *J. Inst. Fuel*, **30**, 13–25.

Cutting, G. W. (1976). 'A practical method for predicting grinding model parameters from laboratory batch tests', *Powder Technol.*, **15**, 21–28.

Epstein, B. (1948). 'Logarithmico-normal distribution in breakage of solids', *Ind. Eng. Chem.*, **40**, 2289–2291.

Gardner, R. P., Aissa, M., and Verghese, K. (1982). 'Determination of ball mill residence time distributions from tracer data taken in closed circuit operation', *Powder Technol.*, **32**, 253–266.

Gardner, R. P., and Austin, L. G. (1962). 'A chemical engineering treatment of batch grinding. Part I: A radioactive tracer technique for the determination of breakage functions', *Proc. 1st Europ. Symp. Zerkleinern*, Frankfurt, VDI Verlag, Dusseldorf, 217–248.

Gardner, R. P., and Sukanjnajtee, K. (1972). 'A combined tracer and back calculation method for determining particulate breakage functions in ball milling. Part I: Rationale and description of the proposed method', *Powder Technol.*, **6**, 65–74.

Gardner, R. P., Verghese, K., and Rogers, R. S. C. (1980). 'On stream determination of large scale mill RTD with short-lived radioactive tracers', *Min. Eng.*, **32**, (4), 422–431.

Gardner, R. P., Verghese, K., Rogers, R. S. C., Yu, B., and Wall, D. (1978). 'Radioactive tracer methods for the study of particulate processes' *Proc. Tech. Progress: Int. Powder Bulk Solids Handling Processing*, Rosemont, Ill, May, Ind. Sci. Conf. Management, Chicago, 479–486.

Goldfarb, D., and Lapidus, L. (1968). 'A rapid method of linearly constrained non-linear optimisation', *Ind. Eng. Chem.*, **7**, 142–151.

Grandy, G. A., and Fuerstenau, D. W. (1970). 'Simulation on non-linear grinding systems: rod mill grinding', *Trans. AIME*, **247**, 348–354.

Gupta, V. K., and Kapur, P. C. (1974). 'Empirical correlations for the effects of particulate mass and ball size on the selection parameters in the discretized batch grinding equation', *Powder Technol.*, **10**, 217—223.

Herbst, J. A., and Fuerstenau, D. W. (1968). 'The zero order production of fine sizes in comminution and its implication in simulation', *Trans. SME/AIME*, **241**, 538–548.

Herbst, J. A., and Fuerstenau, D. W. (1972). 'Influence of mill speed and ball loading on the parameters of the batch grinding equation', *Trans. SME/AIME*, **252**, 169–176.

Herbst, J. A., and Fuerstenau, D. W. (1973). 'Mathematical simulation of dry ball milling using specific power information', *Trans. SME/AIME*, **254**, 343–348.

Herbst, J. A., and Mika, T. S. (1970). 'Mathematical simulation of tumbling mill grinding: an improved method', *RUDY, Proc. IXth Int. Miner, Process. Congr.*, Prague, 3/4, 70–75.

Herbst, J. A., and Mika, T. S. (1973). 'Linearization of tumbling mill models involving non-linear breakage phenomena', *Proc. XIth Int. Symp. Computer Appl. Miner. Ind.*, University of Arizona, E 78–E 124.

Heyes, G. W., Kelsall, D. F., and Stewart, P. S. B. (1973). 'Continuous grinding in a small wet rod mill', 'Pt. I: Comparison with a small ball mill', *Powder Technol.*, **7**, 319–325, 'Pt.II: Breakage of some common ore minerals', *Powder Technol.*, **7**, 337–341.

Himmelbau, D. M., and Bischoff, K. B. (1968). *Process Analysis and Simulation: Deterministic Systems*, Wiley, New York.

Hodouin, D., Bérubé, M. A., and Everell, M. D. (1978). 'Modelling industrial grinding circuits and applications in design', *CIM Bull.*, **73**, (9), 138–146.

Hogg, R., and Rogovin, Z. (1982). 'Mass transport in wet overflow ball mills', *Proc. 14th Int. Miner. Process,. Congr.*, Toronto, 1/7, 1–19.

Hogg, R., Shoji, K., and Austin, L. G. (1975). 'Flow of particles through small continuous dry ball mills', *Trans. AIME*, **258**, 194–198.

Jindal, V. K., and Austin, L. G. (1976). 'Kinetics of hammer milling of maize', *Powder Technol.*, **14**, 35–39.

Kapur, P. C. (1970). 'Kinetics of batch grinding. Part B: An approximate solution to the grinding equation', *Trans. SME/AIME*, **247**, 309–314.

Kapur, P. C. (1982). 'An improved method for estimating the feed size breakage distribution functions', *Powder Technol.*, **33**, 269–275.

Kapur, P. C., and Agrawal, P. K. (1970). 'Effect of feed charge weight on the rate of breakage in batch grinding', *Trans. IMM (London)*, **79**, C269–C274.

Karra, V. K., and Fuerstenau, D. W. (1977). 'Material transport in a continuous rotary drum: Effect of discharge plate geometry', *Powder Technol.*, **16**, 23–28.

Karra, V. K., Swaroop, S. H. R., and Fuerstenau, D. W. (1980). 'Grinding in grate discharge ball mills: Effect of length', *Proc. Europ. Sym. Particle Technol.*, Amsterdam, 493–506.

Kelsall, D. F. (1965). 'A study of breakage in a small continuous open circuit wet ball mill', *Proc. 7th Int. Miner. Process. Congr.*, New York, 1964 (Arbiter, N., Ed.), Gordon & Breach, 33–42.

Kelsall, D. F., and Reid, K. J. (1965). 'The derivation of a mathematical model for breakage in a small continuous wet ball mill', *Proc. A.I.Ch.E./I.Ch.E. Symp. Series* 4, London, Inst. Chem. Eng., 14–20.

Kelsall, D. F., and Reid, K. J. (1969). 'Some effects of a change in environment size distribution on the grinding behaviour in a continuous wet ball mill', *Trans. IMM (London)*, **78**, C198–C205.

Kelsall, D. F., Reid, K. J., and Restarick, C. J. (1966). 'A dynamic investigation of continuous grinding in a small wet ball mill', *Proc. Australian Inst. Min. Metal.*, **220**, 21–23.

Kelsall, D. F., Reid, K. J., and Restarick, C. J. (1967/68). 'Continuous grinding in a small wet ball mill. Part I: A study of the influence of ball diameter', *Powder Technol.*, **1**, 291–300.

Kelsall, D. F., Reid, K. J., and Restarick, C. J. (1968/69). 'Continuous grinding in a small wet ball mill. Part II: A study of the influence of hold-up weight', *Powder Technol.*, **2**, 162–168.

Kelsall, D. F., Reid, K. J., and Restarick, C. J. (1969/70). 'Continuous grinding in a small wet ball mill. Part III: A study of distribution of residence time', *Powder Technol.*, **3**, 170–178.

Kelsall, D. F., Reid, K. J., and Stewart, P. S. B. (1969). 'A study of grinding processes by dynamic modelling', *Inst. Eng. Australia, Elect. Eng. Trans.*, **EE5**, 173–186.

Kelsall, D. F., Stewart, P. S. B., and Weller, K. R. (1973a). 'Continous grinding in a small wet ball mill. Part IV: A study of the influence of grinding media load and density', *Powder Technol.*, **7**, 293–301.

Kelsall, D. F., Stewart, P. S. B., and Weller, K. R. (1973b). 'Continuous grinding in a small wet ball mill. Part IV: A study of the influence of media shape', *Powder Technol.*, **8**, 77–88.

Klimpel, R. R. (1982). 'Laboratory studies of the grinding and rheology of coal-water slurries', *Powder Technol.*, **32**, 267–277.

Klimpel, R. R., and Austin, L. G. (1970). 'Determination of selection for breakage functions in the batch grinding equation by non-linear optimisation', *Ind. Eng. Chem. Fundamentals*, **9**, (2), 230–237.

Klimpel, R. R., and Austin, L. G. (1972). 'The mathematical modelling and optimisation of the industrial rotary cutter milling facility', Dechema Monograph 69, 449–473.

Klimpel, R. R., and Austin, L. G. (1977). 'The back calculation of specific rates of breakage and non-normalised breakage distribution parameters from batch grinding data', *Int. J. Miner. Process.*, **4**, 7–32.

Klimpel, R. R., and Austin, L. G. (1982). 'Chemical additives for wet grinding of minerals', *Powder Technol.*, **31**, 239–253.

Klimpel, R. R., and Austin, L. G. (1984). 'The back calculation of specific rates of breakage from continuous mill data', *Powder Technol.*, **38**, 77–91.

Levenspiel, O. (1962). *Chemical Reaction Engineering*, Wiley, New York.

Luckie, P. T., and Austin, L. G. (1972). 'A review introduction to the solution of the grinding equations by digital computation', *Miner. Sci. Eng.*, **4**, (2), 24–51.

Malghan, S. G., and Fuerstenau, D. W. (1976). 'The scale-up of ball mills using population balance models and specific power input', Dechema Monograph 79, 613–630.

Marchand, J. C., Hodouin, D., and Everell, M. D. (1980). 'RTD and mass transport characteristics of large industrial grinding mills', *Proc. 3rd Int. Fed. Automatic Control Symp.*, Montreal, 295–302.

Mardulier, F. J., and Wightman, D. L. (1971). 'Efficient determination of mill retention time', *Rock products*: Part I June, 74–75 and 90–91; Part II July, 78–79 and 108–110; Part III Aug, 60–61 and 86–88.

Meloy, T. P., and Bergstrom, B. H. (1965). 'Matrix simulation of ball mill circuits considering impact and attrition grinding', *Proc. 7th Int. Miner. Process. Congr.*, New York, 1964, 19–31.

Mika, T. S. (1971). 'Population balance models of a continuous grinding mill as a distributed process', Dr. Eng. thesis, University of California, Berkeley.

Mika, T. S. (1976). 'A solution to the distributed parameter model of a continuous grinding mill at steady state', *Chem. Eng. Sci.*, **31**, 257–262.

Mika, T. S., Berlioz, L. M., and Fuerstenau, D. W. (1967). 'An approach to the kinetics of dry batch ball milling', Dechema Monograph 57, 205–240.

Moreira, R. M., de Castro, J. O. H. M., and Gardner, R. P. (1972). 'A combined tracer and back calculation method of determining particulate breakage functions in ball milling', *Powder Technol.*, **6**, 75–83.

Mori, Y. (1973). 'The RTD and size distribution of products flowing through a continuous type ball mill', *Proc. 1st Int. Conf. Particle Technol.*, 217–223.

Mori, Y., Jimbo, G., and Yamazaki, M. (1964). 'Residence time distribution and mixing characteristics of powders in open-circuit ball mills', *Kagaku Kogaku* (English edition), **2**, 173–178.

Reid, K. J. (1965). 'A solution to the batch grinding equation', *Chem. Eng. Sci.*, **20**, 953–963.

Rogers, R. S. C., and Austin, L. G. (1984). 'Residence time distribution in ball mills', Particulate Sci. Technol., **2**, 191–209.

Rogers, R. S. C., Bell, D. G., and Hukki, A. M. (1982a). 'A short-lived radioactive tracer method for the measurement of closed circuit ball mill RTD', *Powder Technol.*, **32**, 245–252.

Rogers, R. S. C., and Gardner, R. P. (1979a). 'A Monte Carlo method for simulating dispersion and transport through horizontal rotating cylinders', *Powder Technol.*, **23**, 159–167.

Rogers, R. S. C., and Gardner, R. P. (1979b). 'Use of a finite stage transport concept for analysing RTDs of continuous processes', *A.I.Ch.E. J.*, **25**, (2), 229–240.

Rogers, R. S. C., Shoji, K., Hukki, A. M., and Linn, R. J. (1982b). 'The effects of liner design on the performance of a continuous wet ball mill', *Proc. 14th Int. Miner. Process. Congr.*, Toronto, 1/5, 1–20.

Rose, H. E., and Sullivan, R. M. (1958). *Ball, Tube and Rod Mills*, Chemical Publishing Co., New York.

Sedlatschek, K., and Bass, L. (1953). 'Contribution to the theory of milling processes', *Powder Metal Bull.*, **6**, 148–153.

Seebach, von H. M. (1969). 'Die wirkung von dämpfen organischer flüssigkeiton bei der zerkleinerung von zementklinker in trommelmühlen', Res. Inst. Cement Ind., Dusseldorf, Paper 35/1969.

Shoji, K., and Austin, L. G. (1974). 'A model for batch rod milling', *Powder Technol.*, **10**, 29–35.

Shoji, K., Austin, L. G., Smaila, F., Brame, K., and Luckie, P. T. (1982). 'Further studies of ball and powder filling effects in ball mills', *Powder Technol.*, **31**, 121–126.

Shoji, K., Lohrasb, S., and Austin, L. G. (1980). 'The variation of breakage parameters with ball and powder filling in dry ball milling', *Powder Technol.*, **25**, 109–114.

Smith, B. A. (1942). Written discussion on 'Factors influencing rotary drier performance', *Trans. Am. Inst. Chem. Eng.*, **38**, 251–257.

Snow, R. H. (1973). 'Grinding mill simulation and scale-up of ball mills', *Proc. 1st Inst. Conf. Particle Technol.*, IITRI, Chicago, 28–38.

Snow, R. H., and Meloy, T. P. (1972). 'Measurement of breakage and grinding rate functions in ball milling by tracer experiments using nuclear activation analysis', Dechema Monograph 69, 535–555.

Stewart, R. S. B., and Restarick, C. J. (1971). 'A comparison of the mechanism of breakage in full scale and laboratory scale grinding mills', *Proc. Australian Inst. Min. Metall.*, **239**, 81–92.

Weller, K. R. (1980). 'Hold-up and residence time characteristics of full scale grinding circuits', *Proc. 3rd Int. Fed. Auto. Control Symp.*, Montreal, 303–309.

Yang, D. C., Mempel, G., and Fuerstenau, D. W. (1967). 'A laboratory mill for batch grinding experimentation', *Trans. AIME*, **238**, 273–275.

CHAPTER 9*

Modelling of tumbling mill batch processes

NOTATION FOR CHAPTER 9

a	Exponent in expressions for breakage rate function and parameter
a_1, a_2, a_3	Exponents in expression for breakage parameter

*This chapter was written by P. C. Kapur, Department of Metallurgical Engineering, Indian Institute of Technology, Kanpur-208016, India.

A	Constant in expressions for breakage rate function and parameter
b	Exponent in expression for breakage distribution function
$b_{j,j}$	Inner breakage parameter
$b(x, y)$	Breakage distribution function
$b_{i,j}$, $b^*_{i,j}$	Breakage distribution parameters
$B_{i,j}$	Cumulative breakage distribution parameter
$B(x, y)$	Cumulative breakage distribution function
$\mathbf{B}$	$n \times n$ lower triangular matrix of $b_{i,j}$
d_j	Coefficients in expression for breakage rate parameter
$\bar{E}$	Grinding energy
$\boldsymbol{E}(t)$	$n \times 1$ vector of exponential terms
f	Function of $t\Omega(w)$
$F(x, t)$	Distribution function in particle size x at time t
$F_i(t)$	Equal to $F(x_i, t)$
$F_j(n)$	Undersize particles produced in nth cycle
$g(x)$	Proportional to reduced selekage function
$G(x, y)$	Bass function
G_i	Lumped parameter in approximate solution to Bass equation
$h_{i,j}$	Coefficient in Reid solution to Bass equation
H_i	Lumped parameter in approximate solution to Bass equation
$\mathbf{H}$	$n \times n$ lower triangular matrix of $h_{i,j}$
$\mathbf{I}$	Identity matrix
k	Cycles of grinding
K_0, K_1, K_2 etc	Constants
L'	Index of particle fineness
$L(x, y)$	Selekage function or cumulative Bass function
$M(x, t)$	Mass or mass fraction of particles of size x at grinding time t
$M_i(t)$	Mass or mass fraction of particles in ith size interval at time t
$\bar{M}_i(t)$	Uniform distribution of particles in ith size interval at time t
$\boldsymbol{M}(t)$	$n \times 1$ vector of particle distribution in group frequency form
n'	Exponent in Rosin–Rammler distribution
$n_{i,j}$	Elements of $\mathbf{N}$ matrix in equation (9.100)
$\mathbf{N}$	$n \times n$ matrix of $n_{i,j}$
$o^*_{i,j}$	Elements of $\mathbf{O}^*$ matrix
$\mathbf{O}^*$	$n \times n$ lower triangular mill matrix
p	Net power draft
q	Similarity variable

$Q(z)$	Gaussian distribution function
r	Fraction of feed retained on the mesh-of-grind
R	Sieve size ratio
$R(x, t)$	Cumulative distribution larger than particle size x at time t
$R_j(n)$	Amount retained on jth mesh and recycled after nth cycle
$R_i(t)$	Cumulative retained on ith mesh
s	Specific surface area of particles
$S(x)$	Breakage rate function
S_i, S_i^*	Breakage rate parameters
$\mathbf{S}$	$n \times n$ diagonal matrix of S_i
t	grinding time
T	Function of time
u_1	First moment of particle size distribution
U_n	Grinding rate in Bond test in nth cycle
v_1	Expectation of $\Phi(x)$
$v_{i,j}$	Elements of $\mathbf{V}$ matrix in equation (9.101)
$\mathbf{V}$	$n \times n$ diagonal matrix of $v_{i,j}$
w	Operational or mill design variable
W	Hold-up mass of particles in mill
W(n)	Feed mass for nth cycle of bond test
x	Particle size
x_i	Mesh opening of ith sieve
x_{50}	Median particle size
$x_{63.2}$	63.2% passing size
x_{80}	80% passing size
x_{100}	100% passing size
$\bar{x}_i$	Mean size for ith size interval
X, X_1	Factors in expression for breakage parameter
z	Normalized particle size in $Q(z)$
Z	Similarity function
$\bar{Z}$, $\bar{Z}_1$ etc.	Cumulative similarity functions
α	Exponent in Gaudin–Schuhmann distribution
δ_1, δ_2 etc.	Parameters of size distributions Ψ, $\bar{\Psi}$
ε	Distribution factor in expression for $B(x, y)$
θ	Solution to discrete size batch grinding equation
μ	Parameter in $Q(z)$ distribution
σ	Parameter in $Q(z)$ distribution
τ	Scaled time
$\lambda_{i,j}$	Lumped parameter in equation (9.140)
ϕ	Function defined in equation (9.134)
Φ	Function in pseudo-similarity solution
Ψ	Density function in particle size
$\bar{\Psi}$	Cumulative distribution in particle size
Ω	Function of w

9.1 INTRODUCTION

9.1.1 Setting the stage

The current state of the art in mathematical modelling and simulation of the size reduction processes is based essentially on two mechanistic–phenomenological concepts, the selection function or the breakage rate function and the breakage distribution function (Epstein, 1947). These two comminution functions are basic to a realistic model representation and a detailed knowledge of their structure is ordinarily essential for process simulation.

At this point it is expedient to invoke the following two stipulations, otherwise it is simply not possible to make any meaningful progress.

Firstly, the population of particles in the comminution environment is always sufficiently large over the entire size range of interest, and so it is a valid procedure to define appropriate statistical averages of the comminution functions (Bass, 1954; Filippov, 1961). Consequently, even though the crushing and grinding events associated with individual particles are probabilistic in nature (Auer, 1981), it is admissible to operate within an equivalent deterministic framework which is certainly simpler and more pliable as far as modelling and process simulation is concerned.

Secondly, while it is conceded that non-linear models or time-dependent linear models are more realistic and versatile (Austin *et al*, 1982; Kinneberg and Herbst, 1984), as indicated already in Chapter 8, these models are significantly more complex in implementation and their applications have not yet been well developed. Hence this chapter is devoted to linear, time-invariant models whose potential for simulation has been amply demonstrated, as testified by numerous published studies.

9.1.2 Breakage rate function and disappearance kinetics

The specific rate of breakage $S(x)$ is defined as the fraction of particles of point size x broken per unit time. Here the focus is on the rate at which particles of a specified size are broken, ignoring for the time being the rate at which they are produced by the breakage of larger particles. Starting from this definition, the expression for so-called first-order disappearance (or decay) kinetics can be written as:

$$\begin{matrix}\text{Disappearance rate of} \\ \text{particles of size } x\end{matrix} = -S(x)M(x,t) \qquad (9.1)$$

where $M(x, t)\,dx$ is the mass or mass fraction, as the case may be, of particles in size range x to $x + dx$ at time t. Clearly $M(x, t)$ is the absolute or relative (normalized) density function in particle size, and the corresponding absolute or relative cumulative finer and cumulative retained distributions are, respectively:

$$F(x,t) = \int_0^x M(y,t)\,dy \qquad (9.2)$$

$$R(x, t) = \int_x^\infty M(y, t)\,\mathrm{d}y \tag{9.3}$$

It is readily verified that the density function and the two distribution functions are related by:

$$M(x, t) = -\frac{\mathrm{d}R(x, t)}{\mathrm{d}x} = \frac{\mathrm{d}F(x, t)}{\mathrm{d}x} \tag{9.4}$$

9.1.3 Breakage distribution function and appearance kinetics

The breakage distribution function $b(x, y)\,\mathrm{d}x$ is the mass fraction of progeny particles in size range x to $x + \mathrm{d}x$ formed directly, without undergoing any rebreakage, when unit mass of particles of point size y is broken ($y \geqslant x$). The cumulative breakage distribution function $B(x, y)$ is the mass fraction of progeny particles equal to or finer than size x when unit mass of particles of size y is broken. These two forms of the breakage function are related by:

$$B(x, y) = \int_0^x b(w, y)\,\mathrm{d}w \tag{9.5}$$

or

$$b(x, y) = \frac{\mathrm{d}B(x, y)}{\mathrm{d}x} \tag{9.6}$$

Since mass is conserved, it also follows that $b(x, y)$ is a normalised density function in x:

$$\int_0^y b(x, y)\,\mathrm{d}x = B(y, y)$$

$$= 1 \tag{9.7}$$

A class of breakage distribution functions of special interest is the self-similar function whose representation in cumulative form is:

$$B(x, y) = B(x/y) \tag{9.8}$$

In other words, the value that the self-similar breakage function takes depends on the ratio of progeny to parent particle sizes only. The density function form of the self-similar breakage distribution function is:

$$b(x, y) = \frac{\mathrm{d}B(x/y)}{\mathrm{d}x}$$

$$= \frac{1}{y}\,b(x/y) \tag{9.9}$$

An example of the self-similar cumulative breakage distribution function is:

$$B(x/y) = \left(\frac{x}{y}\right)^b, \; b > 0 \tag{9.10}$$

From the definitions of breakage rate and breakage distribution on functions, the expression for the rate of appearance of progeny particles of point size x which are formed by breakage of parent particles of point size y is:

$$\begin{matrix}\text{Appearance rate of}\\ \text{particles of size } x\\ \text{by breakage of}\\ \text{particles of size } y\end{matrix} = S(y)b(x, y)M(y, t) \tag{9.11}$$

The overall appearance rate is obtained by integrating the above term over all parent particles; hence:

$$\begin{matrix}\text{Appearance rate of}\\ \text{particles of size } x\end{matrix} = \int_x^{\infty} S(y)b(x, y)M(y, t)\,\mathrm{d}y \tag{9.12}$$

The breakage rate and breakage distribution product term will be called the Bass function $G(x, y)$; i.e.:

$$G(x, y) = S(y)b(x, y) \tag{9.13}$$

in honour of Bass who first introduced it in his now classic investigation of grinding kinetics (Bass, 1954). It is also convenient to introduce the cumulative Bass function or selekage function (Gupta and Kapur, 1976):

$$L(x, y) = S(y)B(x, y) \tag{9.14}$$

Clearly the two functions are related by:

$$\frac{\mathrm{d}L(x, y)}{\mathrm{d}x} = G(x, y) \tag{9.15}$$

Whereas the Bass function is the specific rate for 'feed forward' of material from parent particle size y to progeny size x, the selekage function is the specific rate for 'feed forward' from parent size y to all particles less than or equal to size x. It is shown in §9.3 that the latter function plays a central role in the derivation of discrete size models of grinding.

9.2 TIME- AND SIZE-CONTINUOUS PARTICLE MASS-BALANCE MODELS

9.2.1 Integro-differential equation of batch grinding

The linear, time-invariant, time- and size-continuous integro-differential equation of grinding in a batch system is readily formulated by taking a mass balance on particles of a point size. This entails combining the expressions for disappearance and appearance kinetics in equations (9.1) and (9.12) respectively. The resulting grinding kinetics equation is (Bass, 1954; Filippov, 1961;

Gaudin and Meloy, 1962; Gardner and Austin, 1962):

$$\frac{\partial M(x,t)}{\partial t} = -S(x)M(x,t) + \int_x^{\infty} S(y)b(x,y)M(y,t)\,\mathrm{d}y \qquad (9.16)$$

To summarize, the more pertinent assumptions which are inherent in the formulation of this grinding model are:

(1) Breakage of a particle is independent of its past history, i.e. the process is memoryless.
(2) Breakage of particles is independent of the overall particulate environment, i.e. particle-to-particle interaction is nil.
(3) The breakage rate function $S(x)$ depends on the particle size x only, and breakage distribution function $B(x, y)$ depends on size y and x only.
(4) Size enlargement of particles by aggregation is absent.
(5) The breakage process is independent of the position of particles in the comminution environment or the particles are fully mixed.

In some modelling and simulation situations it is more advantageous to deal with time evolution of the cumulative distributions directly. The analogous integro-differential equations of grinding in the cumulative retained and cumulative passing distributions can be derived by combining equations (9.16) with (9.2)–(9.6). The resulting expressions are:

$$\frac{\partial R(x,t)}{\partial t} = -S(x)R(x,t) + \int_x^{\infty} \frac{\partial}{\partial y}\,[S(y)B(x,y)]\,R(y,t)\,\mathrm{d}y \qquad (9.17)$$

or

$$\frac{\partial R(x,t)}{\partial t} = \int_x^{\infty} S(y)B(x,y)\,\frac{\partial R(y,t)}{\partial y}\,\mathrm{d}y \qquad (9.18)$$

and

$$\frac{\partial F(x,t)}{\partial t} = \int_x^{\infty} S(y)B(x,y)\,\frac{\partial F(y,t)}{\partial y}\,\mathrm{d}y \qquad (9.19)$$

The general analytical solution to the batch grinding equation (9.16) comprises a convergent series of iterated kernels (Bass, 1954). This solution is complicated and computationally difficult, and it has not found any practical application. Many authors have attempted simpler analytical solutions to the grinding equation for specialized functional forms of breakage rate and distribution functions (Gaudin and Meloy, 1962; Harris, 1968a; Austin *et al.* 1972; Gupta and Kapur 1976). Suppose there exist certain restrictive interrelationships between the two breakage functions such that the selekage function simplifies to:

$$\begin{aligned} L(x,y) &= S(y)B(x,y) \\ &= Ag(x) \end{aligned} \qquad (9.20)$$

where A is a constant and the function g depends on size x only. This reduced selekage function (RSF) possesses the following properties:

$$S(x) = Ag(x) \tag{9.21}$$

$$b(x, y) = \frac{g(x)}{g(y)} \tag{9.22}$$

$$g(0) = 0 \tag{9.23}$$

Moreover, the RSF is a monotonically non-decreasing function in x. Substitution of the RSF into equation (9.17) results in the following batch grinding equation:

$$\frac{\partial R(x,t)}{\partial t} = -Ag(x)R(x,t) \tag{9.24}$$

the solution of which is simply:

$$R(x,t) = R(x,0)\exp[-Ag(x)t] \tag{9.25}$$

Substitution in equation (9.4) gives the density function:

$$M(x,t) = \exp[-Ag(x)t]\left[M(x,0) + At\frac{dg(x)}{dx}R(x,0)\right] \tag{9.26}$$

Further, on specializing the RSF as a power function we have

$$Ag(x) = Ax^a, \qquad a > 0 \tag{9.27}$$

The solution to equation (9.24) becomes:

$$R(x,t) = R(x,0)\exp[-Ax^a t] \tag{9.28}$$

This is the well-known equation of Rosin–Rammler distribution for comminuted particles. Over the years, this distribution has been found reasonably satisfactory for many grinding systems (Rosin and Rammler 1933; Roberts, 1950; Fuerstenau, 1959; Schönert, 1969; Furuya *et al* 1971; Gupta and Kapur, 1974a; Finch and Ramirez-Castro, 1981). To recapitulate, the Rosin–Rammler distribution is generated when:

$$S(x) = Ax^a \tag{9.29}$$

and

$$B(x,y) = \left(\frac{x}{y}\right)^a \tag{9.30}$$

There seems to be no satisfactory explanation why so many comminution systems should seemingly exhibit such severely restricted interrelationships between the comminution functions, which are by definition independent of each other. Several authors have published analytical solutions to the grinding equation under somewhat less restrictive conditions; namely, $S(x)$ is as in equation (9.29) and:

$$B(x,y) = \left(\frac{x}{y}\right)^b, \qquad b \neq a \tag{9.31}$$

Two of these solutions (Filippov, 1961; King, 1972) incorporate hypergeometric functions and series which are computationally inconvenient. Moreover, in both cases it was necessary to assume mono-sized feed particles as the initial condition. Consequently, these solutions are of theoretical interest only. In this context, an approximate analytical solution in terms of an incomplete gamma function has greater merit (Nakajima and Tanaka, 1973).

9.2.2 Similarity solution

9.2.2.1 Preamble

A more general similarity solution to the integro-differential equation of batch grinding has been presented by Kapur (1970b, 1972). Let the breakage rate function be as in equation (9.29), and the breakage distribution function be self-similar as in equation (9.8). Substitution of these functions into equation (9.16) results in:

$$\frac{\partial M(x,t)}{\partial t} = -Ax^a M(x,t) + \int_x^\infty Ay^a \frac{\partial B(x/y)}{\partial x} M(y,t)\, \mathrm{d}y \tag{9.32}$$

A similarity solution to this equation is:

$$M(x,t) = \frac{1}{u_1(t)} Z(q) \tag{9.33}$$

where the similarity function Z depends only on a dimensionless particle size q which is:

$$q = \frac{x}{u_1(t)} \tag{9.34}$$

and the scale factor or the characteristic particle length $u_1(t)$ is the mean of the normalized particle size distribution:

$$u_1(t) = \int_0^\infty M(x,t)x\, \mathrm{d}x \tag{9.35}$$

and it varies with grinding time in the following manner:

$$\frac{\mathrm{d}u_1(t)}{\mathrm{d}t} = -AK_0[u_1(t)]^{a+1} \tag{9.36}$$

where K_0 is a dimensionless constant. In particular, when the breakage distribution function is given by equation (9.31), an explicit expression for the self-similar particle size distribution is (Kapur, 1972):

$$M(x,t) = \frac{K_1}{u_1(t)} \left[\frac{x}{u_1(t)}\right]^{b-1} \exp\left\{-\frac{1}{aK_0}\left[\frac{x}{u_1(t)}\right]^a\right\} \tag{9.37}$$

where K_1 is a normalizing factor.

It turns out that by means of standard transformation techniques it is possible to extract a pseudo-similarity solution for a more general class of the comminution functions. Let:

$$S(x) = A\,[\Phi(x)]^{a} \tag{9.38}$$

$$B(x, y) = \left[\frac{\Phi(x)}{\Phi(y)}\right]^{b} \tag{9.39}$$

where Φ is any monotonically non-decreasing function in particle size and $\Phi(0) = 0$. Now a solution to the grinding equation is (Gupta and Kapur, 1975):

$$M(x, t) = \frac{K_I}{u_1(t)}\left[\frac{\Phi(x)}{u_1(t)}\right]^{b-1} \times \exp\left\{-\frac{1}{aK_0}\left[\frac{\Phi(x)}{u_1(t)}\right]^{a}\right\}\frac{\mathrm{d}\Phi(x)}{\mathrm{d}x} \tag{9.40}$$

where

$$u_1(t) = \int_0^{\infty} M(x, t)\Phi(x)\,\mathrm{d}x \tag{9.41}$$

The density function for comminuted particles in equation (9.40) is the most general closed-form analytical expression that can be derived at present from a formal particle mass-balance model of grinding kinetics.

The similarity solutions in equations (9.33), (9.36) and (9.37) are of considerable importance. For the first time it has become possible to obtain a valuable insight into, as well as a rational basis for, a unifying framework for a number of size reduction phenomena which were hitherto studied, more or less, empirically and in isolation from one another, e.g. size spectra of cominuted particles, their evolution in time and the energy–size reduction relationships. Some of the more important implications of the similarity solution are discussed below.

9.2.2.2 Self-preserving size spectra

Inspection of equation (9.33) shows that the particle size distributions are self-preserving or self-similar. This means that when $u_1(t)M\,(x, t)$ is plotted as a function of $x/u_1(t)$ the distributions generated at different grinding times should collapse on a single curve which is characterized by the similarity function Z. It will be noted that the size spectra in the similarity solution are independent of initial conditions, i.e. influence of the feed size distribution is smoothed out after a sufficiently long grinding time. For this reason the similarity solution is an asymptotic result which is not valid for the initial grinding periods.

Equation (9.33) is not convenient for the analysis of empirical data. Firstly, since the grinding data are invariably acquired grouped into discrete size intervals, the density function $M(x, t)$ is not directly accessible. Secondly, computation of the first moment $u_1(t)$ of comminuted particles is likely to give an uncertain estimate because of the well-known problems encountered in size measurements in the sub-sieve range. These difficulties can be overcome by employing the cumulative distribution form of the similarity solution (Kapur, 1972):

$$F(x, t) = \bar{Z}(q) = \bar{Z}\left(\frac{x}{u_1(t)}\right) \tag{9.42}$$

where $\bar{Z}$ is a cumulative similarity function in q. Moreover, it is easily shown that in a similarity distribution the first moment (u_1), median (x_{50}) or any other percentile size (e.g. 80% passing size x_{80}), and the inverse of the specific surface area (s^{-1}) are all proportional to one another (Kapur, 1971b); i.e.:

$$u_1(t) \propto x_{50}(t) \propto x_{80}(t) \propto s^{-1}(t) \tag{9.43}$$

with the result that the similarity solution to the grinding equation can be written in a number of equivalent forms, e.g.:

$$F(x, t) = \bar{Z}_1(x/x_{50}(t)) \tag{9.44a}$$

$$F(x, t) = \bar{Z}_2(x/x_{80}(t)) \tag{9.44b}$$

$$F(x, t) = \bar{Z}_3(xs(t)) \tag{9.44c}$$

where $\bar{Z}_1$, $\bar{Z}_2$ and $\bar{Z}_3$ are the cumulative similarity functions appropriate to the scale factor employed.

Figure 9.1 shows experimental data for quartz ground in a ball mill, when $F(x, t)$ is plotted as a function of dimensionless size x/x_{80}. The distributions tend to be self-similar or self-preserving and collapse on or in the vicinity of a curve given by the function $\bar{Z}_2$. Examination of many sets of grinding data from diverse sources has shown that an overwhelming majority of the distributions exhibit the self-preserving behaviour. Moreover, the similarity spectra are attained quite rapidly in relatively short grinding times, unless the starting feed is either exceptionally dispersed in size or quite far removed in distribution shape from that of the eventual similarity curve.

The self-preserving character of comminuted particles has a number of interesting implications, as follows.

(i) A knowledge of the similarity distribution from either the empirical collapsed curve in Figure 9.1 or the theoretical function in equation (9.37), along with the variation of the scale factor with time, is all that is needed to map the evolution of the particle size spectra in the course of grinding.

(ii) If a set of size distributions generated at different grinding times are self-preserving, it is then reasonable to conclude that either there has been no change in the mode and the mechanism of size reduction over the range of particle size and grinding time under scrutiny; or at the most, the change can be explicitly incorporated into the time-dependence of the breakage rate function in the following manner (Kapur, 1971a, 1972):

$$S(x, t) = S(x)T(t) \tag{9.45}$$

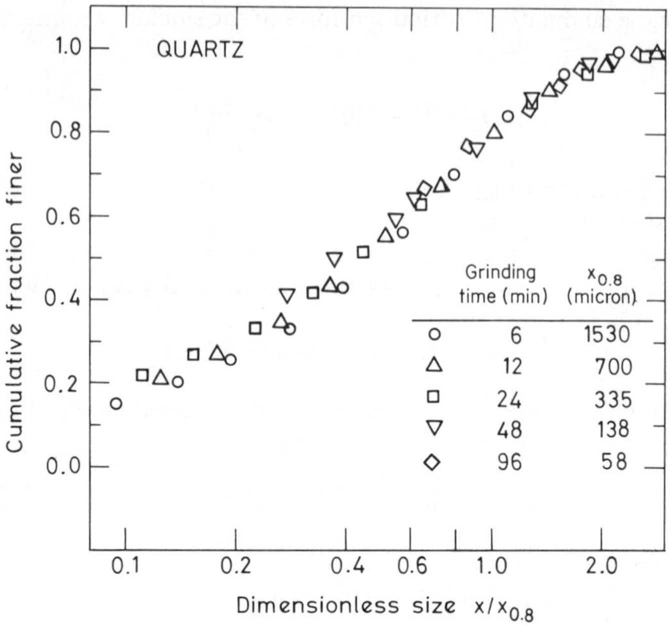

Figure 9.1. Self-similar size distributions of quartz ground in a ball mill (Kapur, 1972). *Reproduced by permission of Pergamon Press Ltd.*

where $T(t)$ is a function of time only, and equation (9.32) is modified to:

$$\frac{\partial M(x,t)}{T(t)\partial t} = -Ax^a M(x,t) + \int_x^\infty Ay^a \frac{\partial B(x/y)}{\partial x} M(y,t)\,\mathrm{d}y \tag{9.46}$$

the solution of which is again the same as in equation (9.33), with the difference that now the scale factor varies as:

$$\frac{\mathrm{d}u_1(\tau)}{\mathrm{d}\tau} = -AK_0[u_1(\tau)]^{a+1} \tag{9.47}$$

where the scaled time τ is given by:

$$\mathrm{d}\tau = T(t)\,\mathrm{d}t \tag{9.48}$$

(iii) If with any changes in design parameters and operational conditions of the grinding mill (e.g. speed and size of mill, density, size and weight of grinding media, charge load etc.), the self-similar character of the size distributions remains invariant, it may then be inferred that the breakage distribution function is not altered, and the breakage rate function is influenced by the imposed change in the following manner only:

$$S(x,w) = S(x)\Omega(w) \tag{9.49}$$

where w is the design or operational variable and Ω is a function of that variable. It can be shown that the scale factor is now a function f of the product term $t\Omega(w)$, or:

$$u_1(t)[\text{or } x_{50}(t) \text{ or } x_{80}(t) \text{ etc.}] = f[t\Omega(w)] \tag{9.50}$$

This result is of some practical importance, because the effect of design or operational variables can be readily quantified by simply examining the variation of the scale factor with time at different levels of the variable under scrutiny. In this manner it is possible to determine the nature of the function Ω and derive useful empirical correlations for the breakage rate function (Gupta and Kapur, 1986).

9.2.2.3 Size distribution equations

Many comminuted particle size distribution equations in the literature were advanced on the strength of practical experience gained from curve fitting to experimental data. It has been pointed out that almost all these equations are specialized cases of the following general equation (Dallavalle, 1948; Harris, 1968b):

$$M(x,t) = \frac{K_1}{L'}\left(\frac{x}{L'}\right)^{b-1} \exp\left[-K_2\left(\frac{x}{L'}\right)^{a}\right] \tag{9.51}$$

where K_1 and K_2 are dimensionless constants and L' is a fineness index or characteristic particle length. Clearly, this expression is equivalent to the similarity distribution in equation (9.37). However, since the latter result was derived from a particle mass balance model of grinding kinetics, it is now possible to interpret the constants and parameters in equation (9.51) in terms of physically identifiable comminution functions. In particular, variation of the fineness index L' with grinding time—which in fact shifts the size distributions toward increased fineness and which could not be explained satisfactorily in the earlier empirical models—is given explicitly in equation (9.36). It is also interesting to note that the two best-known comminuted particle size distribution equations given below are self-preserving, as well as independent of feed size distribution:

Gaudin–Schuhmann:

$$M(x,t) = \frac{\alpha x^{\alpha-1}}{[x_{100}(t)]^{\alpha}} \tag{9.52}$$

$$u_1(t) = \left(\frac{\alpha}{\alpha+1}\right) x_{100}(t) \tag{9.53}$$

Rosin–Rammler:

$$M(x,t) = \frac{n' x^{n'-1}}{[x_{63.2}(t)]^{n'}} \exp\left[\frac{-x}{x_{63.2}(t)}\right] \tag{9.54}$$

$$u_1(t) = \left(\frac{1}{n'}\right)! \, x_{63.2}(t) \tag{9.55}$$

where α and n' are the distribution moduli, and the scale factor or fineness indices x_{100} and $x_{63.2}$ are 100% and 63.2% passing sizes.

9.2.2.4 Energy–size reduction relationships

A number of so-called classical laws of grinding, already discussed in §6.5, purport to describe the functional relationship between a single characteristic parameter of the size distribution of ground particles and the input of grinding energy. In general, these laws were arrived at by empirical and even arbitrary and tenuous arguments, and have remained shrouded in ambiguity and controversy (Austin, 1973). It turns out that, starting from the similarity solution, it is possible to accomplish a unified and integrated derivation of the energy–size reduction laws. This exercise not only leads to a direct synthesis of the classical and modern viewpoints of size reduction phenomena, it also provides a measure of the limitations that are inherent in these laws.

If the total energy input to the mill is proportional to the grinding time, then the classical grinding laws are interchangeable in grinding time and mill energy. Kapur and Fuerstenau (1986) have recently shown that a master energy law, similar to the empirical law of Walker *et al* (1937), can be derived in a more rigorous manner by combining equation (9.36) with an interesting correlation between power to the mill and breakage rate function (Herbst and Fuerstenau, 1973; Malghan and Fuerstenau, 1976). In essence the correlation requires that the breakage rate function is proportional to power p such that in equation (9.36):

$$A \propto p \tag{9.56}$$

Since the specific grinding energy $\bar{E}$ is related to power by:

$$\mathrm{d}\bar{E} = p\,\mathrm{d}t \tag{9.57}$$

it then follows that equation (9.36) can be rewritten as:

$$\frac{\mathrm{d}u_1}{\mathrm{d}\bar{E}} = -K_0'\,[u_1]^{a+1} \tag{9.58}$$

All the well-known energy–size reduction laws existing in the literature are generated by interpreting u_1 in different but equivalent ways, and by specializing, more or less arbitrarily, either the exponent a in equation (9.58) or the similarity function Z in equation (9.33), or both (Kapur, 1971b).

(a) Kick's law (Beke, 1964) When $a = 0$, equation (9.58) expresses Kick's law, namely, grinding energy expended is proportional to the logarithm of the reduction ratio. In view of the interrelationships in equation (9.43), this law can also be formulated in the reduction ratio of specific surface area or any other percentile. Kick's law implies that specific grinding rate $S(x)$ in equation (9.29) is independent of particle size x. From actual measurements it is now known that the breakage rate function for coarse particles, especially when the particle-to-grinding media size ratio is relatively large, may turn out to be

approximately invariant over a limited size range. It is therefore not surprising that in some systems this law is apparently valid for a short grinding period when most of the solids are confined within the coarse size range.

(b) Rittinger's law (Beke, 1964) In equation (9.43) the specific surface area of particles s is inversely proportional to the mean size u_1. Substitution of u_1 by s into equation (9.58) results in:

$$\frac{\mathrm{d}s}{\mathrm{d}\bar{E}} = K_3 s^{1-a} \tag{9.59}$$

where K_3 is a constant. When $a = 1$, this equation becomes a mathematical statement of Rittinger's law, namely an increment in surface area is proportional to an increment in grinding energy. This law requires that the grinding rate function should be proportional to particle size x, which could be the case for a class of fairly hard materials over a limited size range. See §6.5.3.

(c) Holmes' law (Holmes, 1957) Since in equation (9.43), x_{80} is proportional to u_1, integration of equation (9.58) and substitution of x_{80} for u_1 leads to Holmes' law:

$$\bar{E} = K_4 [x_{80}^{-a}(p) - x_{80}^{-a}(f)] \tag{9.60}$$

where K_4 is a constant and $x_{80}(p)$ and $x_{80}(f)$ are 80th percentiles in product and feed, respectively. Strictly speaking, the feed size distribution must also lie in the self-similar regime, otherwise the integration of equation (9.58) is not valid.

(d) Bond's law (Bond, 1952) Bond's law of grinding results immediately if one specializes $a = 0.5$ in equation (9.60). Consequently, this law requires that $S(x)$ is proportional to the square-root of particle size x, which is a much more rational and meaningful interpretation than the one originally proposed by Bond and others. However, it is obvious that this square-root relationship cannot possibly be applicable to all grinding situations. See §6.5.3.

(e) Charles' law (Charles, 1957) Instead of specializing the exponent a, as in the laws of Kick, Rittinger and Bond, one may specialize the similarity distribution function Z in equation (9.33). Assume that the Gaudin–Schuhmann distribution is a sufficiently close approximation to the function Z, at least to the extent that both have nearly identical first moments. Since in equation (9.53) $u_1 \propto x_{100}$, substitution of the latter into equation (9.58), followed by integration, gives for product (p) and feed (f):

$$\bar{E} = K_5 [x_{100}^{-a}(p) - x_{100}^{-a}(f)] \tag{9.61}$$

For large grinding times, $x_{100}(p) \ll x_{100}(f)$, and this equation simplifies to:

$$\bar{E} = K_5 x_{100}^{-a}(p) \tag{9.62}$$

which is the Charles energy–size reduction law. An energy–size reduction expression appropriate to the Rosin–Rammler equation can be derived in a completely analogous manner by substitution of equation (9.55) into (9.58):

$$\bar{E} = K_6 x_{63.2}^{-a}(p) \tag{9.63}$$

where K_6 is a constant.

Some of the more important implications of the classical energy–size reduction grinding laws in the light of the unified derivation presented above are as follows.

(i) The classical laws, all of which are formulated in terms of a single time-varying parameter of the ground particles ensemble, are meaningful if, and only if, the size distribution can be represented as:

$$M(x,t) = \Psi\{x\colon \delta_1, \delta_2, \ldots, \delta_i(t), \ldots, \delta_n\} \tag{9.64}$$

or

$$F(x,t) = \bar{\Psi}\{x\colon \delta_1, \delta_2, \ldots, \delta_i(t), \ldots, \delta_n\} \tag{9.65}$$

where δ represents parameters of the distribution of comminuted particles of which only one parameter is time dependent, and the distribution function Ψ (or $\bar{\Psi}$) does not change its nature with time. Clearly then, in general, the energy–size reduction laws are invalid for short grinding time periods since the chosen feed size distribution would not ordinarily correspond to the function Ψ.

(ii) The similarity solution in equation (9.33) or (9.42) not only guarantees the invariance of the function Ψ, but also ensures interchangeability between the different parameters u_1, s, x_{50}, x_{80} etc. Since a very large body of grinding data is demonstrably in agreement with the similarity solution, it is obvious that the empirical energy–size reduction law of Walker *et al.* (1937) is identical to the relationship in equation (9.58), and the 'particle size' in this law can be interpreted as any one of the distribution parameters mentioned above. Thus, both Rittinger's and Bond's laws are similar except for the value of the exponent a, which is arbitrarily specialized to 1.0 and 0.5 respectively.

(iii) Only Holmes' law is free of any arbitrary and *ad hoc* assumption or specialization which characterize the laws of Kick, Rittinger, Bond, Charles and others. It is, in fact, an alternate form of the similarity solution, and so its validity is as general as the starting grinding model in equation (9.32).

9.3 DISCRETE-SIZE, TIME-CONTINUOUS PARTICLE MASS-BALANCE MODELS

9.3.1 Discrete-size models

Solution to the general integro-differential equation of grinding, as well as estimation of the comminution functions therein, present formidable tasks.

The discrete-size models of grinding kinetics, on the other hand, are much more amenable to model identification, parameter estimation and computation. It is also pertinent to note that the experimental data are invariably acquired in group density form distributed over discrete-size intervals, and therefore it is more logical to employ the discrete-size models which can 'interface' directly with the data. Many authors have discussed the discrete-size mathematical models of batch grinding and their derivations under various implicit or explicit assumptions (Bass, 1954; Reid, 1965; Mika *et al.*, 1967; Freeh *et al.*, 1967, Austin, 1971/72; Olsen, 1972; Whiten, 1974). A mathematically rigorous derivation of the discrete-size models, under clearly stated stipulations, is presented here.

9.3.2 Exact relationships between size-continuous comminution functions and discrete-size comminution parameters

The precursor expression for all the common discrete-size, time-continuous particle mass-balance models of grinding kinetics is the following master set of first-order coupled differential equations:

$$\frac{\mathrm{d}M_i(t)}{\mathrm{d}t} = -S_i(t)M_i(t) + \sum_{j=1}^{i-1} S_j(t)b_{i,j}(t)M_j(t) \quad \text{for} \quad i = 1, 2, 3, \ldots \tag{9.66}$$

or

$$\frac{\mathrm{d}M_i(t)}{\mathrm{d}t} = -S_i^*(t)M_i(t) + \sum_{j=1}^{i} S_j^*(t)b_{i,j}^*(t)M_j(t) \quad \text{for} \quad i = 1, 2, 3, \ldots \tag{9.67}$$

where S_i(or S_i^*) is the breakage rate parameter and $b_{i,j}$(or $b_{i,j}^*$) is the breakage distribution parameter. Clearly S_i is the discrete-size analogue of the breakage rate function and $b_{i,j}$ is the analogue of the breakage distribution function. $M_i(t)$ is the mass or mass fraction of particulate solids in the ith size interval at grinding time t. According to the established convention, the discrete-size interval is bounded by mesh openings of two adjacent sieves in a standard series in which the sieve ratio R is usually $2^{1/2}$ or $2^{1/4}$. For the ith size interval (or size index) the mesh openings of upper and lower sieves are denoted by x_{i-1} and x_i, respectively, and therefore:

$$R = \frac{x_{i-1}}{x_i} \text{ for } i = 1, 2, 3, \ldots \tag{9.68}$$

Moreover:

$$M_i(t) = \int_{x_i}^{x_{i-1}} M(y, t)\, \mathrm{d}y \tag{9.69}$$

and $F_i(t)$, the cumulative amount passing through the mesh opening x_i (i.e.

the ith sieve) is:

$$\begin{aligned} F_i(t) &= F(x_i, t) \\ &= \int_0^{x_i} M(y, t)\,\mathrm{d}y \\ &= \sum_{j > i+1} M_j(t) \end{aligned} \tag{9.70}$$

In equation (9.66) the summation runs from $j = 1$ to $i - 1$, while in equation (9.67) it spans $j = 1$ to i. The two grinding models are in fact completely equivalent, since it can be shown by inspection only that:

$$S_i(t) = S_i^*(t)[1 - b_{i,i}^*(t)] \tag{9.71}$$

and

$$b_{i,j}(t) = \frac{b_{i,j}^*(t)}{1 - b_{i,j}^*(t)} \tag{9.72}$$

Therefore $S_i^*(t)$ and $b_{i,i}^*(t)$ can always be lumped into one parameter $S_i(t)$ and the product term $S_j^*(t)b_{i,j}^*(t)$ can always be replaced by $S_j(t)b_{i,j}(t)$. However, the first model in equation (9.66) is almost invariably employed in practice.

The objective is to derive the master grinding equation in equation (9.66) starting from the integro-differential equation of grinding in equation (9.16), and thereby obtain a mathematical equivalent discrete-size model, and also unambiguous definitions of the discrete-size grinding parameters in terms of the two fundamental size continuous comminution functions, $S(x)$ and $b(x, y)$ (Austin and Bhatia, 1973; Gupta and Kapur, 1976). Integrating equation (9.16) throughout with respect to particle size over the range x_{i-1} to x_i and substituting equation (9.69) on the left-hand side results in:

$$\frac{\mathrm{d}M_i(t)}{\mathrm{d}t} = -\int_{x_i}^{x_{i-1}} S(x)M(x, t)\,\mathrm{d}x + \int_{x_i}^{x_{i-1}} \mathrm{d}x \int_x^{x_0} S(y)b(x, y)M(y, t)\,\mathrm{d}y \tag{9.73}$$

where the upper size limit is x_0, the mesh opening of the topmost sieve. Substitution of equation (9.6), followed by some manipulation and rearrangement, leads to:

$$\begin{aligned} \frac{\mathrm{d}M_i(t)}{\mathrm{d}t} &= -\int_{x_i}^{x_{i-1}} S(y)B(x_i, y)M(y, t)\,\mathrm{d}y \\ &\quad + \sum_{j=1}^{i-1} \int_{x_j}^{x_{j-1}} S(y)[B(x_{i-1}, y) - B(x_i, y)]M(y, t)\,\mathrm{d}y \end{aligned} \tag{9.74}$$

Since breakage rate and breakage distribution functions occur in product terms only, it is convenient to replace these terms by the selekage function in equation (9.14); hence equation (9.74) becomes:

$$\begin{aligned} \frac{\mathrm{d}M_i(t)}{\mathrm{d}t} &= -\int_{x_i}^{x_{i-1}} L(x_i, y)M(y, t)\,\mathrm{d}y \\ &\quad + \sum_{j=1}^{i-1} \int_{x_j}^{x_{j-1}} [L(x_{i-1}, y) - L(x_i, y)]M(y, t)\,\mathrm{d}y \end{aligned} \tag{9.75}$$

A term-by-term comparison of this equation with the master equation (9.66), reveals that:

$$S_i(t)M_i(t) = \int_{x_i}^{x_{i-1}} L(x_i, y)M(y, t)\,\mathrm{d}y \tag{9.76}$$

and the instantaneous breakage rate parameter is:

$$S_i(t) = \frac{\int_{x_i}^{x_{i-1}} L(x_i, y)M(y, t)\,\mathrm{d}y}{\int_{x_i}^{x_{i-1}} M(y, t)\,\mathrm{d}y} \tag{9.77}$$

Similarly the instantaneous breakage distribution parameter is:

$$b_{i,j}(t) = \frac{\int_{x_j}^{x_{j-1}} [L(x_{i-1}, y) - L(x_i, y)]\,M(y, t)\,\mathrm{d}y}{\int_{x_j}^{x_{j-1}} L(x_j, y)M(y, t)\,\mathrm{d}y} \quad \text{with } i \neq j \tag{9.78}$$

If the cumulative breakage distribution parameter is defined in the following manner:

$$B_{i,j}(t) = \sum_{k \geqslant i+1} b_{k,j}(t) \quad \text{with } i \geqslant j \tag{9.79}$$

substitution of equation (9.78) results in:

$$B_{i,j}(t) = \frac{\int_{x_j}^{x_{j-1}} L(x_i, y)M(y, t)\,\mathrm{d}y}{\int_{x_j}^{x_{j-1}} L(x_j, y)M(y, t)\,\mathrm{d}y} \tag{9.80}$$

The relationships in equations (9.77) and (9.78) are quite important. From inspection of equation (9.77) it is evident that the correct definition of breakage rate parameter $S_i(t)$ is the fractional rate of *exit* of particles by breakage from the *i*th size interval at time t. From equation (9.78) the correct definition of breakage distribution parameter $b_{i,j}(t)$ is the fraction of particles *broken out* of size index j that reports directly to the *i*th size interval at time t. Further, the instantaneous inner breakage parameter $b_{j,j}(t)$ is defined as the fraction of material broken in size index j at time t which simply falls back into the same size interval; and it is given by:

$$b_{j,j}(t) = 1 - \frac{\int_{x_j}^{x_{j-1}} L(x_j, y)M(y, t)\,\mathrm{d}y}{\int_{x_j}^{x_{j-1}} S(y)M(y, t)\,\mathrm{d}y} \tag{9.81}$$

A scrutiny of equations (9.77), (9.78) and (9.81) establishes that strictly speaking: (i) the comminution parameters $S_i(t)$, $b_{i,j}(t)$ and $b_{j,j}(t)$ are not independent of one another; all three are functionals and determined by functions $S(x)$, $B(x, y)$ and $M(y, t)$ jointly; (ii) there is no exclusive one-to-one correspondence between $S(x)$ and S_i or between $b(x, y)$ and $b_{i,j}$ and (iii) even if the grinding system is inherently linear and time-invariant, the discrete-size model representation of the system introduces a non-linearity which is implicitly reflected in the time dependence of the grinding parameters. In conclusion, the master grinding equation in equation (9.66) is an exact equivalent of the integro-differential equation in equation (9.16), provided the time-dependent discrete-size grinding parameters are given by equations (9.77) and

(9.78). For practical reasons, however, there exists a powerful motivation for a simpler and more tractable time-invariant parameter model which is derived next.

9.3.3 Bass grinding model

It is assumed that one is justified in approximating the particle size distribution within a relatively narrow discrete-size interval by a uniform distribution $\bar{M}_i(t)$; that is:

$$\bar{M}_i(t) = \frac{M_i(t)}{x_{i-1} - x_i} \quad \text{for } x_i \leqslant x < x_{i-1} \tag{9.82}$$

Substitution of this expression into equations (9.77) and (9.78) results in the following time-invariant discrete-size grinding parameters (Gupta and Kapur, 1976):

$$S_i = \frac{\int_{x_i}^{x_{i-1}} L(x_i, y)\,\mathrm{d}y}{x_{i-1} - x_i} \tag{9.83}$$

and

$$b_{i,j} = \frac{\int_{x_j}^{x_{j-1}} [L(x_{i-1}, y) - L(x_i, y)]\,\mathrm{d}y}{\int_{x_j}^{x_{j-1}} L(x_j, y)\,\mathrm{d}y} \tag{9.84}$$

and the discrete-size model of grinding kinetics becomes:

$$\frac{\mathrm{d}M_i(t)}{\mathrm{d}t} = -S_i M_i(t) + \sum_{j=1}^{i-1} S_j b_{i,j} M_j(t) \tag{9.85}$$

$$\text{for} \quad i = 1, 2, 3, \ldots$$

This then is the model of Bass, (given earlier as equation (8.15)) who was the first to derive it using somewhat different terminology and arguments (Bass, 1954). A few comments are in order:

(1) The Bass model (or one of its derivatives) is the most widely used grinding kinetics model, and it has demonstrably performed satisfactorily in mapping the trajectory of particle size spectra in the time domain in both simulation and prediction modes.

(2) The inner breakage parameter in the Bass model is given by combining equations (9.81) and (9.82):

$$b_{j,j} = 1 - \frac{\int_{x_j}^{x_{j-1}} L(x_j, y)\,\mathrm{d}y}{\int_{x_j}^{x_{j-1}} S(y)\,\mathrm{d}y} \tag{9.86}$$

Numerical simulation shows that $b_{j,j}$ is ordinarily not vanishingly small, not even when $R = 2^{1/4}$. Consequently, $b_{i,j}$ are not actual members of the normalized or relative breakage distribution.

(3) Parameters S_i and $b_{i,j}$ are not uncorrelated or independent of each other in the sense of Epstein's original proposition.

(4) A sufficient condition for the realization of the Bass model is that the particles are piecewise uniformly distributed in discrete-size intervals. This approximation becomes increasingly valid as the width of the size interval decreases. It has been confirmed experimentally that the time dependence of the grinding rate parameters becomes progressively weaker as the sieve size ratio is reduced from 2 to $2^{1/2}$ and to $2^{1/4}$ (Austin and Luckie, 1970/71a). Clearly then the Bass model is an approximation to the size-continuous model in equation (9.16), and not vice versa, even though the latter model has never been explicitly and directly validated with empirical data.

When all the breakage rate parameters are unequal, an analytical solution to equation (9.85) of the Bass model is (Reid, 1965) (see also equation (8.17)):

$$M_i(t) = \sum_{j=1}^{i} h_{i,j} \exp(-S_j t) \tag{9.87}$$

where

$$h_{i,j} = \sum_{k=j}^{i-1} \frac{S_k b_{i,k} h_{k,j}}{S_i - S_j} \text{ for } k \neq i \tag{9.88}$$

$$h_{i,i} = M_i(0) - \sum_{k=1}^{i-1} h_{i,k} \text{ for } i \neq 1 \tag{9.89}$$

$$h_{1,1} = M_1(0) \tag{9.90}$$

In particular, for $i = 1$:

$$M_1(t) = M_1(0) \exp(-S_1 t) \tag{9.91}$$

When two or more of these parameters exhibit the same value, e.g. when $S_i = S_j$, the solution in equation (9.88) and (9.89) must be modified (Olsen, 1972):

$$h_{i,j} = \sum_{k=j}^{i-1} S_k b_{i,k} h_{k,j} t \tag{9.92}$$

and

$$h_{i,i} = M_i(0) - \sum_{\substack{k=1 \\ k \neq j}}^{i-1} h_{i,k} \tag{9.93}$$

When the particle size range of interest is divided into n discrete intervals ($i = 1, 2, \ldots, n$) then the solution in equation (9.87) can be recast in a matrix notation as:

$$\begin{bmatrix} M_1(t) \\ M_2(t) \\ \vdots \\ M_n(t) \end{bmatrix} = \begin{bmatrix} h_{1,1} & & & \bigcirc \\ h_{2,1} & h_{2,2} & & \\ \vdots & \vdots & \ddots & \\ h_{n,1} & h_{n,2} & \cdots & h_{n,n} \end{bmatrix} \begin{bmatrix} e^{-S_1 t} \\ e^{-S_2 t} \\ \vdots \\ e^{-S_n t} = 1 \end{bmatrix} \tag{9.94}$$

or

$$\boldsymbol{M}(t) = \mathbf{H}\boldsymbol{E}(t) \tag{9.95}$$

where $\boldsymbol{M}(t)$ is an $n \times 1$ vector of product size distribution, $\mathbf{H}$ is an $n \times n$ lower triangular matrix whose elements are $h_{i,j}$, and $\boldsymbol{E}(t)$ is an $n \times 1$ vector of exponential terms. Note that conservation of mass requires that $S_n = 0$, and:

$$\begin{aligned} b_{n,j} &= 1 - \sum_{i=j+1}^{n-1} b_{i,j} \\ &= B_{n-1,j} \end{aligned} \tag{9.96}$$

An alternative matrix solution starts with the set of n simultaneous differential equations in equation (9.85) which can always be written compactly in a matrix equation form as:

$$\frac{\mathrm{d}\boldsymbol{M}(t)}{\mathrm{d}t} = -[\mathbf{I} - \mathbf{B}]\mathbf{S}\boldsymbol{M}(t) \tag{9.97}$$

where $\mathbf{I}$ is an $n \times n$ identity matrix, $\mathbf{B}$ is an $n \times n$ lower triangular matrix of $b_{i,j}$, and $\mathbf{S}$ is an $n \times n$ diagonal matrix of breakage rate parameters S_i. The solution to this equation is:

$$\boldsymbol{M}(t) = \exp[-(\mathbf{I} - \mathbf{B})\mathbf{S}t]\boldsymbol{M}(0) \tag{9.98}$$

where $\boldsymbol{M}(0)$ is an $n \times 1$ vector of feed distribution. In the case when no two rate parameters are equal, this equation can be transformed into (Herbst and Mika, 1970):

$$\boldsymbol{M}(t) = \mathbf{N}\mathbf{V}(t)\mathbf{N}^{-1}\boldsymbol{M}(0) \tag{9.99}$$

where the elements of the $n \times n$ matrices $\mathbf{N}$ and $\mathbf{V}$ are, respectively:

$$n_{i,j} = \begin{cases} 0 & \text{for } i < j \\ 1 & \text{for } i = j \\ \displaystyle\sum_{k=j}^{i-1} \frac{b_{i,k}S_k}{S_i - S_j} n_{k,j} & \text{for } i > j \end{cases} \tag{9.100}$$

$$v_{i,j}(t) = \begin{cases} \exp(-S_i t) & \text{for } i = j \\ 0 & \text{for } i \neq j \end{cases} \tag{9.101}$$

The three matrices can be lumped into one by multiplying them together, and the solution in equation (9.99) becomes:

$$\boldsymbol{M}(t) = \mathbf{O}^*(t)\boldsymbol{M}(0) \tag{9.102}$$

or

$$\begin{bmatrix} M_1(t) \\ M_2(t) \\ \vdots \\ M_n(t) \end{bmatrix} = \begin{bmatrix} O^*_{1,1}(t) & & & \bigcirc \\ O^*_{2,1}(t) & O^*_{2,2}(t) & & \\ \vdots & \vdots & \ddots & \\ O^*_{n,1}(t) & \cdot & \cdots & O^*_{n,n}(t) \end{bmatrix} \begin{bmatrix} M_1(0) \\ M_2(0) \\ \vdots \\ M_n(0) \end{bmatrix} \quad (9.103)$$

An important significance of this solution lies in the fact that the $n \times n$ lower triangular matrix $\mathbf{O}^*(t)$ maps the feed vector $\boldsymbol{M}(0)$ directly into the product vector $\boldsymbol{M}(t)$, and for this reason it is called, appropriately enough, the mill matrix. The element $O^*_{i,j}$ is the fraction of material originally in size index j which is found in the ith size interval after undergoing repeated cycles of breakage over a time period t. After two consecutive time intervals t_1 and t_2 $(t = t_1 + t_2)$, the product vector is:

$$\boldsymbol{M}(t) = \mathbf{O}^*(t_2)\mathbf{O}^*(t_1)\boldsymbol{M}(0) \quad (9.104)$$

In particular, after k intervals of duration t each (i.e. so-called k cycles of grinding) the product is given by multiplication of the feed vector with the mill matrix raised to the power k:

$$\boldsymbol{M}(kt) = [\mathbf{O}^*(t)]^k \boldsymbol{M}(0) \quad (9.105)$$

The grinding model in equation (9.102) is an exact discrete-size, discrete-time analogue of the discrete-size but time-continuous Bass model. The mill matrix $\mathbf{O}^*(t)$ is explicitly defined in terms of physically identifiable parameters S_i and $b_{i,j}$, and it is a continuous function of time. Consequently, it provides a more meaningful, versatile and accurate modelling tool than the empirical *ad hoc* mill matrix of the type employed formerly. (Callcott, 1967; Brookes *et al.*, 1970).

In some applications it is more convenient to formulate the Bass model in cumulative distribution mode. Let the cumulative amount of solids retained on mesh opening x_i be:

$$R_i(t) = \sum_{j=1}^{i} M_j(t) \quad (9.106)$$

Hence:

$$M_i(t) = R_i(t) - R_{i-1}(t) \quad (9.107)$$

Combining these two expressions with equation (9.85), the Bass model becomes (Kapur, 1970a):

$$\frac{\mathrm{d}R_i(t)}{\mathrm{d}t} = -S_iR_i(t) + \sum_{j=1}^{i-1}\left[(S_{j+1}B_{i,j+1} - S_jB_{i,j})\right]R_j(t) \text{ for } i = 1, 2, \ldots \quad (9.108)$$

The solution to this equation is again given by equations (9.87) and (9.99) except that M_i is now replaced everywhere by R_i and the term $S_kb_{i,k}$ by $(S_{k+1}B_{i,k+1} - S_kB_{i,k})$.

9.3.4 RSF and SSF grinding models

Inspection of equations (9.77) and (9.78) shows that time-invariant grinding parameters can be generated by an alternative route, that is, by invoking the reduced selekage function (RSF) in equation (9.20). (Gupta and Kapur, 1976) In discrete-size formulation:

$$L(x_i, y) = Ag(x_i) \tag{9.109}$$

Substitution into equations (9.77), (9.78) and (9.80) gives the time-invariant parameters:

$$S_i = Ag(x_i) \tag{9.110}$$

$$b_{i,j} = \frac{g(x_{i-1}) - g(x_i)}{g(x_j)} \tag{9.111}$$

$$\begin{aligned} B_{i,j} &= \frac{g(x_i)}{g(x_j)} \\ &= \frac{S_i}{S_j} \end{aligned} \tag{9.112}$$

It also follows that:

$$\begin{aligned} S_j B_{i,j} &= Ag(x_i) \\ &= S_i \end{aligned} \tag{9.113}$$

Substitutions of equations (9.110) and (9.113) into (9.108) leads to the reduced selekage function model of grinding kinetics:

$$\frac{\mathrm{d}R_i(t)}{\mathrm{d}t} = -Ag(x_i)R_i(t) \tag{9.114}$$

which on integration gives:

$$R_i(t) = R_i(0) \exp[-Ag(x_i)t)] \tag{9.115}$$

This is the discrete-size analogue of the Rosin–Rammler type distribution equation. In fact, when the RSF is specialized to:

$$L(x_i, y) = Ax_i^q \tag{9.116}$$

equation (9.115) becomes the Rosin–Rammler distribution. Further, it can be shown that the exact discrete-size, discrete-time version of the RSF model is (Gupta and Kapur, 1974a):

$$\begin{bmatrix} M_1(t) \\ M_2(t) \\ M_3(t) \\ \vdots \\ M_n(t) \end{bmatrix} = \begin{bmatrix} O^*_{1,1} & & & & \bigcirc \\ O^*_{2,1} & O^*_{2,2} & & & \\ O^*_{3,1} & O^*_{3,2} & O^*_{3,3} & & \\ \vdots & \vdots & \vdots & \ddots & \\ O^*_{n,1} & O^*_{n,2} & O^*_{n,3} & \cdots & O^*_{n,n} \end{bmatrix} \begin{bmatrix} M_1(0) \\ M_2(0) \\ M_3(0) \\ \vdots \\ M_n(0) \end{bmatrix} \tag{9.117}$$

where the elements of the mill matrix are:

$$O^*_{i,i}(t) = \exp[-Ag(x_i)t] \quad \text{for} \quad i = 1, 2, \ldots, n-1 \tag{9.118}$$

and

$$O^*_{n,n}(t) = 1 \tag{9.119}$$

and in any one row all the off-diagonal elements are not only identical but are related to the diagonal elements by

$$O^*_{i,j} = O^*_i = O^*_{ii} - O^*_{i-1,i-1}$$
$$\text{for } i = 2, 3, \ldots, n$$
$$\text{and } j = 1, 2, \ldots, i-1 \tag{9.120}$$

Because of its relative simplicity, the RSF model has been employed widely for simulation purposes. Moreover, the phenomenon of the zero-order production of fines (Arbiter and Bhrany, 1960; Mika *et al.*, 1967; Harris, 1968a; 1968c Herbst and Fuerstenau, 1968) is a direct consequence of this model. Substitution of equations (9.14) and (9.109) into (9.19) and writing $F_i(t)$ in place of $F(x_i, t)$, the expression for rate of production of fines in the RSF model becomes:

$$\frac{dF_i(t)}{dt} = Ag(x_i)[1 - F_i(t)] \tag{9.121}$$

It is now argued that if the mesh opening x_i is sufficiently smaller than the feed particles, then, at least for some intial time, $F_i(t) \ll 1$; and therefore equation (9.121) simplifies to:

$$\frac{dF_i(t)}{dt} = Ag(x_i), \qquad t \to 0^+ \tag{9.122}$$

which implies zero-order (constant) rate of production of fines. Moreover, if the RSF is specialized as in equation (9.116), then (9.122) becomes:

$$\frac{dF_i(t)}{dt} = Ax_i^a \tag{9.123}$$

Combining this equation with equation (9.113) results in the following restrictive interrelationship between breakage rate and breakage distribution parameters as a sufficient condition for zero-order production of fines (Herbst and Fuerstenau, 1968):

$$\frac{dF_i(t)}{dt} = S_j B_{i,j} \tag{9.124}$$

Another grinding model can be generated if one chooses to specialize the selekage function into a separable selekage function (SSF) in the following manner (Gupta and Kapur, 1976):

$$L(y, x) = g_1(y) g_2(x) \tag{9.125}$$

The discrete-size grinding parameters in equations (9.77) and (9.78) now become:

$$S_i(t) = \frac{g_2(x_i)}{M_i(t)} \int_{x_i}^{x_{i-1}} g_1(y)M(y, t)\,dy \tag{9.126}$$

which is of course a function of time, and:

$$b_{i,j} = \frac{g_2(x_{i-1}) - g_2(x_i)}{g_2(x_j)} \tag{9.127}$$

which interestingly is independent of time. The grinding equation for the resulting SSF model is;

$$\frac{dM_i(t)}{dt} = -S_i(t)M_i(t) + \sum_{j=1}^{i=1} S_j(t)b_{i,j}M_j(t) \tag{9.128}$$

This model is of interest for studying the time variability of $S_i(t)$ introduced by the discretization of particle size (Gupta, 1975).

9.4 REDUCTION OF BASS GRINDING MODEL

9.4.1 Reduction of number of parameters

For the purpose of process simulation it is necessary to acquire reasonably accurate values of the parameters in the grinding model. The current understanding of the underlying principles of the comminution mechanics is not adequate to allow purely theoretical computations of these parameters. The crux of the matter is that the parameters must be estimated from experimental grinding data and this is by no means a trivial task. For n size intervals in the Bass model there are $n(n + 1)/2$ breakage rate and distribution parameters (for $n = 13$, the number of parameters to be estimated is 91). For practical reasons this large set of parameters must be extracted from a relatively limited amount of data which are invariably error corrupted. As a consequence, not surprisingly, the estimation exercise can lead to uncertain and misleading results. Many attempts have been made to simplify the model and hence reduce the number of model parameters to a more manageable level. It is of course understood that some accuracy in model representation of the comminution system is inevitably sacrificed, in return (it is hoped) for some tangible gains in convenience in handling the model. It has already been shown that by imposing the RSF model on the system the number of parameters is drastically reduced to just n. Some additional techniques for model reduction are now discussed.

9.4.2 Bass model with difference–similar breakage distribution parameter

An immediate reduction from $n(n + 1)/2$ to $2n - 1$ parameters is realized in the Bass model by stipulating a difference–similar form of breakage distribution parameter (normalization):

$$B_{i,j} \text{ (or } b_{i,j}) = B_{i-j} \text{ (or } b_{i-j}) \tag{9.129}$$

implying thereby that the value of the parameter depends on the difference between size indices of progeny and parent particles only. The Bass model now reduces to;

$$\frac{\mathrm{d}M_i(t)}{\mathrm{d}t} = -S_i M_i(t) + \sum_{j=1}^{i-1} S_j b_{i-j} M_j(t) \tag{9.130}$$

This is the most widely employed model for simulation (Kelsall *et al.*, 1968; Herbst and Fuerstenau, 1968; Klimpel and Austin, 1970; Gumtz and Fuerstenau, 1970; Olsen and Krogh, 1972; Gardner and Sukanjnajtee, 1973). It should be pointed out, however, that first of all there is no *a priori* guarantee that a given grinding system should exhibit a self-similar breakage distribution function in continuous size. Even if a system does exhibit a self-similar function, strictly speaking it does not necessarily follow that the difference–similar breakage distribution parameter prevails in the analogous discrete-size model. In the case of the Bass model, sufficient conditions for the difference–similar parameter are that breakage rate and breakage distribution function be of the forms (Gupta and Kapur, 1976):

$$S(y) = Ay^a \tag{9.131}$$

$$B(x, y) = B(x/y) \tag{9.132}$$

Then the difference–similar breakage distribution parameter in the Bass model is:

$$\begin{aligned} B_{i,j} &= [R^{j-i}]^{a+1} \frac{\phi(R^{j-i-1}, R^{j-i})}{\phi(R^{-1}, 1)} \\ &= B_{i-j} \end{aligned} \tag{9.133}$$

where:

$$\phi(y_2, y_1) = \int_{y_1}^{y_2} y^{-(a+2)} B(y)\,\mathrm{d}y \tag{9.134}$$

In the RSF and SSF models, on the other hand, the difference–similar breakage parameter is guaranteed only if the selekage function conforms to:

$$L(x, y) = Ax^b \tag{9.135}$$

and

$$L(x, y) = g_1(y)x^b \tag{9.136}$$

respectively. Now in both models the expression for this parameter is:

$$\begin{aligned} B_{i,j} &= [R^{j-i}]^b \\ &= B_{i-j} \end{aligned} \tag{9.137}$$

If the above-mentioned conditions are not satisfied, then, strictly speaking, imposition of the difference–similar parameter on a discrete-size model is merely an arbitrary convenience and these parameters are more-or-less disposable curve-fitting constants.

9.4.3 Truncated Bass model

In this model it is assumed that the broken material is fed only to the next four or fewer size intervals (Freeh *et al.*, 1967). Consequently, the Bass model is truncated to:

$$\frac{dM_i(t)}{dt} = -S_i M_i(t) + \sum_{j=k}^{i-1} S_j b_{i,j} M_j(t) \tag{9.138}$$

for

$$i - k = 4, \qquad i > 4$$

and

$$k = 1, \qquad i \leqslant 4$$

Naturally the number of parameters in the model is substantially reduced. However, since no rational procedure exists for truncation of the terms under the summation sign, therefore $b_{i,j}$ which are now termed 'inter-size flow coefficients', no longer possess any physical significance other than that of data-fitting constants.

9.4.4 Lumped-parameter Bass models

It is possible to devise a rational strategy for systematic reduction of, and approximate solutions to, the Bass model. The cumulative form of the Bass model in equation (9.108) can be written as:

$$\frac{dR_i(t)}{dt} = -S_i R_i(t) + \sum_{j=1}^{i-1} \lambda_{i,j} R_j(t) \tag{9.139}$$

where:

$$\lambda_{i,j} = S_{j+1} B_{i,j+1} - S_j B_{i,j} \tag{9.140}$$

In general, for most grinding systems operating under normal conditions, one can justify the following two inequalities:

$$S_j \geqslant S_{j+1} \text{ for } j = 1, 2, \ldots \tag{9.141}$$

and

$$B_{i,j} \leqslant B_{i,j+1} \text{ for } j = 1, 2, \ldots \text{ and } i > j \tag{9.142}$$

As a consequence of these inequalities all $\lambda_{i,j}$ terms are very small in relation to S_i, and these terms may be replaced by an appropriately averaged single-parameter without paying an excessive penalty in loss of accuracy. The resulting lumped-parameter Bass model becomes (Kapur, 1970a):

$$\frac{dR_i(t)}{dt} = -S_i R_i(t) + \lambda_i \sum_{j=1}^{i-1} R_j(t) \tag{9.143}$$

which has $2n - 1$ parameters, the same as in the Bass model with difference–similar breakage parameters in equation (9.130).

The solutions of the Bass model and of simplified models described above are non-linear in their parameters. This is perhaps the major source of difficulty in executing any parameter estimation scheme. It is then desirable to seek approximate solutions to the grinding models in which the parameters, either singly or lumped together, occur in a linear form. In principle, solutions to any degree of accuracy can be extracted by the Cauchy–Picard method of successive iterations. For example, the $(k+1)$th approximate solution to equation (9.139) is given by integration of the following equation (Kapur, 1970c):

$$\frac{\mathrm{d} \ln R_i{}^{(k+1)}(t)}{\mathrm{d}t} = -S_i + \sum_{j=1}^{i-1} \lambda_{i,j} \frac{R_j^{(k)}(t)}{R_i^{(k)}(t)} \tag{9.144}$$

where:

$$R_i^{(k+1)}(0) = R_i(0), \text{ for } k = -1, 0, 1, 2, \ldots \tag{9.145}$$

The first and the second approximate solutions are:

$$\ln \frac{R_i^{(1)}(t)}{R_i(0)} = -S_i t + t \sum_{j=1}^{i-1} \lambda_{i,j} \frac{R_j(0)}{R_i(0)} \tag{9.146}$$

and

$$\ln \frac{R_i^{(2)}(t)}{R_i(0)} = -S_i t + \sum_{j=1}^{i-1} \lambda_{i,j} \frac{R_j(0)}{R_i(0)} \times \frac{\exp[(G_j - G_i)t] - 1}{(G_j - G_i)} \tag{9.147}$$

It turns out that two iterations are sufficient for a reasonably accurate solution provided the grinding duration is not extremely long. Further, since the term $(G_j - G_i)$ is very small, the exponential function may be expanded in a three-term series; equation (9.147) then becomes:

$$\ln \frac{R_i(t)}{R_i(0)} = G_i t + H_i \frac{t^2}{2} \tag{9.148}$$

where:

$$G_i = -S_i + \sum_{j=1}^{i-1} \lambda_{i,j} \frac{R_j(0)}{R_i(0)} \tag{9.149}$$

$$H_i = \sum_{j=1}^{i-1} \lambda_{i,j}(G_j - G_i) \frac{R_j(0)}{R_i(0)} \tag{9.150}$$

Simulations of batch (Kapur, 1970a, 1970c Kapur and Agarwal, 1970) and locked-cycle (Kapur and Rahman, 1971) grinding tests have shown that the G–H solution in equation (9.148) is in good agreement with the exact solution to the Bass equation even for relatively long grinding times. In view of the fact that equation (9.148) is linear in the two lumped parameters G_i and H_i, these can be readily computed from a limited amount of experimental data.

Moreover, it is possible to reduce the order of the time polynomial in equation (9.148) by one in the following manner (Kapur, 1982):

$$\frac{1}{t}\ln\frac{R_i(t)}{R_i(0)} = G_i + \frac{H_i}{2}t \tag{9.151}$$

This equation, as written, can be plotted as a straight line with slope $H/2$ and intercept G. Figure 9.2 shows typical plots based on the dolomite data of Herbst and Fuerstenau (1968). It may be pointed out that at present this plotting format is the only one available for straight-line graphical representation of grinding data over the full size range and time intervals which are normally of interest.

Also for a single size feed, $R_i(0) = 1$ for $i = 1, 2, \ldots$; hence G_i in equation

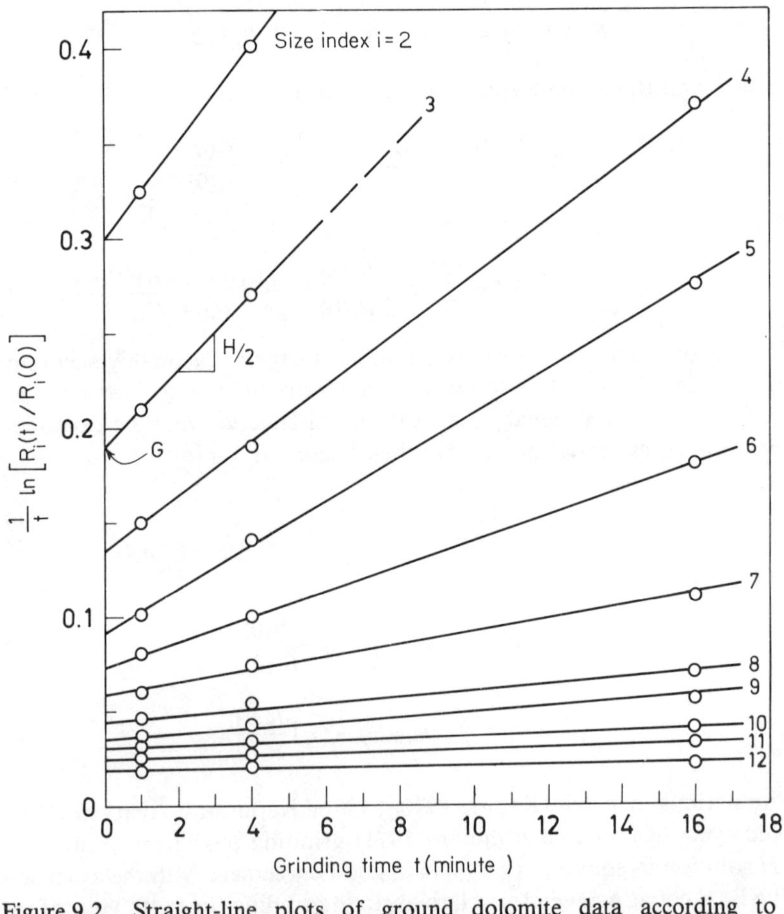

Figure 9.2. Straight-line plots of ground dolomite data according to equation (9.151) (Kapur, 1982). *Reproduced by permission of Elsevier Sequoia*

(9.149) simplifies to:

$$G_i = -S_1 B_{i,1} \text{ for } i = 2, 3, \ldots \tag{9.152}$$

$$G_1 = -S_1 \tag{9.153}$$

By substitutions of these relationships equation (9.148) can be recast in the following alternative forms:

$$\frac{\ln R_i(t)}{-S_1 t} = B_{i,1} - \frac{H_i}{2S_1} t \tag{9.154}$$

or

$$\frac{\ln R_i(t)}{\ln R_1(t)} = B_{i,1} - \frac{H_i}{2S_1} t \tag{9.155}$$

These equations provide a convenient and accurate method for computing the feed size breakage distribution parameters $B_{i,1}$ from the grinding data, which need not be restricted to short grinding times (Kapur, 1982).

9.4.5 Model reduction by functional forms of parameters

In general, both the breakage rate and breakage distribution functions and the corresponding parameters tend to vary gradually and regularly with particle size (or size index). Therefore a perfectly valid approach to model reduction lies in imposing a polynomial or some other appropriate function in particle size for the representation of rate and distribution parameters. The advantage gained is that the parameters are estimated, in fact overdetermined, in terms of a fewer number of constants and coefficients in the chosen function while avoiding possible difficulties with noise corrupted data.

An extreme example of such a reduction for the breakage distribution function is (Broadbent and Callcott, 1956; Cross, 1982):

$$\begin{aligned} B(x, y) &= [1 - \exp(-x/y)]\,[1 - \exp(-1)]^{-1} \\ &= B(x/y) \end{aligned} \tag{9.156}$$

which is not only self-similar but is also free of any unknown constant. A somewhat less restrictive representation is (Kuwahara, 1971):

$$B(x, y) = \left[1 - \exp\left(-\varepsilon \frac{x}{y}\right)\right] [1 - \exp(-\varepsilon)]^{-1} \tag{9.157}$$

where ε is called a distribution factor. Many investigators have used the Gaudin–Schuhmann distribution for breakage distribution parameters (Herbst and Fuerstenau, 1968; Olsen and Krogh, 1972; Moreira *et al.*, 1972; Randolph and Ranjan, 1977):

$$\begin{aligned} B_{i,j} &= \left(\frac{x_i}{x_j}\right)^a \\ &= B_{i-j} \end{aligned} \tag{9.158}$$

A more flexible empirical equation for a difference–similar breakage parameter, which has proved adequate for most applications, is (Austin and Luckie, 1971/72) (see also equation (8.13)):

$$B_{i,j} = X\left(\frac{x_i}{x_j}\right)^{a_1} + (1 - X)\left(\frac{x_i}{x_j}\right)^{a_2} \text{ for } 0 \leqslant X \leqslant 1 \tag{9.159}$$

which reduces to the earlier expression in equation (9.158) when $X = 1$. For non-difference–similar parameters, X is replaced by X_j, which is:

$$X_j = X_1 \left(\frac{x_j}{x_1}\right)^{-a_3} \text{ for } a_3 \geqslant 0 \tag{9.160}$$

where X, X_1, a_1, a_2, a_3 are constants and exponents. A suite of empirical equations for non-normalizable breakage parameters, suitable for complex heterogeneous ores, is also available (Gupta *et al.*, 1981).

A commonly used representation of the breakage rate parameter is (Herbst and Fuerstenau, 1980; Shoji *et al.*, 1980):

$$S_i = A(\bar{x}_i)^a \tag{9.161}$$

where $\bar{x}_i$ is an average particle size for the ith size interval (e.g. geometric mean, arithmetic mean etc), or the lower or upper mesh opening of the size interval. Other functions suggested for this purpose are the polynomials of the following kind (Klimpel and Austin, 1970):

$$S_i = \sum_{j=0}^{m} d_j x_i^j \tag{9.162}$$

$$\log S_i = \sum_{j=0}^{m} d_j (\log x_i)^j \tag{9.163}$$

where d_j are coefficients of the polynomials. In a study of the effect of grinding media ball diameter on the breakage rate parameter, the following functional form was used (Austin *et al.*, 1976):

$$S_i = A x_i^q [Q(z)] \tag{9.164}$$

where A is a function of the ball size, and

$$Q(z) = \frac{1}{\sqrt{\pi}} \int_z^{\infty} \exp\left[-\left(\frac{z'}{\sqrt{2}}\right)^2 d\left(\frac{z'}{\sqrt{2}}\right)\right] \tag{9.165}$$

is the Gaussian distribution function in z, a normalized particle size given by:

$$z = \log\left(\frac{x_i}{\mu}\right) \Big/ \log \sigma \tag{9.166}$$

where μ and σ are parameters of the Gaussian distribution. Thus a plot of x_i versus S_i/Ax_i^q on log-probability graph paper should give a straight line from which the two parameters of this distribution can be determined. A simpler expression which shows a satisfactory fit to the experimentally computed parameters is (Gupta and Kapur, 1974b):

$$S_i = \frac{K_7 x_i}{K_8 + x_i} \tag{9.167}$$

where K_7 and K_8 are constants, and K_7 is also a function of the ball size.

9.5 SIMULATION

9.5.1 Objectives

In the final analysis, the primary goal of mathematical modelling is to provide a practical tool for scale-up, design, simulation, control and optimization of industrial comminution circuits with an acceptable degree of precision and with least possible expenditure of money, labour and time. Chapter 10 indicates the extent to which this goal has been achieved.

In this section, however, the discussion is restricted to laboratory-scale simulation of batch grinding and locked-cycle tests, including the Bond grindability test.

9.5.2 Simulation of batch grinding

There is little doubt that accurate mathematical simulation of batch grinding in tumbling mills is perfectly feasible, as testified by numerous examples in the literature (especially in the work of Austin and co-workers—e.g. Austin *et al.* (1976)). Batch grinding simulation plays an important role in verification of grinding models, parameter estimation and development of empirical correlations. An example of predictive simulation using the Bass model is shown in Figure 9.3.

9.5.3 Simulation of locked-cycle grinding tests

9.5.3.1 Survey

The first practical application of the mathematical modelling techniques in comminution was in predictive simulation of locked-cycle or pseudo-closed-circuit grinding tests, including the well-known Bond grindability test.

In a locked-cycle grinding test, a fixed amount of crushed feed solid is ground in a tumbling mill, and the product is screened on a prescribed 'mesh-of-grind'. The oversize retained component is combined with make-up feed and the grinding cycle is repeated. These cycles are continued until the recycle material, make-up feed, and undersize product become apparently constant; in other words, the system seemingly attains a steady state. The locked-cycle test at equilibrium simulates exactly the steady-state closed-circuit grinding with a sharp classifier and an ideal plug flow transport of the charge in the mill.

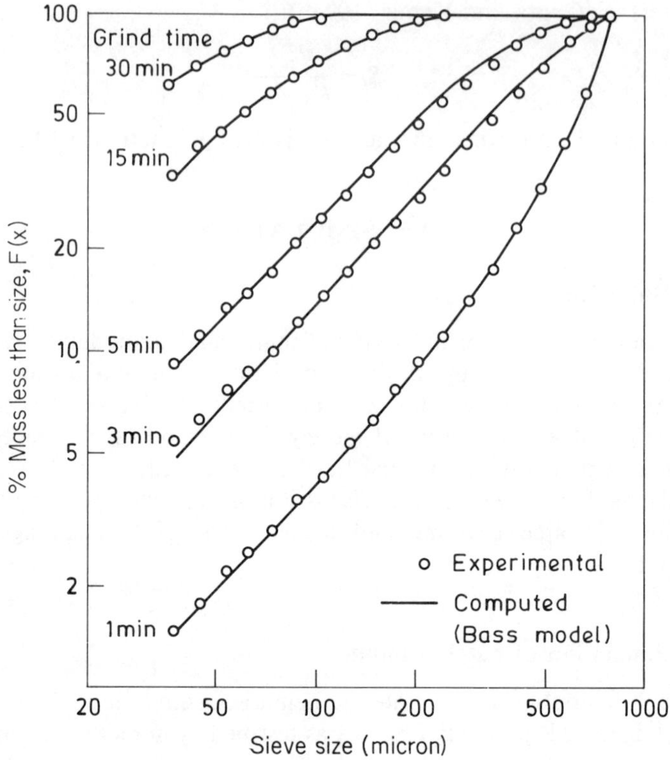

Figure 9.3. Comparison of computed and experimental size distributions for batch dry grinding of quartz in an 8-inch diameter ball mill (Klimpel and Austin, 1977). *Reproduced by permission of Elsevier Science Publishers*

Both the fixed-time locked-cycle test and the Bond grindability test (in the latter the grinding time for each cycle is adjusted) are time-consuming and, in addition, require large amounts of sample feed of assured uniformity in size consist and quality. These tests also demand a rather high degree of experimental precision, otherwise error-induced oscillations may consume a large number of additional cycles to dampen out or, in an extreme case, may even cause the system to settle down in a spurious steady state. Since a great deal of testing of the grindability of ores and other solids as well as laboratory-scale simulation of industrial circuits is done by the tedious locked-cycle tests, a reliable algorithm for computer simulation of these tests with only a limited amount of experimental data input would be highly desirable.

For simulation of the locked-cycle test, first a suitable grinding model is chosen. The exact or approximate solution to this model can always be written as:

$$R_i(t) = \theta_i(t)W \tag{9.168}$$

where R_i is the absolute amount retained on mesh opening x_i, the function θ_i depends on feed size distribution, model parameters and grinding time and W is mass hold-up in the mill. From appropriately designed batch grinding tests the model parameters are estimated by one or more techniques available for this purpose. In order to minimize the experimental component of the simulation, it is clearly advantageous to choose a model with the least possible number of parameters or a model whose parameters can be estimated with greater ease, consistent with the overall objective of a reasonably accurate simulation. The most straightforward approach to simulation entails in essence repetitive application of the solution to the batch grinding equation (9.168). In other words, one duplicates the experimental procedure for the test cycle-wise on a calculator/computer, using an appropriate algorithm.

9.5.3.2 Bond grindability test

Two essentially similar algorithms are available for this test (Kapur, 1970d and Karra, 1981). The Kapur algorithm is:

$$R_j(n) = \sum_{l=1}^{n} \left[W(l)\theta_j\left(\sum_{i=l}^{n} t_i \right) \right] \tag{9.169}$$

$$F_j(n) = R_j(n-1) + rW(n) - R_j(n) \tag{9.170}$$

$$U_n = \frac{F_j(n)}{t_n} \tag{9.171}$$

$$W(n+1) = F_j(n) + (1-r)W(n) \tag{9.172}$$

$$t_{n+1} = \frac{1}{U_n}\,[F_j(e) - (1-r)W(n+1)] \tag{9.173}$$

where at the end of the nth cycle of grinding duration t_n, $F_j(n)$ is the mass of undersize particles produced, $R_j(n)$ is the amount recycled (i.e. retained on the mesh-of-grind 'j'), U_n is the rate of grinding and $W(n+1)$ is the make-up feed for the next cycle. The fraction of original feed retained on the mesh-of-grind is r, and $F_j(e)$ is the mass of total undersize product for the stipulated 250% circulating load in the Bond ball mill test at steady state.

The sequence of U_n ($n = 1, 2, \ldots$) is computed recursively until it converges to the value of Bond grindability. Figure 9.4 shows an example of simulation of the Bond test on a highly heterogeneous copper ore as compared with the actual experimental results. This simulation was carried out with the RSF model in equation (9.115). Moreover, it turns out that an explicit analytical solution to the simulated Bond test is available when the RSF model is employed. Let symbol 'e' denote steady-state values in the test; then for 250% circulating load:

$$U_e = \frac{1}{3.5}\frac{r}{t_e}\,W(1) \tag{9.174}$$

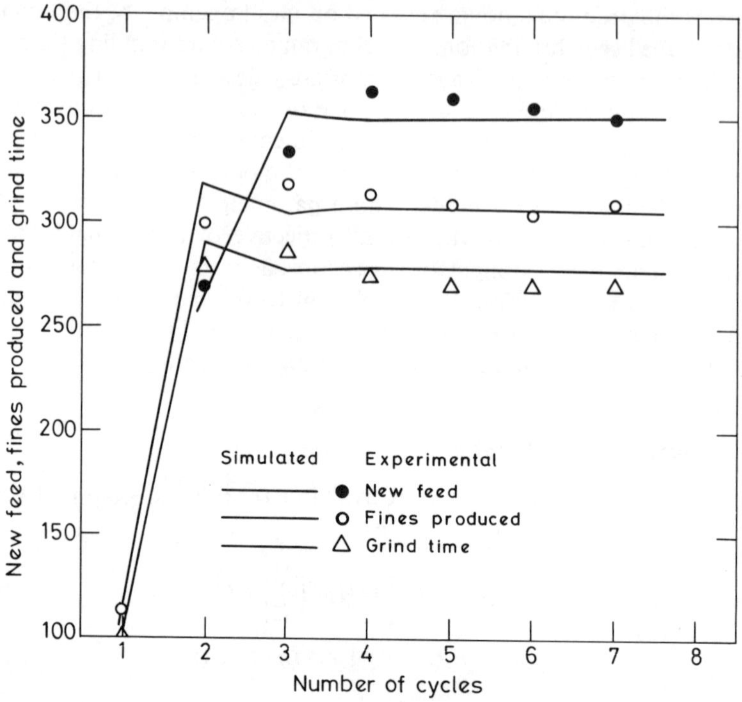

Figure 9.4. Cyclic simulation of Bond test for copper ore using equations (9.169) to (9.173) (Kapur, 1970d). *Reproduced by permission of the Institution of Mining and Metallurgy*

where t_e is solution to the following equation for mass balance in a steady-state cycle:

$$R_j(e) = R_j(e)\theta_j^{\prime}(t_e) + W(e)\theta_j(t_e) \tag{9.175}$$

Here, while θ_j is a function of feed size distribution, $\theta_j^{\prime}$ is a function of size distribution of the circulating load. In the case of the RSF model:

$$\theta_j(t_e) = r \exp(-S_j t_e) \tag{9.176}$$

and

$$\theta_j^{\prime}(t_e) = \exp(-S_j t_e) \tag{9.177}$$

Substitution in equation (9.175) gives:

$$t_e = \frac{1}{S_j} \ln \left[1 + \frac{rW(e)}{R(e)}\right] \tag{9.178}$$

or

$$t_e = \frac{1}{S_j} \ln \left[1 + \frac{r}{2.5}\right] \tag{9.179}$$

Combining these with equation (9.174) gives an explicit expression for Bond grindability in term of breakage rate parameter S_j:

$$U_e = \frac{rS_j w(1)}{3.5 \ln [1 + r/2.5]} \tag{9.180}$$

In other words, the Bond grindability is proportional to the initial batch grinding rate of the oversize since in the RSF model:

$$\begin{aligned} \frac{dR_j(t)}{dt} &= -S_j R_j(t) \text{ as } t \to 0^+ \\ &= -S_j r W(1) \end{aligned} \tag{9.181}$$

The extensive data of Smith and Lee (1968), which compare the Bond and the batch grindabilities, are in conformity with this result.

For more complex models it is normally not possible to obtain an analytical solution for t_e in equation (9.175), and it is necessary to fall back on the sequential simulation algorithm in equations (9.169)–(9.173). Further, if the interest is in computing the entire size distribution of product, then equation (9.169) in the algorithm for the ground product must be modified to:

$$R_k(n) = W(n)\theta_k(t_n) + R_k(n-1)\theta_k^{n-1}(t_n) \text{ for } k = 1, 2, \ldots \tag{9.182}$$

where $\theta_k^{n-1}(t_n)$ is a function of the particle size distribution of the material recycled after the $(n-1)$th cycle.

9.5.3.3 *Fixed-time locked-cycle grinding test*

The fixed-time locked-cycle experiments are employed to simulate on the laboratory scale the probable behaviour of the industrial grinding circuits for the effects of the classifier's efficiency, recirculating load, circuit configuration etc. Hukki and co-workers have carried out literally hundreds of grinding and sieving tests in an extensive programme based on the locked-cycle tests (Hukki and Allenius, 1968). Kapur and Rahman (1971) have shown that these tests can be quite accurately simulated on a computer using experimental data generated by one or two man-days effort only. Their cycle-wise simulation algorithm is similar to the Bond grindability algorithm. Alternately, if the mill matrix $\mathbf{O}(t)$ in equation (9.102) is known, it is possible to obtain the steady-state solution from the mass balance on each size interval, using for this purpose the closed-circuit grinding equations with plug flow regime in the mill (Herbst *et al.*, 1971). Interesting and useful conclusions from the simulated locked-cycle test have been obtained (Gumtz and Fuerstenau, 1970).

9.6 REFERENCES

Arbiter, N., and Bhrany, U. N. (1960). 'Correlation of product size, capacity and power in tumbling mills', *Trans. AIME*, **217**, 245–252.

Auer, A. (1981). 'Das probabilistische modell des kontinuierlichen zerkleinern', *Powder Technol.*, **28** 77–82.
Austin, L. G. (1971/72). 'A review introduction to the mathematical description of grinding as a rate process', *Powder Technol.*, **5**, 1–17.
Austin, L. G. (1973). 'A commentary on the Kick, Bond and Rittinger laws of grinding', *Powder Technol.*, **7**, 315–317.
Austin, L. G., and Bhatia, V. K. (1973). 'Note on conversion of discrete size interval values of breakage parameters *S* and *B* to point values and vice versa', *Powder Technol.*, **7**, 107–110.
Austin, L. G., and Luckie, P. T. (1970/71). 'Note on influence of interval size on the first-order hypothesis of grinding', *Powder Technol.*, **4**, 109–110.
Austin, L. G., and Luckie, P. T. (1971/72). 'The estimation of non-normalized breakage distribution parameters from batch grinding tests', *Powder Technol.*, **5**, 267–271.
Austin, L. G., Luckie, P. T., and Klimpel, R. R. (1972). 'Solutions of the batch grinding equation leading to Rosin–Rammler distributions', *Trans. AIME*, **252**, 87–94.
Austin, L. G., Shoji, K., and Bell, D. (1982). 'Rate equations for nonlinear breakage in mills due to materials effects', *Powder Technol.*, **31**, 127–133.
Austin, L. G., Shoji, K., and Luckie, P. T. (1976). 'The effect of ball size on mill performance', *Powder Technol.*, **14**, 71–79
Bass, L. (1954). 'Zur theorie der mahlvorgänge', *Zeits. Angew Maths. Phys.* **5**, 283–292.
Beke, B. (1964). *Principles of Comminution*, Akadémiai Kiado, Budapest.
Bond, F. C. (1952). 'The third theory of comminution', *Trans. AIME*, **193**, 484–494.
Broadbent, R. A., and Callcott, T. G. (1956). 'Coal breakage processes, *J. Inst. Fuel*, **29**, 524–539.
Brookes, G. F., Hughes, T. H., Watson, D., and Woodley, D. J. A. (1970). 'Control and modelling of a grinding/classifier circuit', *Proc. 9th Commonwealth Min. Metall. Cong.*, 1969, Paper 15, Min. Process. Extractive Metall., **3**, 701–711.
Callcott, T. G. (1967). 'Solution of comminution circuits', *Trans. IMM (London)*, **76**, C1–C11.
Charles, R. J. (1957). 'Energy-size reduction relationships in comminution', *Trans. AIME*, **208**, 80–88.
Cross, M. (1982). 'A method for extracting product size distributions from empirical comminution models', *Powder Technol.*, **31**, 233–237.
Dallavalle, J. M. (1948). *Micromeritics* (2nd Ed.), Pitmans, New York.
Epstein, B. (1947). 'The mathematical description of certain breakage mechanisms leading to the logarithmico-normal distribution', *J. Franklin Inst.*, **244**, 471–477.
Filippov, A. F. (1961). 'On the distribution of the sizes of particles which undergo splitting', *Theory of Probability and its Applications*, **6**, 275–294.
Finch, J. A., and Ramirez-Castro, J. (1981). 'Modelling mineral size reduction in the closed-circuit ball mill at the Pine Point Mines concentrator', *Int. J. Miner. Process.*, **8**, 61–78.
Freeh, E. J., Horst, W. E., and Kellner, R. C. (1967). 'A mathematical model for batch grinding in a ball mill', *Trans. AIME*, **238**, 167–172.
Fuerstenau, D. W. (1959). 'Retention time in continuous vibratory ball milling', *Trans. AIME*, **214**, 1238–1243.
Furuya, M., Nakajima, Y., and Tanaka, T. (1971). 'Theoretical analysis of closed-circuit grinding system based on comminution kinetics', *Ind. Eng. Chem. Process. Des. Devel.*, **10**, 449–456.
Gardner, R. P., and Austin, L. G. (1962). 'A chemical engineering treatment of batch grinding, Part I', *Symposium Zerkleinern*, 217–231.
Gardner, R. P., and Sukanjnajtee, K. (1973). 'A combined tracer and back-calculation

method for determining particulate breakage functions in ball milling. Part III', *Powder Technol.*, **7**, 169–179.

Gaudin, A. M., and Meloy, T. P. (1962). 'Model and comminution distribution equation for repeated fracture', *Trans. AIME*, **223**, 43–50.

Gumtz, G. D., and Fuerstenau, D. W. (1970). 'Simulation of locked-cycle grinding', *Trans. AIME*, **247**, 330–335.

Gupta, V. K. (1975). Ph.D. thesis, IIT, Kanpur.

Gupta, V. K., Hodouin, D., Berube, M. A. and Everell, M. D. (1981). 'The estimation of rate and breakage distribution parameters from batch grinding data for a complex pyritic ore using a back-calculation method', *Powder Technol.*, **28**, 97–106.

Gupta, V. K., and Kapur, P. C. (1974a). 'A simple mill matrix for grinding mills', *Chem. Eng. Sci.*, **29**, 634–637.

Gupta, V .K., and Kapur, P. C. (1974b). 'Empirical correlations for the effects of particulate mass and ball size on the selection parameters in the discretized batch grinding equation', *Powder Technol.*, **10**, 217–223.

Gupta, V. K., and Kapur, P. C. (1975). 'A pseudo-similarity solution to the integro-differential equation of batch grinding', *Powder Technol.*, **12**, 175–178.

Gupta, V. K., and Kapur, P. C. (1976). 'A critical appraisal of the discrete size models of grinding kinetics', *Proc. 4th Europ. Symp. Comminution* (Zerkleinern), Nuremberg, 1975, Dechema Monograph 79, 447–465.

Gupta, V. K., and Kapur, P. C. (1986). 'Effect of unit rod weight on the grinding characteristics of a rod mill', *Proc. Int. Conf. Prog. Met. Res.*, IIT, Kanpur, 1985 (to be published).

Harris, C. C. (1968a). 'Batch grinding kinetics', *Trans. AIME*, **241**, 359–364.

Harris, C.C (1968b). 'The application of size distribution equations to multi-event comminution processes', *Trans. AIME*, **241**, 343–358.

Harris, C. C. (1968c). 'Size reduction-time relationships of batch grinding', *Trans. AIME*, **241**, 449–453.

Herbst, J. A., and Fuerstenau, D. W. (1968). 'The zero order production of fine sizes in comminution and its implications in simulation', *Trans. AIME*, **241**, 538–548.

Herbst, J. A., and Fuerstenau, D. W. (1973). 'Mathematical simulation of dry ball milling using specific power information', *Trans. AIME*, **254**, 343–348.

Herbst, J. A., and Fuerstenau, D. W. (1980). 'Scale-up procedure for continuous grinding mill design using population balance models', *Int. J. Miner. Process.*, **7**, 1–31.

Herbst, J. A., Grandy, G. A., and Mika, T. S. (1971). 'On the development and use of lumped parameter models for continuous open and closed circuit grinding systems', *Trans. IMM (London)*, **80**, C193–C198.

Herbst, J. A., and Mika, T. A. (1970). 'Mathematical simulation of tumbling mill grinding: an improved method', *Rudy*, **18**, 70–75.

Holmes, J. A. (1957). 'A contribution to the study of comminution—a modified form of Kick's law', *Trans. Inst. Chem. Eng.*, **35**, 125–156.

Hukki, R. T., and Allenius, H. (1968). 'A quantitative investigation of the closed grinding circuit', *Trans. AIME*, **241**, 482–487.

Kapur, P. C. (1970a). 'Kinetics of batch grinding—Part A', *Trans. AIME*, **247**, 299–303.

Kapur, P. C. (1970b). 'A similarity solution to the integro-differential equation describing batch grinding', *Chem. Eng. Sci.*, **25**, 899–901.

Kapur, P. C. (1970c). 'Kinetics of batch grinding—Part B', *Trans. AIME*, **247**, 309–313.

Kapur, P. C. (1970d). 'Analysis of the Bond grindability test', *Trans. IMM (London)*, **79**, C103–C108.

Kapur, P. C. (1971a). 'The energy–size reduction relationships in comminution of solids', *Chem. Eng. Sci.*, **26**, 11–16.

Kapur, P. C. (1971b). 'Synthesis of grinding time-size reduction relationships', *Trans. IMM (London)*, **80**, C170–C173.
Kapur, P. C. (1972). 'Self-preserving size spectra of comminuted particles', *Chem. Eng. Sci.*, **27**, 425–431.
Kapur, P. C. (1982). 'An improved method for estimating the feed size breakage distribution functions', *Powder Technol.*, **33**, 269–275.
Kapur, P. C., and Agarwal, P. K. (1970). 'Approximate solutions to the discretized batch grinding equation', *Chem. Eng. Sci.*, **25**, 1111–1113.
Kapur, P. C., and Fuerstenau, D. W. (1986). 'Energy-size reduction 'laws' revisited' *Int. J. Miner. Process*, (in press).
Kapur, P. C., and Rahman, M. (1971). 'Simulation of fixed-time, locked-cycle grinding tests', *Trans. IMM (London)*, **80**, C74–C78.
Karra, V. K. (1981). 'Simulation of Bond grindability test', *CIM Bull.*, **74**, 195–199.
Kelsall, D. F., Stewart, P. S. B., and Reid, K. J. (1968). 'Confirmation of a dynamics model of closed-circuit grinding with a wet ball-mill', *Trans. IMM (London)*, **77**, C120–C127.
King, R. P. (1972). 'An analytical solution to the batch comminution equation', *J. South Afr. Inst. Min. Metal.*, **73**, 127–131.
Kinneberg, D. J. and Herbst, J. A. (1984). 'A comparison of linear and nonlinear models for open-circuit ball mill grinding', *Int. J. Miner. Process.*, **13**, 143–165.
Klimpel, R. R., and Austin, L. G. (1970). 'Determination of selection-for-breakage functions in batch grinding equation by nonlinear optimization', *Ind. Eng. Chem. Fundl.*, **9**, 230–237.
Kuwahara, Y. (1971). 'Experimental study of selection function and distribution function', *J. Chem. Eng. (Japan)*, **4**, 359–363.
Malghan, S. G., and Fuerstenau, D. W. (1976). 'The scale-up of ball mills using population balance models and specific power input', *Proc. 4th Europ. Symp. Comminution* (Zerkleinern), Nuremberg, 1975, Dechema Monograph 79, 613–630.
Mika, T. S., Berloiz, L. M., and Fuerstenau, D. W. (1967). 'An approach to the kinetics of dry batch milling', Zerkleinern, Dechema Monograph 57, 205–240.
Moreira, R. M., De Castro, J. O. N. M., and Gardner, R. P. (1972). 'A combined tracer and back-calculation method for determining particulate breakage functions in ball milling. Part II', *Powder Technol.*, **6**, 75–83.
Nakajima, Y., and Tanaka, T. (1973). 'Solution of batch grinding equation', *Ind. Eng. Chem. Process Des. Devel.*, **12**, 23–25.
Olsen, T. O. (1972). 'Modelling and control of ball mill grinding', Report 72-86-W. Inst. Reguleringsteknikk, Univ. Trondheim.
Olsen, T. O., and Krogh, S. R. (1972). 'Mathematical model of grinding at different conditions in ball mills', *Trans. AIME*, **252**, 452–457.
Randolph, A. D. and Ranjan, R. (1977). 'Effect of a material-flow model in prediction of particle size distributions in open and closed circuit mills', *Int. J. Miner. Process*, **4**, 99–110.
Reid, K. J. (1965). 'A solution to the batch grinding equation', *Chem. Eng. Sci.*, **20**, 953–963.
Roberts, E. J. (1950). 'The probability theory of wet ball milling and its applications', *Trans. AIME*, **187**, 1267–1272.
Rosin, P., and Rammler, E. (1933). 'The laws governing the fineness of powdered coal,' *J. Inst. Fuel*, **7**, 29–36.
Schönert, K. (1969). 'The influence of classifier characteristics on the stability of closed circuit grinding. In: *A Decade of Digital Computing in the Mineral Industry* (Weiss, A., Ed.), SME/AIME, New York, 677–694.
Shoji, K., Lohrasb, S, and Austin, L. G. (1980). 'The variation of breakage parameter with ball and powder loading in dry ball milling', *Powder Technol.*, **25**, 109–114.

Smith, R. W. and Lee, K. H. (1968). 'A comparison of data from Bond type simulated closed-circuit and batch type grindability tests', *Trans. AIME*, **241**, 91–101.
Walker, W. H., Lewis, W. K., McAdams, W. H., and Gilliland, E. R. (1937). *Principles of Chemical Engineering*, McGraw Hill, NY, 255.
Whiten, W. J. (1974). 'A matrix theory of comminution machines', *Chem. Eng. Sci.*, **29**, 589–599.

CHAPTER 10

Simulation of closed circuits

NOTATION FOR CHAPTER 10

a'	Apparent bypass of fines to coarse stream in classifier
a^*	Parameter in equation (8.2)
A_i	Absolute rate of breakage = $S_i W^*$
$A_{i\mathrm{HI}}$	Absolute rate of breakage in the Hardgrove machine
$A_{16 \times 20}$	Absolute rate of breakage in size interval 16 × 20 mesh
$ACFM$	Actual cubic feet per minute
$b_{i,j}$	Breakage distribution function of size interval j into size interval i
$b(B)_{j,k}$	Breakage distribution function for ball breakage of material of size j by balls of size k in a semi-autogenous mill
$b(P)_{i,j,k}$	Breakage distribution function for tumbling pebble breakage of material from size interval j into size interval i by pebbles of size k
$b(S)_{i,j}$	Breakage distribution function for self-breakage of size interval j material into size interval i in a semi-autogenous mill
B'	Fraction of pre-classified feed with leaves in the fine stream
$B_{i,j}$	Cumulative breakage distribution function of size interval j into size interval i
$\mathbf{B}$	Cumulative breakage distribution matrix
$B(x, y)$	Cumulative breakage distribution function from size y to size x
B_1	Crushing zone cumulative breakage distribution function
B_2	Abrasion zone cumulative breakage distribution function
B_3	Transition zone cumulative breakage distribution function
c_i	Classification function for material in size interval i
$\mathbf{c}$	Classification matrix (diagonal)
C'	Recycle ratio
$\bar{C}'$	Combined recycle ratio of two classifiers in parallel
C''	Recycle ratio of one of two classifiers in parallel
CS	Controller setting (subscripts 1, 2 according to type)
d	Ball or roll diameter
d_{50}	Cut size corresponding to 50% of classifier feed to coarse stream
$d_{i,j}$	Mass fraction of material in the size interval j in the mill feed which appears in size interval i in the mill product (subscripts 1, 2 for two-component mixture)

D	Mill internal drum diameter (tumbling mill) or mean track diameter (vertical spindle mill) (subscript T = test mill)
D^*	Mixing factor due to ball tumbling action within a mill
D^{**}	Dimensionless form of $D^* = D^*/\bar{v}L$
$\mathbf{D}'$	Discharge matrix for autogenous mill in equation (10.53)
e	Base of natural logarithms
$e_{i,j}$	Mass fraction of material in the size interval j in the circuit feed which appears in size interval i in the circuit product
$\bar{E}$	Specific energy at mill input drive
$\bar{E}^*$	Converged value of $\bar{E}$
f_i	Fraction of mill feed which lies in size interval i
f_i'	Fraction of pre-feed stream which lies in size interval i
f_i^*	Fraction of feed crushing zone which lies in size interval i (Figure 10.28)
$\boldsymbol{f}$	Mill feed size vector
$\boldsymbol{f}'$	Pre-mill feed size vector in twin classifier circuit
F_1, F_2, F_3, F_4	Factors in statistical tests of variance
F'	Relative feed rate to mill
g_i	Fraction of circuit feed which lies in size interval i
$\boldsymbol{g}$	Circuit feed size vector
G'	Mass per unit feed rate of $f_i s_{1,i}$ in Figure 10.28
$\boldsymbol{h}$	Pre-classifier feed stream size vector in Figure 10.25
HI	Hardgrove index
i_g	Size interval corresponding to size x_g
i_s	Upper size interval corresponding to grate opening x_s
$\mathbf{I}$	Unit matrix
J	Fraction of tumbling mill drum volume occupied by the ball charge, including voids
J_B	Volume filling fraction of steel balls in semi-autogenous mill
J_P	Volume filling of tumbling pebbles in semi-autogenous mill
J_T	Volume of J for test mill
k'	Scale-up factor in equation (10.81)
k_0	Adjusting factor in equation (10.86)
K	Constant
l	Axial distance along ball mill drum from feed end
l^*	Dimensionless size l/L
L	Tumbling mill internal drum length
m	Number of stages, or manipulated variables, etc.
m_k	Mass of balls of size k in a semi-autogenous mill
M_i	Mass fraction of mill, or grinding zone, output in size interval i
$M_i(0)$	Value of M_i at time zero
$M_i(t)$	Value of M_i at time t
M_i^*	$(1 + C')M_i$
$\bar{M}_i^*$	$(1 + \bar{C}')M_i$

$\boldsymbol{M}$	Particle size vector for material within autogenous mill
M_k	Mass fraction of balls in size interval k
M_Σ	$\sum_{i_s-1}^{1} M_i$
n	Number
n_f	Number of free parameters
n_s	Number of size intervals
N	Number of revolutions per minute
N^*	Fraction of theoretical critical speed (subscript T = test mill)
p	Mill power
p_s	Mill specific power
p_i	Fraction of mill output which lies in size interval i
$p_{i\text{calc}}$	Calculated value of p_i
$p_{i\text{obs}}$	Observed value of p_i
p_i^*	$p_i(1 + C')$
p_i'	Fraction of airswept product which lies in size interval i
p_i^+	Fraction of crushing zone output which lies in size interval i in equation (10.94)
$\bar{p}_i$	Arithmetic mean of k observations of p_i
$\boldsymbol{p}$	Mill product size vector
P	Total load on the lower grinding ring of a vertical spindle mill
$\mathbf{q}_i$	Fraction of circuit product which lies in size interval i
$\boldsymbol{q}$	Circuit product size vector
Q	Mill output rate
Q_{HI}	Output rate from the Hardgrove machine
r	Radius of ball or roller
r_i	Fraction of coarse stream from classifier in size interval i
$r_{1,i}$	Fraction of coarse stream from classifier 1 in size interval i
$r_{2,i}$	Fraction of coarse stream from classifier 2 in size interval i
R	Mean track radius in vertical spindle mill
R'	Recycle ratio for a combined closed cycle
s_i	Fraction of classifier fine stream in size interval i which leaves with the coarse stream
$\bar{s}_i$	Mean value of s_i for two classifiers in parallel
$s_{1,i}, s_{2,i}$	Value of s_i for classifier 1, 2
S_i	Specific rate of breakage for material in size interval i
$S_{i,k}$	Specific rate of breakage of particles in size interval i by balls in size interval k
S_i (input)	Specific rate of breakage for material from input size in equation (10.24)
$S(B)_{i,k}$	Specific rate of breakage of size i material by balls of size k in a semi-autogenous mill
$S(P)_{i,k}$	Specific rate of breakage of size i material by tumbling pebbles of size k in a semi-autogenous mill

$S(S)_i$	Specific rate of self-breakage by size i material in a semi-autogenous mill
$\bar{S}_i$	Mean specific rate of breakage of size i material by balls, tumbling pebbles and self-breakage in a semi-autogeneous mill
S_i^*	τS_i
$S_i^*(\varkappa)$	$\varkappa \tau S_i$
$\mathbf{S}$	Specific rate of breakage matrix
SG	Specific gravity
SI	Sharpness index
SSQ	Sum of squares Subscripts: calc calculated lof lack of fit mod model obs observed opt optimal re residual error 1,re residual error associated with parameter 1, etc.
t	Time
T	Track width in vertical spindle mill
u_i	Manipulated variable in equations (10.33), (10.34)
U	Fraction of powder filling of the interstices between the balls of a tumbling mill ball charge (subscript T = test mill)
v	Velocity
$\bar{v}$	Average velocity
$v(r)$	Velocity along a radius r
v_i	Fraction of size i material removed by airsweeping from a ball mill
V	Rate of powder removal by airsweeping $= \sum_1^n v_i$
w_i	Weighting factor in SSQ evaluation dependent on error structure of experimental data
W^*	Mass of hold-up in a continuous process
WF	Feed rate of water to a hydrocyclone
x	Particle size
x_i	Particle size in interval i
x_g	Gap setting in roll crusher
x_s	Effective grate opening size
x_0	Reference particle size
x_{80}	Size corresponding to 80% passing
x^*	Value of x for which the denominator of equation (8.3) $= 0{\cdot}5$
x_{80}^*	Characteristic parameter in equation (10.25)
$\mathbf{X}_j$	Matrix form of breakage event occurring during the jth cycle
y	Particle size $> x$

y_j^*	Controlled variable in equations (10.33), (10.34) (Superscript: sp = set point)
Y	Thickness of material between ball and track in a vertical spindle mill
z	Moisture content of coal in equation (10.77)
Z_1, Z_2 etc	Factors in equation (10.26)
α	Exponent in equation (8.2)
β	Exponent in equation (8.13)
γ	Exponent in equation (8.13)
Δ	Characteristic parameter in equation (10.25)
$\varepsilon_{\mathrm{m,n}}$	Operator in equation (10.34)
ζ	Fraction of hold-up exposed to airstream per revolution of airswept ball mill
η	Electric motor efficiency
θ	Dimensionless time
$\phi(\theta)$	Dimensionless RTD function
$\varkappa$	Slowing down parameter in equation (10.24), (10.25)
$\varkappa_0$	Limiting value of $\varkappa$ for fine distributions in equation (10.25)
λ	Characteristic slope of corrected selectivity curve in equation (10.36), (10.37)
λ^*	Adjustable parameter in equation (10.103)
Λ	Exponent in equation (8.3)
μ_j	Fraction of two component mixture in equation (10.105)
ξ_1, ξ_2, ξ_3	Exponents in equation (10.26)
ξ^*	Adjustable parameter in equation (10.103)
σ^2	Variance
τ	Mean residence time
τ_k	Mean residence time in kth stage in equation (10.43)
τ_i	Mean residence time in ith size interval
υ	Parameter defined by equation (10.41)
χ	Proportion of distance across transition zone in autogenous mill
ϕ	Angle of nip
Φ	Parameter in equation (8.13)
$\psi_{i,j}$	Parameter in equations (10.14), (10.27)
Ω_i	Probability of material being held in the hold-up in ith stage of an airswept mill

10.1 INTRODUCTION

To demonstrate the effects of change from open to closed circuit, Kelsall *et al.* (1969) have produced data which are summarized in Figure 10.1. This graph gives product size distributions for batch, open-cycle and closed-cycle size reduction in a given ball mill, in which has been achieved very similar patterns of breakage functions and even the same x_{80} size. The distributions still differ

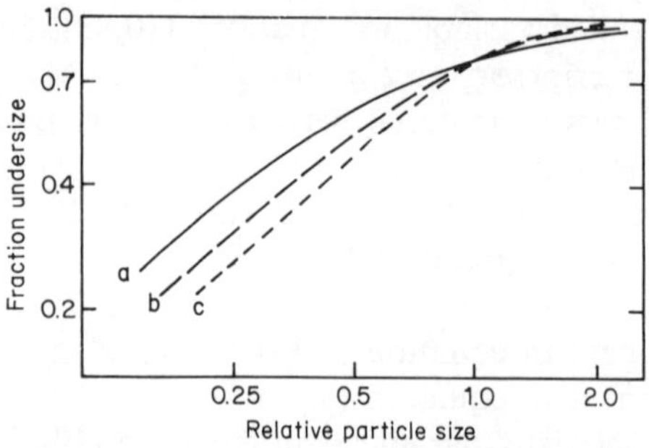

Figure 10.1. Comparison of size distributions from batch, open- and closed-circuit grinding for a ball mill with constant $S_i, B_{i,j}$ and x_{80} for the product: a open-circuit, b batch, c closed-circuit, (Kelsall *et al.*, 1969). *Reproduced by permission of the Institution of Engineers, Australia*

appreciably owing to the differences in flow characteristics from one mode of operation to another. For example, for the same feed make-up rate, the closed-circuit clearly enjoys a much greater flow through the grinding zone than in the case of the open. Again, the classifier loop allows the removal from the grinding zone of a proportion of particles of product size, depending on the sharpness of cut of the classifier. This results in less overgrinding in the closed compared with the open circuit.

These advantages usually more than compensate for the extra complexity of the closed circuit.

The three main objects of simulation of the closed circuit are:

(1) to control or optimize the performance of an existing plant;
(2) to scale-up from a small to a larger plant in a more realistic manner than the 'end-result' scaling of Chapter 7;
(3) to design new plant.

Examples of simulation in these three areas are included in §§10.3–10.8, which are divided according to mill type, with sub-division according to the level of sophistication introduced into the models.

10.2 BASICS OF THE CLOSED CIRCUIT

10.2.1 Types of closed circuit

Three basic types of closed circuit may be distinguished, indicated in Figure 10.2: normal, reverse and combined. In the reverse and combined configurations, the object of the classifier upstream of the pulverizer is to remove product size from the feed, without it having to enter the grinding zone.

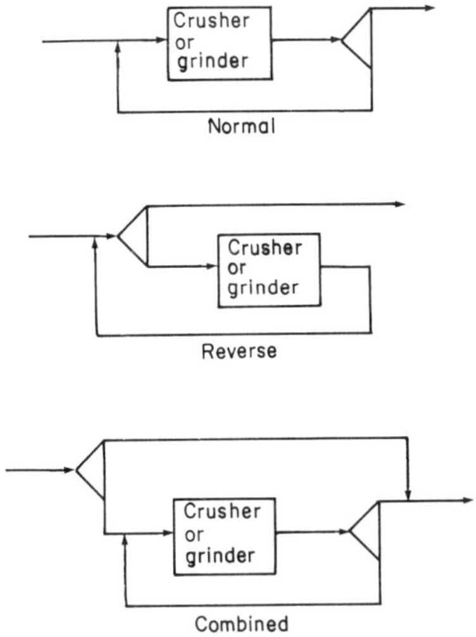

Figure 10.2. Three basic closed-circuit configurations (Rogers, 1983). *Reproduced by permission of Elsevier Sequoia*

In order to achieve the necessary product size distribution, particularly in mineral beneficiation, a two-stage series circuit is called for, shown in Figure 10.3. Luckie and Austin (1972) have recommended a modular treatment of such a circuit. Even more complex arrangements may be necessary to attain the desired end product. Figure 10.4 exemplifies a primary and secondary stage of size reduction, and a further tertiary stage may be needed for difficult ores.

Vertical spindle mills have an internal recirculation loop inherent in their design in addition to the recirculation from the separately provided classifier, and this has to be taken account of in the simulation.

Fluid energy mills are characterized by classification occurring concurrently with size reduction, which demands appropriate treatment in the mathematical model.

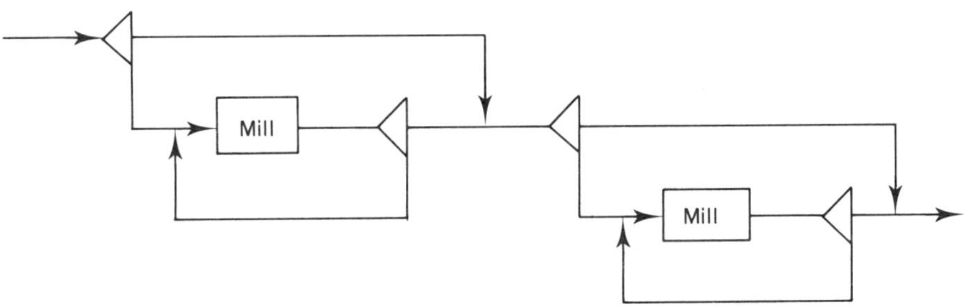

Figure 10.3. General two-stage series circuit

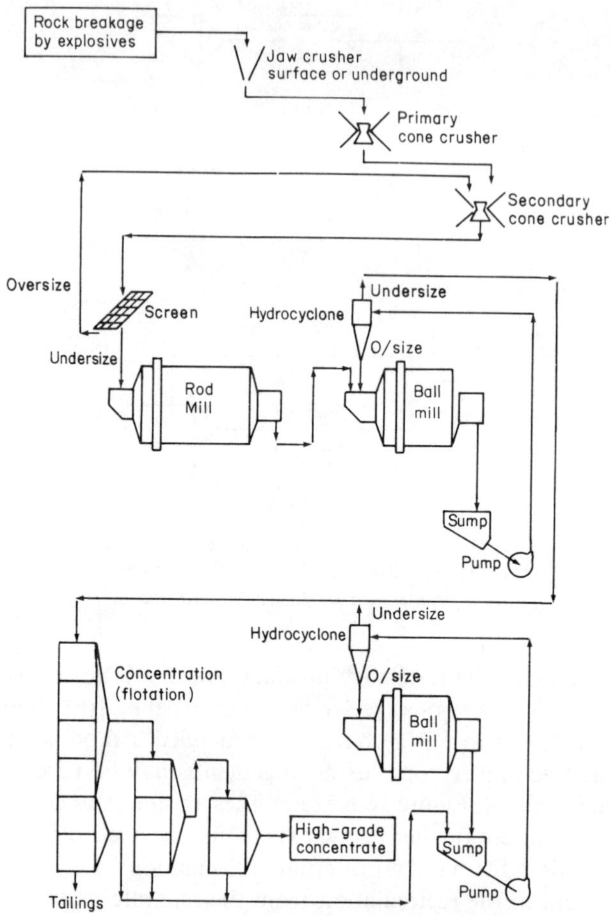

Figure 10.4. Primary and secondary crushers in multiple-loop ore concentration plant (Lynch, 1977). *Reproduced by permission of Elsevier Science Publishers*

10.2.2 Circuit nomenclature and basic equations

Closed circuits involve zones of confluence or of separation of streams in a network. The term 'node' is reserved for zones where a process unit, such as a classifier, occurs: otherwise they are called 'branches'.

For all circuits, relative flow rates are shown in diagrams in rectangular boxes and mass fractions of size i material are placed in circular boxes, and these define size distributions.

The recirculation ratio C' is the ratio of classifier coarse to fine product mass flow rates.

For the normal, reverse and combined configurations, Figures 10.5, 10.6 and 10.7 respectively give the appropriate nomenclature.

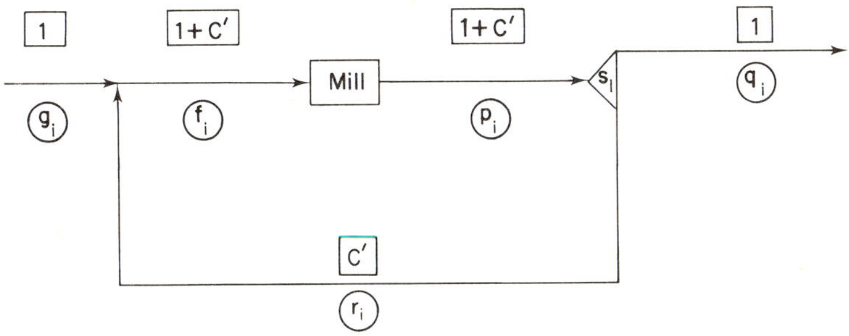

Figure 10.5. Nomenclature for normal closed circuit

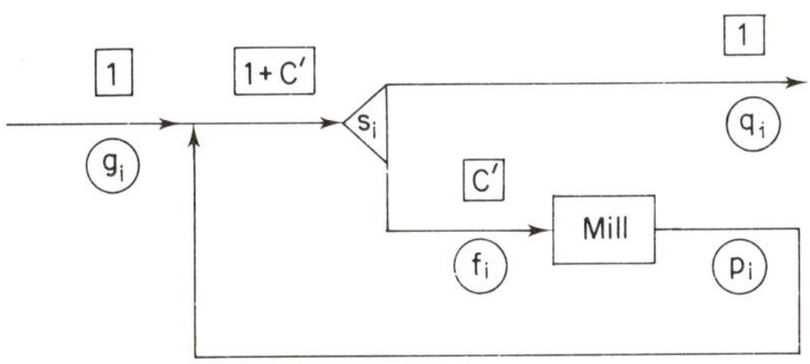

Figure 10.6. Nomenclature for reverse closed circuit

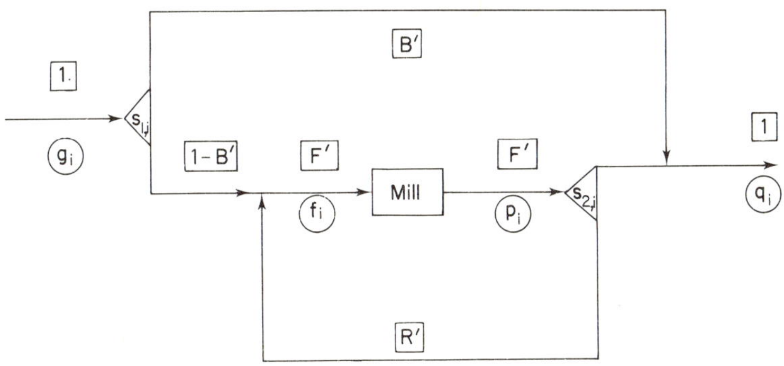

Figure 10.7. Nomenclature for combined closed circuit

In the combined circuit, B' is defined as the fraction of pre-classifier feed which reports to the fine product stream.

When two classifiers operate in parallel with a mill, Figure 10.8 applies.

The classifier selectivity number s_i gives the fraction of size i material in the classifier feed which goes to the coarse stream (see §5.2.4). Hence $(1 - s_i)$ goes

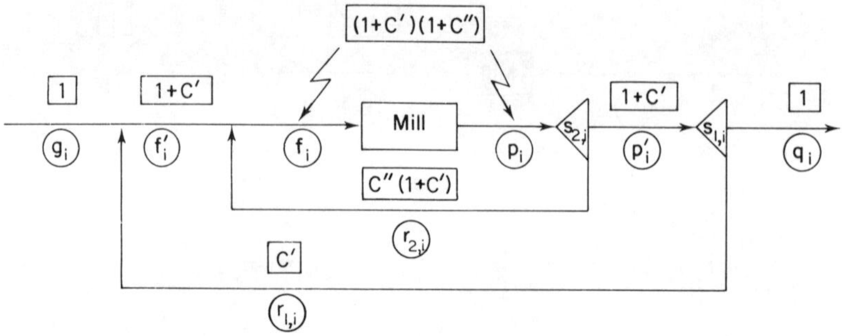

Figure 10.8. Nomenclature for classifiers in parallel

to the fine stream. Thus, for the normal configuration (Figure 10.5):

$$(1 - s_i) = q_i(1 + C')p_i \tag{10.1}$$

It is convenient in simulation exercises to express flow rates in terms of classifier selectivity and of particle size distributions in circuit feed g_i and mill product p_i. Taking the most general case, that of the combined closed circuit (Figure 10.7), the following are relevant (Luckie and Austin, 1972).

By total particulate mass balance on the coarse stream of the pre-classifier:

$$(1 - B') = \sum_i s_{i,j} g_i \tag{10.2}$$

Similarly for the node of classifier 2:

$$R' = F' \sum_i s_{2,i} p_i \tag{10.3}$$

and

$$F' - R' = 1 - B' \tag{10.4}$$

Hence:

$$F' = \frac{\sum s_{1,i} g_i}{1 - \sum s_{2,i} p_i} \tag{10.5}$$

By definition of $s_{1,i}$:

$$(1 - B')f_i' = s_i g_i$$

From classifier 2 coarse stream:

$$R' r_i = F' s_{2,i} p_i$$

A mass balance at the mill feed branch gives:

$$F' f_i = R' r_i + (1 - B') f_i' \tag{10.6}$$

Hence:

$$f_i = s_{2,i}p_i + \frac{s_{1,i}g_i(1 - s_{2,i}p_i)}{\Sigma s_{1,i}g_i} \tag{10.7}$$

and

$$r_i = s_{2,i}p_i/\Sigma s_{2,i}p_i \tag{10.8}$$

Similarly it may be shown that:

$$q_i = (1 - s_{1,i})g_i + \frac{(1 - s_{2,i})p_i\Sigma s_{1,i}g_i}{1 - \Sigma s_{2,i}p_i} \tag{10.9}$$

$$R' = \frac{s_{1,i}g_i\Sigma s_{2,i}p_i}{1 - \Sigma s_{2,i}p_i} \tag{10.10}$$

It is often useful to express the relationships in terms of the recirculation ratio C' for classifer 2. Then, from equation (10.4):

$$C' = \frac{R'}{F' - R'} = \frac{R'}{1 - B'}$$

or

$$(1 + C') = \frac{F'}{F' - R'} = \frac{F'}{1 - B'}$$

Then equation (10.6) may be written as:

$$s_{1,i}g_i = (1 + C')(f_i - s_{2,i}p_i) \tag{10.11}$$

Similarly it may be shown that:

$$q_i = (1 - s_{1,i})g_i + (1 - s_{2,i})p_i(1 + C') \tag{10.12}$$

Corresponding expressions may be obtained for the normal and reverse closed circuits as found in Rogers (1983b) and Gardner *et al.* (1984)

10.2.3 Transfer parameter concept

Luckie and Austin (1972) appear to be the first to have introduced the concept that a mill is a device which acts upon the feed mass–size distribution to give a product mass–size distribution. The set of parameters which convert feed to product is the transfer function for which the symbol $d_{i,j}$ is reserved, and which is defined as the mass fraction of material of size j in the feed to a mill which appears as size i in the product from the mill; or, expressed mathematically for the whole mill product:

$$p_i = \sum_{j=1}^{i} d_{i,j}f_j \text{ for } n \geqslant i > 1 \tag{10.13}$$

Equation (10.13) is only valid if no aggregation of particles occurs, but it does apply under steady- or non-steady state conditions (Rogers, 1983b).

Luckie and Austin (1972), in fact, solved the batch grinding equation (8.15) for the conditions of constant S_i and $b_{i,j}$, as pointed out by Rogers (1983b) in a manner which uses the transfer parameter as follows:

$$M_i(t) = \sum_{j=1}^{i} d_{i,j} M_j(0) \tag{10.14}$$

where

$M_i(0)$ = starting feed size distribution

$$d_{i,j} = \begin{cases} e^{-S_i t} & \text{for } i = j \\ \sum_{k=j}^{i-1} \psi_{i,k}\psi_{j,k}(e^{-S_k t} - e^{-S_i t}) & \text{for } i > j \end{cases}$$

$$\psi_{i,j} = \begin{cases} -\sum_{k=1}^{j-1} \psi_{i,k}\psi_{j,k} & \text{for } i < j \\ 1 & \text{for } i = j \\ \dfrac{1}{S_i - S_j} \sum_{k=j}^{i-1} S_k b_{i,k} \psi_{j,k} & \text{for } i > j \end{cases}$$

In computer software design for a complex circuit the use of sub-programs appropriate to specific modules that can be fitted into a main crushing and grinding programme has obvious advantages. Then the transfer concept allows a different sub-program appropriate to each type of mill to be inserted into a single overall program.

10.2.4 Some simulation problems

10.2.4.1 Types of problem

Because of the complexity of many plant circuits, particularly in the mineral beneficiation industry, simulation problems have arisen in relation to:

(1) steady-state mass balances (see §10.2.4.2)
(2) sensitivity analysis with respect to experimental error and model inaccuracy (see §10.2.4.3)
(3) real and unreal parameter values (see §10.2.4.4)
(4) true mill RTD assessment from closed-circuit data (see §10.2.4.5)
(5) differing requirements for control and design (see §10.2.4.6)
(6) incorporation of the slowing-down effect into a mill model (see §10.2.4.7)

These are briefly discussed in turn, but references are given for further reading where appropriate.

10.2.4.2 Steady-state mass balances

It often happens that streams in a practical circuit cannot be continuously and accurately monitored for flow rate and concentration, and consequently the properties of each stream have to be deduced from available and usually incomplete data by means of system material balances.

Mular *et al.* (1976) have highlighted the mass-balance problem by tabulating seven separate ways of obtaining the ratio of two flows in a particular mill system, and demonstrating that the ratio varied from 3.00 to 5.55 according to the method selected. Clearly the raw data are inconsistent. They go on to discuss systems of search variables whereby the best obtainable values of each flow rate may be arrived at to give a more consistent mass balance overall.

Three main types of solution to the mass flow problem are extant: simultaneous equations; iterative means; and vector balance equations.

Theoretically, it is possible to write a sufficient number of equations based on mass balances at nodes and branches for simultaneous solution to obtain all the unknowns, if adequate basic measurements are made (Hutchinson, 1974). However, it is easy to incorporate redundant equations (overdefined data) and thus not arrive at fully independent relationships. Further, according to Richardson and White (1982), there appears to be no satisfactory published computerized system for assistance in this area.

An alternative to simultaneous equations is an iterative system in which each circuit loop is isolated and solved for unknowns by trial and error, before proceeding to the next. The whole process is repeated to obtain convergence in terms of (guessed – calculated)/calculated quantities. The system is greatly simplified by dividing a complex circuit into modules; but even so, it is most tedious.

Sood *et al.* (1979) and Sood and Reklaitis (1979) have proposed a vector balance method which requires neither simultaneous solution nor iterative convergence. Instead, for each node and branch in the flow sheet, a vector balance equation is developed which contains as unknowns only mixed output streams which are sequenced for solution in a computer program named MBP-I, and an example of its application is given.

Other available computer programs are as follows, and references should be studied for details:

	Computer program
Hodouin and Everell (1980) Hodouin *et al.* (1980)	BILMAT
White *et al.* (1977)	SIMUMIL

10.2.4.3 Sensitivity analysis

(i) Klimpel and Austin (1984) have proposed a preferred statistical sensitivity analysis due to Blau *et al.* (1972) and based on the balance:

variation of observations = Variation of model
+ Variation due to model inaccuracy
+ Variation due to experimental error

Variation is assessed by the sum of squares (SSQ) method. An example of the above balance when applied to mill feed and product size distributions is:

$$\sum_i w_i(p_{i\mathrm{obs}})^2 = (SSQ_{\mathrm{mod}}) + \sum_i w_i(p_{i\mathrm{obs}} - p_{i\mathrm{calc}}) \tag{10.15}$$

In this case, the variance due to model inaccuracy and to experimental error are combined in the final term. The weighting factor w_i depends on the error structure of the experimetnal data and may be $1/\sigma^2$ (where σ^2 is the error variance), or $1/\bar{p}_i$ where $\bar{p}_i$ is the arithmetic mean of k observations of p_i. Then (SSQ_{mod}) is calculated from equation (10.15) knowing w_i, $(p_{i\mathrm{obs}})$ and $(p_{i\mathrm{calc}})$.

(ii) The first statistical test is based on the ratio of the variance of the model to the variance associated with residual error (re). Thus:

$$F_1 = \frac{(SSQ_{\mathrm{mod}})/n_{\mathrm{f}}}{(SSQ_{\mathrm{re}})/(n_{\mathrm{s}} - n_{\mathrm{f}})} \tag{10.16}$$

where n_{s} is the number of size intervals in the size distribution, n_{f} is the number of free parameters, and:

$$SSQ_{\mathrm{re}} = \sum_i w_i(p_{i\mathrm{obs}}) - p_{i\mathrm{calc}})^2$$

Normally F_1 is highly significant as judged by published F tables (Perry, 1963) at a given confidence level, e.g. 95%.

(iii) The second statistical test is to compare the variances of residual error given for the different values of the search parameters. For example, if S_i is being considered, then α, x^*, a^* or Λ (see §8.2.2.3) may be used to find the optimal set of parameters, giving an associated SSQ_{opt}. One selected parameter, say α, is then changed by 5% from the optimal value and held constant at the new value while the search is repeated on the other parameters, giving an associated $SSQ_{1,\mathrm{re}}$. Then:

$$F_2 = \frac{(SSQ_{1,\mathrm{re}})/(n_{\mathrm{s}} - n_{\mathrm{f}} + 1)}{(SSQ_{\mathrm{opt}})/(n_{\mathrm{s}} - n_{\mathrm{f}})} \tag{10.17}$$

If F_2 does not show high significance, this means that the new value of $\alpha(1{\cdot}05\,\alpha_{\mathrm{opt}})$ plus its associated parameters are as good as the optimal values, within the accuracy of the model and the experimental error. This process may then be repeated for another parameter, giving $(SSQ_{2,\mathrm{re}})$ and so on.

(iv) Experimental error may be assessed more completely by means of the lack of fit sum of squares, SSQ_{lof}, given by

$$(SSQ_{lof}) = \left[\sum_{i=1}^{n} \sum_{l=1}^{k} w_i (p_{i,l} - \bar{p}_i)^2 \right] \Big/ (k-1) \tag{10.18}$$

where $\bar{p}_i$ is the arithmetic mean of k observations of $p_{i,l}$. In this case, the number of degrees of freedom is $n_s(k-1)$. An F test may then be applied to the ratio of variance of the equation residual error to the independent experimental error, giving:

$$F_3 = \frac{(SSQ_{re})/(n_s - n_f)}{(SSQ_{lof})/n_s} \tag{10.19}$$

If the F test on this basis with degrees of freedom $(n_s - n_f)$ and $n_s(k-1)$ does not indicate significant difference, i.e.:

$$F_3 < F_{(n_s - n_f), n_s(k-1)}$$

this shows that the optimal model parameters give a model which does not describe the data within the expected accuracy of the experimental data. In this sense, equation (10.19) becomes a measure of model accuracy.

(v) Similarly, to assess sensitivity of model adequacy to the choice of parameters:

$$F_4 = \frac{(SSQ_{1,re})/(n_s - n_f + 1)}{(SSQ_{lof})/n_s} \tag{10.20}$$

When $F_4 > F_{(n_s - n_f + 1), n_s}$ the new parameters give a statistically significant worse match of the model to experiment, and vice versa.

(vi) These four forms of the statistical F test have been found of great value in sensitivity analysis, and Klimpel and Austin (1984) give examples of their use.

(vii) By application of F_2 to a single node separator, in which an entry stream is divided into two streams, Hodouin *et al.* (1982) derived a system from which the following was concluded:

(1) The reliability of the flow rate estimates increases when the concentration ratio of a given component in the two product streams is high and when the overall partition coefficient is far from 0.5 (e.g. 0.9 in one stream and 0.1 in the other).
(2) The reliability of the estimate increases when the number of components analysed increases, but with the proviso that beyond five components, that increase is very slow.
(3) The component which mostly increases the flow rate estimate reliability is that which concentrates in the low flow rate product.

An example is given of the application of this system to a cement clinker grinding circuit.

10.2.4.4 *Real and unreal parameter values*

It was noted in §8.5.3.3(c) in the context of batch grinding that it is quite possible to use a simulation which arrives at a correct product size distribution but breakage functions are not true. Similarly in the context of continuous grinding, Kelsall *et al.* (1968), for example, incorporate measured RTD values but use a single mathematical expression for B values for all ores and minerals. This introduces errors into the back calculated S values, if the B values are not, in fact, true. This type of simulation may be adequate to predict the effect of feed size, feed rate or of classifier behaviour, but is able to give no information on the effect of change of ball size, slurry density, ball load or mill diameter (Austin and Weller, 1982).

Again, if unreal RTD functions are employed in the mathematical model, this has its repercussion on the breakage functions. For a fuller discussion see Klimpel and Austin (1984), where sensitivity analysis is applied to back calculation of breakage functions using real and unreal RTD values. They conclude that any back calculation which assumes tumbling mills are fully mixed, such as in Lynch (1977), gives radically incorrect values of S when applied to data from a mill with an RTD such as those actually measured experimentally. They assert that in the presence of typically full-scale experimental error, this incorrect model cannot be detected by statistical analysis.

This does not mean that it is immaterial that false S values are produced, because product size distributions are incorrectly predicted if a simulation is performed at any other condition than that used in the back calculation. Hence it is recommended that as many as possible size distributions around the closed circuit are taken at steady-state conditions and used in the simulation, because this leads to a smaller range of statistically acceptable values of parameters, and in particular gives α values in equation (8.2) which are more in agreement with those determined by direct experimental in laboratory- or pilot-scale mills. There appears to be a general trend for back calculated α values to be higher than those found experimentally, and Klimpel and Austin (1984) recommend further research to obtain full explanations of this behaviour.

10.2.4.5 *True mill RTD assessment from closed-circuit data*

In §8.6.3, both the single- and double-detector SLRT methods of determination of mill RTD from closed-circuit data are described. These methods are very desirable in that they allow RTD assessment without disturbing the process flows, but they have posed problems arising from the added requirement of a tracer impulse function to an already complex situation. Gardner *et al.* (1982) have addressed themselves to this problem and have produced computer programs which incorporate the impulse function together with either the reactors-in-series model (see §8.6.5.2(d)) or the finite-stage model (see §8.6.5.3(b)) for the RTD function, whichever is most appropriate. Examples are given of the application of each model.

10.2.4.6 *Differing requirements for control and design*

Generally, when the control or optimization of an existing plant is the purpose of mathematical simulation, it may well suffice to work with lumped parameters or unreal values of breakage parameters and the RTD function, if thereby the product size distribution is attained to sufficient accuracy.

By contrast, for new plant design, it is most desirable to obtain true values, as discussed by Klimpel and Austin (1977, 1984). Clearly, if the work involved is justified, control or optimization would also be better performed by the use of true values, than otherwise.

Mill design programs exist, of course, which use lumped parameters, and examples are given in §10.3; but these are classed at a lower level than those which aim at individual and true values of the variables.

10.2.4.7 *Incorporation of the slowing-down effect into a mill model*

Austin *et al* (1984) report preliminary work on the incorporation into a mill model of the slowing-down effect discussed in §8.2.2.8.

Equation (8.15) is the starting point, but it is not possible to use equation (8.56) based on the dimensionless RTD function, because the condition of first-order breakage does not apply with fines build-up. Hence recourse must be made to the convective diffusion equation (8.44).

Austin *et al.* visualize a steady-state mass transfer along the mill length with a superimposed forward–backward mixing caused by the tumbling of mill contents. This mixing, however, gives no net mass transfer because equal masses of all materials are thrown forward and backward with equal probability, and no discrimination in respect of particle size is assumed.

Then combining equations (8.15) and (8.44), and using the parameters $l^* = l/L$, $D^{**} = \bar{v}L$ and $S_j^* = S_j\tau$, it is possible to write:

$$0 = D^{**}\frac{\mathrm{d}^2}{\mathrm{d}l^{*2}} M_i - \frac{\mathrm{d}}{\mathrm{d}l^*} M_i - M_i S_i^* + \sum_{\substack{j=1 \\ i>1}}^{i-1} b_{i,j} S_j^* M_j \qquad (10.21)$$

Using the boundary conditions at feed entry ($l^* = 0$) we have:

$$f_i = M_i(0) - D^{**}\frac{\mathrm{d}}{\mathrm{d}l^*} M_i \qquad (10.22)$$

Using the boundary condition at mill exit ($l^* = 1$) we have:

$$\left.\frac{\mathrm{d}}{\mathrm{d}l^*} M_i\right|_{l^*=1} = 0 \qquad (10.23)$$

Then (10.21), (10.22) and (10.23) are the relevant equations for numerical solution, and it is assumed that all particles move through the mill at the same velocity $\bar{v}$, and that D^* is constant along the mill and independent of particle size. Further it is assumed that the hold-up level along the mill is approximately constant, i.e. that only small gradients in hold-up are required to give bulk movement of powder or slurry through the mill.

The appendix to Austin *et al*. (1984) gives the steps in the numerical solution to equations (10.21), (10.22) and (10.23).

Iteration starts from the coarsest possible size distribution, the mill feed. At the *k*th iteration, the values of $M_i(k)$ are used to calculate $S_i^*(k)$ from:

$$S_i^*(k) = \varkappa \tau S_i(\text{input}) \tag{10.24}$$

where $\varkappa$ is the slowing-down parameter, given by;

$$\varkappa = \varkappa_0 + \frac{(1 - \varkappa_0)}{1 + (x_{80}/x^*_{80}) - \Delta} \tag{10.25}$$

with $\varkappa_0$ the limiting value of $\varkappa$ for fine distributions. The larger the value of x^*_{80}, the coarser the size distribution at which the slowing down is observed; and the larger the value of Δ the more rapidly $\varkappa$ decreases as the size distribution becomes finer.

The work has been applied only to open circuit so far, and Figure 10.9 shows the relation between $\varkappa$ and x_{80}/x^*_{80} for various values of $\varkappa_0$ and Δ using equation (10.25).

The effect of the slowing-down effect on mill capacity is given in Figure 10.10 for 40 and 54 volume % solids in the wet grinding of quart, with D^* taken as 0.2.

In closed circuit, the slowing down is clearly more complex to analyse because it depends on the degree of recycle and the efficiency of the classifier.

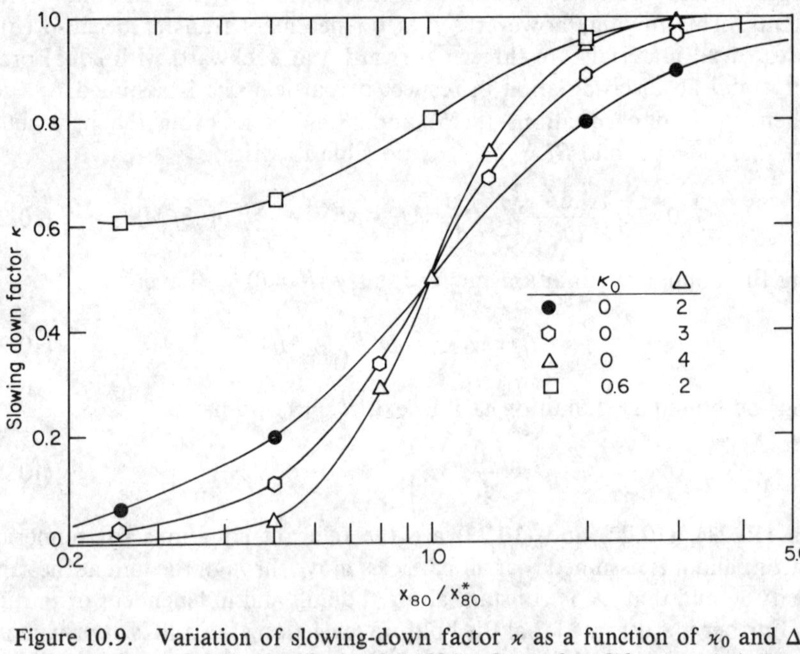

Figure 10.9. Variation of slowing-down factor $\varkappa$ as a function of $\varkappa_0$ and Δ (see equation (10.25)) (Austin *et al*., 1984). *Reproduced by permission of Hemisphere Publishing Corporation*

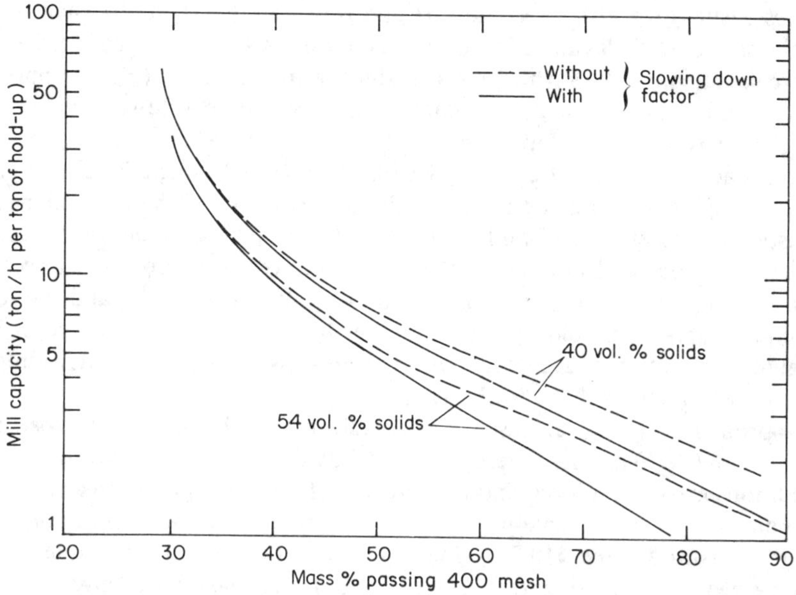

Figure 10.10. Effect of slowing-down factor on mill capacity—open circuit (Austin *et al.*, 1984). *Reproduced by permission of Hemisphere Publishing Corporation*

However, the effect is less than in open circuit because the size distributions being acted upon are much coarser owing to the recycle from the classifier. On the other hand, a closed circuit implies higher mass flow rates through the mill, which increases the overfilling effect. (See §8.2.2.7 and Figure 8.13.)

10.2.5 Approaches to simulation

With regard to the use of full mathematical models, Austin and Weller (1982) discuss three general approaches to the simulation of size reduction in large-scale, steady-state continuous tumbling mill circuits:

(1) Size distributions and RTD values are taken on the large-scale plant itself and breakage functions are back calculated from them.
(2) Breakage and RTD functions are measured on a small continuous mill of the same type and under similar conditions as the large-scale mill. The values so found are scaled-up to the larger mill size.
(3) Breakage functions are measured in a laboratory batch mill. These, scaled-up to the large mill, are combined with RTD functions assessed from the large mill itself.

Approach 1 is exemplified in Stewart and Restarick (1971), Lynch (1977) and Marchand *et al.* (1980). However, as indicated in §8.6.6.1, it is most difficult to obtain breakage functions from full-scale mill data because unreal values ensue. (See §10.2.4.4) Stewart and Restarick (1971) and Lynch (1977) used a single set of B values for all ores and minerals, which introduced errors into calculated S values when B values did not match the true. This approach is valid only when prediction of changes in feed size or rate is contemplated or when classifier behaviour is studied.

Approach 2 has been adopted by Kelsall *et al.* (1968) using a laboratory mill and by Rogers (1979) using a pilot-scale mill, because of the convenience of investigation of change of mill conditions. However, the values of S have to be back calculated assuming B values unless tracer tests are used, and even then it is not always easy to separate B and S values accurately, and the RTD function varies with hold-up. On the other hand, a major advantage of this technique is that breakage occurs as particles pass through mill environments which are similar to those in the large-scale mill.

Approach 3 has been used by Austin *et al.* (1976) and Herbst and Fuerstenau (1980) among others. It is certainly the easiest when a range of mill conditions is to be explored and it is capable of non-ambiguous measurements of S and B. However, scale-up from laboratory- to full-scale conditions is an essential feature. (See §10.2.6) The simplest experimental technique uses a single-sized feed fraction (see §8.5.2.2), but this has the disadvantage of precluding breakage to be in the true environment of the continuously operated mill.

Some workers have combined these approaches. For example, Austin and Weller (1982) apply features of approaches 2 and 3 above for the special conditions associated with an ore having three discrete components being ground in an overflow wet mill. (see §10.3.2.4(b))

10.2.6 Scale-up of essential parameters

10.2.6.1 Specific rate of breakage

It is assumed in the following that scale-up of tumbling mills is performed at constant N^*, J and U, with constant linear geometry and with grinding balls of the same density and hardness. Then the remaining variables to consider in the scale-up of specific rate of breakage S are listed in Table 10.1, together with the appropriate equations and section references.

When the above constant conditions do not apply, except for uniform ball diameter and linear mill geometry, Austin *et al.* (1982a) propose that the following equation may be used for scale-up of S_i:

$$S_i = \frac{a^*(x_i/x_0)^\alpha}{[1 + x_i/Z_1 x^*]^\Lambda} Z_2 Z_3 Z_4 Z_5 \tag{10.26}$$

where

$$Z_1 = \left(\frac{D}{D_T}\right)^{\xi_1} \left(\frac{d}{d_T}\right)^2$$

Table 10.1. Variables to consider in the scale-up of the specific rate of breakage S_i for constant N^*, J and U, linear geometry and ball density

Variable	Relation	Section reference
x	$S_i = a^*(x_i/x_0)^\alpha/[1 + x_i/x^*]^\Lambda$	8.2.2.3 equation (8.3)
D	$S_i \propto D^{0.5}$ if $D < 12.5$ ft (3.81 m) $S_i \propto D^{0.3}$ if $D > 12.5$ ft (3.81 m)	8.2.2.5 (a)
L	S_i assumed invariant	8.2.2.5 (b)
Ball size (mixture)	$\bar{S}_i = \sum_k S_{i,k} M_k$	8.2.2.6 equation (8.6)

with $0.1 > \xi_1 > 0.2$, from equation (8.4a) and §8.2.2.3

$$Z_2 = \left(\frac{d_T}{d}\right)^{\xi_2} \text{ with } \xi_2 \text{ usually} = 1.0 \text{ from equation (8.5)}$$

$$Z_3 = \left(\frac{D}{D_T}\right)^{\xi_3} \text{ with} \begin{cases} \xi_3 = 0.5 \text{ for } D > 12.5 \text{ ft} \\ \xi_3 = 0.3 \text{ for } D < 12.5 \text{ ft from §8.2.2.5(a)} \end{cases}$$

$$Z_4 = \left(\frac{1 + 6.6 J_T^{2.3}}{1 + 6.6 J^{2.3}}\right) \exp[-K_1(U - U_T)] \text{ from equation (8.8)}$$

$$\text{where } K_1 \begin{cases} = 1.32 \text{ for wet milling} \\ = 1.20 \text{ for dry milling} \end{cases}$$

$$Z_5 = \left(\frac{N^* - 0.1}{N_T^* - 0.1}\right) \left\{\frac{1 + \exp[15.7(N_T^* - 0.94)]}{1 + \exp[15.7(N^* - 0.94)]}\right\} \text{ from equation (8.4b)}$$

where the suffix T denotes the test mill parameter.

If a mixture of balls is used, then equation (8.6) must be applied to obtain a mean value $\bar{S}_i$ applicable to the particular ball mix of the large-scale mill.

10.2.6.2 Breakage distribution function

The breakage distribution function calculated from equations (8.13) or (8.14) is regarded as invariant in scale-up apart from the effect of change of ball mix, by the application of equation (8.14a).

10.2.6.3 RTD function

Where the RTD function is assessed in a laboratory or pilot plant, as distinct from the full-scale plant itself (see §10.2.5), then scale-up of the RTD function is necessary. However, for constant linear geometry and constant N^*, the dimensionless RTD function $\phi(\theta)$ appears to be invariant over a range of mill sizes, as long as overfilling of the mill is avoided (Austin *et al.*, 1982a).

For a detailed study of scale-up of the RTD function in tumbling reactors, the following papers should be studied: Abouzeid *et al.* (1980), and Abouzeid and Fuerstenau (1980).

10.2.6.4 Classifiers

Classifier data are normally found from the full-size classifier, and §10.2.2 gives the information required.

10.2.7 Equation for circuit product size distribution

The basis for the calculation of the circuit product size distribution is equation (10.13) which applies to continuous flow of material through a mill.

Combining equations (10.11), (10.12) and (10.13) and putting $p_i(1 + C') = p_i^*$, Austin *et al.* (1982a) obtain:

$$p_i{}^* = \frac{d_{i,i}s_{1,i}g_i + \sum_{j=1}^{i-1} d_{i,j}(s_{1,j}g_i + s_{2,j}p_j^*)}{1 - d_{i,i}s_{2,i}} \quad (10.27)$$

$$\text{for } n > i > j > 1$$

where:

$$d_{i,j} = \begin{cases} e_j & \text{for i = j} \\ \sum_{k=j}^{i-1} \psi_{i,k}\psi_{j,k}(e_k - e_i) & \text{for } i > j \end{cases}$$

$$\psi_{i,j} = \begin{cases} \sum_{k=1}^{j-1} \psi_{i,k}\psi_{j,k} & \text{for } i < j \\ 1 & \text{for } i = j \\ \frac{1}{S_i - S_j} \sum_{k=j}^{i-1} S_k b_{i,k}\psi_{k,j} & \text{for } i > j \end{cases}$$

$$e_j = \int_0^\infty e^{-S_j t}\phi(t)\,\mathrm{d}t$$

Note that the RTD function $\phi(t)$ is introduced into the expression for e_j, and hence the form of e_j depends on the particular model chosen for the RTD function. For example, if the 'one large and two small fully mixed reactors in series' model is selected (see §8.6.5.2(d)), then the equivalent RTD leads to:

$$e_j = 1/(1 + S_j\tau_1)(1 + S_j\tau_2)^2 \quad (10.28)$$

and corresponding values of e_j can be found for other expressions of the RTD function.

Then the set of equations (10.27) with the appropriate e_j equation may be used to calculate the mill product size distribution p_i from a known make-up feed distribution g_i for a set value of τ (i.e. a set value of $d_{i,j}$) proceeding in a sequential manner, starting with $i = 1$. The recirculation ratio C' of classifier

2 may then be found from:

$$1 + C' = \sum_{j=1}^{n} p_j^* \tag{10.29}$$

and the values of p_i follow from:

$$p_i = p_i^*/(1 + C') \tag{10.30}$$

Similarly, the circuit product distribution q_i follows from equation (10.12), and the mill feed distribution f_i from equation (10.11).

The mill output rate Q may be derived from:

$$Q = \frac{W^*}{(1 + C')\tau} \tag{10.31}$$

where τ is the mean residence time in the mill.

For set mill conditions in which $S_i, b_{i,j}, s_{1,i}, s_{2,i}$ and g_i are specified, the value of τ can be varied to give a desired one-point match (e.g. 75% minus 75 micron) in the circuit product size distribution, and it is usual to include in the computer program designed to handle equation (10.27) a search routine to find the required value of τ. C' and W^* then follow from equations (10.29), (10.30) and (10.31).

Similarly, if the classifier values are entered into equation (10.27) in a functional form, having parameters a', d_{50} and SI, it is possible to vary a' and/or d_{50}, as well as τ, to obtain a one-point match at a desired value of C'.

10.3 CLOSED CIRCUITS INVOLVING TUMBLING MILLS

10.3.1 Survey: degrees of sophistication

Examples are taken at each of four levels of sophistication which have been incorporated into the simulation of closed circuits:

(a) Level I No SMB equation is used in the model, but only the minimum essential system elements are retained, which are regarded as adequate for optimization or control of an existing plant.

(b) Level II An SMB mathematical model is used with the following simplifying assumptions:

(1) A constant form of breakage distribution function is used. An early popular form was that due to Broadbent and Callcott (1956) given by:

$$B(x, y) = (1 - e^{-x/y})/(1 - e^{-1}) \tag{10.32}$$

which is now known not to be fully representative of B variations.

(2) An inadequate RTD function is applied. For example, the mill may be

assumed to operate in a fully mixed mode, which is only an approximation as indicated in §8.6.2.

(3) Classification may not be adequately represented.

Such models have been applied to plant optimization or control, but in view of the assumptions made, there remain reservations about the validity of the models (Klimpel and Austin, 1984).

(c) Level III An SMB mathematical model is employed in which

(1) lumped breakage parameters may occur, such as $S_i b_{i,j}$ (linear breakage assumed);
(2) RTD function assessment is such that the values may wander from reality;
(3) simplifying classifier assumptions are made.

Many of these models have demonstrated sufficiently accurate forecast of product size distribution for limited variation of such variables as feed size, feed rate or classifier behaviour, and some have been advanced for mill design.

(d) Level IV Into the SMB equation are introduced:

(1) discrete breakage functions;
(2) true RTD functions;
(3) valid classifier selectivity numbers;
(4) provision for linear or non-linear breakage as applicable.

Such models lend themselves to design of new mill circuits and to better optimization or control of existing plant than those at lower levels of sophistication, and they apply over a wider range of operational variables.

In the following sections, arranged according to mill type, each of the above levels is illustrated, except that only one example is taken of level I simulation, that in the context of a wet ball mill, since it is marginal to the theme of the handbook.

Clearly, still more sophisticated models could be conceived; but for each rise in level, penalties are paid in greater complexity of equations and in an increase in the number of equations.

10.3.2 Wet ball mill

10.3.2.1 Level I simulation

The following is geared to wet mill operation, but is really one facet of the control of grinding equipment in general. For a good summary of control in the context of crushing and grinding, see Marshall (1974), and for a specific example for a closed circuit, see Forman *et al.* (1977).

Table 10.2. Process matrix showing response of controlled variables y_1^*, y_2^* etc. to change in manipulated variables u_1, u_2 etc.

Manipulated variables	Controlled variables: y_1^* = product fineness	y_2^* = circulating load	y_3^* = sump level	y_4^* = % mill solids
u_1 = sump water addition rate	+ fast	+ fast	+ fast	+ − slow
u_2 = fresh water feed solids rate	− slow	+ fast	+ slow	+ fast
u_3 = cyclone feed pumping rate	+ fast	+ fast	− fast	+ fast
u_4 = feed water addition rate	+ − slow	+ slow	+ slow	− fast

Reproduced by permission of the American Institute of Mining, Metallurgical and Petroleum Engineers.

Now a detailed SMB model for a multi-module circuit may involve many scores of differential equations which are quite beyond the capacity of, say, a microcomputer. Hence Herbst and Rajamani (1980) have advanced a pragmatic approach to closed-circuit milling systems in which the basic variables y_1^*, y_2^* for control are defined together with appropriate manipulating variables u_1, u_2 etc. A process matrix may then be developed from plant experience, exemplified in Table 10.2 for wet ball mill operation.

Classical controller methods have used a feedback control law which connects u and y^* in the relation:

$$u_i = CS_1(y_j^* - y_j^{*\text{sp}}) + CS_2 \int_0^t (y_j^* - y_j^{*\text{sp}})\, \mathrm{d}t \tag{10.33}$$

where $(y^* - y_j^{*\text{sp}})$ is the deviation from the set point and CS_1, CS_2 are control settings.

Equation (10.33) is geared to a strategy in which one or more of the variables are held at or near preselected set-point values and provision is made for single-input/single-output control loop linking.

More recently, to allow feedback control in a multivariable sense, u and y^* have been related by:

$$u_i = (y_j^* - y_j^{*\text{sp}}) \begin{bmatrix} \varepsilon_{1,1} & \varepsilon_{1,2} & \cdots & \varepsilon_{1,n} \\ \varepsilon_{2,1} & \varepsilon_{2,2} & \cdots & \varepsilon_{2,n} \\ \vdots & \vdots & & \\ \varepsilon_{m,1} & \varepsilon_{m,2} & \cdots & \varepsilon_{m,n} \end{bmatrix} \tag{10.34}$$

where n is the number of measured variables, m is the number of manipulated variables, and ε are constants or operators as required.

Several examples of the application of the above are given in Herbst and Rajamani (1980).

10.3.2.2 Level II simulation

Equation (4.37a) has been applied by Lynch (1977) to a single mill. For the normal closed circuit it is modified to

$$q = g(\mathbf{I} - \mathbf{c})(\mathbf{BS} + \mathbf{I} - \mathbf{S})[(\mathbf{I} - \mathbf{S})(\mathbf{BS} + \mathbf{I} - \mathbf{S})]^{-1} \tag{10.35}$$

Lynch (1977) shows the way equation (10.35) is used in his example 4.3. Olsen and Krogh (1975) also give a comparison of simulated and experimental product size distributions by a similar method. Further, Hodouin *et al.* (1980) describe a modified level II simulation which uses the BILMAT program for material balance round a three-stage grinding circuit for a fine-grained heterogeneous ore. The circuit consists of a rod mill followed by two ball mill/hydrocylone stages in series. Matrix notation is employed for the breakage equations. The authors admit that the weakest element in their system is the modelling of the hydrocyclone and that the description of the grinding kinetics within the mill is in need of improvement. However, good predictions of circuit product size distribution were achieved over a limited range of feed.

10.3.2.3 Level III simulation

(a) Mill design Herbst *et al.* (1971) have advanced a model involving constant net specific energy for grinding and based on the following assumptions: linear breakage, a lumped breakage function $S_i B_{i,j}$ and a classifier performance independent of its feed distribution. Matrix notation is used and expressions are presented for both normal and reverse closed circuits.

Herbst and Rajamani (1982) maintain these principles in a computer program ESTIMILL (Herbst *et al*; 1977) for parameter estimation together with a further program SCALEMILL for scaling-up from laboratory to commercial mills operated in both open and closed circuit. The assumed constant specific energy $\bar{E}$, however, would appear to be valid only if J and U are also maintained constant in the scale-up, because $\bar{E}$ has been shown to vary with J and U according to the pattern of Figure 10.11 (Shoji *et al.*, 1982).

ESTIMILL and SCALEMILL are available at the University of Utah, USA. Steps in the closed-circuit process are:

(1) Choose two initial mill input energy levels $\bar{E}_1$ and $\bar{E}_2$ which are ± 0.5 kWh/ton over that given by a Bond test in the material being ground.
(2) Compute the feed rate of water to the hydrocyclone WF to make up to a preselected percentage of solids in the cyclone feed.

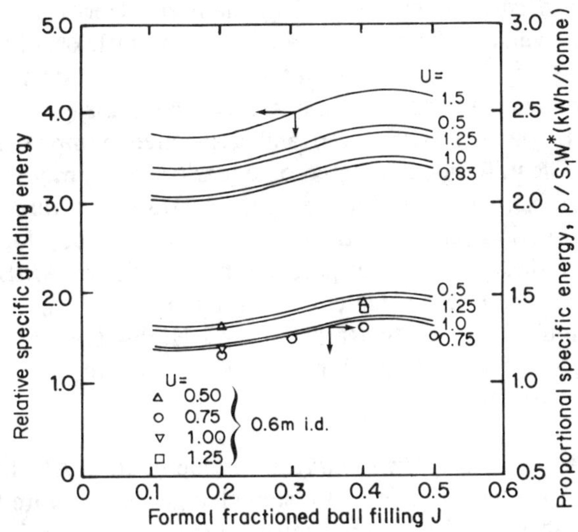

Figure 10.11. Variation of ball mill relative specific grinding energy with ball and powder filling (Shoji *et al.*, 1982). *Reproduced by permission of Elsevier Sequoia*

Relative specific energy is proportional to:

$$\frac{1 - 0.9375\,J}{1 + 5.95(J)^5} \Bigg/ \frac{Ue^{-1.2U}}{1 + 6.6\ J^{2.3}}$$

(3) Calculate the classifier matrix from the hydrocyclone model equations.
(4) Solve the closed-circuit grinding model equations.
(5) Interpolate between the computed circulating loads from step 4 for the energy input $\bar{E}^*$ which would result in the desired circulating load.
(6) Solve the closed-circuit grinding model equations again with the interpolated value of $\bar{E}$ from step 5. If the result is not within the specified convergence limit, repeat steps 5 and 6.
(7) Find x_{80} from the hydrocyclone overflow product distribution. If this is not within the specified convergence limit, adjust *WF* accordingly. If convergence is obtained, omit step 8.
(8) Set two new energy levels 0.5 kWh/ton above and below the previously converged $\bar{E}^*$ from step 6. Repeat steps 3–7.
(9) With final $\bar{E}^*$, compute the mill power required.
(10) From the mill power, derive mill diameter and length from appropriate scale equations.

For more details, see Herbst and Rajamani (1982).

For further reading with particular reference to methods of data acquisition, the following papers should be consulted: Mular (1970, 1971).

(b) Evaluation of control strategies Rajamani and Herbst (1980) have produced at the University of Utah a computer program DYNAMILL with the special purpose of assessment of alternative control strategies prior to experimental implementation. Basically a closed circuit is envisaged including a wet ball mill, together with sump and associated pump in series with a hydrocyclone. The mill model is based on equation (8.15) modified to accommodate the n mixers-in-series type of particle transport through the mill, together with equations covering the sump and the hydrocyclone, which are placed in sub-routines. This makes possible the use of DYNAMILL for either open circuit or normal or reverse closed circuit.

Reference should be made to Rajamani and Herbst (1980) for full details and for examples of the application of the program to circuit tuning and variation of control strategy.

(c) Dynamic response to system changes Inoue *et al.* (1980) have used the SMB approach to study the effect of a step change, in feed rate for example, on the system variables. Their paper must be referred to for details, but the kind of changes their program forecasts are given in Figure 10.12 for closed-circuit grinding. Such dynamic responses have been observed in actual plant operation.

10.3.2.4 Level IV simulation

(a) Homogeneous feed The SMB method already outlined in §10.2.7 following Austin *et al.* (1982) is in fact a level IV simulation. Austin and Brame (1983) have amplified certain aspects of the method, including the slowing-down effect at long times of grind (see §8.2.2.8), but for the most part they verify the earlier paper and underline the advantages of the level IV simulation over the Bond method of mill design.

(b) Heterogeneous feed For a heterogeneous feed, Austin and Weller (1982) find that modifications to the procedures of §10.2.7 are necessary, chiefly with respect to classification.

The example taken is the treatment of an ore containing marmatite ($SG = 4.0$) and galena ($SG = 7.5$) in a silicaceous gangue of $SG = 3.2$. Statistical examination of the data from classifier tests revealed that it did not follow the normal pattern when judged by a form of equation (5.41), but in fact by the equation:

$$S_i = a' + (1 - a')\left[\frac{1}{1 + (d_{50}/x_i)^{\lambda}}\right] \qquad (10.36)$$

where λ is the characteristic slope of the corrected selectivity curve, and the sharpness index is given by:

$$SI = \exp(-2.1792/\lambda) \qquad (10.37)$$

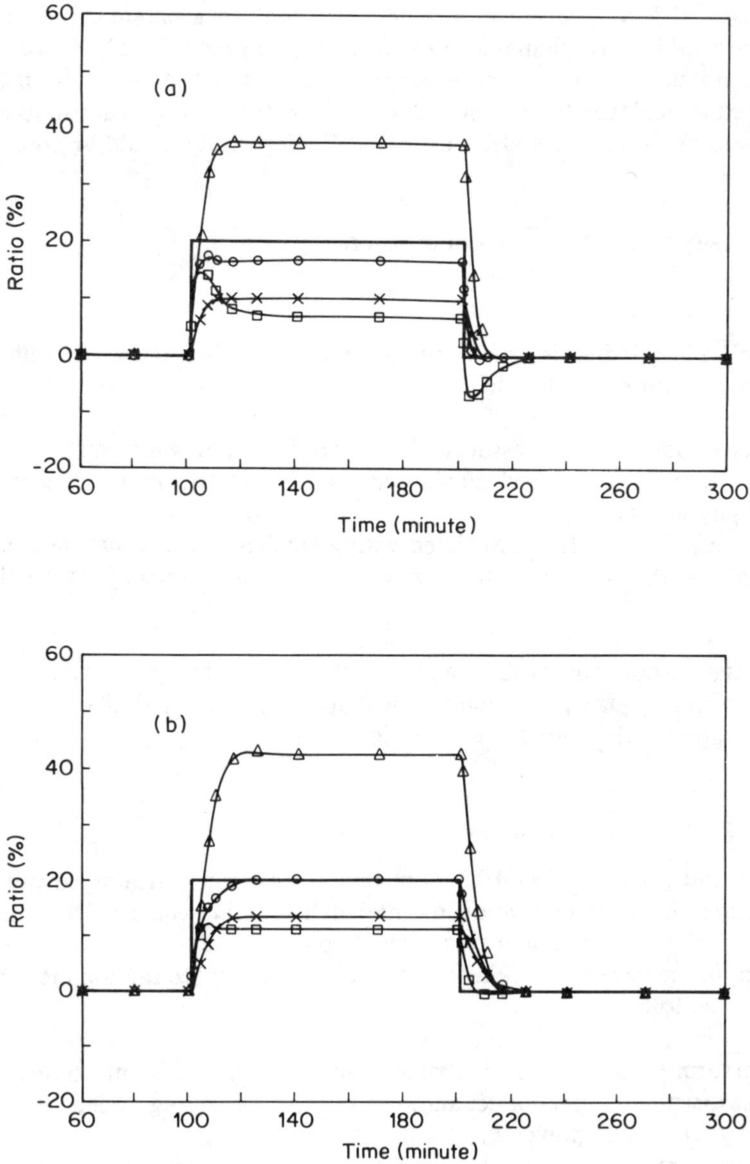

Figure 10.12. For wet grinding in a ball mill, the simulated effect of a stepwise change in feed rate on circulating load and product flow rate (at three finenesses) at (a) low and (b) high recirculation ratio (Inoue *et al.*, 1980). *Reproduced by permission of DECHEMA, Frankfurt*

× circulating load
□ fine-product flow rate
○ medium-product flow rate
△ coarse-product flow rate

The value of λ for the gauge, at 1.04, was found by separate experiment to be considerably lower than that for marmatite and galena (1.35). Hence, when for the mixture the normal procedure was applied at back calculation of S_i, it proved difficult to obtain valid values in the presence of unreal values of the variables. For further details, Austin and Weller (1982) should be consulted.

10.3.3 Dry ball mill—airsweeping ignored

10.3.3.1 Preview

Dry ball mill grinding is usually simpler than wet grinding for the following reasons (Austin *et al.*, 1975):

(1) Wet grinding is often associated with ore liberation which may involve a very heterogeneous feed, having components of widely differing specific gravity and hardness.
(2) Dry grinding is often associated with a smaller range of hardness in the feed and the effect of pulp density, so important a factor in wet milling, is absent.

Hence there would appear to be a greater chance that deductions made on one clinker grinding plant, for example, will apply to another similar plant than for two apparently similar ore liberation plants.

10.3.3.2 Level III simulation

Herbst and Fuerstenau (1980) adopt matrix notation and lumped breakage parameters in their application of the batch grinding equation to a normal closed-circuit dry grind mill in a cement context.

Steps in the scale-up of data from the laboratory mill to the commercial size mill are as follows:

(1) Perform batch grinding experiments in a reference ball mill to obtain the size distribution of product and, via a torque measuring device, obtain the net mill specific power p_s.
(2) Estimate the feed size breakage rate function via the slope of a line on a semi-log plot as in Figure 8.2.
(3) Estimate the feed size breakage distribution function by the method of §8.5.3.2(a).
(4) Estimate the non-feed size kinematic parameters by equation (8.24).
(5) Calculate the specific breakage rate function as the product of S_i from step 4 and p_s from step 1.
(6) Calculate the specific energy to satisfy design conditions.
(7) Assess the mill size from the mill power calculated in step 6.

This system assumes constant net specific energy between laboratory and commercial sized mills and therefore requires the same qualification with regard to ball and powder filling as that of §10.3.2.3(a).

A rather different level III simulation is suggested by Onuma *et al.* (1976) for a normal closed dry grinding circuit in the cement industry, but for which airsweeping is not taken into account. Matrix notation is employed in the lumped-parameter equation for breakage within the mill and also in the treatment of the classifier performance. Only a fair degree of correlation is demonstrated between calculated and observed production rates of fine product.

10.3.3.3 Level IV simulation

Austin *et al.* (1975) commenced a study on a normal closed-circuit cement clinker grinding mill, which was further pursued by Austin *et al.* (1976), with the limited objective of determination of the optimum product size distribution for strength development of concrete. The pattern of simulation generally followed that of §§10.2.6 and 10.2.7 with modifications for the mill having two compartments, with larger balls in the first than in the second. The RTD function was measured by the fluorescein technique and simulated by ten fully mixed reactors in series. The breakage function was found to be non-normalizable and adjustment made accordingly.

The model produced simulations of output size distribution which agreed with experimentally found values, and was further used to predict the size distribution of cements which could be produced from closed-circuit operation with a more efficient classifier. It was concluded that a substantial increase in output and reduction in energy could be achieved with a separator having a sharper cut size. This has been verified in practice more recently with the advent of the O'Sepa classifier. (see §5.4.2.2(iii))

10.3.4 Dry ball mill with airsweeping

10.3.4.1 Effect of airsweeping

Airsweeping is introduced into a cement tube mill primarily to cool the clinker during grinding. However, as pointed out by Austin *et al.* (1980), airsweeping has three additional functions which affect the grinding process:

(1) removal of fines from the mill hold-up, thus giving an advantageous internal classifying action;
(2) fines removal also reduces the severity of the slowing down effect (see §8.2.2.8);
(3) the distribution of organic vapours, used as grinding aids, is thereby improved.

As a dry ball mill rotates, a free flowing layer of powder and balls rolls down the inclined surface of the tumbling mass and the air picks up powder from the layer at a rate which depends on the air velocity and particle size. Fine particles are carried forward with the air stream, and in a short mill they will be carried from the mill. (In a long mill provision may need to be made for large particles falling back into the mill.)

10.3.4.2 Level IV simulation

A relatively short mill, as used for coal grinding, is discussed by Austin *et al.* (1984) in the context of a normal closed circuit. The mill is conceived as m stages in series, in the ith stage of which a fraction υ_i is removed by the air after a residence time τ_i. Figure 10.13 summarizes the model.

Then the total rate of powder removed by airsweeping V is given by $V = \Sigma_i^m \upsilon_i$ and the batch grinding equation becomes:

$$\upsilon_i + p_i(F' - V) = f_i F' + W^* \sum_{\substack{j=1 \\ i>1}}^{i-1} b_{i,j} S_j M_j - S_i M_i W^* \tag{10.38}$$

Assuming that the suspended powder is not subject to breakage action, $M_i = p_i$.

Further, if ζ is the fraction of the hold-up W^* exposed to the airstream per mill revolution, the amount of size i material exposed to the airstream per unit time is $p_i \zeta N W^*$.

If Ω_i is the probability of size i material being retained in the hold-up, then the fraction swept up will be given by:

$$\upsilon_i = p_i \zeta N W^* (1 - \Omega_i) \tag{10.39}$$

Using $\tau = W^*/F'$:

$$p_i = \frac{f_i + \tau \sum_{j=1}^{i-1} b_{i,j} S_j p_j}{(1 - \upsilon) + S_i \tau + \zeta N (1 - \Omega_i) \tau} \tag{10.40}$$

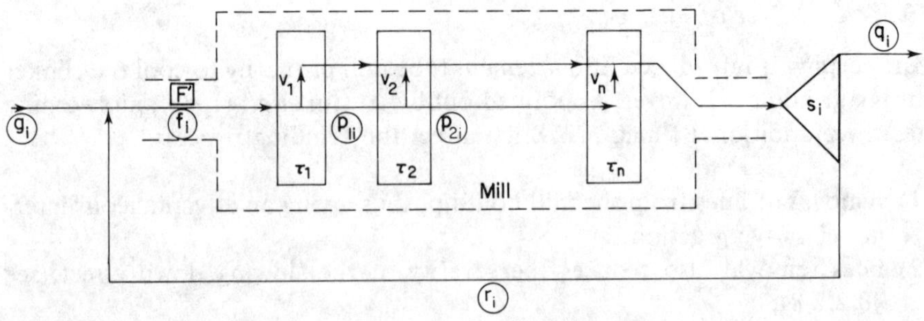

Figure 10.13. Illustration of an airswept ball mill as a series of stages (Austin *et al.*, 1984). *Reproduced by permission of Elsevier Sequoia*

where

$$v = \frac{V}{F'} = \tau \zeta N \sum_{1}^{n} (1 - \Omega_j) p_j \tag{10.41}$$

The treatment is extended to m fully mixed stages in series on the assumption that particles do not fall back into the bed of powder. Then for the first stage:

$$p_{i,1} = \frac{f_i + \tau_1 \sum_{j=1}^{i-1} b_{i,j} S_j p_{j,1}}{(1 - v_1) + S_i \tau_1 + \zeta N (1 - \Omega_i) \tau_1} \qquad \text{for } 1 \leqslant i \leqslant n$$

$$\tau_1 = (W^*/m)/F'$$

$$v_1 = \tau_1 N \sum_{j=1}^{n} (1 - \Omega_j) p_{j,i}$$

For the second stage:

$$p_{2,i} = \frac{p_{i,1} + \tau_2 \sum_{j=1}^{i-1} b_{i,j} S_j p_{j,2}}{(1 - v^2) + S_1 \tau_2 + \zeta N (1 - \Omega_i) \tau_2} \qquad \text{for } 1 \leqslant i \leqslant n$$

$$\tau_2 = (W^*/m)/(F' - V_1) = \tau_1/(1 - v_1)$$

$$v_2 = \tau_2 \zeta N \sum_{1}^{n} (1 - \Omega_j) p_{j,2}$$

Hence in general:

$$p_{i,k} = \frac{p_{i,(k-1)} + \tau_k \sum_{j=1}^{i-1} b_{i,j} S_j p_{i,k}}{(1 - v_k) + S_i \tau_k + \zeta N (1 - \Omega_i) \tau_k} \tag{10.42}$$

$$\text{for } 1 \leqslant i \leqslant n \quad \text{and} \quad 1 \leqslant k \leqslant m$$

$$\tau_k = \frac{\tau/m}{(1 - v_1)(1 - v_2) \ldots (1 - v_{k-1})} = \frac{\tau}{\mathrm{m}} \tag{10.43}$$

$$\text{for } 1 < k \leqslant m$$

$$\tau_1 = \tau/m$$

where τ is the overall mean residence time defined by W^*/F', and:

$$v_k = \tau_k \zeta N \sum_{1}^{n} (1 - \Omega_i) p_{j,k} \text{ for } 1 \leqslant k \leqslant m \tag{10.44}$$

Since there is no powder flow out of the mill reactor, $v_m = 1$.

For a given set of $(1 - \Omega_i)$ values and given ζ, W^*, N and f_i, the value of F' must be such as to lead to $v_m = 1$.

The size distribution of the airswept product is given as $p_1' F' = v_{i,1} + v_{i,2} + \ldots + v_{i,m}$, or:

$$p_i' = \frac{\tau \zeta N(1 - \Omega_i)}{m} \sum_{k=1}^{m} p_{i,k} \text{ for } 1 \leqslant i \leqslant n \tag{10.45}$$

If the values of ζ and S_j are assumed constant along the mill, a mass balance on size i over the total mill gives:

$$\sum_1^m v_{i,k} = f_i F' + \frac{W^*}{m} \sum_{\substack{j=1 \\ i>1}}^{i-1} b_{i,j} S_j \sum_1^m p_{k,i} - S_i \frac{W^*}{m} \sum_1^m p_{k,i}$$

or

$$\begin{aligned} \bar{v}_i &= \zeta N W^*(1 - \Omega_i) M_i \\ &= f_i F' + W^* \sum_{\substack{j=1 \\ i>1}}^{i-1} b_{i,j} S_j M_j - S_i W^* M_i \end{aligned} \tag{10.46}$$

where $M_i = \Sigma_1^m p_{i,k}/m$ is the mean size distribution in the mill, $\bar{v}_i$ is the total flow of size i in the airstream, and W^* is the total mass of powder instantaneously in the mill. Hence:

$$M_i = \frac{f_i + \tau \sum_{j=1}^{i-1} b_{i,j} S_j M_j}{S_i \tau + \zeta N(1 - \Omega_i)\tau} \quad \text{for } 1 \leqslant i \leqslant n \tag{10.47}$$

$$p_i' = \tau \zeta N(1 - \Omega_i) M_i \quad \text{for } 1 \leqslant i \leqslant n \tag{10.48}$$

The important conclusion is thus reached that, for the assumptions made, the mill behaves as if it were a fully mixed mill containing a mean overall size distribution.

Since the rate of flow of material presented to the internal air classification is $\zeta N W^*$ and the rate of fine material leaving is $\zeta N W^*(1 - \Omega_i) M_i$, the equivalent internal circulating load is:

$$1 / \sum_i (1 - \Omega_i) M_i$$

Such a mill model as described above may be inserted into equations for the closed circuit, and in Appendix I to Austin *et al.* (1984) it is shown that the circuit product size distribution q_i may be obtained from:

$$q_i = (1 + C') p_i' (1 - s_i) \tag{10.49}$$

and an expression is developed for $(1 + C')$ in terms of measurable quantities including Ω_i. It is found, in fact, that the external classifier parameter s_i dominates over the internal classification parameter Ω_i, and hence the value of q_i is not very sensitive to Ω_i. Experiment verifies that the model gives reasonable prediction of circuit output grade, despite its relatively simple nature. The assumption that the particles will not fall back into the bed, once airborne, is clearly an over-simplification, which appears to be justified in the case of a short mill, associated with coal grinding. Austin *et al.* (1980b) have attempted to set down equations which take account of particle fall-back, studied only in a laboratory mill, and comparison with the effect in a long mill (cement context) seems not yet to have been accomplished.

10.3.5 Rod mill

10.3.5.1 Level II simulation

Lynch (1977) regards the rod mill as a series of stages in which both breakage and screening occur (Callcott and Lynch, 1964). Each stage is thought of as that interval required to eliminate the largest size fraction from the previous stage, commencing with the largest size fraction in the feed in the first instance. On this basis the following equation is developed (Lynch *et al.*, 1967) in transfer parameter form for n stages:

$$\mathbf{g} = \left[\prod_{j=0}^{j=n} \mathbf{X}_j\right] g \qquad (10.50)$$

where:

$$\mathbf{X}_j = (\mathbf{I} - \mathbf{c})(\mathbf{BS} + \mathbf{I} - \mathbf{S})[\mathbf{I} - \mathbf{c}(\mathbf{BS} + \mathbf{I} - \mathbf{S})]^{-1}$$

Lynch (1977) gives an illustration of the use of equation (10.50) in his example 4.2.

10.3.5.2 Level III simulation

Work in this area appears not to have been developed very far as yet. The application of equation (8.15) to a batch rod mill has been considered in §8.7, where the differences between rod and ball mill operation are discussed, but mass transfer through a rod mill appears to await investigation.

10.3.6 Autogenous mill

10.3.6.1 Level II simulation

Whiten (1974) derived a theoretical model for autogenous mills by dividing the mill interior into segments in each of which perfect mixing was assumed to be achieved. Stanley (1974) simplified the Whiten model by taking the whole mill as a perfectly mixed unit, giving:

$$\frac{\mathrm{d}}{\mathrm{d}t} M = (\mathbf{BS} - \mathbf{S})M + f - p \qquad (10.51)$$

The relationship between mill contents and products is:

$$p = \mathbf{D}'M \tag{10.52}$$

where $\mathbf{D}'$ is the discharge rate matrix.

Thus, for steady-state conditions, equation (10.51) may be written as;

$$0 = (\mathbf{BS} - \mathbf{S} - \mathbf{D}')M + f \tag{10.53}$$

Hence if $\mathbf{B}$ is assessed, as below, and f, M and $\mathbf{S}$ are known, both f and $\mathbf{D}'$ may be found

Now experiment revealed that three zones of size reduction could be identified according to particle size, seen in Figure 10.14 in the context of the specific rate of breakage function S against particle size in $\sqrt{2}$ intervals. Large particles were taken to be reduced in size by abrasion and small particles by crushing, with a transition zone between.

For the crushing zone, the breakage distribution function used was a modified form of equation (10.32):

$$B_1 = B(x, y) = [1 - e^{(-x/y)^2}]/(1 - e^{-1}) \tag{10.54}$$

For the abrasion breakage zone, a novel method of assessment of the breakage distribution function B_2 was used which was related to the wear rates of the particles, and Stanley (1974) must be consulted for details.

Then the transition zone breakage distribution function B_3 was taken as hav-

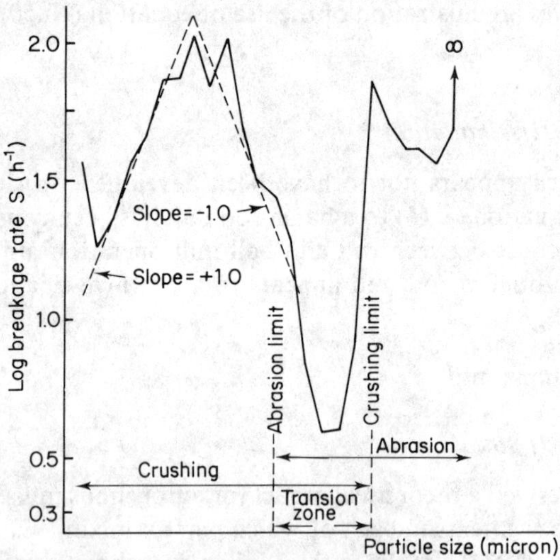

Figure 10.14. Typical breakage rate function for an autogenous mill (Stanley, 1974). *Reproduced by permission of the South African Institute of Mining and Metallurgy*

ing a linear relation between B_1 and B_2 as follows:

$$B_3 = (1 - \chi)B_1 + \chi B_2 \tag{10.55}$$

where χ is the proportion of the distance (in size interval terms) across the transition zone. Based on these concepts, an overall breakage distribution matrix **B** was arrived at.

Examples of the application of the system are given, but the fit of predicted to experimental output size distributions is only approximate.

10.3.6.2 Level III simulation

Austin *et al.* (1977) have made a first approach to simulation of a grate discharge autogenous mill using a discrete-size, discrete-time form of the SMB equation (see Table 8.1), assuming first-order breakage and a fully mixed mill. For size i material ($i = 1, 2, \ldots, n$) they give:

$$F' p_i = F' f_i + W^* \left(\sum_{\substack{j=i-1 \\ i>1}}^{1} b_{i,j} S_j M_j \right) - W^* S_i M_i \tag{10.56}$$

Defining τ as W^*/F' we have:

$$p_i = f_i + \tau \left(\sum_{\substack{j=i-1 \\ i>1}}^{1} b_{i,j} S_j M_j \right) - \tau S_i M_i \tag{10.57}$$

It is assumed that the grate with effective opening size x_s passes an upper size corresponding to interval i_s. Then p_i is zero for all sizes larger than $i_s (i < i_s)$, whereas the size distribution in the mill material less than $x_s (i > i_s)$ is the same as the product size distribution.

Knowing the values of S_i, $b_{i,j}$ and f_i, the product size distributions may be calculated for a given value of τ, starting with size 1. Since $p_1 = 0$, $f_1 = \tau S_1 M_1$, which gives M_1; then for $p_2 = 0$, $f_2 + b_2 S_1 M_1 \tau = \tau S_2 M_2$, which gives M_2; and so on down to size interval i_s. At i_s or smaller interval sizes, from equation (10.57) we have:

$$p_i = f_i + \tau \left(\sum_{j=i-1}^{1} b_{i,j} S_j M_j \right) - \tau S_i M_i \qquad \text{for } i \geqslant i_s$$

But:

$$p_i = M_i / \sum_{i_s - 1}^{1} M_j$$

Hence:

$$p_i = \frac{f_i + \tau \left(\sum_{j=i-1}^{1} b_{i,j} S_j M_j \right)}{1 + \tau S_i (1 - M_\Sigma)} \tag{10.58}$$

where:

$$M_\Sigma = \sum_{i_s-1}^{1} M_j$$

which is known.

In this study, semi-autogenous grinding is assumed, which implies that some steel balls of diameter k are used.

Three types of breakage are then envisaged:

(1) self-breakage by falling material, associated with a specific rate of breakage $S(S)_i$;
(2) breakage of small material by tumbling larger lumps called pebbles associated with $S(P)_{i,k}$ which stands for the breakage of size i material by size k pebbles summed over all pebble sizes. It is asssumed that M_k is the fraction of pebble mass made up of size k pebbles. The total pebble volume filling is J_p;
(3) breakage caused by steel balls, associated with $S(B)_{i,k}$, which stands for the breakage of size i material by size k balls, summed over all ball sizes. It is assumed that m_k is the fraction of the ball mass made up of size k balls with a total volume filling of J_B ($J = J_P + J_B$).

Further, assuming linear additivity, we have:

$$\bar{S}_i = S(S)_i + \sum_k S(P)_{i,k}M_k + \sum_k S(B)_{i,k}m_k \tag{10.59}$$

Clearly $S(S)_i$ depends on J.

Then equation (10.56) becomes:

$$F' p_i = F' f_i + W^* \sum_{\substack{j=i-1\\ i>1}}^{1} M_j \Big[b(S)_{i,j}S(S)_j + \sum b(P)_{i,j,k}S(P)_{j,k}M_k + \sum b(B)_{i,j,k}S(B)_{j,k}m_k \Big] - W^*M_i \Big[S(S)_i + \sum S(P)_{i,k}M_k + \sum S(B)_{i,k}m_k \Big] \tag{10.60}$$

where $b(S)_{i,j}$, $b(P)_{i,j,k}$, $b(B)_{j,k}$ are the breakage distribution functions for self breakage, tumbling pebble breakage and ball breakage respectively. The solution to equation (10.60) follows as that for equation (10.56).

The experimental problems encountered and the means for overcoming the same are not detailed, but may be obtained from the Austin *et al.* (1977) paper. In view of the preliminary nature of the work, only the main conclusions associated with the grinding of a molybdenum ore are given:

(1) for the largest particles, self-breakage makes a dominant contribution;
(2) for the fine sizes, the dominant breakage mechanism is by tumbling pebbles;
(3) for the intermediate sized particles, the steel balls are necessary, as the particles are too small for self-breakage and too large for tumbling pebble breakage.

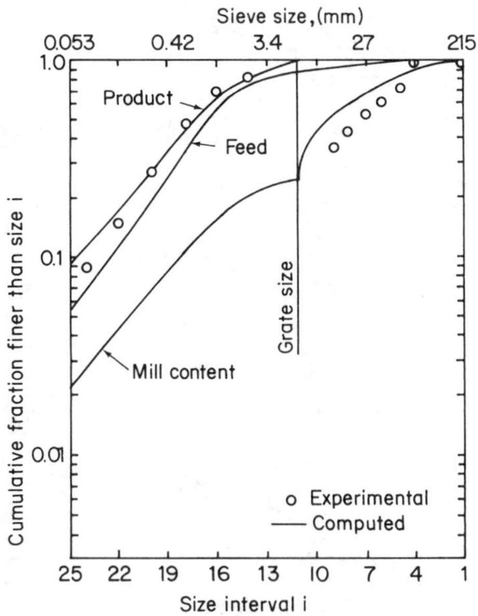

Figure 10.15. Computed and experimental size distributions for the product and mill contents of a 1.83 m × 0.6 m diameter continuous semi-autogenous mill ($J_B = 0.08$, pulp density = 70% solids by weight)(Austin *et al.*, 1977). *Reproduced by permission of the American Institute of Mining, Metallurgical and Petroleum Engineers*

Figure 10.15 compares computed and experimental distributions of both mill content and of product for a typical set of conditions. The fit is seen to be only approximate, but encouraging in view of the problems encountered. It is noted that there is a relatively small difference between feed and product distributions. This is because this particular feed size distribution was selected to match that used in a continuous test with a classifier, involving a circulating load of 850%. Since a classifier model was not incorporated into the overall system simulation, computation using make-up feed size distribution was not possible.

Clearly much more work is necessary on the simulation of autogenous grinding.

10.4 CIRCUITS INVOLVING VERTICAL SPINDLE MILLS: LEVEL IV SIMULATION

10.4.1 Circuit flow rates and size distributions

As has been noted in §5.3.4 and Figure 5.22, the vertical spindle mill is characterized by two recycle loops in parallel, for which relative flow rates and

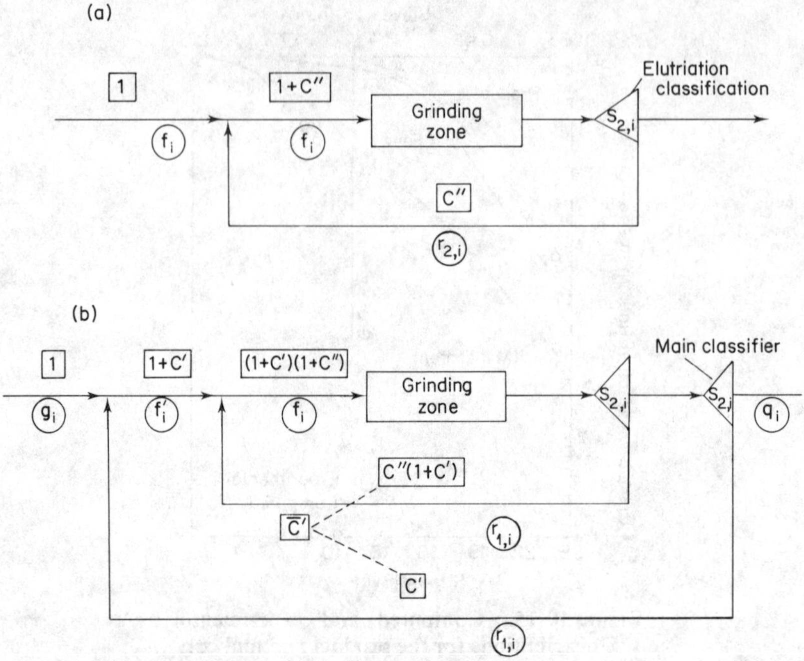

Figure 10.16. Nomenclature for vertical spindle mill circuits: (a) grinding zone + elutriation classification only, (b) complete mill circuit (After Austin *et al.*, 1982b). *Reproduced by permission of Elsevier Sequoia*

size distributions are given in Figure 10.16, where C' and C'' are the recirculation ratios for classification processes numbered 1 and 2 respectively. s_2 represents the elutriation classification and s_1 the main classification.

Now it is very difficult to assess C'' and $r_{2,i}$, as has been indicated in §7.4, and hence Austin *et al.* (1982b) overcome the problem in the following manner.

By application of the SMB principle, for the elutriation recycle only (Figure 10.16(a)):

Rate of removal of size i material from the grinding zone
= rate of addition of size i in the grinding zone
+ rate of production of size i from larger size j
− rate of breakage of size i

or:

$$(1 + C'')M_i = (1 + C'')f_i + \sum_{j=1}^{i-1} b_{i,j}S_jM_jW^* - S_iM_iW^* \tag{10.61}$$

By mass balance at the point of recycle return to the feed to the grinding zone:

$$(1 + C'')f_i = f_i' + C''r_{2,i} \tag{10.62}$$

By definition of $s_{2,i}$:

$$C''r_{2,i} = (1 + C'')M_i s_{2,i} \tag{10.63}$$

Combining (10.62) and (10.63) we have:

$$(1 + C'')f_i = f'_i + (1 + C'')M_i s_{2,i} \tag{10.64}$$

Combining (10.61) and (10.64) we have:

$$(1 + C'')M_i = f'_i + (1 + C'')M_i s_{2,i} + \sum_{j=1}^{i-1} b_{i,j}S_j M_j W^* - S_i M_i W^* \tag{10.65}$$

Let:

$$M_i(1 + C'') = M_i^* \text{ and } S_i W^* = A_i$$

Then, collecting terms in M_i^* we have:

$$M_i^* = \frac{f'_i + [1/(1 + C'')] \sum_{j=1}^{i-1} b_{i,j}S_j A_j M_j^*}{(1 - s_{2,i}) + A_i/(1 + C'')} \tag{10.66}$$

The set of equations represented by equation (10.66) may be computed sequentially starting with $i = 1$. Then:

$$M_1^* = f'_1/\{(1 - s_{2,1}) + [A_1/(1 + C'')]\} \tag{10.67}$$

This value may then be used to calculate M_2^*, and so on, and the value of C'' may be obtained from:

$$\sum_i M_i^* = \sum_i (1 + C'')M_i \tag{10.68}$$

or, since $\sum_i M_i = 1$:

$$(1 + C'') = \sum_i M_i^* \tag{10.69}$$

and the values of M_i follow.

Knowing C'', we may assess the mass flow of material leaving the grinding zone.

The above principles may be equally applied to the case of the combination of elutriation with the main classifier, using Figure 10.16(b), and regarding the combination as a single classifier represented by the single parameter $\bar{s}_i$ and a combined recirculation ratio $\bar{C}'$. Then it may be shown that:

$$(1 - \bar{s}_i) = (1 - s_{1,i})(1 - s_{2,i}) \tag{10.70}$$

Similarly:

$$(1 + \bar{C}') = (1 + C')(1 + C'') \tag{10.71}$$

Putting:

$$\bar{M}_i^* = (1 + \bar{C}')M_i \tag{10.72}$$

and following the pattern of equation (10.66), it is possible to write:

$$\bar{M}_i^* = \frac{g_i + [1/(1+\bar{C}')] \sum_{j=1}^{i-1} b_{i,j} S_j A_j \bar{M}_j^*}{(1-\bar{s}_i) + [A_i/(1+\bar{C}')]} \tag{10.73}$$

and the sequential treatment for $i = 1, 2$ etc. follows as before to obtain $\bar{C}'$ and $\bar{M}_i^*$, and hence M_i from equation (10.72). Then if C' is measured for the main classifier loop by the method of Figure 7.15, C'' may be calculated from equation (10.71). Similarly, if $s_{1,i}$ is measured for the main classifier, then $s_{2,i}$ may be assessed via equation (10.70).

Following equation (10.69):

$$(1+\bar{C}') = \sum_i \bar{M}_i^* \tag{10.74}$$

Further, by definition of $\bar{s}_i$:

$$(1-\bar{s}_i) = \frac{q_i}{(1+C')(1+C'')M_i} = \frac{q_i}{(1+\bar{C}')M_i}$$

or

$$q_i = (1-\bar{s}_i)(1+\bar{C}')M_i \tag{10.75}$$

which gives the circuit output size distribution.

Similarly, all other flows and size distributions may be obtained.

It is to be noted that, in fact, Austin *et al.* (1982b) did not use the method of Figure 7.15. Instead, to find the elutriation recirculating load and size distribution they adopted an open-circuit crash shut-down, opened the mill and sampled at strategic points. Again, in their closed-circuit pilot plant tests, they found it impossible to obtain samples of the feed to the main classifier or the recycle from that classifier, and hence had to estimate these from a similar classifier, separately tested. The use of the concept of Figure 7.15, however, obviates such separate tests or crash shut-down and helps to reduce scatter in the test data.

10.4.2 Laboratory-found breakage functions and their relation to Hardgrove index

The laboratory batch mill corresponding to the continuously operated vertical spindle mill is the Hardgrove machine, in which breakage functions for coal were measured.

Early work by Gardner and Austin (1962) on the Hardgrove machine has been noted in §8.5.2.4, and Figures 8.23 and 8.24 give values, for three particular coals, of S and B respectively, by radioactive tracer methods.

More recently, Austin *et al.* (1981a) and Austin *et al.* (1982c) have studied in greater depth the grinding of coal in the Hardgrove machine. They conclude as follows:

(1) Breakage is linear until accumulation of fines cushions the grinding action.
(2) The specific rate of breakage S_i may be represented by equation (8.2) for the smaller particle sizes, and may be expressed as:

$$A_{i\text{HI}} = A_{16\times20}(x/1180)^{\alpha} \tag{10.76}$$

Both $A_{16\times20}$ and α are related to the Hardgrove index as given in Figure 10.17 and in the following empirical expressions:

$$A_{16\times20} = 0.054(HI)^{0.7} \times 1.18^{\alpha} \times 50(1-z) \tag{10.77}$$

and:

$$\alpha = 0.7\left\{1 - \exp\left[-\left(\frac{HI}{90}\right)^{-1.9}\right]\right\} \tag{10.78}$$

where z is the moisture content of coal (thus $1 - z$ corrects back to dry coal).
(3) The breakage distribution function $B_{i,j}$ may be represented by equation (8.13) and values of Φ, γ and β may be measured for several coals over a range of Hardgrove index. It is found that β may be constant at a value of 5; Φ and γ are related to HI as shown in Figure 10.18, and in the following empirical relations:

$$\Phi = 0.71\left\{1 - \exp\left[-\left(\frac{HI}{42.5}\right)^{1.8}\right]\right\} \tag{10.79}$$

$$\gamma = 1.35 - 0.0056\, HI \tag{10.80}$$

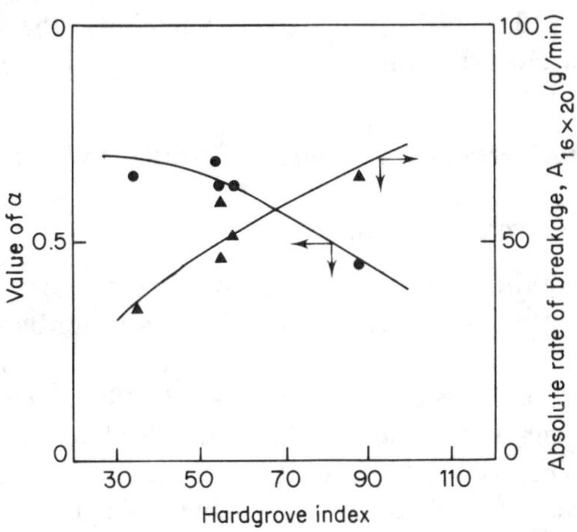

Figure 10.17. Correlation between the Hardgrove index and each of $A_{16\times20}$ and α in equation (10.77) (Austin *et al.*, 1982a)., *Reproduced by permission of Elsevier Sequoia*

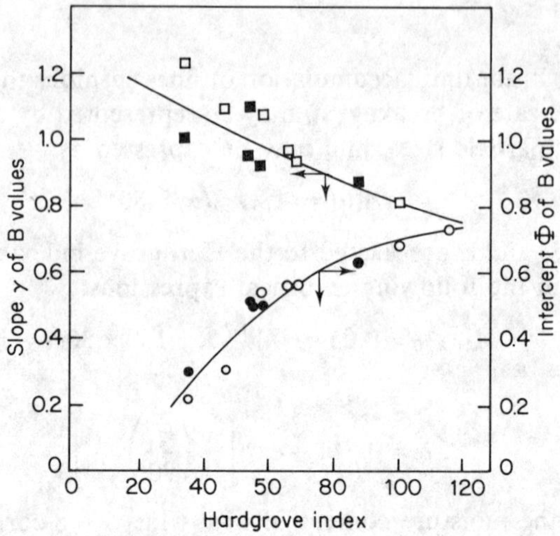

Figure 10.18. Variation of Hardgrove index with Φ and γ for constant $\beta = 5$ in equation (8.13) for several coals (Austin *et al.*, 1981a). *Reproduced by permission of Elsevier Sequoia*

Thus for any coal with a given Hardgrove index and moisture content it is possible to assess the value of $A_{i\mathrm{HI}}$ from equation (10.76) knowing $A_{16\times20}$ and α, and similarly $B_{i,j}$ from equation (8.13) and hence $b_{i,j}$ using equation (8.12), since $B_{i,j}$ was found to be normalizable.

Some manufacturers of vertical spindle mills are collecting Hardgrove data for coals in terms of α, x^* and Λ for S_i in equation (8.3), and β, γ and Φ for $B_{i,j}$ in equation (8.13). This is proving very valuable in the design of mills for a coal of a given Hardgrove index.

10.4.3 Scale-up of breakage functions to a pilot mill

10.4.3.1 Absolute rate of breakage A_i

In order to explore the scale-up of Hardgrove machine breakage functions to a pilot scale mill, it was necessary to modify the mill for batch grinding, since it is not possible to apply laboratory batch techniques to a continuous process.

A Babcock 1.7E mill was used by Austin *et al*. (1982c), having a mean track diameter of 17 inches and 10 balls each 5 inches diameter to which was applied a load of 8400 lb (Figure 10.19). The classifier and feed pipe were removed and replaced by a cone centrally above the rotating table. To retain the coal in the grinding zone, a wall was built on the periphery of the rotating table. The test was performed without air sweeping and at a reduced rotational speed from normal, because otherwise the coal broke so rapidly it was impossible to obtain accurate size distribution of samples. The result is given in Figure 8.17.

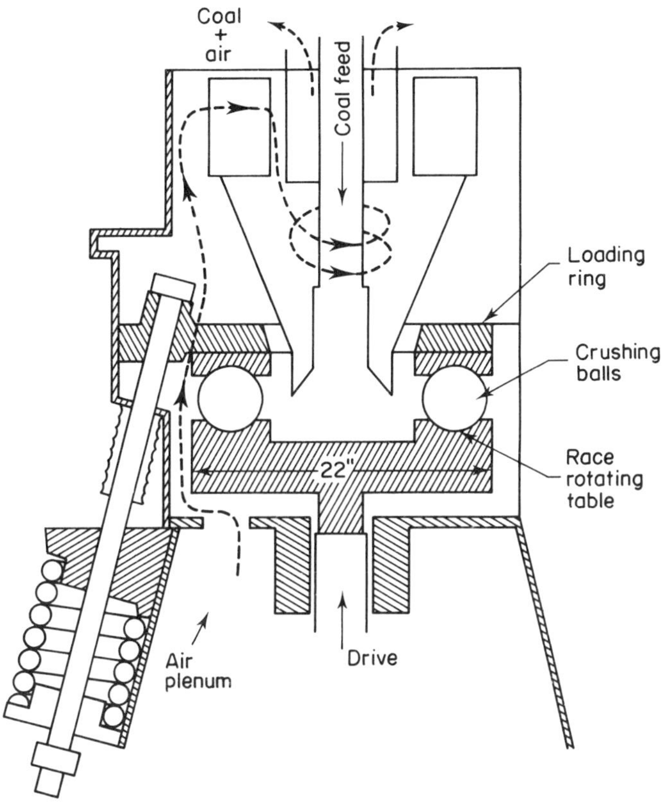

Figure 10.19. Cross-section of the Babcock 1.7E mill (Austin *et al.*, 1982b). *Reproduced by permission of Elsevier Sequoia*

The slope α of the straight-line portion of the graph coincides with that of the straight-line portion of the Hardgrove data, shown on the same plot. The non-linear portion of the graph is explained by the inability of the balls to nip the larger particles.

Neglecting the non-linear portion, A_i for the pilot mill could be related to $A_{i\mathrm{HI}}$ by:

$$A_i = k' A_{i\mathrm{HI}} \tag{10.81}$$

10.4.3.2 Breakage distribution function $B_{i,j}$

Using the Austin and Luckie method II for the assessment of $B_{i,j}$ (see §8.5.2.2(b)) in the pilot scale mill it was possible to obtain Figure 10.20 due to Austin *et al.* (1982b), which shows direct correspondence with values from the Hardgrove machine.

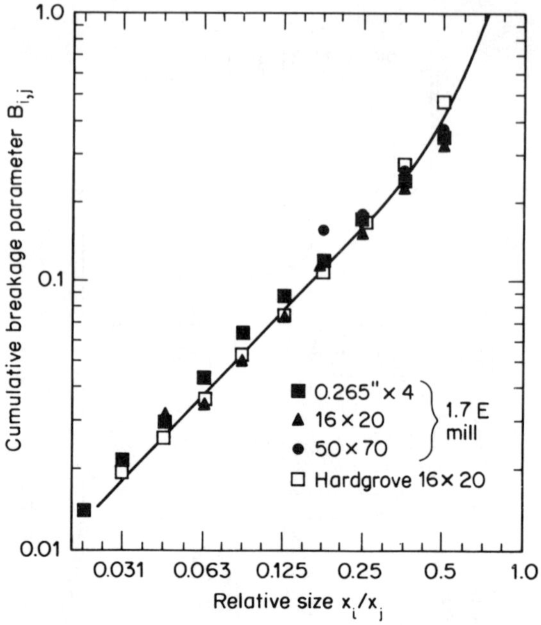

Figure 10.20. $B_{i,j}$ values determined by batch test in the Babcock 1.7E mill and in the Hardgrove machine for anthracite (Austin *et al.*, 1982b). *Reproduced by permission of Elsevier Sequoia*

Equation (8.13) is clearly applicable, and values of Φ and γ may be obtained over a range of coals, and for each, β is constant at 5. Thus the Hardgrove data applies directly to the pilot mill without any scaling.

10.4.4 Application of the simulation model to change of Hardgrove index

The above pilot plant values of A_i and $b_{i,j}$ may be inserted in equation (10.73) from which M_i may be calculated via equation (10.72), and hence the system output size distribution from equation (10.75).

One test of the model would be to forecast the size distribution for a coal of one Hardgrove index from a test on a coal having a different Hardgrove index. The result of such an exercise is given in Figure 10.21 for a coal of $HI = 88$ based on Hardgrove data for a coal of $HI = 55$. The simulation is reasonably good.

10.4.5 Limits of feed rate, product size distribution and air–coal ratio

Variation of size distribution with airflow rate and air–coal mass ratio is given typically for a 1.7E mill grinding coal in Figure 10.22. Capacity is greater for

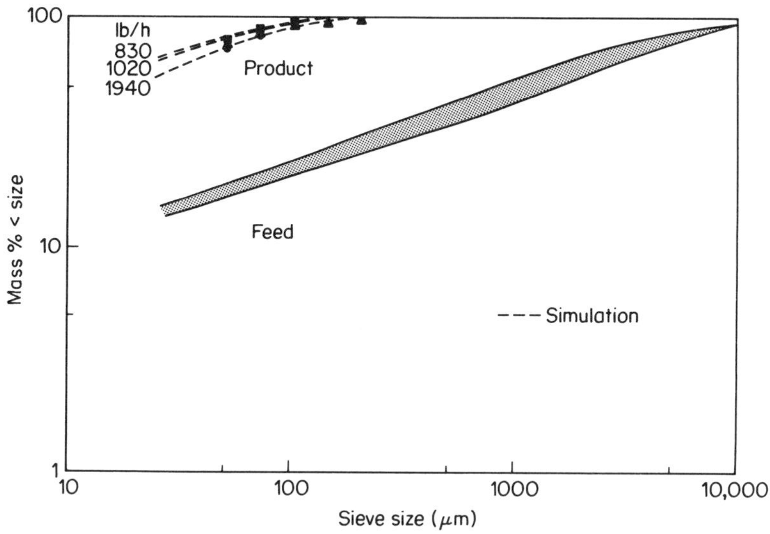

Figure 10.21. Product size distribution for coal of Hardgrove index = 88 ground in the Babcock 1.7E mill (Austin *et al.*, 1982b). *Reproduced by permission of Elsevier Sequoia*

Coal (lb/h)	*ACFM*	*Air/coal*
830	820	3.8
1020	820	3.1
1940	825	1.7

curve 3 than for curves 1 or 2 (all being 70% minus 75 micron) because of the low air–coal ratio, which gives a higher recirculation load and a steeper size distribution. Normal full-load operation is designed for an air–coal ratio of approximately 1.73 lb/lb, which falls to a value about 3 lb/lb at minimum load. Controls are incorporated in the mill system to prevent further reduction, as a potentially explosive mixture of pulverized fuel and air is then entered into. Hence an air–coal ratio boundary X exists, drawn at 4 lb air/lb coal, and superimposed on Figure 10.22

Assuming sufficient mill power is available, increase in flow rate beyond the normal full-load operation leads to unstable operation, indicated by high and fluctuating air pressure loss over the mill (mill differential) and eventually mill chokage. Hence there exists a mill chokage boundary at high flow rates, shown on Figure 10.23 at Y.

Because of twin cone classifier limitations, there is similarly a fineness barrier Z, shown on Figure 10.23 at about 85% minus 75 micron.

Thus a 'window of desirable operation' exists for the vertical spindle mill associated with limiting conditions of the interacting variables. It must be stressed that the positions of the boundaries shown in Figure 10.23 are illustrative only and are not absolute for any one mill.

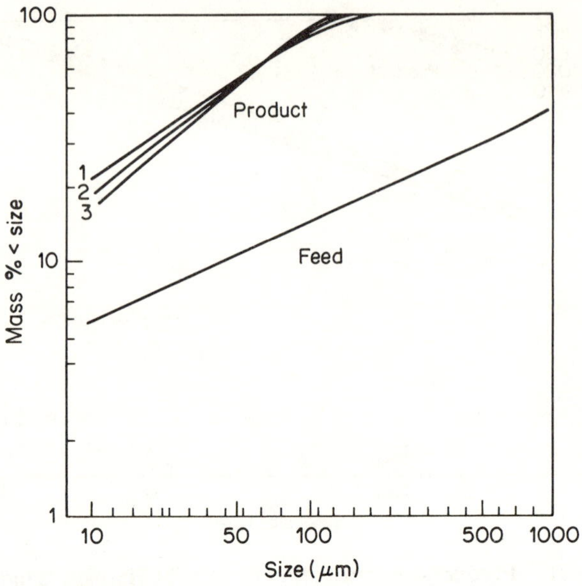

Figure 10.22. Variation of product size distribution with airflow rate for the Babcock 1.7E mill (Austin *et al.*, 1982c). *Reproduced by permission of Elsevier Sequoia*

ACFM	Q (t/h)	$\bar{C}$	*Air/coal*
(1) 900	0.32	1.2	5.5
(2) 700	0.57	2.3	2.4
(3) 500	0.71	6.0	1.4

10.4.6 Scaling of capacity

10.4.6.1 Ring ball mill

Following Austin *et al.* (1982c), and assuming constant geometry in a series of different sized mills, the mean track diameter D may be taken as characteristic of the mill size. Figure 10.24 gives a cross-section of a ring-ball mill grinding zone, in which the coal, entering centrally, accelerates radially outward through the race. However, it is assumed that over the track width T the mean velocity $\bar{v}$ is constant.

Let the depth of coal in the race be Y, and the rotational speed of the lower grinding ring be N revolutions per unit time. Consider particles moving along a radius with velocity $v(r)$. The acceleration along r is:

$$\frac{\mathrm{d}}{\mathrm{d}t}\,v(r) = \frac{\mathrm{d}}{\mathrm{d}r}\,v(r)\,\frac{\mathrm{d}r}{\mathrm{d}t} = v(r)\,\frac{\mathrm{d}v}{\mathrm{d}r}$$

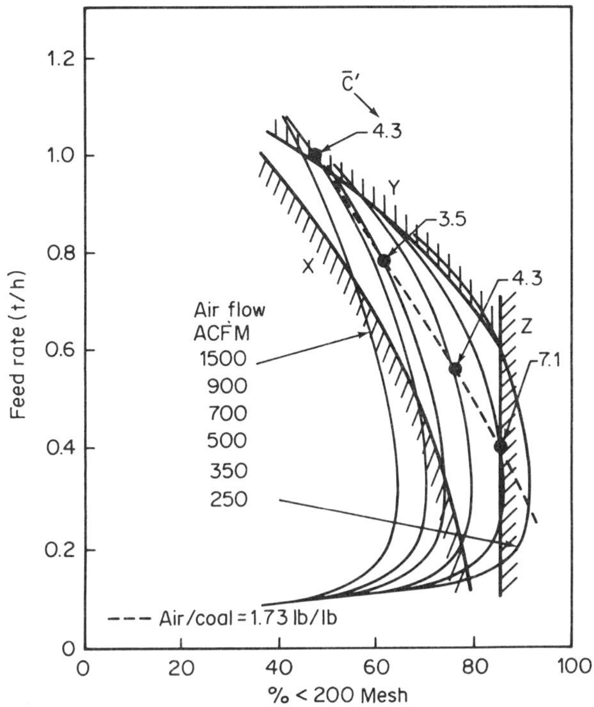

Figure 10.23. Effect of airflow rate on capacity and product fineness for the Babcock 1.7E mill with limiting boundaries. Parameter $\bar{C}'$ = overall recirculation ratio (Austin *et al.*, 1982c). *Basic plot reproduced by permission of Elsevier Sequoia*

X = Air–coal ratio boundary
Y = Chokage boundary
Z = Fineness boundary

Separating and integrating, assuming that $v(r)$ is small at small values of r, gives:

$$\bar{v} = NR = ND/2 \tag{10.82}$$

The time particles spend in the track is $T/\bar{v}$. Hence the number of times material is rolled over in this time interval by n balls moving at a rotational speed of $N/2$ is:

$$\frac{N}{2}\frac{T}{\bar{v}}n = \frac{N}{2}Tn\frac{2}{ND} = \frac{Tn}{D}$$

This is an interesting result, because it shows that the number of crushing actions in the time for material to pass through the race is independent of N.

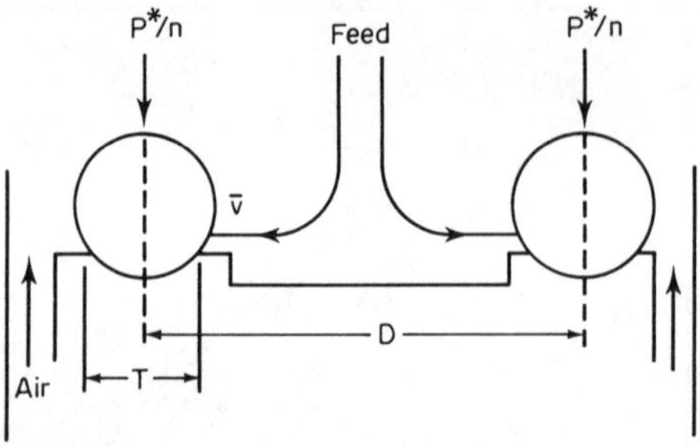

Figure 10.24. Cross-section of ring-ball mill grinding zone (Austin *et al.*, 1982c). *Reproduced by permission of Elsevier Sequoia*

The breakage potential of particles along the radius under a ball will clearly increase with the load per ball P^*/n and decrease with the amount of material which is proportional to TY. Combining these effects, the severity of breakage will be proportional to P^*/nTY. Further, the rate at which material is moving outwards through the track is proportional to $DY\bar{v}$, i.e. to D^2YN by equation (10.82).

Assuming that all particles in the track are of size i, then the absolute rate of breakage A_i would be given by the product of the number of times material is rolled over in moving through the track, the severity of breakage, and the rate material is moving through the track; i.e.:

$$A_i \propto \frac{Tn}{D}\frac{P^*}{nTY}D^2YN \propto P^*DN \tag{10.83}$$

10.4.6.2 Hardgrove machine

Applying the same principles to the batch process of the Hardgrove machine, we have that the number of times the material is rolled over in unit time is nN, the severity of breakage is P^*/nTY, and the material acted on in unit time is πDTY. Hence:

$$A_{i\mathrm{HI}} = nN \times \frac{P^*}{nTY} \times \pi DTY \propto P^*DN \tag{10.84}$$

where P^*, D and N apply to the Hardgrove machine.

Tests on the Hardgrove machine (Austin *et al.*, 1981a) did show that $A_i \propto P^*$ and $A_i \propto \mathrm{N}$ with considerable accuracy, and hence equation (10.84) is confirmed by experiment.

10.4.6.3 Scaling up from the Hardgrove machine

Combining equation (10.81) with equations (10.83) and (10.84) gives:

$$k' = P^*DN/(P^*DN)_{\mathrm{HI}} \tag{10.85}$$

Now:

$$(P^*DN)_{HI} = 64 \times \frac{3}{12} \times 20 \text{ (see §6.6.2)}$$

$$= 320 \text{ ft lb/minute} = 434 \text{ Nm}$$

for operation under standard conditions.

Further, k' may not be expressed quite as simply as given in equation (10.85) because tests have shown that make-up feed rate and feed size are factors as well as P^*, D and N. To take cognisance of these factors, equation (10.85) is therefore written as:

$$k' = k_0 P^* DN/434 \quad (10.86)$$

with P^* in newtons, D in metres and N in rev/minute. Hence if Q_{HI} is the capacity of the Hardgrove machine corresponding to A_{iHI}, then the pilot-scale mill capacity at the same product fineness is given by:

$$Q = k' Q_{HI}$$

or:

$$Q = k_0 P^* DNQ_{HI}/434 \quad (10.87)$$

which is one form of the required scale-up equation.

Now, from §7.2.3.2, P^*/n is held constant at the optimum value, and $T \propto D$. Thus $P^* \propto D^2$. Again $N \propto D^{-0.5}$. Hence, from equation (10.87):

$$Q \propto D^{2.5} Q_{HI} \quad (10.88)$$

which is in keeping with Figure 7.6 for vertical spindle mills.

Based on their work on the 1.7E mill, Austin *et al.* (1982c) give 1.33×10^4 is the proportionality constant in equation (10.88) when D is measured in metres, and Q is mill capacity in tonnes per hour at 70% minus 75 micron.

10.4.7 Scaling of mill power

From equation (7.13) it would be expected that the power to grind would be a function of $D^{2.5}$.

From tests with the 1.7E mill, Austin *et al.* (1982c) give, in fact:

$$p = \frac{29.5}{\eta} D^{2.5} \text{ kW} \quad (10.89)$$

where η is electrical motor efficiency.

Hence from equations (10.88) and (10.89) the specific energy is independent of D, which is borne out in practice, as noted in §7.2.3.3.

In the scaling of mill power, it is clearly necessary to subtract the mill idle power at zero capacity, as it is only *net* power which may be scaled from one mill size to another, as pointed out by Austin *et al.* (1982c).

10.5 CIRCUITS INVOLVING SWING HAMMER MILLS: LEVEL II SIMULATION

Very little work seems to have been attempted in this area. The only known relevant paper is that of Tschorbadjiski (1980), who applies SMB principles to a model beater mill, and demonstrates that the same principles apply in an industrial closed-circuit plant. Further study is clearly necessary.

10.6 CIRCUITS INVOLVING CRUSHERS

10.6.1 Level II simulation

Referring to Figure 10.25, a mass balance at the branch upstream of the classifier gives, in matrix notation:

$$\boldsymbol{g} + \mathbf{Bc}\boldsymbol{h} = \boldsymbol{h} \tag{10.90}$$

A mass balance over the classifier node gives:

$$\boldsymbol{h} = \boldsymbol{q} + \mathbf{c}\boldsymbol{h} \tag{10.91}$$

Eliminating $\boldsymbol{h}$ between equations (10.82) and (10.83) we obtain:

$$\boldsymbol{q} = \boldsymbol{g}[\mathbf{I} - \mathbf{c}]\,[\mathbf{I} - \mathbf{Bc}]^{-1} \tag{10.92}$$

which gives the output product size distribution vector (Whiten, 1972). Lynch (1977) gives his example 4.1 to illustrate the use of equation (10.92) for a cone crusher, including the setting out of the matrices for **B** and **c** and their product by matrix multiplication.

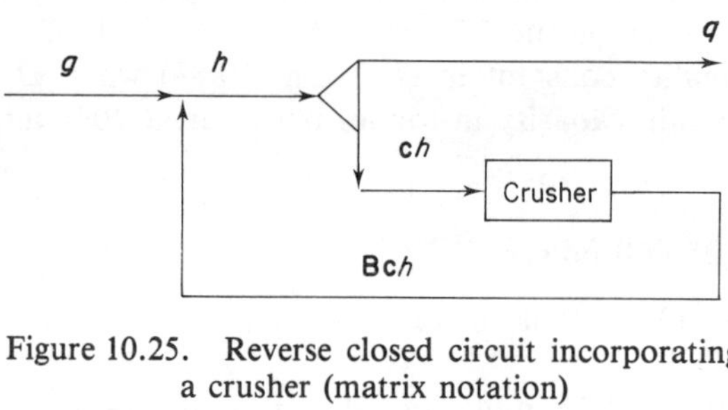

Figure 10.25. Reverse closed circuit incorporating a crusher (matrix notation)

10.6.2 Level IV simulation

10.6.2.1 Double roll crusher in open circuit

(a) Theory Smooth roll crushers in open circuit have been investigated by Austin *et al.* (1980a) and Austin *et al.* (1981b). Roller action is summarized in Figure 10.26, which also gives the symbols used.

Assuming each size fraction breaks independently of other size fractions, the relationship between feed and product size distributions may be expressed

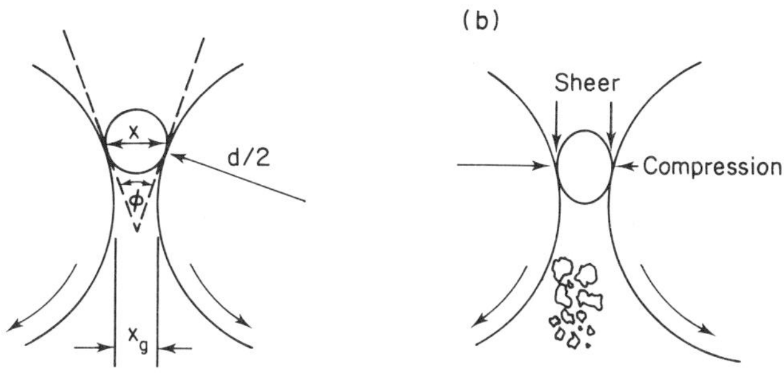

Figure 10.26. Illustration of (a) nip angle in rolls, (b) compression and shear forces (Austin *et al.*, 1980a). *Reproduced by permission of Elsevier Science Publishers*

as:

$$p_i = \sum_{j=1}^{i} f_j d_{i,j} \tag{10.93}$$

where $d_{i,j}$ is defined as the mass fraction of size j feed which when passed through the rolls appears in size i.

Now it is important to realize that for each starting size j in the feed, a certain proportion $d_{j,j}$ is found in the product within the same starting size, and these even include sizes greater than the gap size x_g, as seen in Figure 10.27.

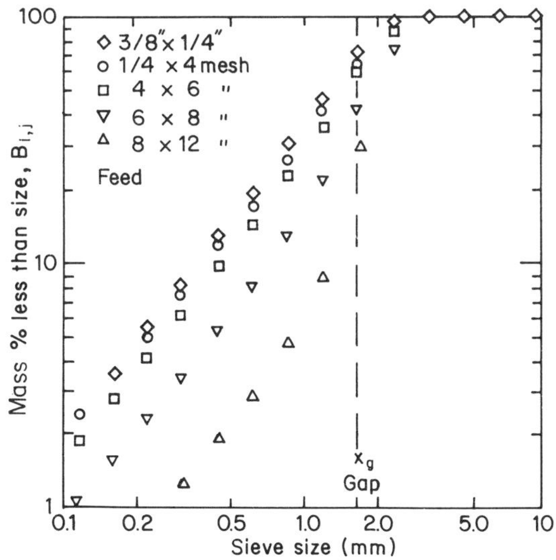

Figure 10.27. Products from various coal feed sizes crushed through a gap corresponding to 12 mesh between 200 mm diameter smooth rolls (2 : 1 speed ratio) (Austin *et al.*, 1980a). *Reproduced by permission of Elsevier Science Publishers*

Defining breakage as that process which leads to size reduction, $d_{j,j}$ can be considered as material which has bypassed through the rolls, i.e. $d_{j,j} = (1 - s_{i,j})$ bypasses without breakage and a fraction $s_{i,j}$ is broken. The fraction $s_{i,j}$ for a given size j is referred to as the primary bypass parameter for size j.

It is further assumed that, on being nipped in the rolls, prior to passing through, the first fracture leads to a set of primary daughter fragments, denoted by $b_{i,j}$. In other words, $b_{i,j}$ symbolizes the mass fraction of size j material which is broken to size i as a result of this primary breakage. Of the daughter fragments, there will still be those larger than the gap setting, which can in turn either bypass through the rolls or be selected for another fracture, with a fraction $s_{2,i}$ broken. The fraction $s_{2,i}$ for a given size i is referred to as the secondary bypass parameter for size i. These values are a function of the ratio of particle size to gap setting.

Clearly classifier principles may be applied to both primary and secondary bypass processes, resulting in Figure 10.28, a diagram of the crusher model here analysed.

Let G' be the mass per unit feed of the $f_i s_{1,i}$ component and C' the mass per unit feed of rebreakage component. Then breakage of a mass $(G' + C')$ of size f_i^+ produces a size distribution p_i^+ given by:

$$p_i^+ = \sum_{j=1}^{i=1} f_i^+ b_{i,j} \qquad \text{for } n \geqslant i > 1$$
$$p_1^+ = 0 \tag{10.94}$$

At steady state, the mass balance at the confluence of the rebreakage stream with the product from $s_{1,i}$ is taken. Then, by definition of $s_{1,i}$ and $s_{2,i}$ we have:

$$(G' + C')f_i^+ = f_i s_{1,i} + (G' + C')p_i^+ s_{2,i} \tag{10.95}$$

Let $f_i^* = f_i^+(G' + C')$ and $p_i^* = p_i^+(G' + C')$. Then:

$$f_i^* = f_i s_{1,i} + p_i^* s_{2,i} \tag{10.96}$$

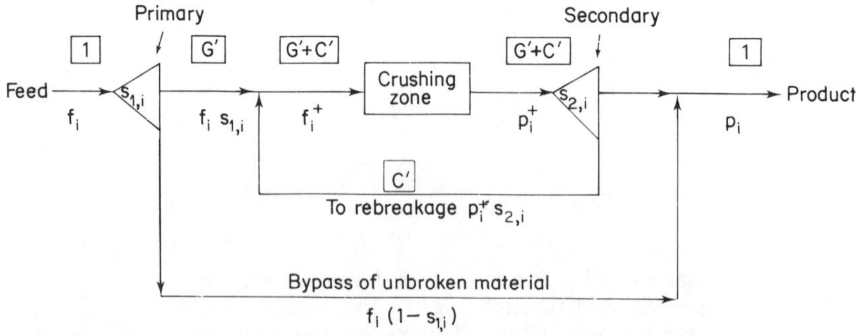

Figure 10.28. Schematic of roll-crusher model (Based on Austin *et al.*, 1980a). *Reproduced by permission of Elsevier Science Publishers*

Multiplying each size of equation (10.94) by $(G' + C')$ and substituting from equation (10.96), gives:

$$p_i^* = \sum_{j=1}^{i-1} b_{i,j}(p_j^* s_{2,j} + f_j s_{1,j})$$
$$p_1^* = 0 \tag{10.97}$$

The set of equations given by (10.97) is computed sequentially, starting with size 1 to give p_2^*, which is used to obtain p_3^* and so on. Since $\Sigma_i p_i = 1$:

$$(G' + C') = \sum_{i=1}^{n} p_i^* \tag{10.98}$$

and then:

$$p_i^+ = p_i^* \bigg/ \sum_{j=1}^{n} p_j^* \tag{10.99}$$

Taking a mass balance at the confluence of the unbroken material bypass with the final product stream results in:

$$p_i = f_i(1 - s_{1,i}) + p_i^+(1 - s_{2,i}) \sum_{j=1}^{n} p_j^* \tag{10.100}$$

which gives the size distribution of the double roll crusher product.

If, as is often the case for hard rocks and many coals, the primary daughter fragment distributions can be normalized with respect to the breakage size (see §8.2.3.3), then b_{i-j} may replace $b_{i,j}$, and equation (10.97) becomes:

$$p_i^* = \sum_{j=1}^{i-1} b_{i-j}(p_j^* s_{2,i} + f_j s_{1,j}) \text{ for } n \geqslant i > 1$$
$$p_1^* = 0 \tag{10.101}$$

This can be arranged to give:

$$b_{i-1} = \left[p_i^* - \sum_{\substack{j=2 \\ i>2}}^{i-1} b_{i-j}(p_j^* s_{2,i} + f_i s_{1,i}) \right] \bigg/ f_1 s_{1,i} \tag{10.102}$$

from which b_{i-j} can be estimated when the product set p_i is from a single feed $f_1 = 1$, if values of $s_{1,i}$ and $s_{2,i}$ are known. Otherwise the normalized form of equation (8.13) may be used to obtain B_{i-j}, and hence b_{i-j}.

(b) Estimation of $s_{1,i}$ and $s_{2,i}$ $s_{1,i}$ was found by feeding into the roll crusher, in succession, each of the relevant size intervals. The fraction left in the product for each size interval is the bypass fraction for that interval. Figure 10.29 gives a typical result for both coal and greenstone.

The curve of Figure 10.29 was found in preliminary work to give a straight line on log-probability paper, and hence $s_{1,i}$ was assessed in terms of the median and the standard deviation as given in §4.3.8.3(a). Further, $s_{2,i}$ was obtained by trial and error means in the calculation of b_{i-j} using equation (10.100).

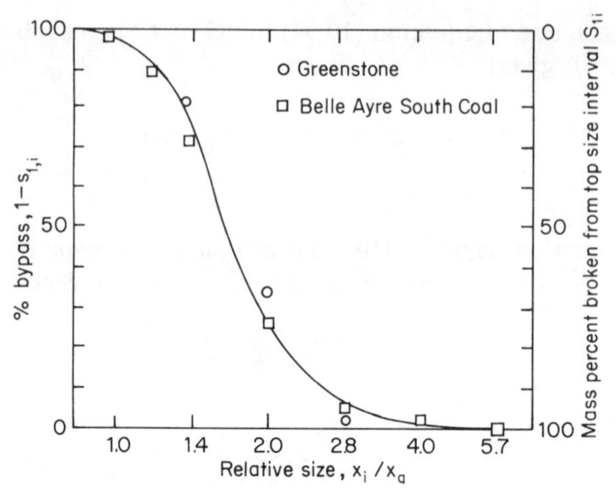

Figure 10.29. Primary bypass fraction $s_{1,i}$ for greenstone and coal (Austin *et al.*, 1980a). *Reproduced by permission of Elsevier Science Publishers*

Later studies by Austin *et al.* (1981b) indicated that a better correlation for the primary bypass is;

$$s_{1,i} = 1 - 1\Big/\left[1 + \left(\frac{x_i/x_g}{\xi^*}\right)^{\lambda^*}\right] \tag{10.103}$$

where λ^* and ξ^* are adjustable parameters and x_g is the gap setting in millimetres. Then the secondary bypass is given by the empirical expression:

$$s_{2,i} = \begin{cases} s_{2,i-1} & \text{for } i < i_g - 1 \\ \frac{1}{2}(s_{2,i_g-1} + s_{2,i_g-2}) & \text{for } i = i_g - 1 \\ s_{2,i} & \text{for } i > i_g \end{cases} \tag{10.104}$$

where i_g is the size interval number corresponding to x_g.

Typical values of $s_{2,i}$ are given in Figure 10.30, and all values are seen to be les than the corresponding $s_{1,i}$.

(c) Comparison of experimental and simulated data Preliminary comparison of product size distribution from industrial-scale coal crushers with those scaled from laboratory equipment via equation (10.100), with a knowledge of b_{i-j}, $s_{1,i}$ and $s_{2,i}$ only gave a fair agreement. However, by making a screening correction in a similar manner to that described in §8.5.2.2(a), the agreement was improved. Rogers and Shoji (1983), by this method, were able to produce the good comparison given in Figure 10.31 for a relatively small gap setting over a range of closely sized feeds.

For larger gap settings the agreement was less good, as indicated in Figure 10.32.

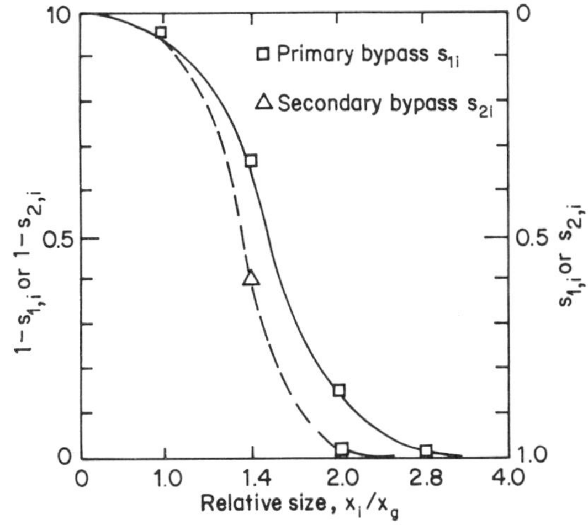

Figure 10.30. Primary and secondary bypass fractions $s_{1,i}$ and $s_{2,i}$ for coal (Austin *et al.*, 1981b). *Reproduced by permission of Elsevier Sequoia*

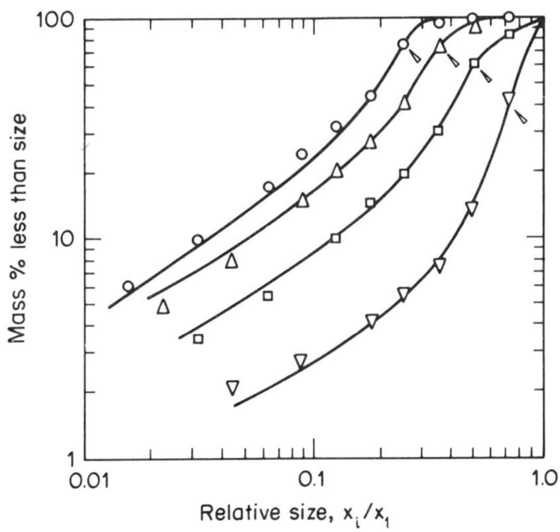

Figure 10.31. Actual and simulated products from KVS 3020 double smooth roll crusher with 4.76 mm gap setting and roll speeds of 170 and 85 rev/minute using closely sized coal feeds (Rogers and Shoji, 1983). *Reproduced by permission of Elsevier Sequoia*

○ 19.1 × 13.5 mm
△ 13.5 × 9.52 mm
□ 9.52 × 6.73 mm
▽ 6.73 × 4.76 mm
◁ Gap — Predicted

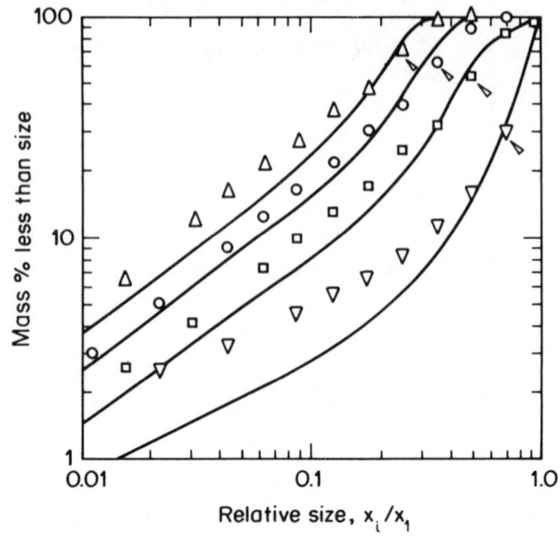

Figure 10.32. Actual and simulated products from the KVS 3020 double smooth roll crusher with 9.52 mm gap setting and roll speeds of 170 and 85 rev/minute using closely sized coal feeds (Rogers and Shoji, 1983). *Reproduced by permission of Elsevier Sequoia*

△ 38.2 × 27.0 mm
○ 27.0 × 19.1 mm
□ 19.1 × 13.5 mm
▽ 13.5 × 9.52 mm
◁ Gap — Predicted

With a distributed feed size and a relatively small gap setting, again agreement between simulated and experimentally found points was close, shown in Figure 10.33.

Overall, the simulated size distributions for industrial-scale equipment derived from laboratory-scale experiments proved to be very encouraging over a wide range of feed sizes and crusher operating conditions, for a single feed material, namely Pittsburgh No. 8 coal. It is an outstanding exercise to discover if the agreement is as good for a wider range of materials.

(d) Two-component feed Rogers (1983a) has commenced the study of materials other than coal by considering a two-component feed to the crushing rolls consisting of gravel and limestone. On the assumption that each would break independently of the other, a model was produced as follows:

$$p_i = \sum_{j=1}^{i} \mu_j f_j d_{1,i,j} + \sum_{j=1}^{i} (1 - \mu_j) f d_{2,i,j} \qquad (10.105)$$

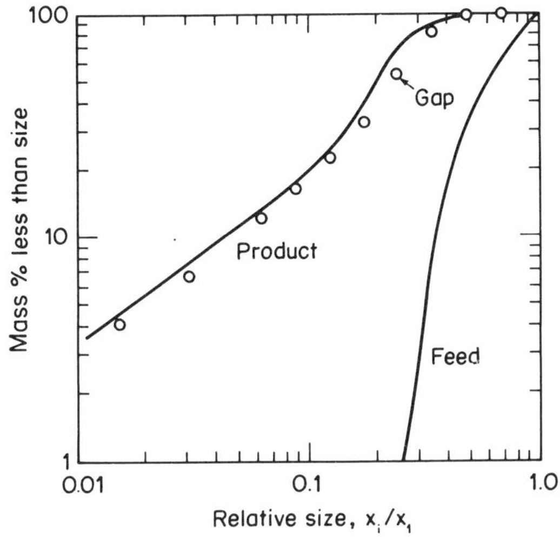

Figure 10.33. Actual and predicted products from the KVS 3020 double smooth roll crusher with 4.76 mm gap setting and roll speeds of 170 and 85 rev/minute using a distributed feed size (coal) (Rogers and Shoji, 1983). *Reproduced by permission of Elsevier Sequoia*

where μ_j is the fraction of component 1, $d_{1,i,j}$ is the mass fraction of component 1 in the size interval j in the feed which appears as size i in the product, and $d_{2,i,j}$ is the mass fraction of component 2 in the size interval j in the feed which appears as size i in the product.

The principles of §10.6.2.1(a) were applied and good agreement obtained between simulated and experimental points on the product size distributions, as given in Figure 10.34.

10.6.2.2 Double roll crushers in closed circuit

Austin *et al.* (1981b) have extended the crusher model to incorporate closed-circuit grinding. Figure 10.35 gives the circuit and symbols used.

With c_i as the fraction of classifier feed returned to the coarse stream, the mass balances become:

$$p_i = \sum_{j=1}^{i} d_{i,j} f_i$$

$$(1 + C')f_i = C' r_i + g_i \text{ and } C' r_i = (1 + C')p_i c_i$$

$$p_i^* = (1 + C')p_i = \frac{d_{i,j} g_i + \sum_{\substack{j=1 \\ i>1}}^{i-1} d_{i,j}(p_j^* c_j + g_i)}{1 - d_{i,i} c_i} \qquad (10.106)$$

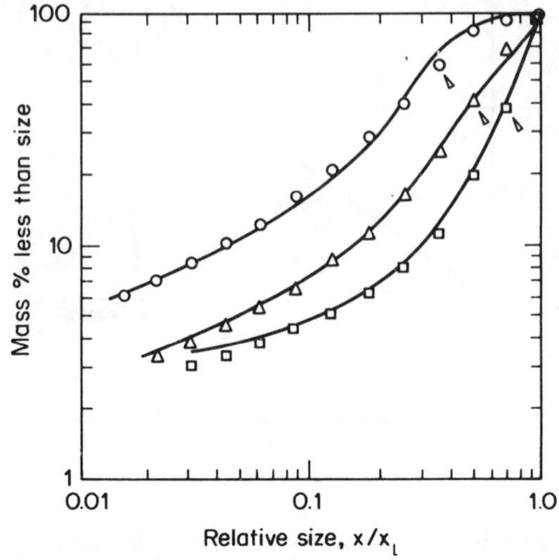

Figure 10.34. Actual and predicted products from a gravel–limestone mixture in a 200 mm smooth double roll crusher with a gap setting of 6.73 mm using closely sized feeds (Rogers, 1983a). *Reproduced by permission of Elsevier Sequoia*

	Feed size (mm)	μ_j	x_1 (mm)
○	19.1 × 13.5	0.8	19.1
△	13.5 × 9.52	0.8	13.5
□	9.52 × 6.73	0.09	9.52
▷	Gap setting		— Simulated

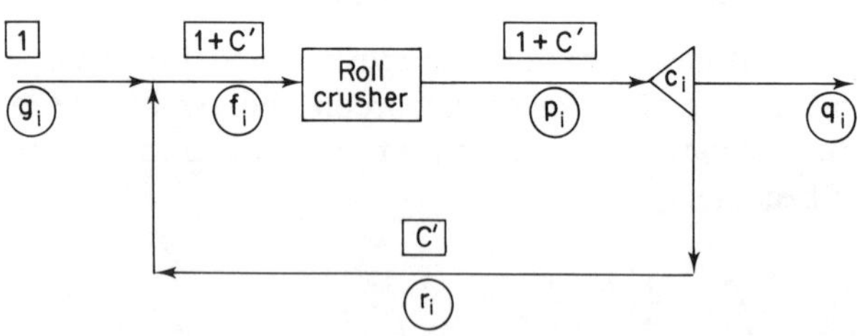

Figure 10.35. System and symbolism for closed roll crusher–classifier circuit (After Austin *et al.,* 1981b). *Reproduced by permission of Elsevier Sequoia*

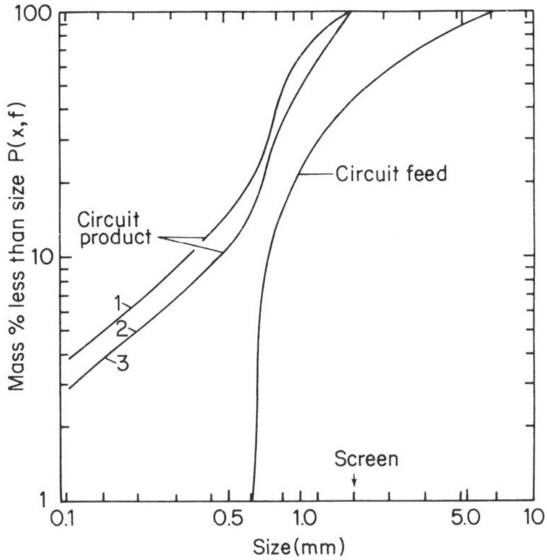

Figure 10.36. Simulated circuit product size distribution for 3 × 30 mesh coal as a function of gap setting and ideal screening of 12 mesh (Austin *et al.*, 1981b). *Reproduced by permission of Elsevier Sequoia*

	x_g(mm)	C'
1	1.19	0.05
2	1.68	0.83
3	2.38	10.50

where:

$$1 + C' = \sum_{i=1}^{n} p_i^* \tag{10.107}$$

$$p_i = p_i^*/(1 + C') \tag{10.108}$$

$$q_i = p_i(1 + C')(1 - c_i) \tag{10.109}$$

A computer program embodying equations (10.106)–(10.109) was used to show the effect of increase of gap setting on a closed-circuit product, as seen in Figure 10.36 for ideal screening ($c_i = 1$)

The conclusion reached is that closing the circuit with a screen gives a significant reduction in the product fines compared with open-circuit grinding. A gap setting slightly higher than the screen is sufficient to reach the desirable minimum fines, with a low circulating load, and increasing the gap setting to obtain a high circulating load gives negligible improvement in performance.

Rogers (1982) has produced the mathematical models corresponding to normal, reverse and combined closed circuits incorporating roll crushers, and reference should be made to his paper for their forms.

10.7 CIRCUITS INVOLVING FLUID ENERGY MILLS: LEVEL II SIMULATION

Pitchumani and Venkateswarlu (1980) have given the products from the grinding of calcite in a vertical fluid energy mill in matrix notation.

Hendry (1967) proceeded further by attempting the SMB simulation of a horizontal fluid energy micronizer type of mill shown diagrammatically in Figure 7.10. For a mill having six jets, the micronizer operation may be visualized from Figure 10.37. At each jet, feed material is reduced in size and a fraction of cut size is peeled off to the product outlet: the remainder is recirculated for further size reduction.

Typical proportions of this recycle are seen in Figure 10.38, both for individual circuits and for the total, which in this instance shows an overall

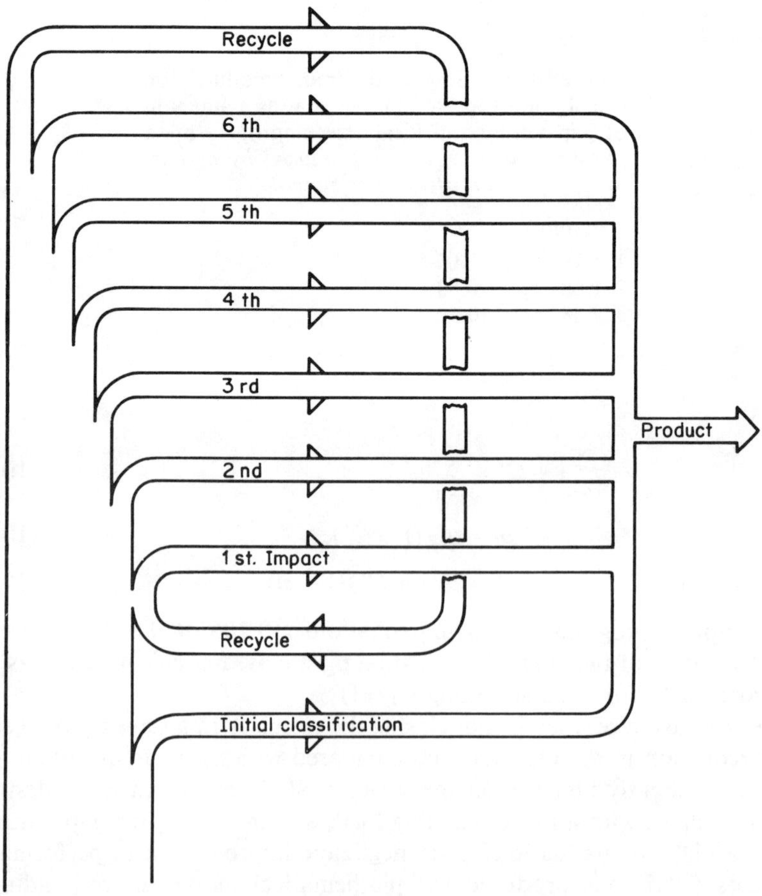

Figure 10.37. Diagram of horzontal micronizer fluid energy mill operation (Hendry, 1967). *Reproduced by permission of DECHEMA, Frankfurt*

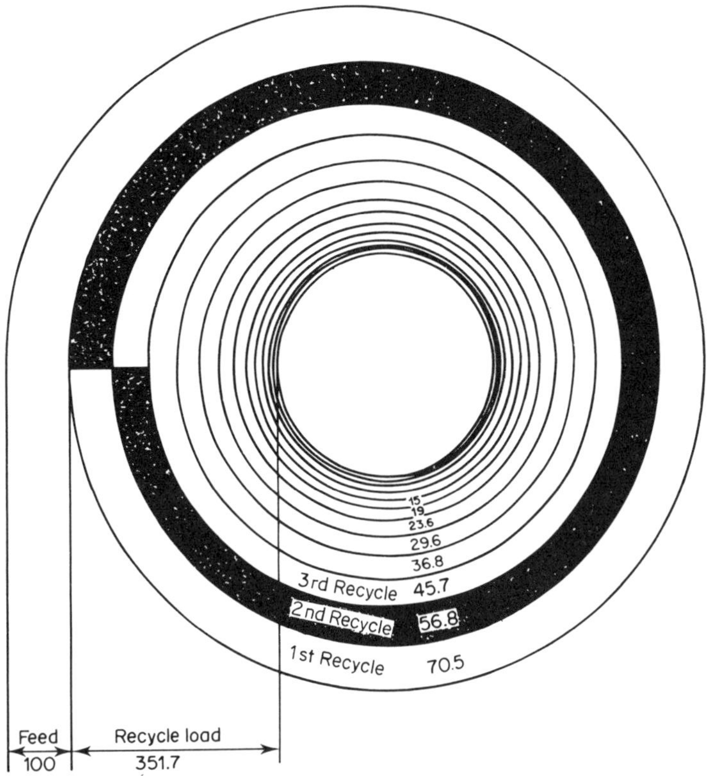

Figure 10.38. Recycling in the horizontal micronizer fluid energy mill (Hendry, 1967). *Reproduced by permission of DECHEMA, Frankfurt*

recycle ratio of 3.517 : 1. The greater this ratio, the greater is the difficulty in grinding the material concerned.

Thus breakage and classification proceed concurrently in the fluid energy mill, and hence S_i, $b_{i,j}$ and c_i are incorporated into the mathematical model as presented by Hendry (1967). He sets out data in matrix form for S_i and c_i and uses the Broadbent and Callcott (1956) equation (10.36) for the evaluation of $B_{i,j}$, and hence $b_{i,j}$, assuming normalization holds. Then:

$$p_i = \sum_{i=1}^{n} f_i S_i b_{i,j} c_i \qquad (10.110)$$

gives the product size distribution from a given feed size distribution f_i. A computer program was written incorporating these variables.

To obtain S_i, a special apparatus was devised which eliminated the classifying action. Then with assumed $b_{i,j}$ and measured p_i, it was possible to assess c_i from equation (10.110) over a range of feed rate.

This work is clearly preliminary and needs to be followed up to allow prediction of product size distributions.

10.8 CIRCUITS INVOLVING OTHER MILLS

10.8.1 Attrition mill: level III simulation

Herbst and Sepulveda (1978) have written an SMB equation for an attrition mill (stirred ball mill) to be used in conjunction with the computer program ESTIMILL (see §10.3.2.3) and have demonstrated good correlation between predicted and observed size distributions when grinding limestone, for short times of grind.

This simulation has been applied only to open circuit so far, but could be modified for closed circuit if required.

10.8.2 Rotary cutter: level III simulation

Klimpel and Austin (1972) have applied the circuit of Figure 10.16(b) to the case of a rotary cutter as used for the disintegration of plastics, having an internal screen, which they treat as the classifier s_2 and a parallel external classifier s_1. The principles of circuits containing two classifiers in parallel are outlined in §10.4, but simplification is introduced in the case of the rotary cutter, because the screen is treated as a perfect classifier.

The full equations for the closed circuit and the steps in the iterative solution of the same are fully set out in the Appendix to Klimpel and Austin (1972).

10.8.3 Planetary mill (Szego): level IV simulation

Koka and Trass (1985) have successfully used the SMB method with the planetary (Szego) mill in multi-pass operation. For a mill description, see §11.2.2.2 and Figure 11.7.

The basis of the simulation is equation (8.15) for which the solution given in equation (8.17) is applicable. The RTD function is modelled on the m fully mixed reactors in series, equation (8.40), and introduced into equation (8.56) for open-circuit steady grinding. The transfer parameter concept expressed in equations (10.13) and (10.14) is used to good advantage.

For multi-pass grinding, adopting matrix notation, the first pass may be represented by:

$$p_1 = \mathbf{D}_1 f_1$$

where $\mathbf{D}_1$ is the transfer matrix corresponding to the first pass. The steady-state first-pass product becomes feed for the second pass. Hence for two passes:

$$p_2 = [\mathbf{D}_2\mathbf{D}_1] f_1$$

and for three passes:

$$p_3 = [\mathbf{D}_3\mathbf{D}_2\mathbf{D}_1]f_1$$

Hence for n passes:

$$p_n = [\mathbf{D}_n\mathbf{D}_{n-1} \ldots \mathbf{D}_1]f_1 \qquad (10.111)$$

Hence, if it can be demonstrated that $\mathbf{D}_n = \mathbf{D}_{n-1} = \ldots = \mathbf{D}_1$, then equation (10.11) may be expressed as:

$$p_n = [\mathbf{D}_1]^n f_1 \qquad (10.112)$$

The assumptions made in this model are:

(1) grinding is at steady state;
(2) particles in their size intervals are not heterogeneous;
(3) breakage is first-order and linear;
(4) all particle sizes have the same RTD in the mill.

Breakage functions were measured on the Szego mill itself modified as follows. For S, the top size fraction method was adopted for each pass using a single short roller. S was found to be well represented by equation (8.2) and the correction of equation (8.3) was not required. Thus values of a^* and α are listed. For B, equation (8.13) was found to be valid by using single-sized feed

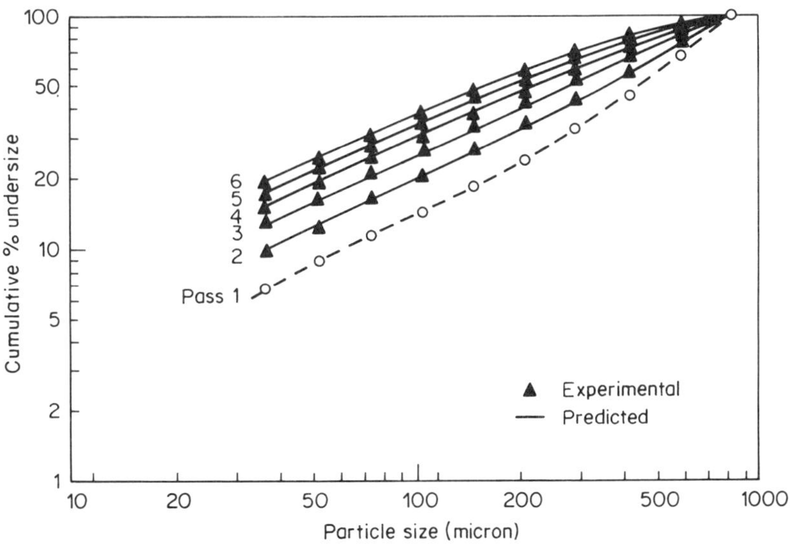

Figure 10.39. Comparison of experimental and predicted size distributions for multi-pass grinding of a coal with the Szego mill at 500 rev/minute and a feed rate of 31 g/s (Koka and Trass, 1985). *Reproduced by permission of the Clean Energy Research Institute*

fractions at low rotational speed on the modified mill, and values of Φ, γ and β are listed.

The effect of the following variables on the breakage function proved significant:

(1) S in the first pass differed from that in subsequent passes because a large proportion of irregularly shaped particles with high aspect ratio was formed in the first pass, which Austin *et al.* (1973) showed would lead to abnormal breakage.
(2) B values changed considerably with mill speed.
(3) The RTD function was found to be constant, except for the first pass.

Hence, equation (10.112) proved to be valid when the first pass was omitted, and on this basis good correlation was demonstrated (Figure 10.39) between predicted and experimentally found product size distributions at 500 rev/minute. At higher speeds the agreement was not so good because of the change of B values as indicated in effect 2 above.

10.9 SUMMARY

Table 10.3 summarizes the up-to-date situation for mill simulation by SMB methods.

Table 10.3. Summary of mill simulation by SMB methods

Mill type	Mill sub-type	Level of sophistication II	III	IV
Tumbling	Wet ball	✓	✓	✓
Tumbling	Dry ball, airsweeping ignored		✓	✓
Tumbling	Dry ball, with airsweeping			✓
Tumbling	Rod	✓	✓	
	Autogenous	✓	✓	
Vertical spindle	Ring-ball			✓
Swing hammer	Beater	✓		
Crusher	Cone	✓		
Crusher	Double roll			✓
Fluid energy	Horizontal micronizer	✓		
Attrition	Stirred ball		✓	
Rotary cutter	—		✓	
Planetary	Szego			✓

As would be expected, most work has been done for mills in greatest use—namely the tumbling mills. Future studies will no doubt fill the gaps, particularly at level IV simulation and at even higher levels for all mills.

10.10 REFERENCES

Abouzeid, A.-Z. M., and Fuerstenau, D. W. (1980). 'Scale-up of particulate hold-up in rotary drums', *Powder Technol.*, **25**, 65–70.

Abouzeid, A.-Z. M., Feurstenau, D. W., and Sastry, K. V. S. (1980). 'Transport behaviour of particulate solids in rotary drums: scale-up of RTD using the axial dispersion model', *Powder Technol.*, **27**, 241–250.

Austin, L.G., Bell, D., and Rogers, R. S. C. (1984). 'Incorporation of the slowing-down effect into a mill model', *Particulate Sci. Technol.*, **2**, 135–145.

Austin, L. G., and Brame, K. (1983). 'A comparison of the Bond method for sizing wet tumbling ball mills with a size-mass balance simulation model', *Powder Technol.*, **34**, 261–274.

Austin, L. G., Klimpel, R. R., Luckie, P. T., and Rogers, R. S. C. (1982a). 'Simulation of grinding circuits for design'. In: *Design and Installation of Comminution Circuits* (Mular, A. L., and Jergensen, G. V., (Eds.), AIME, NY, 301–324.

Austin, L. G., Luckie, P. T., and Seebach, H. M. (1976). 'Optimization of a cement milling circuit with respect to particle size distribution and strength development by simulation models', Dechema Monograph 79, 519–537.

Austin, L.G., Luckie, P. T., and Shoji, K. (1982b). 'An analysis of ball and race milling. Part II: The Babcock 1.7E mill', *Powder Technol.*, **33**, 113–125.

Austin, L. G., Luckie, P. T., and Shoji, K. (1982c). 'An analysis of ball and race milling. Part III: Scale-up to industrial mills', *Powder Technol.*, **33**, 127–134.

Austin, L. G., Luckie, P. T., Shoji, K., Rogers, R. S.C., and Brame, K. (1984), 'A simulation model for an airswept ball mill grinding coal', *Powder Technol.*, **38**, 255–266.

Austin, L. G., Luckie, P. T., and Wightman, D. (1975). 'Steady state simulation of a cement milling circuit', *Int. J. Miner. Process.*, **2**, 127–150.

Austin, L. G., Shah, J., Wang, J., Gallacher, E., and Luckie, P. T. (1981a). 'An analysis of ball and race milling. Part I: The Hardgrove mill', *Powder Technol.*, **29**, 263–275.

Austin, L. G., Shoji, K., and Everell, M. D. (1973). 'An explanation of abnormal breakage of large particles in laboratory mills', *Powder Technol.*, **7**, 3–7.

Austin, L. G., Van Orden, D., McWilliams, B., Pérez, J. W., and Shoji, K. (1981b). 'Breakage parameters of some materials in smooth roll crushers', *Powder Technol.*, **28**, 245–251.

Austin, L. G., Van Orden, D. R., and Pérez, J. W. (1980a). 'A preliminary analysis of smooth roll crushers', *Inst. J. Miner. Process.*, **6**, 321–336.

Austin, L. G., and Weller, K. R. (1982). 'Simulation and scale-up of wet ball milling', *Proc. 14th Int. Miner. Process. Congr.*, Toronto, 1–8/1–24.

Austin, L. G., Weymont, N. P., and Knobloch, O. (1980b). 'The simulation of airswept cement mills', *Proc. Europ. Symp. Particle Technol.*, Amsterdam, 640–655.

Austin, L. G., Weymont, N. P., Prisbrey, K. A., and Hoover, M. (1977). 'Preliminary results on the modelling of autogenous grinding', *Proc. 14th Symp. Appln. Computer Methods Miner. Ind.* (Ramani, R. V., Ed.), SME/AIME, 207–226.

Blau, G., Klimpel, R., and Steiner, E. (1972). 'Non-linear parameter estimation and model distinguishability of physico-chemical models at chemical equilibrium', *Can. J. Chem. Eng.*, **50**, 399–409.

Broadbent, R. A., and Callcott, T. G. (1956). 'A matrix analysis of processes involving particle assemblies', *Phil. Trans. Royal Soc.*, **A249**, 99–123.

Callcott, T. G., and Lynch, A. J. (1964). 'An analysis of breakage processes within mills', *Proc. Australian Inst. Min. Metall.*, **209**, 109–131.

Forman, I., Kreysa, K., and Cigánek, A. (1977). 'Regelung einer zementumlaufmahlage', *ZKG*, **30**, (2), 79–83; English trans., *ZKG*, **30**, (4), 87–89.

Gardner, R. P., Aissa, M., and Verghese, K. (1982). 'Determination of ball mill RTD from tracer data taken in closed circuit operation', *Powder Technol.*, **32**, 253–266.

Gardner, R. P., Aissa, M., and Verghese, K. (1984). 'A model with closed form analytical solution for steady state closed circuit linear comminution processes based on the crossflow finite stage RTD model', *Powder Technol.*, **39**, 143–145.

Gardner, R. P., and Austin, L. G. (1962). 'A chemical engineering treatment of batch grinding. Part I: A radioactive tracer technique for the determination of breakage functions', *Proc. 1st Europ. Symp.* (Zerkleinern), Frankfurt, Verlag-Chemie, Dusseldorf, 217–231.

Hendry, R. (1967). 'A mathematical model for fluid energy mills', Dechema Monograph 57, 695–727.

Herbst, J. A., and Fuerstenau, D. W. (1980). 'Scale-up procedure for grinding mill design using population balance methods, *Int. J. Miner. Process.*, **7**, 1–31.

Herbst, J. A., Grandy, G. A., and Mika, T. S. (1971). 'On the development and use of lumped parameter models for continuous open and closed circuit grinding systems', *Trans. IMM (London)*, **80**, C193–C198.

Herbst, J. A., and Rajamani, K. (1980). 'Control strategies: control of grinding circuits'. In: *Computer Methods for the 80s* (Weiss, A., Ed.), AIME, NY, 770–786.

Herbst, J. A., and Rajamani, K. (1982). 'Developing a simulator for ball mill scale-up: a case study'. In: *Design and Installation of Comminution Circuits* (Mular, A. L., and Jergensen, G. V., Eds.), SME/AIME, NY, 325–342.

Herbst, J. A., Rajamani, K., and Kinneberg, D. J. (1977). 'ESTIMILL: a program for grinding simulation and parameter estimation with linear models', University of Utah, Metallurgical Eng. Dept. Report.

Herbst, J. A., and Sepulveda, J. L. (1978). 'Fundamentals of fine and ultrafine grinding in a stirred ball mill', *Proc. Int. Conf. Powder Bulk Handling Process.*, Rosemont, Illinois, Ind. Sci. Conf. Management, Chicago, 452–470.

Hodouin, D., and Everell, M. D. (1980). 'A hierarchical procedure for adjustment and material balancing of mineral processes data', *Int. J. Miner. Process.*, **7**, 91–116.

Hodouin, D., Gelpe, T., and Everell, M. D. (1982). 'Sensitivity analysis of material balance calculations—an application to a cement clinker grinding process', *Powder Technol.*, **32**, 139–153.

Hodouin, D., McMullen, J., and Everell, M. D. (1980). 'Mathematical simulation of the operation of a three stage grinding circuit for a fine grained Zn/Pb/Cu ore', *Proc. Europ. Symp. Particle Technol.*, Amsterdam, 686–702.

Hutchinson, H. P. (1974). 'Plant simulation by linear methods', *Trans. Inst. Chem. Eng.*, **52**, 287.

Inoue, T., Okaya, K., and Imaizumi, T. (1980). 'A dynamic simulation model for the closed circuit grinding process', *Proc. Europ. Symp. Particle Technol.*, Amsterdam, 656–671.

Kelsall, D. F., Reid, D. J., and Stewart, P. S. B. (1969). 'The study of grinding processes by dynamic modelling', *Inst. Eng. Australia Elect. Eng. Trans.*, **EE5**, 173–186.

Kelsall, D. F., Stewart, P. S. B., and Reid, K. J. (1968). 'Confirmation of a dynamic model of closed circuit grinding with a wet ball mill', *Trans. IMM (London)*, **77**, C120–C127.

Klimpel, R. R., and Austin, L. G. (1972). 'The mathematical modelling and optimisation of the industrial rotary cutter milling facility', Dechema Monograph 69, 449–473.

Klimpel, R. R., and Austin, L. G. (1977). 'The back-calculation of specific rates of breakage and non-normalized breakage distribution parameters from batch grinding data', *Int. J. Miner. Process.*, **4**, 7–32.

Klimpel, R. R., and Austin, L. G. (1984). 'The back-calculation of specific rates of breakage from continuous mill data', *Powder Technol.*, **38**, 77–91.

Koka, V. R., and Trass, O. (1985). 'Determination of breakage parameters and modelling of coal breakage in the Szego mill', Fine Particle Society Meeting, Miami.

Luckie, P. T., and Austin, L. G. (1972). 'A review introduction to the solution of the grinding equations by digital computation', *Miner. Sci. Eng.*, **4**, (2), 24–51.

Lynch, A. J. (1977). *Mineral Crushing and Grinding Circuits*, Elsevier, Amsterdam.

Lynch, A. J., Whiten, W. J., and Draper, B. E. (1967). 'Developing the optimum performance of a multi-stage grinding circuit', *Trans. IMM (London)*, C169–C179.

Marchand, J. C., Hodouin, D., and Everall, M. D. (1980). 'RTD and mass transport characteristics of large industrial grinding mills', *Proc. 3rd Int. Fed. Automatic Control Symp.*, Montreal, 295–302.

Marshall, V. C. (Ed.) (1974). *Comminution: a Report of the Inst. Chem. Eng. Working Party*.

Mular, A. L. (1970). 'The determination of selection elements and lumped parameters for grinding mills', *CIM Bull.*, **63**, (699), 821–826.

Mular, A. L. (1971). 'Mathematical models for optimum design of grinding circuits', *CIM Bull.*, **64**, (708), 34–41.

Mular, A. L., Bradburn, R. G., Flintoff, B. C., and Larson, C. R. (1976). 'Mass balance of a grinding circuit', *CIM Bull.*, **69**, (776), 124–129.

Olsen, T., and Krogh, S. R. (1975). 'A low order model of continuous ball mill grinding', *Proc. 11th Int. Miner. Process. Congr.*, Cagliari, 119–137.

Onuma, E., Asai, N., and Jimbo, G. (1976). 'Analysis and operating characteristics of steady state closed circuit ball mill grinding', Dechema Monograph 79, 559–573.

Perry, J. H. (Ed.) (1963—4th edn.). *Chemical Engineers' Handbook*, McGraw-Hill, New York, Table 1-18, page 1-41.

Pitchumani, B., and Venkateswarlu, D. (1980). 'Matrix representation of batch and continuous size reduction processes', *Proc. Int. Symp. Fine Particle Process.*, Las Vegas, Nevada, AIME, 148–163.

Rajamani, K., and Herbst, J. A. (1980). 'A dynamic simulator for the evaluation of grinding circuit control strategies', *Proc. Europ. Symp. Particle Technol.*, Amsterdam, 64–81.

Richardson, J. M., and White, J. W. (1982). 'Mass balance calculations for comminution circuits'. In: *Design and Installation of Comminution Circuits* (Mular, A. L., and Jergensen, G. V., Eds.), SME/AIME, NY, 150–175.

Rogers, R. S. C. (1979). 'Application of the segregated flow concept to the analysis and simulation of continuous ball mills by the mechanistic approach', Ph.D. thesis, North Carolina State University, Raleigh, NC.

Rogers, R. S. C. (1982). 'Closed form analytical solutions for models of closed circuit roll crushers', *Powder Technol.*, **32**, 125–127.

Rogers, R. S. C. (1983a). 'The double roll crusher model applied to a two-component feed material', *Powder Technol.*, **35**, 131–134.

Rogers, R. S. C. (1983b). 'A generalised transfer parameter of crushing and grinding circuit simulation', *Powder Technol.*, **36**, 137–143.

Rogers, R. S. C., and Shoji, K. (1983). 'A double roll crusher model applied to a full scale crusher', *Powder Technol.*, **35**, 123–129.

Shoji, K., Austin, L. G., Smaila, F., Brame, K., and Luckie, P. T. (1982). 'Further studies of ball and powder filling effects in ball milling', *Powder Technol.*, **31** 121–126.

Sood, M. K., and Reklaitis, G. V. (1979). 'Solution of material balances for flowsheets modelled with elementary modules: the constrained case', *A.I.Ch.E.J*, **25**, (2), 220–229.

Sood, M. K., Reklaitis, G. V., and Woods, J. M. (1979). 'Solution of material balances for flowsheets modelled with elementary modules: the unconstrained case', *A.I.Ch.E. J*, **25**, (2)., 209–219.

Stanley, G. G. (1974). 'Mechanisms in the autogenous mill and their mathematical representation', *J. South African Inst. Min. Metall.*, **75**, 77–78.

Stewart, P. S. B., and Restarick, C. J. (1971). 'A comparison of the mechanism of breakage in full scale and laboratory scale grinding mills', *Proc. Australian Inst. Min. Metall.*, **239**, 81–92.

Tschorbadjiski, I. (1980). 'Expermentelle prüfung eines mathematischen modells der kombination schlagradmühle-sichter', *Proc. Europ. Symp. Particle Technol.*, Amsterdam, 595–607.

White, J. W., Winslow, R. L., and Rossiter, G. J. (1977). 'A useful technique for metallurgical mass balances—applications in grinding', *Int. J. Miner. Process.*, **4**, 39–49.

Whiten, W. J. (1972). 'The simulation of crushing plants with models developed using multiple spline regression', *J. South African Inst. Min. Metall.*, **72**, (10), 257–264.

Whiten, W. J. (1974). 'A matrix theory of comminution machines', *Chem. Eng. Sci.*, **29**, 589–599.

CHAPTER 11

Specialized methods of size reduction

NOTATION FOR CHAPTER 11

D	Diameter of cylinder
g	Acceleration due to gravity
G	Diameter of gyration
HI	Hardgrove index
M	Mass
N	Rotational, speed of gyration
N_T	Rotational speed of cylinder about its own axis

R Ratio of sun to planetary rotational speeds $= N/N_T$ (Subscript c = at critical condition)

11.1 INTRODUCTION

In comminution-oriented industry, as in all industry, changes occur with time as new demands arise. Then fresh concepts of size reduction machinery must be advanced to meet these demands.

Currently, examples are found in:

(1) cryogenic crushing which has assumed new importance with the greatly expanded use of frozen products (pin mills and rotary cutters);
(2) the reduction of size and weight of crushers so that they may be used (a) down a mine near the source of coal, with subsequent air-lift of fine particles to the surface, or (b) as a portable device on an open-cast coal site, with subsequent hydraulic transport of fine product to the site where it is required (planetary mills);
(3) the production of stable coal slurries which can be pumped and atomized in the same way as oil (attrition mills);
(4) the cleaning of coal by removal of ash and pyritic impurity for which a superfine poduct is advantageous (planetary mills and ultrasonic systems of comminution);
(5) the elimination of the need of separate removal of sulphur and ash from coal (chemical comminution);
(6) the production of close size distributions of materials such as coke and sinter, advantageous for use in blast-furnace operation (electrohydraulic crushers);
(7) advantageous concurrent size reduction and liberation of ores (tower mills).

In general, it is agreed with Bilby (1979) that minds should remain constantly alert to the harnessing of all possible methods of crushing and grinding, because the demands of, and conditions in, tomorrow's industry may be different from those of today.

Most of the size reduction methods considered in this handbook rely on mechanical forces to produce stress within a particle, leading to breakage, and most of the applications in this chapter concern mechanical devices. However, thermal and chemical forces can be large compared with mechanical (Rumpf, 1965), hence thermal decrepitation and chemical comminuton are discussed. Similarly, shock waves from conventional chemical explosion are known to be effective in quarrying, but their use in comminution requires a more controlled force, found in electrohydraulic crushing. Again, the explosive release of steam under pressure, as used in the Snyder process, has reached pilot scale investigation for a range of minerals.

As noted above, in the realm of superfine grinding, ultrasonics has been advanced for the laboratory production of micron-sized coal with a strong recommendation for scale-up to a pilot size experiment.

Other energy sources involving electrothermal or laser or even atomic devices have been proposed in a quarrying context (Cook and Harvey, 1974; Maurer, 1967). These appear not to show economic advantage at present in a comminution context, but they should be kept under review in the future in the light of any new industrial demand on the one hand and of new advances in the processes themselves on the other.

11.2 MECHANICAL APPLICATIONS

11.2.1 Mills for cryogenic crushing

The size reduction of non-brittle materials such as plastics has been achieved in rotary cutting machines such as that shown diagrammatically in Figure 11.1

In their study of this machine Jimbo *et al.* (1980) have found that the replacement of sharp cutters by blunt ones made very little difference to the performance, and hence it was natural that the pin mill, depicted in Figure 11.2, should also be examined.

As temperature is reduced, both the rotary cutter and the pin mill show increased energy utilization. Figure 11.3 indicates there is little to choose between either mill in the size reduction of maize at optimum conditions.

When a wider range of foodstuffs is examined, it becomes clear that reduction of specific energy is only attainable when the moisture content is below 15%. When the water content of potato, for example, is increased to 75%, a reverse trend is noticed: reduction of temperature increases the specific energy consumption.

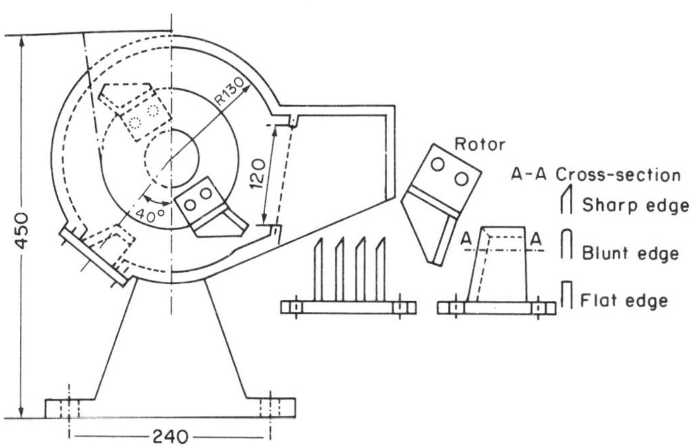

Figure 11.1. Diagram of a rotary cutting mill (Jimbo *et al.*, 1980). *Reproduced by permission of DECHEMA, Frankfurt*

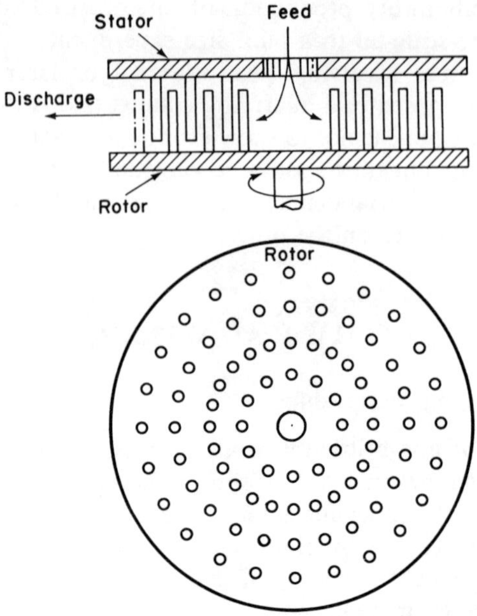

Figure 11.2. Diagram of a typical pin mill

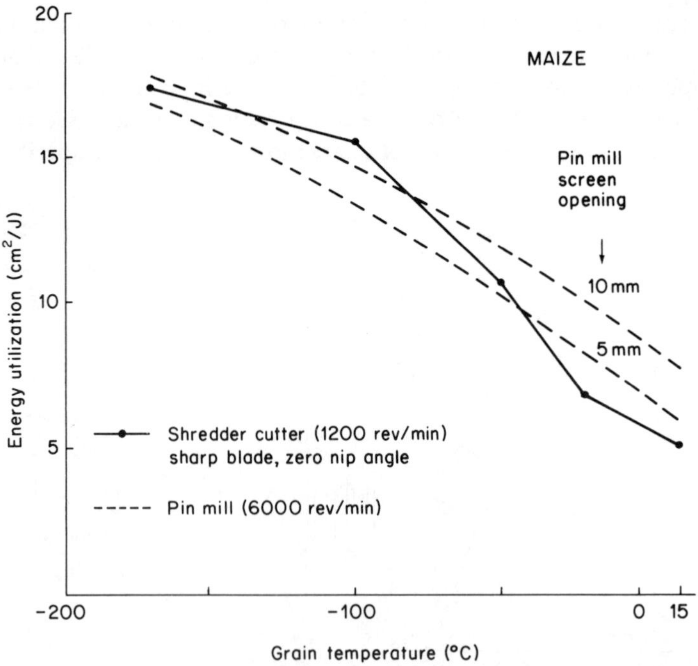

Figure 11.3. Improved energy utilization with reduced grain temperature for a rotary cutter and a pin mill grinding maize (Jimbo *et al.*, 1980). *Reproduced by permission of DECHEMA, Frankfurt*

Typical size distributions for rotary cutters operating on plastics at room temperature are given in Figure 11.4, which indicates that the distributions may be represented by straight lines on a log–linear chart. For low rotational speeds and high Charpy values of the plastic (less brittle) a fairly coarse product results, as curve 5, characterized by two straight lines of different slope (bimodal distribution). With either increased speed or reduced Charpy value (more brittle) or both, the curves shift to a position of greater fineness, as would be expected, and also lose their bimodal character (Jimbo *et al.*, 1976). See §4.3.5.2.

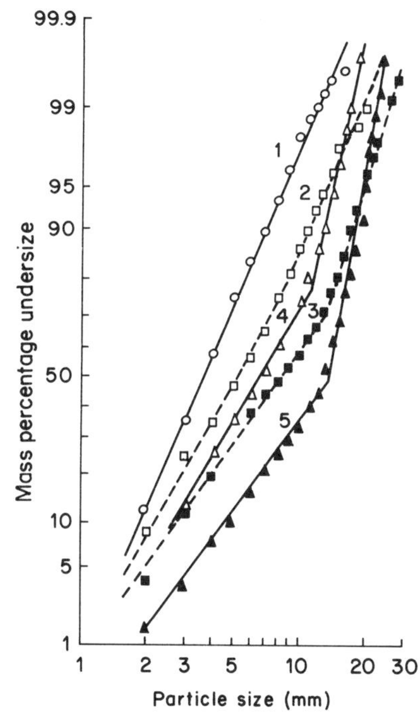

Figure 11.4. Size distributions of the product from a rotary cutter operating on a plastic (PVC) material (Jimbo *et al.*, 1976). *Reproduced by permission of DECHEMA, Frankfurt*

1 3395 rev/minute A
2 1832 rev/minute A
3 1832 rev/minute C
4 995 rev/minute A
5 995 rev/minute D

with the following percentage ratios of impact modifier added to the plastic:
A = 0 (brittle), C = 7.5, and D = 10.0

11.2.2 Planetary mills

11.2.2.1 General theory

The concept of a planetary mill appears to have arisen during the last century (Herzfeld, 1896) but not until recently has it been developed and that in the context of gold mining (Bradley *et al.*, 1972, 1980).

The geometry of the planetary mill, also called the centrifugal mill, is given in Figure 11.5. It shows a cylinder of diameter D (in practice there are several) which simultaneously rotates about a centre of gyration and about its own axis. The diameter of gyration is G. Then a particle within the cylinder will be subject to three forces (Schönert, 1984):

(1) a centrifugal force due to rotation about the centre of gyration ($=\frac{1}{2}\mathrm{MGN}^2$);
(2) a centrifugal force due to rotation about its own axis ($=\frac{1}{2}MDN_T^2$);
(3) the Coriolis force ($= \pm \mathrm{MDNN_T}$, the minus sign applying when the rotations are in opposite directions).

If $R = N_T/N$ is the sun to planetary gear ratio, for a particle on the wall the Coriolis force will not apply, and neglecting gravity, which is relatively small at high rotational speeds, the critical condition is given when the centrifugal force due to gyration at a frequency of N and the centrifugal force due to rotation about the cylinder axis at a frequency of $(R+1)N$, are equal. Thus

$$D(R+1)^2N^2 = GN^2, \text{ or:}$$

$$(R+1)^2 = G/D \tag{11.1}$$

Thus there are two critical values of the gear ratio given by:

$$R = -1 \pm \sqrt{(G/D)} \tag{11.2}$$

The fraction of critical speed is then given by:

$$\% \text{ critical} = 100(R+1)/(R_c+1) \tag{11.3}$$

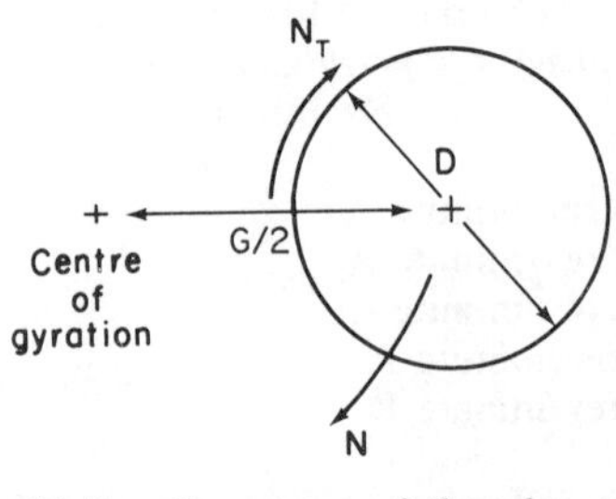

Figure 11.5. Geometry of the planetary mill (Bradley *et al.*, 1980). *Reproduced by permission of DECHEMA, Frankfurt*

where:

$$R_c = -1 + \sqrt{(G/D)} \text{ when } (R+1) > 0$$
$$= -1 - \sqrt{(G/D)} \text{ when } (R+1) < 0$$

From equation (11.3) for $R = -1$, the operation of the mill will be non-critical for any finite value of $\sqrt{(G/D)}$. Kitschen *et al.* (1982) have demonstrated that:

(1) Optimum G/D lies between 0.4 and 0.6 for minimum specific energy.
(2) The optimum cylinder filling is 50–60%.
(3) Rotational speeds giving accelerations up to 15g are practical, with an optimum at 13g for minimum specific energy.
(4) The facility may be used for wet or dry grinding, either with a grinding charge or autogenously.
(5) Because of the small space requirement and relatively low weight, the mill is well suited to (a) use underground, associated with an air-lift to transport powder to the surface (particularly valuable in South African gold mines which have very deep shafts), and (b) use in open-cast mining as a portable size reducer, allowing powder to be transported hydraulically directly from the mining area to the location required for use (e.g. a boiler in the case of coal).
(6) When operating with iron ore and giving a product with a specific surface of 2500 cm^2/g, the specific energy is of the order of 20–25 kWh/tonne (dry grinding).

11.2.2.2 Szego planetary mill

(a) Mill description One particular variant of the planetary mill is that conceived by Szego and described by Gandolfi *et al.* (1984). A cross-section of the mill is given in Figure 11.6. It consists of a stationary vertical cylinder (stator) and a composite rotor comprising three or more helically grooved rollers suspended by flexible steel shafts from flanges connected to a central drive shaft. Feed material enters at the top of the mill and the product is removed from the bottom.

The crushing force is created by the radial acceleration of the rotors and the high velocity gradients that are generated induce a high degree of shear. The grooves in the rotors, the contours of which are vitally important design features, aid transport of even pasty materials through the mill. In fact, pre-crushed coal to minus 4 mesh, at a concentration in water of 60–70% by mass, has such a pasty consistency. Further, the machine can handle wood chippings and bark.

Adjustments are made to roller diameter, roller length and mill speed to suit the requirements of a given material and product characteristics.

(b) Coal slurry context Stable slurries of coal have been sought in both oil and water, which may be pumped and sprayed like oil for use in utility and

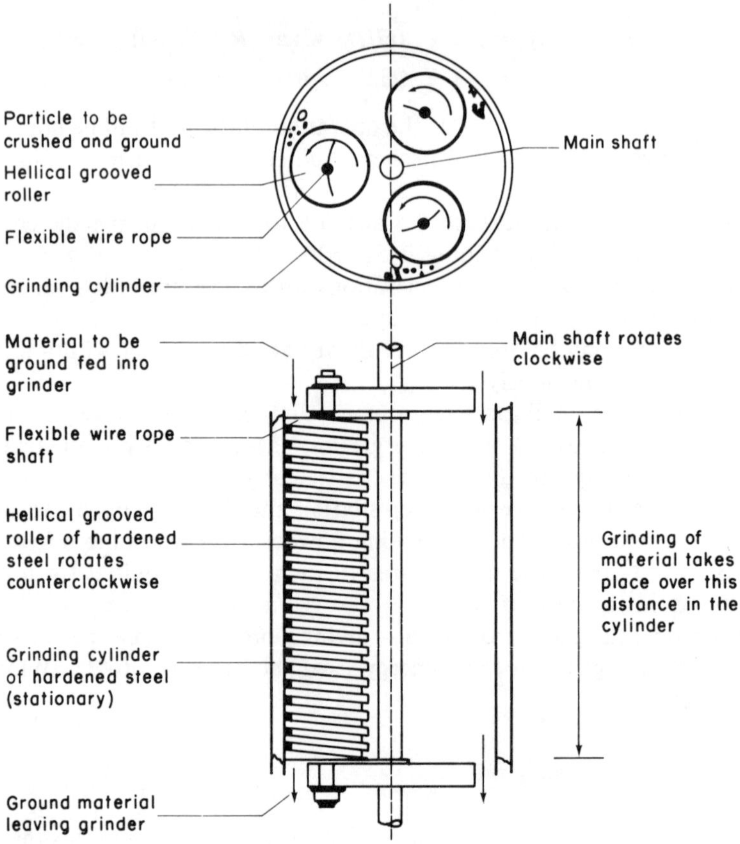

Figure 11.6. Cross-section of Szego mill (Gandolfi *et al.*, 1984). *Reproduced by permission of Elsevier Sequoia*

industrial furnaces. Coal–water slurries find more attraction than coal–oil slurries because they obviate the use of high-cost oil altogether, and the disadvantage of reduced boiler efficiency because of the presence of water is found not to be prohibitive.

To obtain a coal–water slurry both dry and wet grinding routes are available. However, the former demands a drying process in an airswept mill followed by disengaement of solids from the airstream using a cyclone and bag filter, in which inert gas blanketing is essential as a safety measure. This is further followed by water admixture. Such complexity is avoided in the wet grinding route, which is clearly preferable.

The first obvious choice of mill is the wet tumbling ball mill, but this has proved to be economic only for median particle sizes down to 45 microns, approximately 80% minus 75 micron (Gandolfi *et al.*, 1984). For very fine slurries with a median size below 10 microns, the USBM attrition mill (see

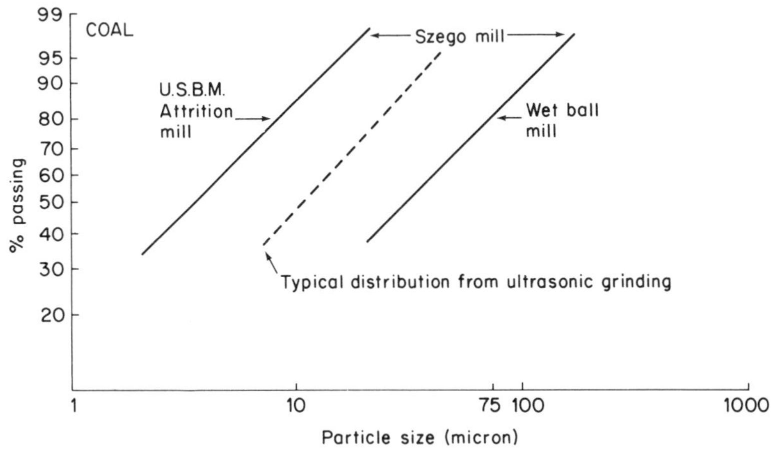

Figure 11.7. Operational product size ranges in coal slurry grinding for the USBM attrition mill, the Szego mill and the wet ball mill compared with a typical distribution from an ultrasonic grinding system

§11.2.3) has been found suitable. The Szego mill appears to fill the gap in the median particle size range from 10 to 45 microns, as indicated in Figure 11.7.

Gandolfi *et al.* (1984) measured the slurry product size distribution by sieving down to 44 microns and used a CMB 60 sensor below 60 microns. Agreement between the two sets of data over the common size range between 44 and 60 microns confirmed consistency. Typical distributions are given in Figure 11.8.

For this particular 8-roller mill, four passes were required to obtain 85% minus 75 micron, when operating at 600 rev/minute; hence a recycle system is essential.

Typically, energy consumption to grind to 80% minus 75 micron is of the order of 20–25 kWh/tonne, depending on the Hardgrove index, when grinding coal at a capacity of 3 tonne/hour for an 8-roller mill. Scale-up to capacities of more than 20 tonne/hour is envisaged at similar product fineness with the same specific energy.

(c) Coal cleaning context Trass *et al.* (1985) and Trass and McCracken (1985) have reported the use of the planetary (Szego) mill for coal beneficiation by simultaneous grinding and oil agglomeration. The agglomeration technique was developed by the National Research Council of Canada (NRCC). Capes and Germain (1982) describe the process: finely ground coal, dispersed in water, is contacted with oil by intensive mixing. The oil phase bridges between the hydrophobic coal particles to form agglomerates, while the hydrophilic ash particles remain in water suspension. The agglomerates are then separated from the aqueous phase by screening or some other suitable method.

The process improvement by using the Szego mill comes from simultaneously grinding in water and adding appropriate amounts of oil, which

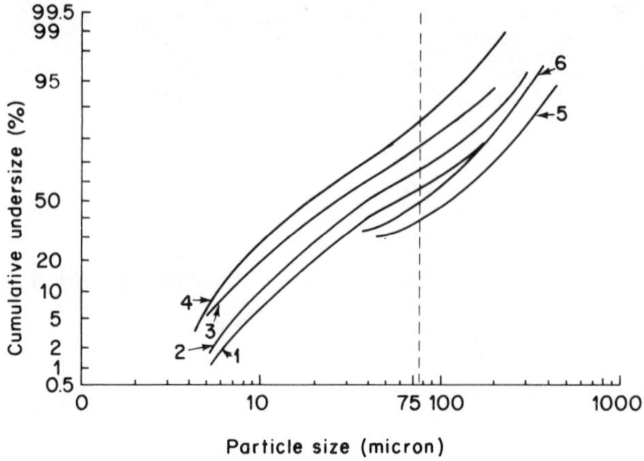

Figure 11.8. Typical particle size distributions for coal ground with water in the Szego mill (Gandolfi *et al.*, 1984). *Reproduced by permission of Elsevier Sequoia*

	No of Passes	*rev/minute.*
1	1	600
2	2	600
3	3	600
4	4	600
5	1	900
6	1	1200

gives the desired products in a single operation, and considerable savings over the NRCC method in equipment cost and power consumption are possible.

It is reported (Trass and McCracken, 1985) that a coal containing 20% ash may be reduced to 10% by the Szego mill technique, with an energy consumption of the order of 30 kWh/tonne.

This system is radically different from the ultrasonic method of pyritic ash removal described in §11.3.1, but the order of magnitude of energy consumption appears to be the same.

(d) Context of wood chips and bark As an indication of the versatility of the Szego mill, Trass (1984) has reported its ability to grind wood chips and bark. For the former, high rotational speeds (>1000 rev/minute) and high feed rates are the most desirable operating conditions. For bark, however, lower speeds (<600 rev/minute) and high feed rate are optimal. In each case the moisture content is an important factor, but barks with more than 70% moisture require a two-stage process: initial low rotational speed followed by screening and followed again by grinding the coarse fraction from the screen at high speed.

The Szego mill at 15 t/h can produce the same product as that from a hammer mill (long used for the grinding of wood) with the expenditure of only 435 horsepower, compared with 600 horsepower for the hammer mill.

11.2.3 Attrition mills

11.2.3.1 Mill applications

The attrition mill is not new, as it has been used in several forms in the paint, pigment and pharmaceutical industries, including the sand (Perl) mill (Figure 7.11). (see §7.2.7.)

In the context of the rapid production of sub-micron ceramic oxides suitable for sintering, the US Bureau of Mines (USBM) developed a particular type of attrition mill containing a rotor within a stator (Stanley *et al.,* 1973). It was also demonstrated for silicon carbide that this mill could be operated autogenously, that is, replacing the normally used sand by rounded silicon carbide grains so avoiding product contamination that occurred with sand (Stanley *et al.*, 1974).

It is this mill, with sand as the grinding medium, which has been used for coal slurry production. (See §11.2.3.2.)

Another attrition mill, the tower mill, has been developed in Japan in the context of advantageous concurrent comminution and liberation for uranium and gold ores. (See §11.2.3.3.)

11.2.3.2 USBM attrition mill

Figure 11.9 shows the stator and rotor cages having relatively small clearances at the sides and at the bottom of the mill. The cages are mounted in a water-cooled container because of the considerable heat evolved during rapid size

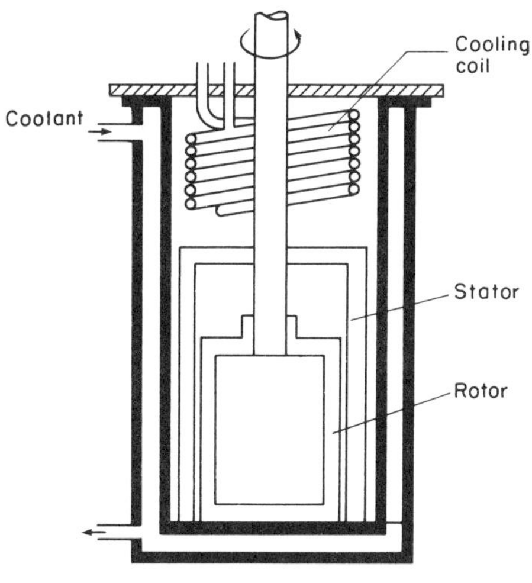

Figure 11.9. Cross-section of USBM attrition mill for batch coal slurry production (Davies *et al.,* 1980). *Reproduced by permission of the American Institute of Mining, Metallurgical and Petroleum Engineers*

Table 11.1. Summary of performance data for the attrition micro-mill grinding industrial minerals (Davis et al., 1980)

Mineral	Feed (wt %) Minus 10 μm	Feed (wt %) Minus 2 μm	Product (wt %) Minus 10 μm	Product (wt %) Minus 2μm	Energy consumed per tonne of feed GJ	Energy consumed per tonne of feed kWh
Mica	0	0	62	54	2.09	580
Sericite	38	3.6	96.8	40	1.07	621
Fluorite	<5	0	96.5*	88.7	2.0	555
Anthracite coal	41.2	15.5	87.8	34.5	†	—
Lignite	38.6	14.1	92.7	84.8	3.4	944
Talc	12	7	97	60	2.0	555
Calcium carbonates:						
Texas limestone‡	10	2	98	85	2.5	694
Calif. calcite	3.6	1	100	91	2.4	666
Calif. whiting	36	0	98.5	90	1.2	333
English whiting	90	47.3	98	93	1.5	417
Canadian calcite	10	0	100	95.6	2.03	563
Alabama marble	0.2	0	98	87	1.2	333

*Minus 5 μm material
†No data available
‡Surface area 11.68 m^2/g which calculates to 0.2 μm average particle size

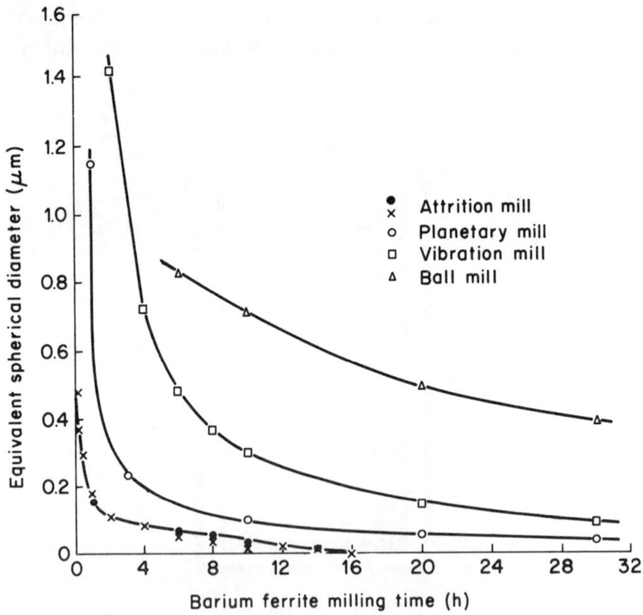

Figure 11.10. Comparison of grinding times for barium ferrite in four mill types (Davies et al., 1980). *Reproduced by permission of the American Institute of Mining, Metallurgical and Petroleum Engineers*

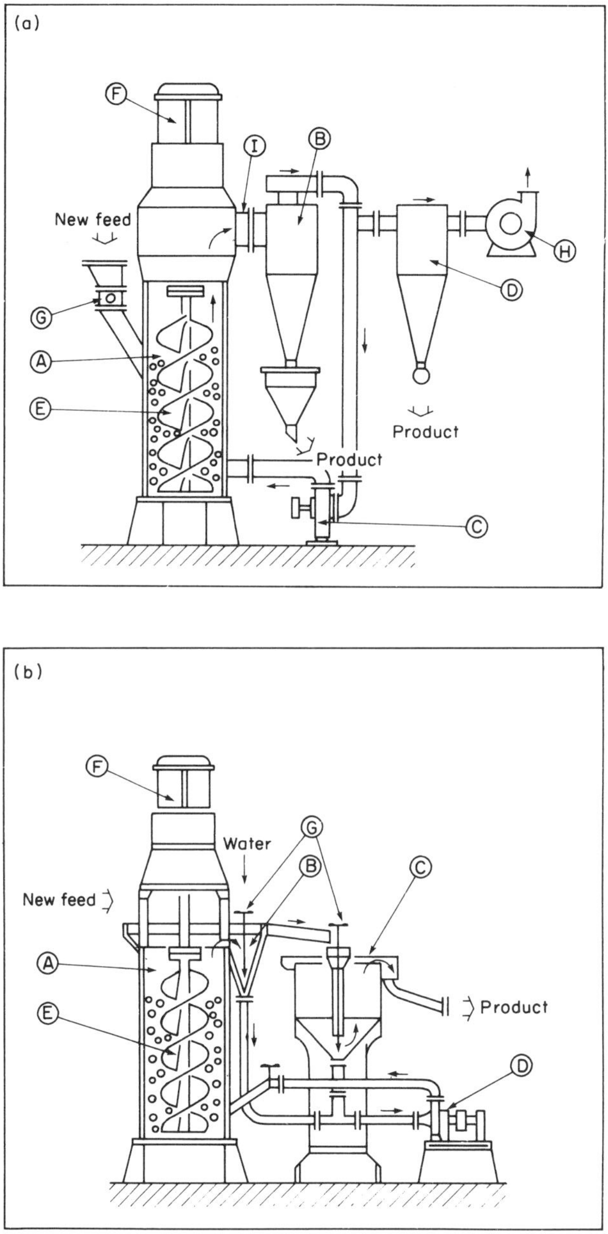

Figure 11.11. Cross-section of tower mill and associated equipment for (a) dry grinding, (b) wet grinding (Japan Tower Mill Co. Ltd., 1984)
(a): A grinding chamber, B cyclone, C blower, D bag filter, E screw, F main motor, G rotary feeder, H vent fan, I sensor
(b); A grinding chamber, B coarse classifier, C cone classifier, D sand pump, E screw agitator, F main motor, G cone valve

reduction. Typically, the grinding medium occupies 76% of the volume of the mill. Initial experiments reveal that the really critical feature is the type, size and shape of the grains of the grinding medium. Spherical particles prove to be superior to angular. Other important variables are the mass ratio of grinding medium to feed hold-up, rotor speed and the angular disposition of the stator and rotor blades. These have more influence than the clearance between the rotor and the base of the container, or the pulp level.

For a description of a 20-inch (50-cm) diameter version of the USBM attrition mill, see Davis *et al.* (1973).

Table 11.1 summarizes typical performance data. Note that coal requires the highest specific energy of all the minerals tested.

The advantage of the USBM attrition mill over the planetary, vibration and tumbling ball mills is seen for barium ferrite in Figure 11.10 (Davis *et al.*, 1980).

11.2.3.3 Tower mills

Figure 11.11 shows a cross-section of the tower mill which consists of a vertical cylinder containing a screw agitator mounted on a central vertical shaft and a grinding charge consisting of steel balls or ceramic pebbles (or alternatively autogenous grinding is possible). The associated equipment for dry grinding or wet grinding is also depicted. Both the cylinder interior and the agitator are rubber-covered to counteract wear. The rotational speed of the agitator may be in the range 35–85 rev/minute depending on mill size.

A typical performance on minus 0.625 inch limestone is to produce 22.5 t/h product with a fineness of 95% passing 325 mesh, at 20 kWh/ton (wet grinding) for a flue gas desulphurization system.

However, the greatest advantage of the tower mill appears to be its potential for simultaneous size reduction and beneficiation of ores.

Two examples are given:

(1) Simultaneous size-reduction and leaching of uranium ore. Wet grinding of minus 3 mm yellow cake uranium ore may be performed to give 60% leaching of minus 200 mesh product in half an hour.
The corresponding process in a ball mill would involve initial grinding together with subsequent leaching, requiring 72 hours and twice the energy consumption of the tower mill.
(2) Simultaneous size reduction and gold ore extraction. Much more effective extraction has been demonstrated in shorter time than by the use of a ball mill.

11.3 ULTRASONIC SYSTEMS

11.3.1 Preferential liberation of pyrite and ash

Early experiments by the US Department of Energy (DoE), using ultrasonic systems, demonstrated that coal particles of less than 10 microns in size

Table 11.2. Pyritic iron and ash analysis of ultrasonically ground dry coal (Tarpley *et al.*, 1981)

Size category	% total pyritic iron	% ash
−200 + 325 mesh	0.416	4.6
−325 mesh	0.456	4.5
−10 micron	0.100	3.2
Control (bulk unground)	0.300	4.1

contained a smaller proportion of pyritic iron and ash than those in immediately larger size ranges, as indicated in Table 11.2 for coal of $HI = 86$.

This trend encouraged the development of an improved ultrasonic comminution system with an even finer product, for which the most advantageous preferential proportions of ash and pyritic iron (for moist coal) were found to be in the minus 7 micron band.

11.3.2 Dry versus moist coal grinding

The advantage of the use of moist rather than dry coal in ultrasonic grinding is summarized in Table 11.3. It is seen to give a 7:1 increase in minus 7 micron material over dry grinding and a 13:1 increase over moist grinding without ultrasonics.

Table 11.3. Coal comminution with and without ultrasonics and with and without moisture (Tarpley *et al.*, 1981)

	% in size range		
Size range	Dry grind + ultrasonics	Moist grind + ultrasonics	Moist grind without ultrasonics
+200 mesh	57.70	35.00	73.00
−200 + 325 mesh	13.99	17.00	8.00
−325 mesh + 30 micron	3.23	11.23	14.82
−30 + 15 micron	18.82	12.43	1.84
−15 + 7 micron	4.62	12.24	1.43
−7 micron	1.64	12.10	0.91

11.3.3 Description of system

Figure 11.12 shows the final form of ultrasonic comminution equipment for moist coal at the conclusion of the laboratory development programme. (Tarpley *et al.*, 1981). The curved plate gave improved performance compared with a flat plate because it:

(1) allowed a longer residence time of coal in the grinding nip, which was subject to intense ultrasonic vibrations;

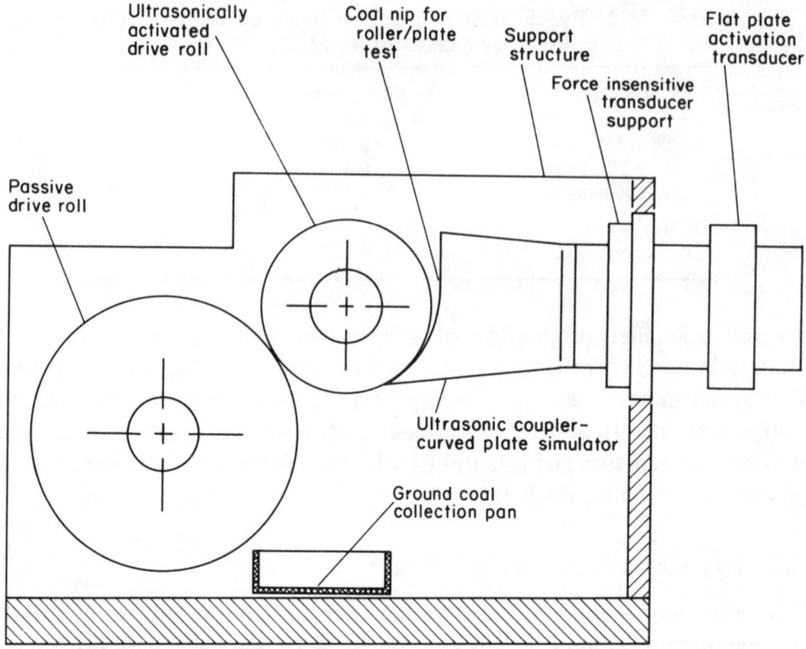

Figure 11.12. Ultrasonic coal grinding apparatus (Tarpley *et al.*, 1981). *Reproduced by permission of the Energy and Minerals Research Co.*

(2) increased the active working area;
(3) provided a gradually decreasing passage gap, so increasing the proportion of the all important minus 7 micron product to 37% with a coal of $HI = 56.7$.

Figure 11.7 gives a typical size distribution from the equipment shown in Figure 11.12. It is seen that it falls into the range of the Szego mill, developed in the context of coal slurry production, and hence ultrasonics could well be valuable in the context of coal slurry production.

11.3.4 Energy requirements

Compared with some other mills, ultrasonics showed a distinct specific energy advantage, as summarized in Table 11.4

No valid reason has been advanced why the laboratory equipment described above should not be scaled-up to industrial size, with similar specific energy requirements, and this, in fact, is strongly recommended. Meanwhile laboratory work proceeds on a modified apparatus (Aber, 1985).

Table 11.4. Comparison of specific energy required for coal grinding to minus 10 micron size by four methods (Tarpley *et al.*, 1981)

Comminution method	Specific energy* to produce −10 micron coal (kWh/ton)
Fluid energy mill	330–1000
Ball mill	Circa 330
Attrition mill (Figure 7.11)	Up to 326
Ultrasonic grinding	Circa 37

*Depends on coal grindability

11.4 ELECTROHYDRAULIC CRUSHING

Figure 11.13 gives a typical arrangement of an electrohydraulic crushing apparatus, the original concept of which is attributed to Yutkin (1955).

The active region is the underwater discharge generated by the release of stored energy from a high-voltage capacitor via an air-gap switch. A unidirectional water pulse may also be applied from a separate vessel using compressed air with the object of removal of undersized material as it is formed, thus assisting in the production of a close sized distribution. The length of the spark gap is typically between 1.5 and 8.0 cm in this equipment according to the material to be crushed, as indicated in Table 11.5.

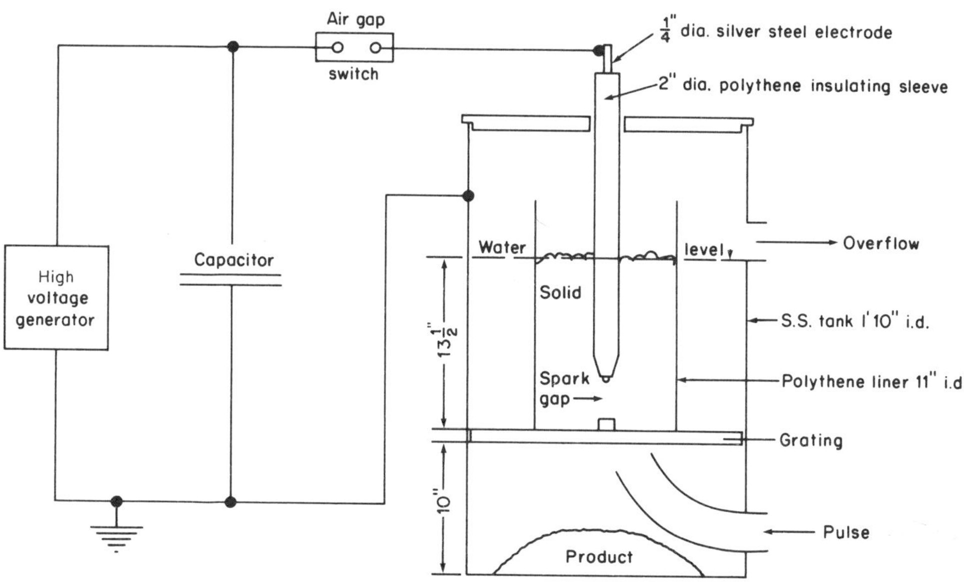

Figure 11.13. Diagram of electrohydraulic crushing equipment (Hitchon and Carley-Macauly, 1969). *Reproduced by permission of the British Steel Corporation*

Table 11.5. Typical results from electrohydraulic crushing

Material	Feed x_{100} (mm)	Product x_{80} (mm)	Spark parameters: Gap (cm)	Capacitance (μF)	Voltage kV	Energy (J)	Yield (kg/kWh)	Specific energy (kWh/tonne)
Alumina*	10	0.2	1.5	0.01	80	32	5	200
		0.2					15	67
Quartz*	10	0.5	1.5	0.05	60	9	40	25
		1.5					110	9
Quartz*	200	30	6	1.5	40	1290	1000	1
Magnetite†	100	12	6	1.5	40	1200	93	11
			6	3.0	40	2400		
Hematite†	100	15	8	0.3	70	735	322	3.1
Coke†	150	50	8	0.3	70	735	673	1.5
	150	50	4	0.1	70	245	338	3.0
Sinter†	100	25	8	0.3	70	735	1350	0.6
	100	25	4	0.1	70	245	880	1.1

*Carley-Macauly (1968) †Hitchon and Carley-Macauly (1969)

The advantages of electrohydraulic crushing may be summarized as:

(1) There is the ability to crush very hard materials without excessive machine wear and with corresponding control of contamination.
(2) A less flaky product may well be produced than that from mechanical crushing, i.e. a cubiform shape (§4.2.6).
(3) A close product size distribution is possible, which is of particular interest to blast-furnace operation.
(4) A relatively confined crushing zone is used, allowing both economical use of equipment volume and portable plant without heavy vessels.

Cost comparison has been made available between electrohydraulic and conventional crushing in Table 11.6 for coke and sinter as used in blast furnaces.

It is deduced that electrohydaulic crushing is unlikely to be widely used, but could find applications where a cubiform closely sized product is required, free from iron contamination or in conditions where a conventional crusher is not allowable because of its bulk and weight.

Table 11.6. Comparison of crushing costs by electrohydraulic and conventional means for coke and sinter—1969 prices (Hitchon and Carley-Macauly, 1969)

	Electrohydraulic		Conventional	
Factor	Coke	Sinter	Coke	Sinter
Capital cost (g/hour)	£0.80	£0.33	£0.26	£0.07
Crushing cost (£/ton)	0.066	0.0275	0.0185	0.005
Relative crushing cost	13.2	5.5	3.7	1.0
Yield (kg/kWh)	500	1200	670	2600
Specific energy (kWh/tonne)	2.0	0.83	1.49	0.38

11.5 THERMODYNAMIC SYSTEMS

11.5.1 Survey

Two similar avenues of exploration of explosive shattering by thermodynamic means are reported. The first, a laboratory unit was developed by the US Bureau of Mines (Gross, 1938). The second, a pilot-plant unit, embodied the Snyder process (Cavanaugh and Rogers, 1974). Each used steam as the heating and pressurizing medium, on the sudden release of which particle disintegration occurred.

11.5.2 US Bureau of Mines method

Of several prototypes developed, that shown in Figure 11.14 proved to be both simple and effective. The cycle of operations, which are synchronized in

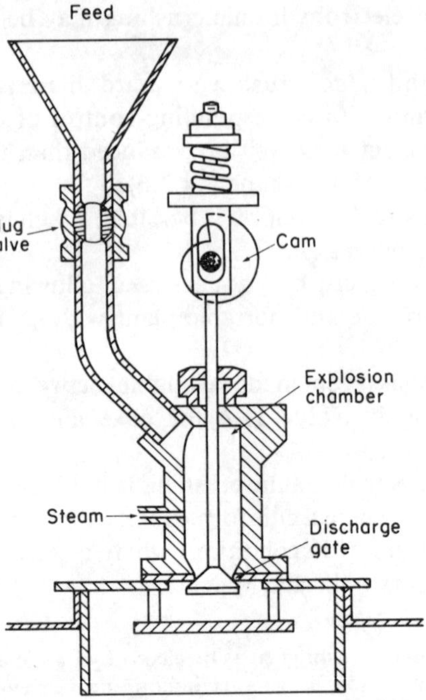

Figure 11.14. Cross-section of US Bureau of Mines explosive shatter device (Gross, 1938). *Courtesy Bureau of Mines, US Department of the Interior*

practice, are as follows:

(1) The cam lifts to close the discharge gate.
(2) The plug valve opens.
(3) The feed delivers the charge.
(4) The plug valve closes.
(5) Steam enters the explosion chamber.
(6) Steam is shut off.
(7) The cam falls to open the discharge gate.
(8) The product is received in the collecting chamber.

The effect of increased steam pressure on the product size distribution is given in Figure 11.15 for dolomite, and that of increased temperature in Figure 11.16. Increase in the size of the charge up to the limit of the capacity of the chamber increases the product mass consistently, as shown in Figure 11.17 for quartz, sphalerite and galena.

Gross (1938) gives data which show that with 250 lb/in^2 steam, 0.28 lb of minus 48 mesh product required 70 Btu, i.e. 250 Btu/lb or 164 kWh/ton, a most unattractive specific energy.

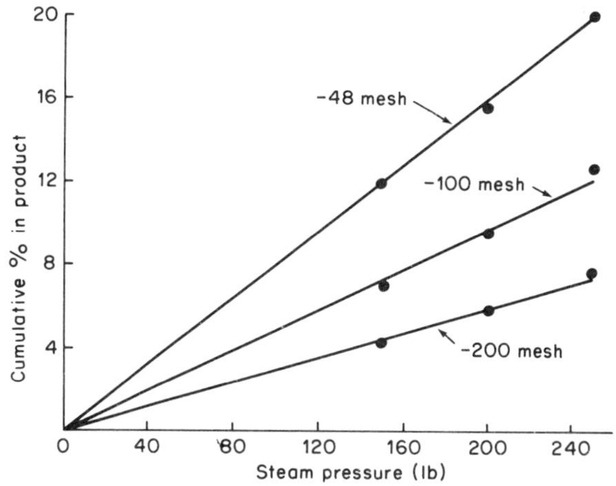

Figure 11.15. Effect of steam pressure on product size distribution in an explosive shatter device using dolomite (Gross, 1938). *Courtesy Bureau of Mines, US Department of the Interior*

Increase in time for steam to soak into the charge bed has practically no effect on product size distribution, which seems to indicate that the mechanism is not that of penetration by steam into the pores and fissures of the material. Rather it would appear to be the result of shock waves generated by the sudden loss of pressure together with particle to particle collision in the opposed jet discharges from the explosion chamber.

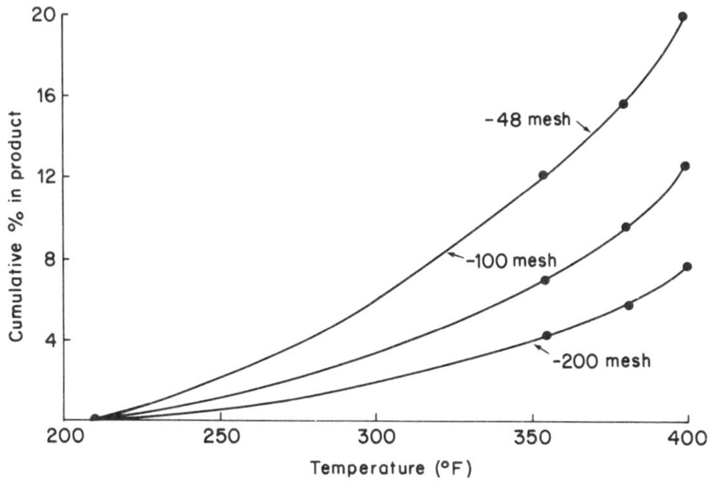

Figure 11.16. Effect of steam temperature on product size distribution in an explosive shatter device using dolomite (Gross, 1938). *Courtesy Bureau of Mines, US Department of the Interior*

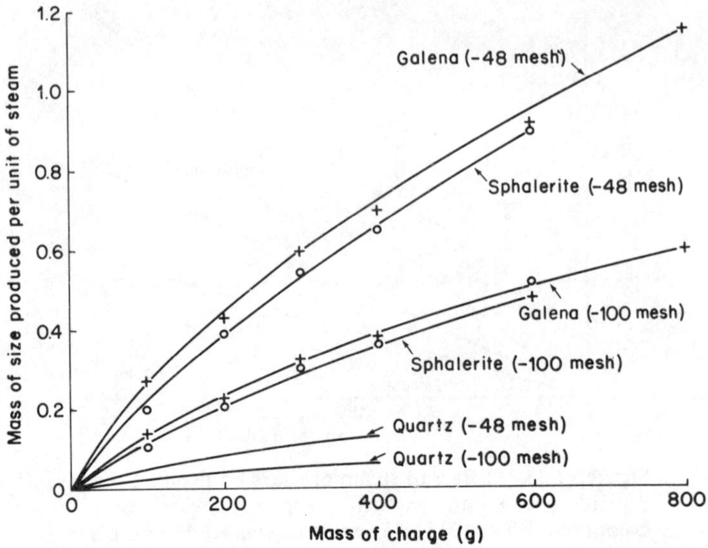

Figure 11.17. Effect of charge mass on explosive shattering of quartz, sphalerite and galena (Gross, 1938). *Courtesy Bureau of Mines, US Department of the Interior*

11.5.3 The Snyder process

Figure 11.18 depicts the twin-gun Snyder system (US Patent 3257 080 1966) which in principle differs little from the US Bureau of Mines system, except that opposed firing ducts are incorporated, somewhat as in some forms of fluid energy mill.

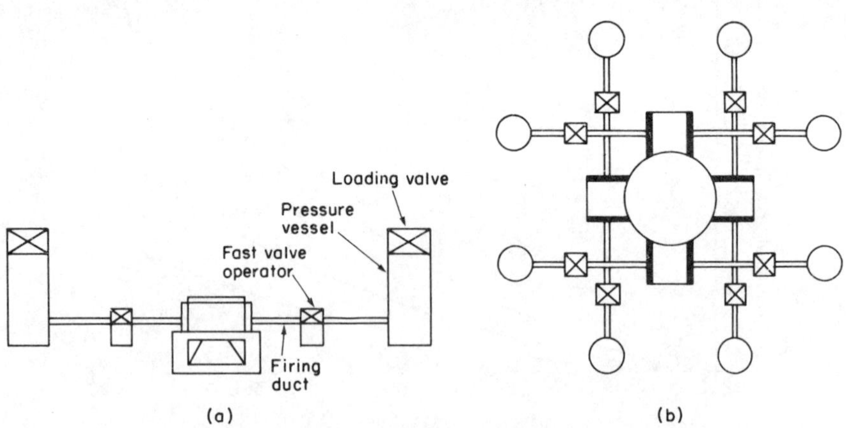

Figure 11.18. Twin-gun Snyder system: (a) layout of a single unit, (b) layout of four units feeding a comon receiving chamber (Cavanaugh and Rogers, 1974). *Reproduced by permission of the Institution of Mining and Metallurgy.*

Comparative test data are tabulated by Cavanaugh and Rogers (1974) to demonstrate improved metallurgical separation of ores by the Snyder process than with conventional tumbling mill equipment; but no data are produced on steam consumption or the cost of fuel to produce the steam, so it is difficult to judge the efficiency of the process, but it is likely to be low.

11.6 THERMAL DECREPITATION

Decrepitation, the spontaneous shattering of a mineral on heating, has long been known, but explanations differ widely, including differential thermal expansion, evaporation of volatile occlusions, phase change and dehydration by heating. Fluorites and byrites are particularly susceptible to decrepitation, but Turba (1962) reported for byrites that the process was uneconomic compared with mechanical methods of size reduction. Chakravati and Jowett (1967) have made a broader study, including quenching as well as heating, which reduced the specific energy for crushing compared with heating alone. However, even with quenching, the process was not economic and is much less likely to be today in the present climate of high fuel costs.

11.7 CHEMICAL COMMINUTION

11.7.1 The process for coal

When coal is contacted by a low-molecular-weight substance (e.g. ammonia) a penetration is effected into the laminated structures and pores, which penetration is facilitated by irregularities and discontinuities in the carbon structure of the coal. Ammonia has been demonstrated to be the most effective known comminution agent (Datta, 1976)

Possible mechanisms for the process have been advanced as follows (Kruyt, 1982):

(1) a difference in the swelling characteristics of macerals during the absorption of the comminution agent;
(2) expansion and contraction as a result of adsorption at the pore walls;
(3) partial dissolution of the coal;
(4) stresses generated by rapid local temperature changes (evaporation and condensation may be responsible for these changes);
(5) capillary pressure of the liquid phase in the pores, directly after condensation.

Insufficient research has been possible, to date, to decide which of these is of major importance in comminution with concurrent liberation of inorganic constituents. However, comparative tests with mechanical size reduction have claimed that chemical comminution, at the same particle size, does liberate more pyritic sulphur and a similar amount of ash after size reduction (Datta and Howard, 1978; Datta *et al.,* 1976).

11.7.2 Recent experimental progress with coal

Ansems and Bloebaum (1984) have reported work in the Netherlands on three types of bituminous coal: low-, medium- and high-volatile. Results so far may be summarized as follows:

(1) The degree of comminution increases with:
 (a) increase in carbon content;
 (b) decrease in content of volatile constituents;
 (c) increase in content of vitrinites;
 (d) decrease in content of exinites;
 (e) decrease in content of inertites.
(2) The efficiency of comminution depends on the history of the coal and the location where it is mined.
(3) The relative size reduction decreases as the initial size of particles becomes smaller, and at an initial size of 1 mm, no further fracturing occurs in most cases.
(4) Treatment of coal with liquid ammonia results in a higher degree of comminution than treatment with gaseous ammonia.
(5) The contact time between ammonia and coal can be small as comminution mainly takes place during the first few minutes.
(6) Pressure has a negligible effect on comminution in liquid ammonia at the boiling point condition.
(7) Pressure has some influence on comminution in gaseous ammonia: no comminution at 10^5 Pa and a moderate degree of comminution at 8×10^5 Pa.
(8) Ultrasonic vibration has a negligible effect.

11.7.3 Conclusions for coal

Ansems and Bloebaum (1984) see the up-to-date position in respect of chemical comminution for coal as follows:

(1) Because the process is coal selective, it has little attraction to a country such as the Netherlands which has to import a wide variety of coals from overseas.
(2) Large lumps of coal can be fractured by the process which could well be applied to run-of-mine coal, but the economics of the advantage over mechanical methods have not been clearly stated.
(3) Although claims have been made in the literature that beneficial liberation of pyrite and ash is possible, these claims should be critically viewed in the light of the lack of suitable analysis techniques for judging the liberation. This is an area for continued research.
(4) Further research is also required to ascertain the chief mechanism(s) of chemical comminution.

11.8 SUMMARY

Table 11.7 brings together on a comparative basis the superfine ($x_{80} = 1$ micron) and the very fine ($x_{80} = 25$–75 microns) product data found in this chapter.

It would appear that for superfine grinding the attrition mill is unique, but for very fine products, a choice of mill is available.

For unconventional size reduction, data are only available for electro-hydraulic crushing as given in Table 11.5. No energy data appears to have been given for thermal decrepitation, and that found for explosive shattering rather indicates a high specific energy. Hence these methods of size reduction are unattractive at the present time.

Table 11.7. Comparative data for fine grinding mills

System	Material	Product x_{80} (microns)	Specific energy (kWh/tonne)	Reference
USBM attrition mill	Lignite	1	944	Davis *et al.* (1980)
	Limestone	1–2	694	
	Fluorite	1	555	
	Whiting	1	333	
Tower mill	Limestone	15	20	Japan Tower Mill Co. Ltd. Brochure 1984-01-01
Planetary mill (Szego)	Coal slurry	75	20–25	Gandolfi *et al.* (1984)
Planetary mill	Iron ore (dry)	(Surface only quoted —see text)	20–25	Kitschen *et al.* (1982)
Ultrasonic system	Coal (dry)	5	37	Tarpley *et al.* (1981)

11.9 REFERENCES

Aber, R. C. (1985). 'Ultrasonically enhanced size reduction of coal—Phase II', Energy and Minerals Research Co., PO Box 389, Exton, Pa., USA. DoE contract DE-AC 22-83PC 63049.

Ansems, A., and Bloebaum R. K. (1984). 'Chemical comminution', *Power Technol.*, **40**, 265–268.

Bilby, B. A. (1979). 'Putting fracture to work', *Proc. 4th Tewksbury Symp.*, Melbourne, Australia, 1.1–1.44.

Bradley, A. A. (1972). 'Some principles of centrifugal milling', Dechema Monograph 69, 781–803.

Bradley, A.A., Hindle, A. L., Lloyd, P. J. D., and Schymura, K. (1980). 'Development in centrifugal milling', *Proc. Europ. Symp. Particle Technol.*, Amsterdam, 153–170.

Bradley, A.A., Lloyd, P. J. D., White, D. A., and Willows, P. W. (1972). 'High speed centrifugal milling and its potential use in the mining industry', *South African Mech. Eng.*, 129–134.

Capes, C. E., and Germain, R. F. (1982). 'Selective oil agglomeration in fine coal beneficiation'. In: *Physical Cleaning of Coal, Present and Developing Methods* (Lin, Y. A., ed.), Marcell-Dekker, NY.

Carley-Macauly, K. W. (1968). 'Electrohydraulic crushing', *Chem. Proc. Eng.*, Sept., 87–96.

Cavanaugh, W. J., and Rogers, D. J. (1974). 'Applications of the Snyder process', *Proc. 10th Int. Miner. Process Congr.*, 1973 (Jones, M. J., Ed.), IMM (London), **83**, 91–122; Discussion 170–175.

Chakravarti, A., and Jowett, A. (1967). 'Aspects of comminution by heating', Dechema Monograph 57, 583–604.

Cook, N. G. W., and Harvey, V. R. (1974). 'An appraisal of rock excavation by mechanical, hydraulic, thermal and electromagnetic means: advances in rock mechanics', *Proc. 3rd Congr. Int. Soc. Rock Mech.* Themes 4–5, Denver, Colorado, 1599–1615.

Datta, R. S. (1976). 'Feasability study of precombustion coal cleaning using chemical comminution', NTIS Report FE-2520-4.

Datta, R. S., and Howard, P. H. (1978). 'Characterization of the chemical comminution of coal', FE-2520 4, Final Report, ERDA Contract EX 76-C-01-2520 (March).

Datta, R. S., Howard, P. H., and Hanchett, A. (1976). 'Feasability study of precombustion coal cleaning using chemical comminution', FE-1777-4, Final Report, ERDA Contract EX-76-C-01-1777 (November).

Davis, E. G., Collins, E. W., and Feld, I. L. (1973). 'Large scale continuous attrition grinding of coarse kaolin', Report of Investigations 7771, US Bureau of Mines.

Davis, E. G., Hansen, J. P., and Sullivan, G. V. (1980). 'Attrition microgrinding', *Proc. Int. Symp. Fine Particle Process.*, Las Vegas, Nevada, AIME, NY, 74–95.

Gandolfi, E. A. J., Papachristodoulou, G., and Trass, O. (1984). 'Preparation of coal slurry fuels with the Szego mill', *Powder Technol.*, **40**, 269–282.

Gross, J. (1938). 'Crushing and grinding', US Bureau of Mines, Bulletin 402.

Herzfeld, A. (1896). US Patent 569828.

Hitchon, J., and Carley-Macauly, K. W. (1969). 'Electrohydraulic crushing of blast furnace feed materials', British Steel Corporation Report PE/D/39/69.

Japan Tower Mill Co. Ltd. (1984). Brochure 01-01.

Jimbo, G., Saito, Y., and Kosaka, M. (1980). 'The energy efficiency of cryogenic crushing of grain and vegetable particles', *Proc. Europ. Symp. Particle Technol.*, Amsterdam, 125–136.

Jimbo, G., Sakurai, Y., Kimura, K., and Oshitani, S. (1976). 'The crushing mechanism of plastic materials in a shredder type pulveriser', Dechema Monograph 79, 371–384.

Kitschen, L. P., Lloyd, P. J. D., and Hartmann, R. (1982). 'The centrifugal mill: experience with a new grinding system and its applications', *Proc. 14th Int. Miner. Process. Congr.*, Toronto, I-9.1 to I-9.13.

Kruyt, J. (1982). 'Chemisch vergruizen van steenkool', Afstudeerverslag, Technische Hogeschool, Delft Afdeling Chemische Technologie (August).

Maurer, W. C. (1967). 'The state of rock mechanics knowledge in drilling', *Proc. 8th Symp. Rock Mech.*, 1966, University of Minnesota, AIME, NY, 355–395.

Rumpf, H. (1965). 'Die einzelkornzerkleinerung als grundlage einer technischen zerkleinerungswissenschaft', *Chemie. Ing. Techn.*, **37**, 187–202 (English translation, US Special Libraries Association, 66-13859).

Schönert, K. (1984). 'Grinding media mills', Section III Comminution and air classification Conf., Brighton, UK.

Stanley, D. A., Sadler, L. Y., and Brooks, D. R. (1973). 'Size reduction of ceramic powders by attrition milling', *Proc. 1st Int. Conf. Powder Technol.*, IITRI, Chicago, 279–281.

Stanley, D. A., Sadler, L. Y., Brooks, D. R., and Schwartz, M. A. (1974). 'Attrition milling of ceramic oxides', *Ceram. Bull.*, **53**, (11), 813–815, 829.

Tarpley, W. B., Howard, P. L., and Moulder, G. R. (1981). 'Efficient ultrasonic grinding: a new technology for micron-sized coal', US Department of Energy Report DOE/ER/10466-1.

Trass, O. (1984). 'Grinding of wood chips and barks to fine powders', Bioengineering Projects Meeting, Ottowa.

Trass, O., Koka, V. R., and Papachristodoulou, G. (1985). 'Beneficiation and fine grinding of coal for slurry fuel preparation', *Proc. 7th Int. Symp. Coal Slurry Technol.,* New Orleans.

Trass, O., and McCracken, T. W. (1985). 'Grinding and oil agglomeration of coal in the Szego mill', *Proc. 4th Int. Symp. Agglomeration,* Toronto, Ontario, June.

Turba, E. (1962). 'Thermische zerkleinerung von schwerspat', *Proc. 1st Europ. Symp.* (Zerkleinern), Frankfurt, VDI-Verlag, Dusseldorf, 160–169.

Yutkin, L. A. (1955). 'The electrohydraulic effect', *Moscow Mashgiz* (English translation AD 267722).

CHAPTER 12

Appendices

12.1 NOMENCLATURE FOR MILL TYPES

It is convenient to classify mills into families and the following are recognized:

	Figure
Crushers	12.1
Grinders	
Tumbling mills	12.2
Vertical spindle mills	12.3
Impact mills	12.4
Vibration mills	12.5
Attrition mills	12.6
Fluid energy mills	12.7

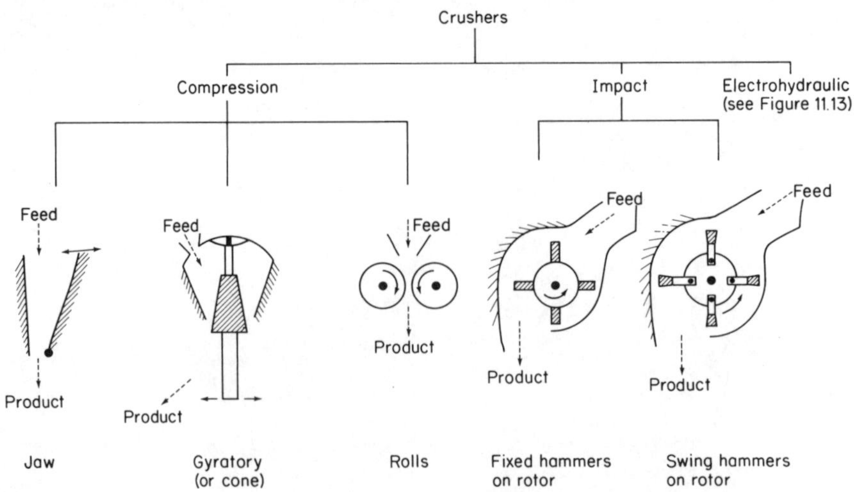

Figure 12.1. Family of crushers (not to scale)

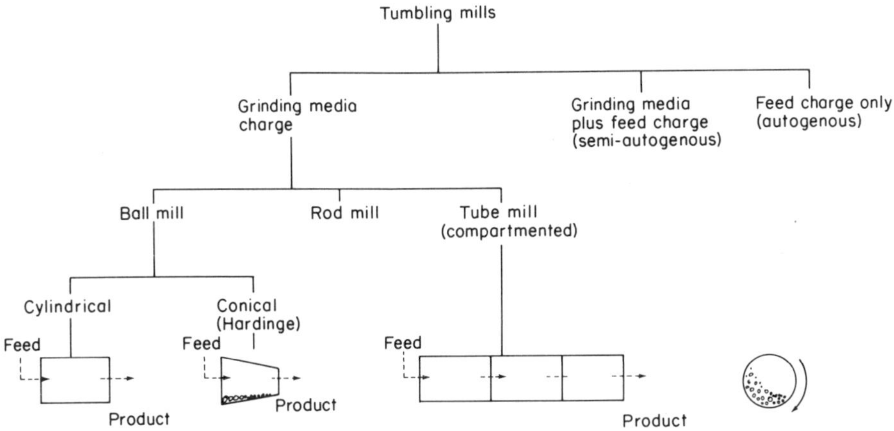

Figure 12.2. Family of tumbling mills (not to scale)

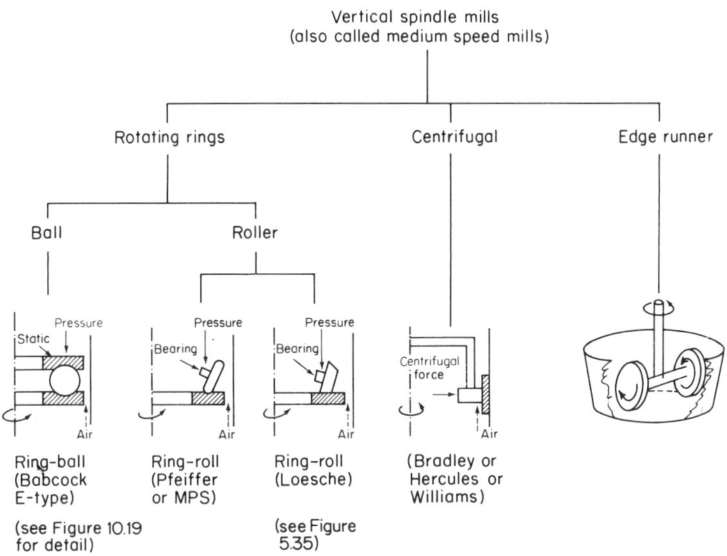

Figure 12.3. Family of vertical spindle mills (not to scale)

The mills within each of these families are given in the figures referred to, with alternative nomenclature, where appropriate.

It must be emphasized that no attempt has been made to include all known mill types. Behrens (1962)* and Prem and Prior (1979)† describe many types

*Behrens, D. (1962). 'Einige neuentwicklungen von fein- und feinst–prallzerkleinerungsmaschinen', *Proc. 1st Europ. Symp.* (Zerkleinern) Frankfurt, 474–502.
†Prem, H., and Prior, J. H. (1979). 'Impact mills for fine reduction', *Powtech. 6th Int. Conf. Powder Technol.*, Birmingham, UK, 630–652.

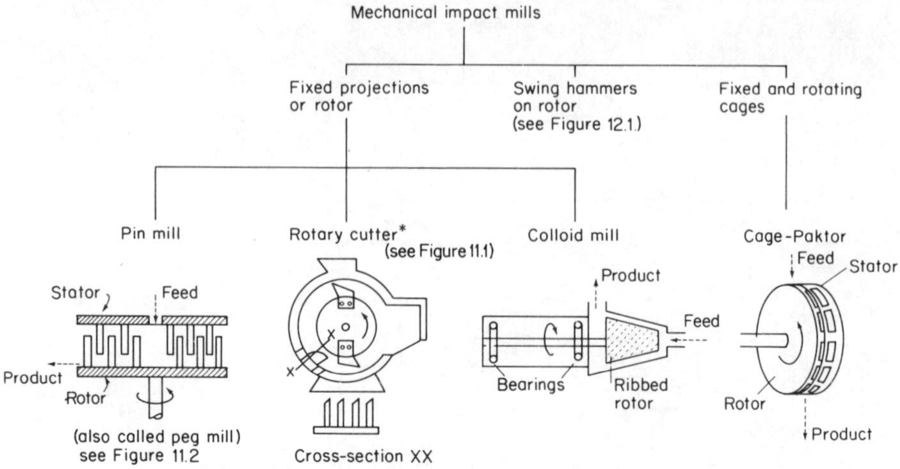

Figure 12.4. Family of impact mills (not to scale)

*The rotary cutter works as well with blunt blades as with sharp blades when grinding plastics

of impact mills, for example, but only a few are well known. The list is therefore confined to mills in common use.

12.2 US TRANSLATION CENTRES

National Translation Centre (NTC)
John Crerar Library
35 West 33rd Street
Chicago, Ill. 60616, USA

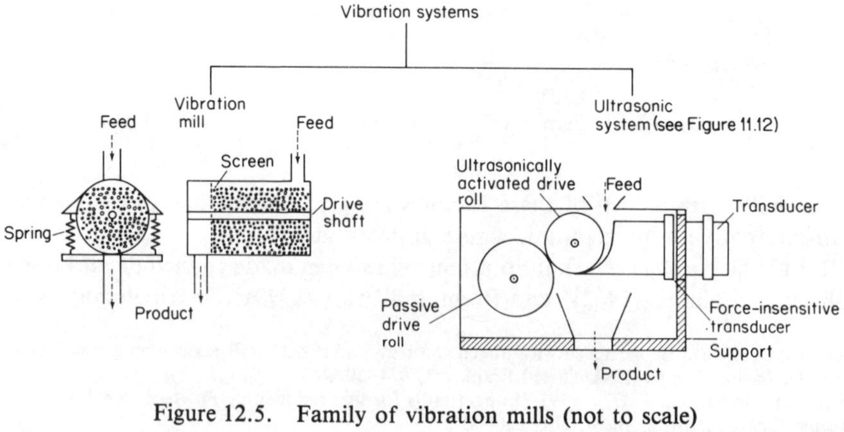

Figure 12.5. Family of vibration mills (not to scale)

Special Libraries Association (SLA)
Translation Centre
235 Park Avenue South
New York, NY 10003, USA

12.3 UNITS AND CONVERSION FACTORS

12.3.1 British, metric and SI units

Reference is made to BS 350 (Part 1) 1974), *Conversion Factors and Tables*, for a complete list of units and conversion factors. A useful summary is Table 12.1 for conversion between British, metric and SI units.

12.3.2 Conversion of national mesh sizes to metric units

Table 12.2 gives standard sieve sizes according to USA, Tyler, Canadian, British, French and German specifications.

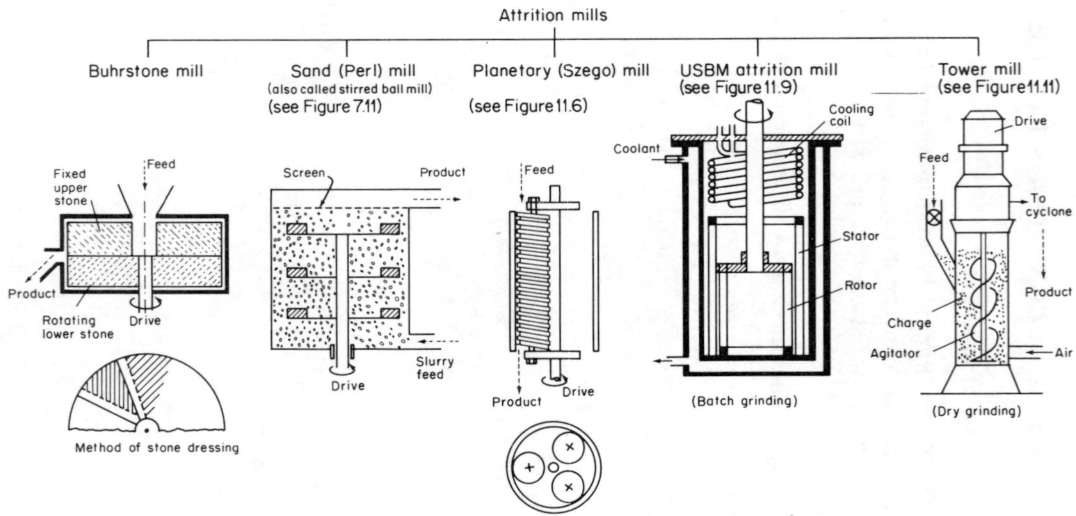

Figure 12.6. Family of attrition mills (not to scale)

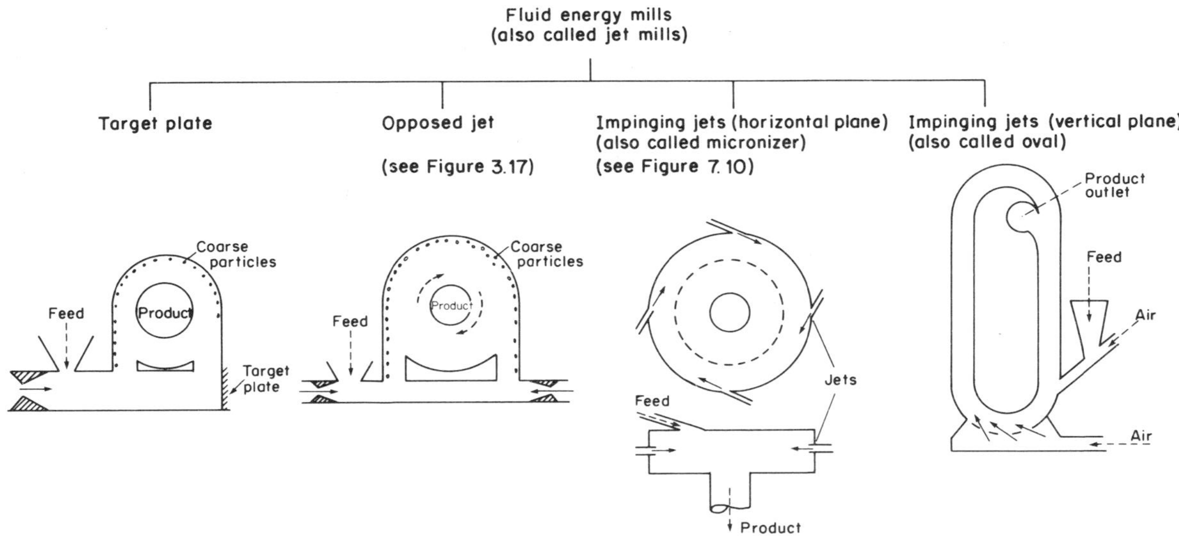

Figure 12.7. Family of fluid energy mills (not to scale)

Table 12.1. Conversion between British, metric and SI units

Concept	British	Metric	SI
Length	1 ft	0.3048 m*	0.3048 m*
	1 in	25.4 mm*	25.4 mm*
Area	1 ft^2	0.09290304 m^2*	0.09290304 m^2*
Volume	1 ft^3	0.02831685 m^3	0.02831685 m^3
Mass	1 lb	0.45359237 kg*	0.45359237 kg*
Mass flow	1 lb/h	0.45359237 kg/h*	0.0001259979 kg/s
Mass flux	1 lb/hft^2	0.001356230 kg/m^{2}s	0.001356230 kg/m^{2}s
Specific volume	1 ft^3/lb	0.06242796 m^3/kg	0.06242796 m^3/kg
Density	1 lb/ft^3	16.01846 kg/m^3	16.01846 kg/m^3
H_2O max at 3.98°C & 1 atm		0.999972 g/cm^3	999.972 kg/m^3
H_2O Ideal at g_n		1. g/cm^3	1000. kg/m^3
Hg at g_n		13.5951 g/cm^3	13595.1 kg/m^3
Acceleration (g_n)		9.80665 m/s^2*	9.80665 m/s^2*
Force	1 lbf	0.45359237 kgf*	4.448222 N
Pressure	1 lbf/in^2	0.07030696 kgf/cm^2	0.06894757 bar
(1 MPa = 10 bar = 10 × 10^5 N/m^2 = 1 N/mm^2)	1 lbf/in^2 = 2.306659 ft H_2O = 0.06804596 atm 1 lbf/in^2 = 27.679905 in H_2O = 703.06958 mm H_2O 1 lbf/in^2 = 2.036021 in Hg = 51.714925 mm Hg 1 in H_2O = 2.54 gf/cm^2 = 0.249089 kPa		
	1 atm	1.0332275 kgf/cm^2	1.01325 bar*
Specific energy	1 Btu/lb	0.5555556 kcal/kg	2.326 kJ/kg*
Spec. energy (vol.)	1 Btu/ft^3	8.8991463 kcal/m^3	37.258946 kJ/m^3
Specific heat	1 Btu/lb°F	1. kcal/kg°C*	4.1868 kJ/kg°C*
Energy	1 Btu	0.25199576 kcal	1.0550559 kJ
	1 ft-lbf	0.13825495 m-kgf	1.3558179 J
	1 Btu = 107.585756 m-kgf = 778.169262 ft-lbf		
Power (1W = 1 J/s)	1 Btu/h	0.25199576 kcal/h	0.29307107 W
	1 Brit.hp = 550. ft-lbf/s* = 76.040225 m-kgf/s		
	1 Brit.hp	1.0138697 metric hp	745.6998716 W
Heat transfer coef.	1 Btu/hft^2°F	4.8824276 kcal/h°Cm^2	5.6782633 W/m^2°C
Therm. conductivity	1 Btu/hft°F	1.4881639 kcal/hm°C	1.7307347 W/m°C
Dynamic viscosity	1 lb/fts	14.881639 gm/cms(P)	1.4881639 kg/ms(Pas)
Kinematic visc.	1 ft^2/s	929.0304 cm^2/s(St)*	0.09290304 m^2/s*
Heat flux	1 Btu/hft^2	2.7124598 kcal/hm^2	3.1545907 W/m^2
Volume heat rate	1 Btu/hft^3	8.8991463 kcal/hm^3	10.349707 W/m^3

Concept	Metric	SI	British
Length	1 m	1.0 m	3.280840 ft
	1 mm	1.0 mm	0.0393701 in
Area	1 m^2	1.0 m^2	10.763910 ft^2
Volume	1 m^3	1.0 m^3	35.314667 ft^3
Mass	1 kg	1.0 kg	2.204623 lb
Mass flow	1 kg/h	0.0002777778 kg/s	2.204623 lb/h
Mass flux	1 kg/m^{2}s	1.0 kg/m^{2}s	737.33812 lb/hft^2
Specific volume	1 m^3/kg	1.0 m^3/kg	16.01846 ft^3/lb
Density	1 kg/m^3	1.0 kg/m^3	0.06242796 lb/ft^3
H_2O max at 3.98°C & 1 atm	Molar volume of perfect gas at NTP (0°C & 1 atm) = 22.4138277 m^3/kg mol = 359.0350781 ft^3/lb mol		62.426213 lb/ft^3
H_2O Ideal at g_n			62.427961 lb/ft^3
Hg at g_n			848.714367 lb/ft^3
Acceleration (g_n)			32.1740486 ft/s^2
Force	1 kgf	9.80665 N*	2.204623 lbf
Pressure	1 kgf/cm^2	0.980665 bar*	14.223343 lbf/in^2
(1 MPa = 10 bar = 10 × 10^5 N/m^2 = 1 N/mm^2)	1 kgf/cm^2 = 0.9678411 atm = 32.80840 ft H_2O 1 kgf/cm^2 = 10,000 mm H_2O* = 393.70079 in H_2O 1 kgf/cm^2 = 735.55914 mm Hg = 28.959021 in Hg 1 in Hg = 0.03386389 bar = 0.4911541 lbf/in^2 1 atm = 759.9998917 mm Hg = 29.921256 in Hg		
Specific energy	1 kcal/kg	4.1868 kJ/kg*	1.8 Btu/lb*
Spec. energy (vol.)	1 kcal/m^3	4.1868 kJ/m^3*	0.11237033 Btu/ft^3
Specific heat	1 kcal/kg°C	4.1868 kJ/kg°C*	1. Btu/lb°F*
Energy	1 kcal	4.1868 kJ*	3.9683207 Btu
	1 m-kgf	9.80665 J*	7.2330139 ft-lbf
	1 kcal = 426.934784 m-kgf = 3088.02521 ft-lbf		
Power (1W = 1 J/s)	1 kcal/h	1.163 W*	3.9683207 Btu/h
	1 metric hp = 75 m-kgf/s* = 0.986320070 Brit.hp		
	1 metric hp	735.4987499 W	542.47604 ft-lbf/s
Heat transfer coef.	1 kcal/h°Cm^2	1.163 W/m^2°C*	0.20481614 Btu/hft^2°F
Therm. conductivity	1 kcal/hm°C	1.163 W/m°C*	0.67196897 Btu/hft°F
Dynamic viscosity	1 poise(P)	0.1 kg/ms(Ns/m^2, Pas)*	0.06719690 lb/fts
Kinematic visc.	1 Stoke(St)	0.0001 m^2/s*	0.0010763910 ft^2/s
Heat flux	1 kcal/hm^2	1.163 W/m^2*	0.36866906 Btu/hft^2
Volume heat rate	1 kcal/hm^3	1.163 W/m^3*	0.11237033 Btu/hft^3

Concept	SI	British	Metric
Length	1 m	3.280840 ft	1.0 m
	1 mm	0.0393701 in	1.0 mm
Area	1 m^2	10.763910 ft^2	1.0 m^2
Volume	1 m^3	35.314667 ft^3	1.0 m^3
Mass	1 kg	2.204623 lb	1.0 kg
Mass flow	1 kg/h	7936.6414 lb/h	3600 kg/h*
Mass flux	1 kg/m^{2}s	737.33812 lb/hft^2	1.0 kg/m^{2}s
Specific volume	1 m^3/kg	16.01846 ft^3/lb	1.0 m^3/kg
Density	1 kg/m^3	0.06242796 lb/ft^3	1.0 kg/m^3
H_2O max at 3.98°C & 1 atm	1 litre = 0.21996916 gal = 1000 cm^3*		
H_2O Ideal at g_n	1 gal (UK) = 4.5460919 cm^3		
Hg at g_n	Solids concentration 1 grain/ft^3 = 2.288 g/m^3		
Acceleration (g_n)		416975669 ft/h^2	
Force	1 N	0.2248089 lbf	0.1019716 kgf
Pressure	1 bar	14.503774 lbf/in^2	1.019716 kgf/cm^2
(1 MPa = 10 bar	1 bar	33.455256 ft H_2O	0.9869233 atm
= 10 × 10^5 N/m^2	1 m bar	0.401463076 in H_2O	10.197162 mm H_2O
= 1 N/mm^2)	1 bar	29.529983 in Hg	750.06158 mm Hg
	1 m H_2O = 1.4223343 lbf/in^2 = 0.0980665 bar* 1 atm = 14.695949 lbf/in^2 = 33.8985384 ft H_2O		
Specific energy	1 kJ/kg	0.4299225 Btu/lb	0.23884590 kcal/kg
Spec. energy (vol.)	1 kJ/m^3	0.026839192 Btu/ft^3	0.23884590 kcal/m^3
Specific heat	1 kJ/kg°C	0.2388459 Btu/lb°F	0.23884590 kcal/kg°C
Energy	1 kJ	0.94781712 Btu	0.23884590 kcal
	1 J	0.73756215 ft-lbf	0.10197162 m-kgf
	1 kWh	3412.142 Btu	859.84523 kcal
Power (1W = 1 J/s)	1 W	3.41214163 Btu/h	0.85984523 kcal/h
	1 kW	737.56215 ft-lbf/s	101.97162 m-kgf/s
	1 kW	1.34102209 Brit.hp	1.35962162 metric hp
Heat transfer coef.	1 W/m^2°C	0.17611018 Btu/hft^2°F	0.85984523 kcal/h°Cm^2
Therm. conductivity	1 W/m°C	0.57778952 Btu/hft°F	0.85984523 kcal/hm°C
Dynamic viscosity	1 kg/ms	0.6719690 lb/fts	10.0 gm/cms(Poise)*
Kinematic visc.	1 m^2/s	10.76391 ft^2/s	10000. cm^2/s(St)*
Heat flux	1 W/m^2	0.31699833 Btu/hft^2	0.85984523 kcal/hm^2
Volume heat rate	1 W/m^3	0.096621091 Btu/hft^3	0.85984523 kcal/hm^3

Notes: 1) Boxed items supply additional information. 2) 'atm' refers to Standard Atmosphere and should not be confused with Technical Atmosphere (at), where 1 at = 1 kgf/cm^2. 3) *Exact.

Table 12.2. Conversion of national mesh sizes to metric units

U.S.A. (1)		Tyler (2)	Canadian (3)		British (4)		French (5)		German (6)
Standard*	Alternative	Mesh designation	Standard	Alternative	Nominal aperture	Nominal mesh no.	Opg. mm	No.	Opg.
125 mm	5″		125 mm	5″					
106 mm	4.24″		106 mm	4.24″					
100 mm	4″		100 mm	4″					
90 mm	$3\frac{1}{2}$″		90 mm	$3\frac{1}{2}$″					
75 mm	3″		75 mm	3″					
63 mm	$2\frac{1}{2}$″		63 mm	$2\frac{1}{2}$″					
53 mm	2.12″		53 mm	2.12″					
50 mm	2″		50 mm	2″					
45 mm	$1\frac{3}{4}$″		45 mm	$1\frac{3}{4}$″					
37.5 mm	$1\frac{1}{2}$″		37.5 mm	$1\frac{1}{2}$″					
31.5 mm	$1\frac{1}{4}$″		31.5 mm	$1\frac{1}{4}$″					
26.5 mm	1.06″	1.05″	26.5 mm	1.06″					
25.0 mm	1″		25.0 mm	1″					25.0 mm
22.4 mm	$\frac{7}{8}$″	0.883″	22.4 mm	$\frac{7}{8}$″					
19.0 mm	$\frac{3}{4}$″	0.742″	19.0 mm	$\frac{3}{4}$″					20.0 mm
									18.0 mm
16.0 mm	$\frac{5}{8}$″	0.624″	16.0 mm	$\frac{5}{8}$″					16.0 mm
13.2 mm	0.530″	0.525″	13.2 mm	0.530″					
12.5 mm	$\frac{1}{2}$″		12.5 mm	$\frac{1}{2}$″					12.5 mm
11.2 mm	$\frac{7}{16}$″	0.441″	11.2 mm	$\frac{7}{16}$″					
									10.0 mm
9.5 mm	$\frac{3}{8}$″	0.371″	9.5 mm	$\frac{3}{8}$″					
8.0 mm	$\frac{5}{16}$″	$2\frac{1}{2}$	8.0 mm	$\frac{5}{16}$″					8.0 mm
6.7 mm	0.265″	3	6.7 mm	0.265″					
6.3 mm	$\frac{1}{4}$″		6.3 mm	$\frac{1}{4}$″					6.3 mm
5.6 mm	No. $3\frac{1}{2}$	$3\frac{1}{2}$	5.6 mm	No. $3\frac{1}{2}$	5.60 mm	3			
					4.75 mm	$3\frac{1}{2}$	5.000	38	5.0 mm

Table 12.2. (*continued*)

U.S.A. (1)		Tyler (2)	Canadian (3)		British (4)		French (5)		German (6)
Standard*	Alternative	Mesh designation	Standard	Alternative	Nominal aperture	Nominal mesh no.	Opg. mm	No.	Opg.
4.75 mm	4	4	4.75 mm	4					
4.00 mm	5	5	4.00 mm	5	4.00 mm	4	4.000	37	4.0 mm
3.35 mm	6	6	3.35 mm	6	3.35 mm	5			
							3.150	36	3.15 mm
2.80 mm	7	7	2.80 mm	7	2.80 mm	6			
2.36 mm	8	8	2.36 mm	8	2.36 mm	7	2.500	35	2.5 mm
2.00 mm	10	9	2.00 mm	10	2.00 mm	8	2.000	34	2.0 mm
1.70 mm	12	10	1.70 mm	12	1.70 mm	10	1.600	33	1.6 mm
1.40 mm	14	12	1.40 mm	14	1.40 mm	12			
							1.250	32	1.25 mm
1.18 mm	16	14	1.18 mm	16	1.18 mm	14			
1.00 mm	18	16	1.00 mm	18	1.00 mm	16	1.000	31	1.0 mm
850 μm	20	20	850 μm	20	850 μm	18			
							0.800	30	800 μm
710 μm	25	24	710 μm	25	710 μm	22			
							0.630	29	630 μm
600 μm	30	28	600 μm	30	600 μm	25			
500 μm	35	32	500 μm	35	500 μm	30	0.500	28	500 μm
425 μm	40	35	425 μm	40	425 μm	36			
							0.400	27	400 μm
355 μm	45	42	355 μm	45	355 μm	44			
							0.315	26	315 μm
300 μm	50	48	300 μm	50	300 μm	52			

250 μm	60	60	250 μm	60	250 μm	60	0.250	25	250 μm
212 μm	70	65	212 μm	70	210 μm	72			
							0.200	24	200 μm
180 μm	80	80	180 μm	80	180 μm	85			
							0.160	23	160 μm
150 μm	100	100	150 μm	100	150 μm	100			
125 μm	120	115	125 μm	120	125 μm	120	0.125	22	125 μm
106 μm	140	150	106 μm	140	106 μm	150			
							0.100	21	100 μm
90 μm	170	170	90 μm	170	90 μm	170			90 μm
							0.080	20	80 μm
75 μm	200	200	75 μm	200	75 μm	200			
									71 μm
63 μm	230	250	63 μm	230	63 μm	240	0.063	19	63 μm
									56 μm
53 μm	270	270	53 μm	270	53 μm	300			
							0.050	18	50 μm
45 μm	325	325	45 μm	325	45 μm	350			45 μm
							0.040	17	40 μm
38 μm	400	400	38 μm	400	38 μm	400			

(1) USA Sieve Series—ASTM Specification E-11-70
(2) Tyler Standard Screen Scale Sieve Series.
(3) Canadian Standard Sieve Series 8-GP-1d.
(4) British Standards Institution, London BS-410-76.
(5) French Standard Specifications, AFNOR X-11-501.
(6) German Standard Specification DIN 4188.
*These sieves correspond to those recommended by ISO (International Standards Organization) as an International Standard and this designation should be used when reporting sieve analysis intended for international publication.

Index